Traveling Wave Analysis of Partial Differential Equations

Traveling Wave Analysis of Partial Differential Equations

Numerical and Analytical Methods with MATLAB® and Maple™

Graham W. Griffiths

City University, London, UK

William E. Schiesser

Lehigh University, Bethlehem, PA, USA

AMSTERDAM • BOSTON • HEIDELBERG • LONDON
NEW YORK • OXFORD • PARIS • SAN DIEGO
SAN FRANCISCO • SINGAPORE • SYDNEY • TOKYO
Academic Press is an imprint of Elsevier

Academic Press is an imprint of Elsevier
30 Corporate Drive, Suite 400, Burlington, MA 01803, USA
The Boulevard, Langford Lane, Kidlington, Oxford, OX5 1GB, UK

Library of Congress Cataloging-in-Publication Data
Application submitted

British Library Cataloguing-in-Publication Data
A catalogue record for this book is available from the British Library.

ISBN: 978-0-12-384652-5

For information on all Academic Press publications
visit our Web site at *www.elsevierdirect.com*

Typeset by: diacriTech, India

Dedication

To our teachers, with respect and appreciation.

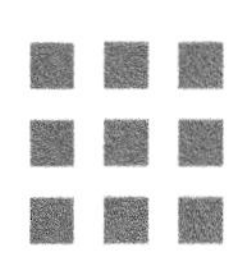

Table of Contents

Preface xi

1. Introduction to Traveling Wave Analysis 1

Traveling Wave Solutions 1

Residual Function Solutions 4

References 6

2. Linear Advection Equation 7

Smooth Solutions 7

Solutions with Sharp Gradients or Discontinuities 20

Appendix 37

References 44

3. Linear Diffusion Equation 47

Reference 55

4. A Linear Convection Diffusion Reaction Equation 57

Reference 65

5. Diffusion Equation with Nonlinear Source Terms 67

Appendix 1 102

Appendix 2 106

References 109

6. Burgers–Huxley Equation 111
Appendix 118
References 121

7. Burgers–Fisher Equation 123
Appendix 128
Reference 133

8. Fisher–Kolmogorov Equation 135
Appendix 141
References 146

9. Fitzhugh–Nagumo Equation 147
Appendix 164
References 171

10. Kolmogorov–Petrovskii–Piskunov Equation 173
Appendix 179
References 183

11. Kuramoto–Sivashinsky Equation 185
Appendix 192
References 195

12. Kawahara Equation 197
Appendix 1 217
Appendix 2 234
References 237

13. Regularized Long Wave Equation 239
Appendix 254
References 260

14. Extended Bernoulli Equation 261
Appendix 271
References 273

15. Hyperbolic Liouville Equation 275
Appendix 284
References 292

16. Sine-Gordon Equation 293
Appendix 301
References 307

17. Mth-Order Klein–Gordon Equation 309
Appendix 336
References 338

18. Boussinesq Equation 339
Appendix 370
References 374

19. Modified Wave Equation 377
Appendix 387
Reference 389

Appendix: Analytical Solution Methods for Traveling Wave Problems 391
A.1 Introduction 391
A.2 Tanh Method 391
A.3 Exp Method 411
A.4 Riccati Equation Method 418
A.5 Direct Integration 429
A.6 Factorization 430

A.7 Additional Solutions by Addition of Arbitrary Constants 435
A.8 Other Methods 435
A.9 Maple Built-In Procedure TWSolutions 436
References 437

Index 441

Preface

Partial differential equations (PDEs) have been developed and used in science and engineering for more than 200 years, yet they remain a very active area of research because of both their role in mathematics and their application to virtually all areas of science and engineering. This research has been spurred by the relatively recent development of computer solution methods for PDEs. These have extended PDE applications such that we can now quantify broad areas of physical, chemical, and biological phenomena. The current development of PDE solution methods is an active area of research that has benefited greatly from advances in computer hardware and software, and the growing interest in addressing PDE models of increasing complexity.

A large class of models now being actively studied are of a type and complexity such that their solutions are usually beyond traditional mathematical analysis. Consequently, numerical methods have to be employed. These numerical methods, some of which are still being developed, require testing and validation. This is often achieved by studying PDEs that have known exact analytical solutions. The development of analytical solutions is also an active area of research, with many advances being reported recently, particularly for systems described by nonlinear PDEs. Thus, the development of analytical solutions directly supports the development of numerical methods by providing a spectrum of test problems that can be used to evaluate numerical methods.

This book surveys some of these new developments in analytical and numerical methods and is aimed at senior undergraduates, postgraduates, and professionals in the fields of engineering, mathematics, and the sciences. It relates these new developments through the exposition of a series of *traveling wave* solutions to complex PDE problems. The PDEs that have been selected are largely *named* in the sense that they are generally closely linked to their original contributors. These names usually reflect the fact that the PDEs are widely recognized and are of fundamental importance to the understanding of many application areas. Each chapter follows the general format:

- The PDE and its associated auxiliary conditions (initial conditions (ICs) and boundary conditions (BCs)) are stated.
- A series of routines is discussed with detailed explanations of the code and how it relates to the PDE. They are written in Matlab but have been specifically programmed so that they can be easily converted to equivalent routines in other languages. The routines have the following common features:
  - The numerical procedure is the method of lines (MOL) in which the boundary value (spatial) partial derivatives are replaced with algebraic approximations, in the present case finite differences (FDs), although other approximations such as finite elements (FEs), finite volumes (FVs), and spectral methods (SMs) could be used. The FD approximations are implemented in a series of library routines; the details of how these routines were developed are given as an introduction to facilitate the development of new routines that may be required for particular PDE applications.

  - The resulting system of ordinary differential equations (ODEs) in an initial value variable, typically time in an application, is then integrated numerically using an initial value ODE integrator from the Matlab library.
  - The displayed numerical output also includes the analytical solution and the difference between the numerical and analytical solutions. The agreement between the two solutions is displayed numerically and graphically as a way of demonstrating the validity of the numerical methods.
- An analytical solution for the PDE is stated, including a reference to the original source of the solution, and in some cases, a verification (proof) of the solution by substitution into the PDE and auxiliary conditions.
- Additionally, in several chapters, the analytical solution is derived by relatively new techniques such as the tanh-, exp-, Riccati- or factorization-based methods. The derivation is either by direct application of the analytical method or through the use of the *computer algebra system* (CAS), Maple. Where Maple is used, the associated code is included in the text along with a description of its main functional elements. This code usually demonstrates the use of our new Maple procedures, which implement various analytical methods that are described in the text. Graphical output from these Maple applications is provided, including a 2D animation (to facilitate insight into and understanding of the solution) and a plot in 3D perspective. Maple is also used in other chapters to confirm analytical solutions from the literature. Where appropriate, the code is provided in the *mws* file format as well as the *mw* format so that it will also run in early versions of Maple.
- The form of the analytical solution is considered, with particular emphasis on traveling wave analysis by which the PDE (in an Eulerian or fixed frame) is converted to an ODE (in a Lagrangian or moving frame). An analytical solution to the ODE is then derived and the inverse coordinate transformation is applied to provide an analytical solution to the PDE.
- A second approach to a PDE analytical solution, the method of residual functions, is also used in some of the chapters to derive an analytical solution to a PDE that is closely related to the original PDE.
- The basic approach of traveling wave analysis, whereby a PDE is transformed to an associated ODE, is also reversed in two chapters. These start with ODEs that are then restated as PDEs that are first and second order in the initial value variable. The analytical solution to the initial ODE is then provided as an analytical solution to the PDE.
- The structure of the PDE is usually revisited briefly with regard to its form, such as whether it is first or second order in the initial value variable, the order of the boundary value derivatives, the features of nonlinear terms, and the form of the BCs. In this way, the intention of the final summary is to suggest concepts and computational approaches that can be applied in new PDE applications.
- Each chapter concludes with a discussion of the numerical solution, particularly how it conforms to the initial statement of the PDE and its auxiliary conditions; also the numerical solution is evaluated with regard to the magnitude of the errors and how these errors might be reduced through additional computation.
- In Chapter 2 we discuss the linear advection equation, one of the simplest PDEs, and show that solutions involving steep gradients or discontinuities can be difficult to achieve numerically. We then illustrate how flux limiters can be employed to improve the fidelity of the numerical solution. A short appendix to this chapter is also included, which briefly discusses some of the background to the ideas behind flux limiters.
- A general appendix details the tanh-, exp-, Ricatti-, direct integration-, and factorization-based methods. Maple implementation, by way of newly developed procedures, is included for the tanh-, exp- and Ricatti-based analytical methods. As

referred to above, these general features are then referenced for specific applications in appendices to individual chapters.

In summary the major focus of this book is the numerical MOL solution of PDEs and the testing of numerical methods with analytical solutions, through a series of applications. The origin of the analytical solutions through traveling wave and residual function analysis provides a framework for the development of analytical solutions to nonlinear PDEs that are now widely reported in the literature. Also in selected chapters, procedures based on the tanh, exp, and Ricatti methods that have recently received major attention are used to illustrate the derivation of analytical solutions. References are provided where appropriate to additional information on the techniques and methods deployed.

Our intention is to provide a set of software tools that implement numerical and analytical methods that can be applied to a broad spectrum of problems in PDEs. They are based on the concept of a traveling wave and the central feature of these methods is conversion of the system PDEs to ODEs. The discussion is limited to one-dimensional (1D) PDEs and complements our earlier book *A Compendium of Partial Differential Equation Models: Method of Lines Analysis with Matlab*, Cambridge University Press, 2009.

Finally all the code discussed in this book, along with a set of the MOL DSS library routines, is available for download from www.pdecomp.net.

Graham W. Griffiths
Nayland, Suffolk, UK

William E. Schiesser
Bethlehem, PA, USA

June 1, 2010

referred to above, these general features are then referenced for specific applications in appendices to individual chapters.

In summary the major focus of this book is the numerical MOL solution of PDEs and the testing of numerical methods with analytical solutions, through a series of applications. The origin of the analytical solutions through traveling wave and residual function analysis provides a framework for the development of analytical solutions to nonlinear PDEs that are now widely reported in the literature. Also in selected chapters, procedures based on the tanh, exp, and Riccati methods that have recently received major attention are used to illustrate the derivation of analytical solutions. References are provided where appropriate to additional information on the techniques and methods deployed.

Our intention is to provide a set of software tools that implement numerical and analytical methods that can be applied to a broad spectrum of problems in PDEs. They are based on the concept of a traveling wave and the central feature of these methods is conversion of the system PDEs to ODEs. The discussion is limited to one-dimensional (1D) PDEs and complements our earlier book *A Compendium of Partial Differential Equation Models: Method of Lines Analysis with Matlab*, Cambridge University Press, 2009.

Finally, all the code discussed in this book, along with a set of the MOL DSS library routines is available for download from www.pdecomp.net

Graham W. Griffiths
Nayland, Suffolk, UK

William E. Schiesser
Bethlehem, PA, USA

June 1, 2010

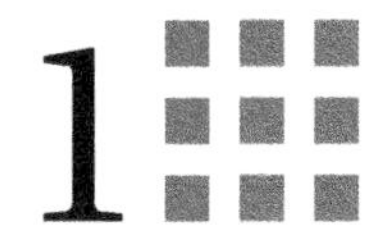

1 Introduction to Traveling Wave Analysis

Most applications of partial differential equations (PDEs) in science and engineering require numerical solutions, since the equations are typically too complicated, both in number and form, to admit analytical solutions. However, numerical procedures (methods, algorithms) are available to compute numerical solutions to most problems. In this book, we introduce the method of lines (MOL), a general numerical procedure that can be applied to all the major classes of PDEs.

In order to test MOL algorithms and software, which gives us some assurance, that the methods are correct, we utilize analytical (exact) solutions for comparison with the numerical solutions. In the subsequent discussion, we present two distinct methodologies with regard to the derivation of analytical solutions that have been widely used and reported extensively. They are (1) the *traveling wave* method and (2) the *residual function* method. The approach we have followed is, for each chapter, to illustrate the use of these methods through example applications. Thus, typically a MOL numerical solution is presented for an important PDE through a parallel discussion of the equations and Matlab routines. The particular focus of the application is emphasized, e.g., calculation of the PDE spatial derivatives, implementation of the boundary conditions, and extension of the application to other cases that require a numerical solution. In addition, an analytical solution is derived using either the associated traveling wave or residual function method. The traveling wave analytical solutions are derived with Maple procedures and scripts that are also presented. Through this approach, we hope to convey the essence of MOL and analytical analysis as applied to a series of applications that illustrate a spectrum of important concepts and details.

We start with a brief introduction to the methods of traveling wave solutions and residual functions to provide analytical solutions that can be used to test the numerical MOL procedures.

Traveling Wave Solutions

We consider a general PDE

$$\frac{\partial u}{\partial t}=f\left(u,\frac{\partial u}{\partial x},\frac{\partial^2 u}{\partial x^2},\frac{\partial^2 u}{\partial x\partial t},\frac{\partial^3 u}{\partial x^2\partial t},\dots\right) \tag{1.1}$$

which can be analyzed through a change of variables $u(x,t) = U(\xi)$, where $\xi = \xi(x,t)$ is a function to be specified. Then, eq. (1.1) can be written as

$$\begin{aligned}\frac{\partial u}{\partial t} = \frac{dU}{d\xi}\frac{\partial \xi}{\partial t} &= f\left(U, \frac{dU}{d\xi}\frac{\partial \xi}{\partial x}, \frac{\partial}{\partial x}\left(\frac{dU}{d\xi}\frac{\partial \xi}{\partial x}\right), \frac{\partial}{\partial t}\left(\frac{dU}{d\xi}\frac{\partial \xi}{\partial x}\right), \frac{\partial}{\partial t}\left[\frac{\partial}{\partial x}\left(\frac{dU}{d\xi}\frac{\partial \xi}{\partial x}\right)\right], \ldots\right) \\ &= f\left(U, \frac{dU}{d\xi}\frac{\partial \xi}{\partial x}, \frac{dU}{d\xi}\left(\frac{\partial^2 \xi}{\partial x^2}\right) + \frac{d^2U}{d\xi^2}\left(\frac{\partial \xi}{\partial x}\right)^2, \frac{dU}{d\xi}\left(\frac{\partial^2 \xi}{\partial x \partial t}\right) + \frac{d^2U}{d\xi^2}\left(\frac{\partial \xi}{\partial x}\right)\left(\frac{\partial \xi}{\partial t}\right), \ldots\right)\end{aligned} \tag{1.2}$$

For the linear case $\xi(x,t) = k(x - ct)$, the partial derivatives in eq. (1.2) are $\frac{\partial \xi}{\partial t} = -kc$, $\frac{\partial \xi}{\partial x} = k$, $\frac{\partial^2 \xi}{\partial x^2} = \frac{\partial^2 \xi}{\partial x \partial t} = \cdots = 0$. This case ($\xi(x,t) = k(x-ct)$) is generally termed a *traveling wave,* since it corresponds to a linear translation along the x axis with respect to t; k and c are arbitrary constants generally termed the *wavenumber* and *wave velocity,* respectively. For this case, eq. (1.2) reduces to

$$(-kc)\frac{dU}{d\xi} = f\left(U, k\frac{dU}{d\xi}, k^2\frac{d^2U}{d\xi^2}, -k^2c\frac{d^2U}{d\xi^2}, -k^3c\frac{d^3U}{d\xi^3}, \ldots\right)$$

or in *canonical form*

$$\frac{dU}{d\xi} = f\left(U, \frac{dU}{d\xi}, \frac{d^2U}{d\xi^2}, \frac{d^3U}{d\xi^3}, \ldots\right) \tag{1.3}$$

where the constants c and k are included in f. Equation (1.3) is an *ordinary differential equation* (ODE) in ξ (which illustrates a principal advantage of a traveling wave solution, i.e., a PDE is reduced to an ODE). If a solution to eq. (1.3), $U(\xi)$, can be found, then the solution to eq. (1.1) follows as $u(x,t) = U(\xi)$. The extension to other derivatives in eq. (1.1), such as $\frac{\partial^3 u}{\partial x^3}, \frac{\partial^3 u}{\partial x \partial t^2}, \frac{\partial^4 u}{\partial x^4}, \ldots$, follows in the same way as the preceding analysis.

The solution process for eq. (1.3) is often based on the auxiliary conditions that the dependent variable and its first, second, and higher spatial derivatives tend to zero as $\xi \to \infty$, i.e.,

$$U(\xi \to \pm\infty) = 0, \quad \frac{dU(\xi \to \pm\infty)}{d\xi} = 0, \quad \frac{d^2U(\xi \to \pm\infty)}{d\xi^2} = 0, \ldots, \text{etc.} \tag{1.4}$$

Consequently, constants of integration produced during the solution of eq. (1.3) are taken as zero.

Analytical solutions of eq. (1.3) have typically been achieved using many approaches. We discuss in detail the following methods in the main Appendix and give examples throughout the various chapters:

- *Direct integration method,* see appendix 3 of [6], which applies standard calculus techniques to transform the problem into one that can be integrated.

- *Factorization method* [1, 5] which factors the problem ODE (transformed PDE) into smaller problems that can be solved more easily.
- *Expansion methods* based on tanh [4], exp [3], and Riccati [7] expansions of elements of the ODE (transformed PDE). These techniques lead to systems of nonlinear equations that can be solved analytically.

All the above methods benefit from the use of a *computer algebra system* (CAS), such as Maple, which facilitates algebraic operations, such as the solution of simultaneous nonlinear equations.

The preceding discussion applies as well to PDEs that are second and higher order in t. For analytical traveling wave solutions, we would use the expansion method as this generally presents no difficulty, and solutions fall out naturally. Numerical methods can also be extended naturally to higher-order PDEs in t. For example, if eq. (1.1) was second order in t (rather than first order in t) by the standard procedure of defining additional dependent variables, we set $u_1 = u$ and $u_2 = \dfrac{\partial u}{\partial t}$, which transforms eq. (1.1) into two first-order PDEs. Thus, we have

$$\frac{\partial^2 u}{\partial t^2} = f\left(t, u, \frac{\partial u}{\partial x}, \frac{\partial^2 u}{\partial x^2}, \ldots\right) \tag{1.5}$$

expressed as

$$\frac{\partial u_1}{\partial t} = u_2 \tag{1.6a}$$

$$\frac{\partial u_2}{\partial t} = f\left(t, u_1, \frac{\partial u_1}{\partial x}, \frac{\partial^2 u_1}{\partial x^2}, \ldots\right), \tag{1.6b}$$

and the numerical procedures for first order (in t) PDEs to be discussed subsequently can be applied to eqs. (1.6). This approach is applied in several of the following chapters to PDEs that are second order in t. In general, a PDE that is nth order in t can be reduced to n PDEs that are first order in t.

Equation (1.1) is a one-dimensional (1D) PDE in the sense that x corresponds to a single spatial direction or dimension. In principle, the numerical methods to be discussed can be applied to systems of PDEs in 1D, 2D, and 3D. We will consider only numerical solutions to 1D problems in this book. Examples of higher-dimensional MOL analysis are given in [6]. However, some 2D analytical solution examples using the tanh method are provided here.

Equations (1.3) and (1.4) essentially constitute an *initial value problem* in the sense that boundary conditions for eq. (1.1) are not required other than through the conditions $\xi \to \pm\infty$. Thus, we refer to eq. (1.1), subject to the initial condition

$$u(x, t = 0) = f(x) \tag{1.7}$$

with $f(x \to \pm\infty) = 0, \dfrac{df(x \to \pm\infty)}{dx} = 0, \ldots$ (higher-order derivatives in x are homogeneous) as a *Cauchy problem*. Traveling wave solutions and the corresponding Cauchy (initial value) problems are illustrated by several of the problems that follow. Also, some of the subsequent problems are based on boundary conditions at finite values of x that follow from an analytical solution to eq. (1.1) (so that the initial conditions, boundary conditions, and PDE solution are consistent).

To summarize the method of traveling waves, if a solution to the PDE is assumed to be a function of a linear combination of x and t such as $\xi = k(x - ct)$, the PDE is transformed to an ODE which, hopefully, will be easier to solve than the original PDE. A number of methods have been developed for the solution of the ODE. The transformation from a PDE to an ODE can be considered as going from a fixed *Euler coordinate system* in x and t to a moving *Lagrangian coordinate system* in ξ. This change to a moving coordinate system can often lead to a significant simplification as we have observed here.

Residual Function Solutions

A second general approach to the solution of a PDE such as eq. (1.1) is to assume a solution, then determine an associated PDE that, in fact, is solved by the assumed solution. This procedure is best explained through an example. If we start with PDE (1.8) [2], (the *quadratic Klein–Gordon equation*),

$$\frac{\partial^2 u}{\partial t^2} + \alpha \frac{\partial^2 u}{\partial x^2} + \beta u + \gamma u^2 = 0 \tag{1.8}$$

we then assume a solution to eq. (1.8) such as

$$u_a(x,t) = x \cos t, \quad -1 \leq x \leq 1 \tag{1.9}$$

for the particular values $\beta = 0, \gamma = 1$, and any α in eq. (1.8). At the end of this discussion, we give some guidelines for assuming the solution.

If eq. (1.9) is substituted in eq. (1.8), as we might expect, eq. (1.9) will not satisfy eq. (1.8) exactly. In other words, there will be a function remaining that is not part of the original PDE, in this case, eq. (1.8). We refer to this remaining part as the *residual function*, denoted as $f(x,t)$. This process is summarized in the following table

Term in eq. (1.8)	Term from eq. (1.9)
$\frac{\partial^2 u}{\partial t^2}$	$-x\cos t$
$\alpha\frac{\partial^2 u}{\partial x^2}$	0
βu	$\beta x \cos t$
γu^2	$\gamma x^2 \cos t^2$
Sum of terms	Sum of terms
0	$\neq 0$

(1.10)

Note that the terms in the right column do not sum to zero (and in fact, they sum to $-x\cos t + x^2\cos^2 t$ for $\alpha = -1, \beta = 0, \gamma = 1$). However, if we define a residual function as

$$f(x,t) = -x\cos t + x^2\cos^2 t, \tag{1.11}$$

then eq. (1.9) is a solution of the PDE

$$\frac{\partial^2 u}{\partial t^2} + \alpha\frac{\partial^2 u}{\partial x^2} + \beta u + \gamma u^2 = f(x,t). \tag{1.12}$$

Equation (1.12), of course, is not the PDE we started with, eq. (1.8). However, eq. (1.12) can be used as a test problem, since it has an analytical solution, eq. (1.9). Also, eq. (1.12) is similar to eq. (1.8) in the sense that the only difference between the two PDEs is the residual function $f(x,t)$ (which is also termed a *nonhomogeneous* or *inhomogeneous* term). Thus, eq. (1.12) includes the essential terms of eq. (1.8), namely, the same partial derivatives.

Note that this procedure is general in the sense that no matter what we assume for an analytical solution, if it can be substituted into the PDE (i.e., the various partial derivatives exist and can be derived), an analytical solution will result by including the residual function in the original PDE to arrive at a related PDE (with the assumed solution as an analytical solution). In other words, this method can be generally applied to produce an analytical solution to a related PDE.

Finally, some guidelines for selecting an analytical solution are as follows:

- The assumed analytical solution can be of essentially any form so long as it can be substituted in the original PDE. Of course, if we make exactly the right choice of an analytical solution, the assumed solution will satisfy the original PDE, but generally this is unlikely; that is, we in fact will not know the analytical solution to the original PDE.
- We could also add a logical extension that the closer the assumed analytical solution is to the true (but unknown) analytical solution of the initial PDE, the better will be the resulting test problem based on the modified PDE.
- If the assumed solution is "smooth" in the sense that it and its derivatives are well behaved, the resulting modified PDE (with the residual function) will be relatively easy to solve numerically. For example, the functions of x and t in eq. (1.9) are smooth (well behaved), as well as all of their derivatives.
- If a PDE test problem is required to have specified initial conditions (ICs) and/or boundary conditions (BCs) and if these are included in the assumed solution, the method of residual functions can still be applied (including the required ICs and/or BCs).

This completes the discussion of the methods of traveling waves and residual functions. We will consider a series of example PDE applications for which the analytical solutions could be derived by either of these two methods.

References

[1] O. Cornejo-Perez, H.C. Rosu, Nonlinear second order ODE: factorizations and particular solutions, *Prog. Theor. Phys.* 114 (3) (2005) 533–538.

[2] M. Dehghan, A. Shokri, Numerical solution of the nonlinear Klein-Gordon equation using radial basis functions, *J. Comp. App. Math.* 230 (2009) 400–410.

[3] J.-H. He, X.-H. Wu, Exp-function method for nonlinear wave equations, *Chaos, Solitons and Fractals*, 30 (2006) 700–708.

[4] W. Malfliet, Solitary wave solutions of nonlinear wave equation, *Am. J. Phys.* 60 (7) (1992) 650–654.

[5] H.C. Rosu, O. Cornejo-Perez, Supersymmetric pairing of kinks for polynomial nonlinearities: <http://arxiv.org/PS_cache/math-ph/pdf/0401/0401040v3.pdf>, last accessed 23 April, 2010.

[6] W.E. Schiesser, G.W. Griffiths, *A Compendium of Partial Differential Equation Models*, Cambridge University Press, UK, 2009.

[7] A.M. Wazwaz, The tanh-coth method combined with the Riccati equation for solving the KdV equation, *AJMMS* 1 (1) (2007) 27–34.

Linear Advection Equation

We begin our discussion of particular PDEs with one that might be considered the simplest PDE, the *linear advection equation*, but which somewhat paradoxically, is one of the most difficult to solve numerically; this point will be illustrated in the subsequent discussion.

The linear advection equation is

$$\frac{\partial u}{\partial t} = -c\frac{\partial u}{\partial x} \tag{2.1}$$

with the initial condition (IC)

$$u(x, t = 0) = f(x) \tag{2.2}$$

The solution to eqns. (2.1) and (2.2) is

$$u(x, t) = U(\xi) = f(x - ct); \; \xi = x - ct \tag{2.3}$$

Note that the function of eq. (2.3) is a traveling wave solution since it has an independent variable $\xi = x - ct$ (as discussed in Chapter 1).

Equations (2.1) and (2.2) are an example of an *initial value problem*, also termed a *Cauchy problem*. If $f(x)$ has a finite discontinuity (usually at $x = 0$), then this is termed a *Riemann problem*. For example, if $f(x)$ is the Heaviside unit step function, eq. (2.3) indicates this step function propagates left to right in x with increasing t; this propagating discontinuity makes the numerical solution of even a simple PDE such as eq. (2.1) difficult computationally. This aspect will be discussed subsequently.

To confirm that eq. (2.3) is a solution to eqns. (2.1) and (2.2), we substitute it as follows:

$$\frac{dU}{d\xi}\frac{\partial \xi}{\partial t} = -c\frac{dU}{d\xi}\frac{\partial \xi}{\partial x}$$

$$\frac{dU}{d\xi}(-c) = -c\frac{dU}{d\xi}(1)$$

Thus, eq. (2.3) satisfies eq. (2.1). Equation (2.3) also satisfies IC (2.2), so it is a complete solution.

Smooth Solutions

We now consider a specific example with a smooth solution and choose the IC function of eq. (2.2) to be the Gaussian function

$$f(x) = e^{-\lambda x^2} \tag{2.4}$$

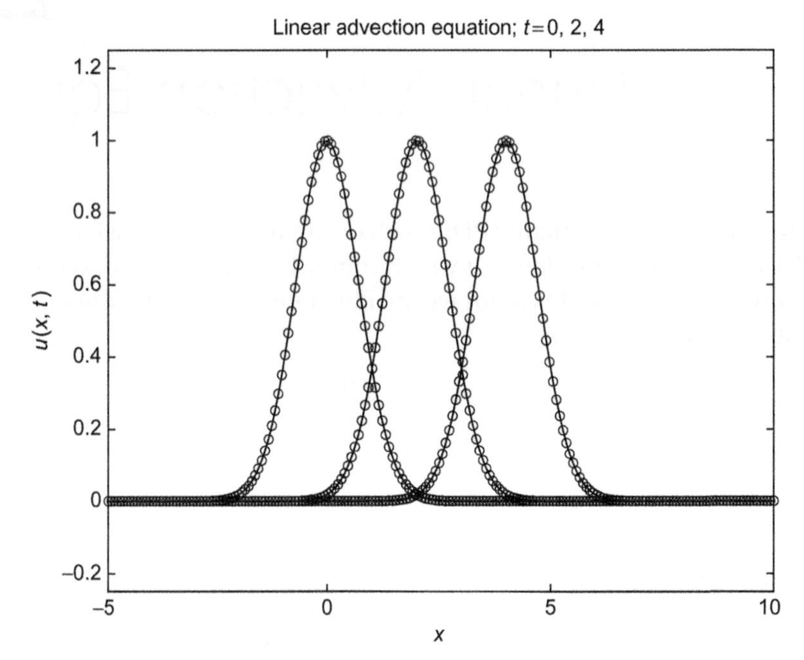

FIGURE 2.1: Numerical solution to eq. (2.1) (lines) with the analytical solution superimposed (circles) using five-point FD approximations in `dss004` [20].

Then, from eq. (2.3), the analytical (exact) solution to eq. (2.1) is

$$u(x,t) = e^{-\lambda(x-ct)^2} \tag{2.5}$$

The numerical and analytical solutions produced by the computer code that follows are in Fig. 2.1.

The close agreement between the numerical and analytical solutions is also apparent from the tabulated numerical output reproduced in Table 2.1.

We can note the following points about this output:

- t varies through the values $t = 0, 2, 4$.
- Within each of these values of t, x varies through the values $x = -5, -4.250, \ldots, 10$.
- The numerical solution, `u(it,i)`, changes with an index `it=1,2,3` (`it` is part of the computer code to follow) for the three values of t (including $t = 0$). This solution also changes with an index `i=1,2,3,...,201` for 201 values of x (`i` is part of the computer code; only every fifth value appears in the output).
- The analytical solution (from eq. (2.5)), `u_anal(it,i)`, is in close agreement with the numerical solution. The difference between the two solutions is `err(it,i)`.

The computer code also produces a plot of the solution in 3D perspective.

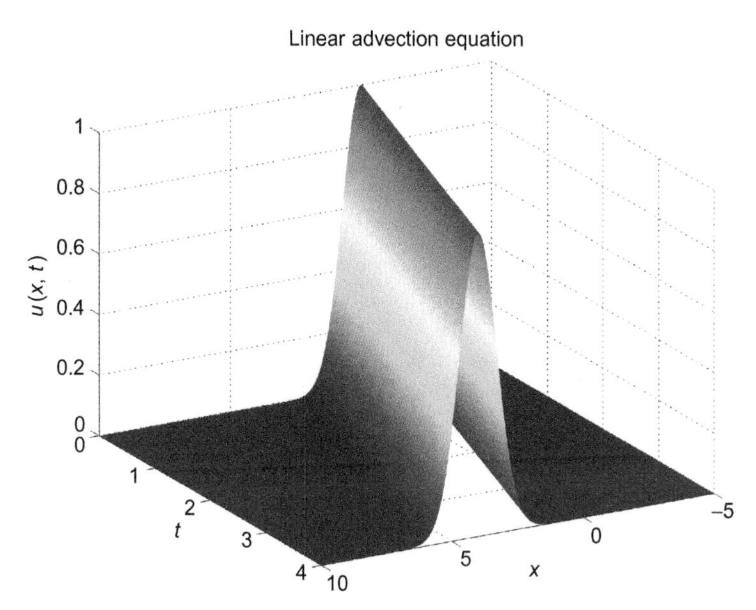

FIGURE 2.2: Numerical solution to eq. (2.1), $u(x,t)$, as a function of x and t using four-point FD approximations in dss004.

The Matlab routines for the numerical method of lines (MOL) integration of eqns. (2.1) and (2.2) that produced the output in Figs. 2.1 and 2.2, and the tabulated output in Table 2.1 are considered next, starting with the main program, pde_1_main.m

```
%
% Linear advection equation
%
% Clear previous files
  clear all
  clc
%
% Parameters shared with other routines
  global xl xu x n ncall
%
  global c lambda
  c=1; lambda=1;
%
% Initial condition
  t0=0.0;
  u0=inital_1(t0);
%
% Independent variable for ODE integration
  tf=4;
  tout=[t0:2:tf]';
  nout=3;
  ncall=0;
%
```

```
% ODE integration
  mf=2;
  reltol=1.0e-06; abstol=1.0e-06;
  options=odeset('RelTol',reltol,'AbsTol',abstol);
%
% Explicit (nonstiff) integration
  if(mf==1)[t,u]=ode45(@pde_1,tout,u0,options); end
%
% Implicit (sparse stiff) integration
  if(mf==2)
    S=jpattern_num_1;
%   pause
    options=odeset(options,'JPattern',S)
    [t,u]=ode15s(@pde_1,tout,u0,options);
  end
%
% Store analytical solution, errors in numerical solution
  for it=1:nout
    for i=1:n
      u_anal(it,i)=ua_1(x(i),t(it));
      err(it,i)=u(it,i)-u_anal(it,i);
    end
  end
%
% Display selected output
  for it=1:nout
    fprintf('\n     t        x          u(it,i)   u_anal(it,i)       err(it,i)\n');
    for i=1:10:n
      fprintf('%6.2f%8.3f%15.6f%15.6f%15.6f\n',...
            t(it),x(i),u(it,i),u_anal(it,i),err(it,i));
    end
  end
  fprintf('     ncall = %4d\n\n',ncall);
%
% Plot numerical and analytical solutions
  figure(2)
  plot(x,u,'-',x,u_anal,'o')
  axis([-5 10 -0.25 1.25]);
  xlabel('x')
  ylabel('u(x,t)')
  title('Linear advection equation; t = 0, 2, 4; solid - numerical;
         o - analytical')
  figure(3)
  surf(x,t,u)
  xlabel('x'); ylabel('t'); zlabel('u(x,t)');
  title('Linear advection equation');
```

LISTING 2.1: Main program pde_1_main.m.

We can note the following points about this listing:

- After clearing previous files, some *global* parameters are declared that can be shared with other routines.

Table 2.1: Tabular numerical and analytical solutions

t	x	u(it,i)	u_anal(it,i)	err(it,i)
0.00	−5.000	0.000000	0.000000	0.000000
0.00	−4.250	0.000000	0.000000	0.000000
0.00	−3.500	0.000005	0.000005	0.000000
0.00	−2.750	0.000520	0.000520	0.000000
0.00	−2.000	0.018316	0.018316	0.000000
0.00	−1.250	0.209611	0.209611	0.000000
0.00	−0.500	0.778801	0.778801	0.000000
0.00	0.250	0.939413	0.939413	0.000000
0.00	1.000	0.367879	0.367879	0.000000
0.00	1.750	0.046771	0.046771	0.000000
0.00	2.500	0.001930	0.001930	0.000000
0.00	3.250	0.000026	0.000026	0.000000
0.00	4.000	0.000000	0.000000	0.000000
0.00	4.750	0.000000	0.000000	0.000000
0.00	5.500	0.000000	0.000000	0.000000
0.00	6.250	0.000000	0.000000	0.000000
0.00	7.000	0.000000	0.000000	0.000000
0.00	7.750	0.000000	0.000000	0.000000
0.00	8.500	0.000000	0.000000	0.000000
0.00	9.250	0.000000	0.000000	0.000000
0.00	10.000	0.000000	0.000000	0.000000
t	x	u(it,i)	u_anal(it,i)	err(it,i)
2.00	−5.000	0.000000	0.000000	0.000000
2.00	−4.250	0.000000	0.000000	0.000000
2.00	−3.500	0.000000	0.000000	0.000000
2.00	−2.750	0.000000	0.000000	0.000000
2.00	−2.000	0.000000	0.000000	0.000000
2.00	−1.250	0.000026	0.000026	0.000000
2.00	−0.500	0.001934	0.001930	0.000004
2.00	0.250	0.046756	0.046771	−0.000014
2.00	1.000	0.367881	0.367879	0.000002
2.00	1.750	0.939459	0.939413	0.000046
2.00	2.500	0.778736	0.778801	−0.000064
2.00	3.250	0.209640	0.209611	0.000028
2.00	4.000	0.018316	0.018316	0.000001
2.00	4.750	0.000518	0.000520	−0.000002
2.00	5.500	0.000005	0.000005	−0.000000
2.00	6.250	0.000000	0.000000	−0.000000
2.00	7.000	0.000000	0.000000	−0.000000
2.00	7.750	0.000000	0.000000	−0.000000
2.00	8.500	0.000000	0.000000	−0.000000
2.00	9.250	0.000000	0.000000	−0.000000
2.00	10.000	0.000000	0.000000	−0.000000

(Continued)

Table 2.1: *(Continued)*

t	x	u(it,i)	u_anal(it,i)	err(it,i)
4.00	−5.000	0.000004	0.000000	0.000004
4.00	−4.250	0.000002	0.000000	0.000002
4.00	−3.500	0.000001	0.000000	0.000001
4.00	−2.750	0.000000	0.000000	0.000000
4.00	−2.000	0.000000	0.000000	0.000000
4.00	−1.250	0.000000	0.000000	0.000000
4.00	−0.500	0.000000	0.000000	0.000000
4.00	0.250	0.000001	0.000001	0.000000
4.00	1.000	0.000126	0.000123	0.000002
4.00	1.750	0.006336	0.006330	0.000006
4.00	2.500	0.105346	0.105399	−0.000053
4.00	3.250	0.569869	0.569783	0.000086
4.00	4.000	0.999978	1.000000	−0.000022
4.00	4.750	0.569725	0.569783	−0.000058
4.00	5.500	0.105447	0.105399	0.000047
4.00	6.250	0.006323	0.006330	−0.000007
4.00	7.000	0.000122	0.000123	−0.000002
4.00	7.750	0.000001	0.000001	−0.000000
4.00	8.500	0.000000	0.000000	−0.000000
4.00	9.250	0.000000	0.000000	−0.000000
4.00	10.000	0.000000	0.000000	−0.000000

```
%
% Linear advection equation
%
% Clear previous files
  clear all
  clc
%
% Parameters shared with other routines
  global xl xu x n ncall
```

- Two model parameters, c in eq. (2.1) and λ in eq. (2.4), are assigned numerical values.

```
%
  global c lambda
  c=1; lambda=1;
```

- The IC of eq. (2.2) is set numerically through a call to function `inital_1(t0)` (to be discussed subsequently); this function basically uses the IC function of eq. (2.4).

```
%
% Initial condition
  t0=0.0;
```

```
  u0=inital_1(t0);
%
% Independent variable for ODE integration
  tf=4;
  tout=[t0:2:tf]';
  nout=3;
  ncall=0;
```

The output sequence of t (in eq. (2.1)) is then defined as $t = 0, 2, 4$ (a total of three outputs as reflected in Table 2.1).

- The grid in x, defined in `inital_1(t0)` for 201 points, is the basis for 201 ODEs in t (as programmed subsequently in function `pde_1.m`). These ODEs are integrated by a call to the nonstiff integrator `ode45` or the stiff integrator `ode15s` as specified by `mf`. The error tolerances for the ODE integration are specified before the call to the ODE integrator.

```
%
% ODE integration
  mf=2;
  reltol=1.0e-06; abstol=1.0e-06;
  options=odeset('RelTol',reltol,'AbsTol',abstol);
%
% Explicit (nonstiff) integration
  if(mf==1)[t,u]=ode45(@pde_1,tout,u0,options); end
%
% Implicit (sparse stiff) integration
  if(mf==2)
    S=jpattern_num_1;
%   pause
    options=odeset(options,'JPattern',S)
    [t,u]=ode15s(@pde_1,tout,u0,options);
  end
```

For `ode15s`, the *sparse option* is specified which requires a function `jpattern_num_1` to determine the structure of the ODE Jacobian matrix (discussed subsequently). As a result of the ODE integration, the solution of the 201 ODEs is returned in array `u` at the values of $t = 0, 1, 2$ in array `t`.

- The analytical solution of eq. (2.5) computed in function `ua_1.m` (discussed subsequently) is subtracted from the numerical solution to give the difference `err`. The two solutions (numerical and analytical) and their difference are then displayed through the `fprintf` statement, which produced the output in Table 2.1.

```
%
% Store analytical solution, errors in numerical solution
  for it=1:nout
    for i=1:n
      u_anal(it,i)=ua_1(x(i),t(it));
      err(it,i)=u(it,i)-u_anal(it,i);
    end
  end
%
% Display selected output
```

```
  for it=1:nout
    fprintf('\n     t       x          u(it,i)    u_anal(it,i)       err(it,i)\n');
    for i=1:10:n
      fprintf('%6.2f%8.3f%15.6f%15.6f%15.6f\n',...
              t(it),x(i),u(it,i),u_anal(it,i),err(it,i));
    end
  end
  fprintf('     ncall = %4d\n\n',ncall);
```

The number of calls to the ODE routine `pde_1.m` is displayed as a measure of the computational effort to produce the numerical solution. The value is rather modest (`ncall = 399`) indicating the computational efficiency of the ODE integrator `ode15s`.

- The plots of Figs. 2.1 and 2.2 are then produced by calls to the Matlab utilities `plot` and `surf`.

```
%
% Plot numerical and analytical solutions
  figure(2)
  plot(x,u,'-',x,u_anal,'o')
  axis([-5 10 -0.25 1.25]);
  xlabel('x')
  ylabel('u(x,t)')
  title('Linear advection equation; t = 0, 2, 4; solid - numerical;
         o - analytical')
  figure(3)
  surf(x,t,u)
  xlabel('x'); ylabel('t'); zlabel('u(x,t)');
  title('Linear advection equation');
```

- The main program of Listing 2.1 produces two additional outputs. The first is a list of the options selected for the sparse matrix version of `ode15s`.

```
options =

              AbsTol: 1.0000e-006
                 BDF: []
              Events: []
         InitialStep: []
            Jacobian: []
           JConstant: []
            JPattern: [201x201 double]
                Mass: []
        MassConstant: []
        MassSingular: []
            MaxOrder: []
             MaxStep: []
         NormControl: []
           OutputFcn: []
           OutputSel: []
              Refine: []
              RelTol: 1.0000e-006
               Stats: []
```

In the present case, the error tolerances and the size of the ODE Jacobian matrix are indicated. In particular, the Jacobian matrix is $201 \times 201 = 40{,}401$ elements. Since this matrix has a size equal to the number of ODEs squared, it increases rapidly with the number of ODEs, which is an important reason for using a sparse matrix integrator that uses only the nonzero elements of the Jacobian matrix.

- This point is illustrated by the map of the Jacobian matrix in Fig. 2.3, produced by the call to `ode15s`.

  We can note the following points about this map:

  - The ODEs are numbered vertically, whereas the ODE-dependent variables are numbered horizontally. For example, an entry at $x = 101, y = 100$ indicates dependent variable 101 that appears in the RHS of ODE 100.
  - The nonzero elements are clustered around the main diagonal, that is, the Jacobian matrix is *banded*. In the present case, the band is five elements wide, and thus the Jacobian matrix is *pentadiagonal*. This bandwidth of five results from the use of five-point finite differences (FDs) in the ODE routine `pde_1.m` is discussed subsequently.
  - Of the 40,401 elements, only 808 are nonzero, which is 2.0% of the total. In other words, 98% of the elements are essentially zero (and using them in any calculations would simply waste computer time).

  In conclusion, the effectiveness (efficiency) of the sparse option of `ode15s` is clear.

To conclude this discussion of the main program in Listing 2.1, the numerical solution is validated through the use of eq. (2.3). This example illustrates two features for most of the numerical solutions that follow:

- The Matlab routines can be considered as templates that will be modified for particular PDE applications in the subsequent chapters.
- The analytical solution (eq. (2.3)) is used to evaluate the errors in the numerical solution. Of course, we attempt to achieve a numerical solution in which the errors are considered acceptable (small enough to be essentially negligible).

We now go on to a discussion of the routines (functions) that were called during the execution of the main program. The ODE routine, `pde_1.m`, called by `ode15s` is listed next.

```
  function ut=pde_1(t,u)
%
% Function pde_1 computes the t derivative vector for the linear
% advection equation
%
  global xl xu x n ncall
%
  global c lambda
%
% ux
  ux=dss004(xl,xu,n,u);
%
```

```
% PDE
%
%   Procedural
%   for i=1:n
%      ut(i)=-c*ux(i);
%   end
%   ut=ut';
%
%   Vectorized
    ut=-c*ux';
%
% Increment calls to pde_1
  ncall=ncall+1;
```

LISTING 2.2: ODE function `pde_1.m`.

We can note the following points about this listing:

- The function and some global parameters are defined.

```
  function ut=pde_1(t,u)
%
% Function pde_1 computes the t derivative vector for the linear
% advection equation
%
  global xl xu x n ncall
%
  global c lambda
```

The first set of global parameters is defined numerically by the call to `inital_1.m` in main program `pde_1_main.m`. The second set of parameters is defined mumerically in `pde_1_main.m`. Thus, all of these parameters are available (numerically) for use in `pde_1.m`.

- The input arguments of `pde_1.m` are the ODE-independent variable, `t`, and the vector of ODE-dependent variables, `u` (of length `n`). `u` is the same vector displayed horizontally in Fig. 2.3.
- The output argument of `pde_1.m` is the vector of ODE derivatives, `ut`, which is the same vector displayed vertically in Fig. 2.3. The programming requirement in `pde_1.m` is to compute all `n` = 201 derivatives before exiting from the routine. Note that to satisfy `ode15s` which calls `pde_1.m` from the main program `pde_1_main.m`, a transpose must be included (to transpose `ut` from a row to a column vector).
- The first derivative $\frac{\partial u}{\partial x}$ in eq. (2.1) is computed by a call to `dss004` ([20]); `u` is an input to `dss004` and the resulting derivative `ux`, a vector of length `n`, is the output. `ux` is computed by five-point FDs in `dss004`, which is the origin of the pentadiagonal system mapped in Fig. 2.3.

```
%
% ux
  ux=dss004(xl,xu,n,u);
```

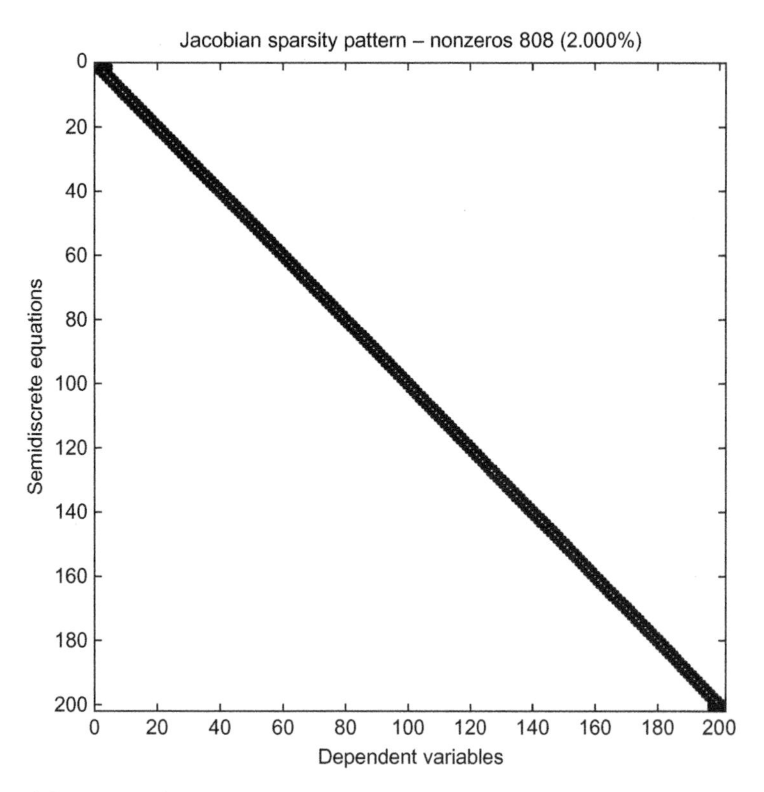

FIGURE 2.3: Map of the ODE Jacobian matrix.

- The PDE, eq. (2.1), is then programmed in either of two ways:

```
%
% PDE
%
%   Procedural
%   for i=1:n
%      ut(i)=-c*ux(i);
%   end
%   ut=ut';
%
%   Vectorized
    ut=-c*ux';
```

In the first approach (commented), a `for` loop is used to program eq. (2.1); this is termed *procedural* and is commonplace in other programming languages such as *Fortran*. The second is termed *vectorized* and uses the features specific to Matlab, which operates on vectors and matrices in much the same way as with scalars. In either case, the close resemblance of the programming to eq. (2.1) is clear, which is one of the advantages of the MOL (a second advantage is the use of ODE library integrators, such as `ode15s` and `ode45`). In both cases, a transpose is used to return the derivative ut as a column vector (required by ode45 and ode15s). Finally, the counter for the calls to pde_1.m is incremented each time pde_1.m is called.

The routine that defines IC (2.4) is considered next.

```
  function u0=inital_1(t0)
%
% Function inital_1 sets the initial condition for the linear
% advection equation
%
  global xl xu x n
%
% Spatial domain and initial condition
  xl=-5;
  xu=10;
  n=201;
  dx=(xu-xl)/(n-1);
%
% IC from analytical solution
  for i=1:n
    x(i)=xl+(i-1)*dx;
    u0(i)=ua_1(x(i),0.0);
  end
```

LISTING 2.3: IC function inital_1.m.

We can note the following points about inital_1.m:

- The function is defined and global parameters are included for use by inital_1.m.

```
  function u0=inital_1(t0)
%
% Function inital_1 sets the initial condition for the linear
% advection equation
%
  global xl xu x n
```

- The spatial domain in x is defined as $-5 \leq x \leq 10$ with 201 points. The interval in x is determined by the characteristics of the solution. In this case, since the Gaussian pulse of eq. (2.4) initially ($t = 0$) is centered at $x = 0$ (see Fig. 2.1), it moves left to right in x at velocity $c = 1$. Thus, at $t = 0, 2, 4$, the Gaussian pulse is centered at $x = 0, 2, 4$, respectively, and the interval $-5 \leq x \leq 10$ includes the moving pulse (it does not interact with the boundaries at $x = -5, 10$). The selection of 201 points provides good resolution of the solution as indicated in Figs. 2.1 and 2.2, but the computer run times are quite acceptable.

```
%
% Spatial domain and initial condition
  xl=-5;
  xu=10;
  n=201;
  dx=(xu-xl)/(n-1);
```

- The IC, eq. (2.4), is then implemented by a call to the routine that calculates the analytical solution to eq. (2.5) with $t = 0$, i.e. ua_1.m (this routine is considered next).

```
%
% IC from analytical solution
  for i=1:n
    x(i)=xl+(i-1)*dx;
    u0(i)=ua_1(x(i),0.0);
  end
```

Note the `for` steps through the 201 values of x to define IC (2.4).

Function `ua_1.m` for IC (2.4) follows.

```
  function uanal=ua_1(x,t)
%
% Function uanal computes the exact solution of the linear advection
% equation for comparison with the numerical solution
  global c lambda
%
% Analytical solution
  uanal=exp(-(lambda*(x-c*t))^2);
```

LISTING 2.4: Function `ua_1.m` for IC (2.4).

The last line of `ua_1.m` is a straightforward programming statement for eq. (2.4).

Two other routines have been used, but they will not be listed and discussed here to conserve space. Briefly,

- `jpattern_num_1.m` calls `pde_1.m` to map the ODE Jacobian matrix (as expected since `pde_1.m` has the programming for the ODEs). The two statements in `jpattern_num_1.m` that call `pde_1.m` are

  ```
  ytbase=pde_1(tbase,ybase);

  [Jac,fac]=numjac(@pde_1,tbase,ybase,ytbase,thresh,fac,vectorized);
  ```

  `numjac` is a Matlab utility that maps the ODE Jacobian matrix (and therefore produced Fig. 2.3).
- The other routine that was used is `dss004` to compute `ux` in `pde_1.m`. The "dss" denotes *Differentiation in Space Subroutine* and "004" indicates *fourth-order FDs* (five-point FDs) that are used in `dss004`. A set of spatial differentiation routines, ranging from second to tenth order are used for MOL analysis in subsequent chapters . These routines form the DSS library ([20]) which is provided with the downloads for this book.

There are two additional details we can consider concerning the preceding MOL solution.

- Equation (2.1) has a first-order derivative in x, which indicates a BC in x is required. But in `pde_1.m` of Listing 2.2, a BC has not been specified. One possibility is to recognize that $u(x,t)$ remains zero at the left boundary, $x = x_l$, and therefore could be used as a BC $u(x = x_l, t) = 0$ which could be coded as `u(1)=0` just before the call to `dss004`. When this is done, the solution does not change, since $u(x = x_l, t)$ remains at its initial value $u(x = x_l, t = 0) = 0$ defined in `inital_1.m`. Thus, in this particular case, the required BC has no effect. Of course, this will not always be the case (depending on the circumstances of the problem), and in fact, the use of a BC is illustrated in the subsequent discussion of eq. (2.1).

- From eqns. (2.3) and (2.5), we see that the solution does not change shape as it moves along the x axis. In other words, the area under the traveling Gaussian pulse remains constant. Thus, an *invariant* or *integral constraint* can be stated mathematically as

$$I = \int_{-\infty}^{\infty} u(x,t)dx = \text{const}$$

I could be computed numerically for the numerical solution (in Fig. 2.1) using a quadrature such as *Simpson's rule*. In other words, an invariant can be computed as a test of the numerical solution (it should remain constant). We do not carry out this calculation here, since we would expect (from Fig. 2.1) that I remains essentially constant, but the use of an integral constraint as a test of the numerical solutions for some PDE applications is discussed in subsequent chapters.

This completes the discussion of the MOL solution of eqns. (2.1) and (2.2) for the Gaussian pulse IC of eq. (2.4). The agreement between the numerical solution and the analytical solution of eq. (2.5) in Fig. 2.1 and Table 2.1 indicates that the numerical solution is quite acceptable.

Solutions with Sharp Gradients or Discontinuities

Unfortunately, it is not always the case that solutions are smooth, and we now consider a variant of the preceding problem that illustrates this point. Equations (2.1) and (2.2) will again be analyzed, but IC (2.4) will be changed to a *square pulse*

$$f(x) = \begin{cases} 0, & x < 0 \\ 1, & 0 \le x < 1 \\ 0, & x \ge 1 \end{cases} \tag{2.6}$$

Equation (2.3) indicates that this pulse should move left to right with velocity $c = 1$. We now consider what the numerical solution actually looks like. To handle this second case of eq. (2.6), we will use the preceding routines with only the changes required to include eq. (2.6). However, in order to keep the routines separated and organized, we will change the `1` designation in the routine names to `2`, i.e., `pde_1_main.m`, `pde_1.m`, `inital_1.m`, and `ua_1.m` will become `pde_2_main.m`, `pde_2.m`, `inital_2.m`, and `ua_2.m`, respectively. In order to conserve space, we will indicate only the changes in the preceding code in going from IC (2.4) to IC (2.6).

First, we note that eq. (2.3) is the analytical solution of eqns. (2.1) and (2.2) for $f(x)$ of eq. (2.6). We can then put this solution into the routine `ua_2.m`.

```
  function uanal=ua_2a(x,t)
%
% Function uanal computes the exact solution of the linear advection
% equation for comparison with the numerical solution
  global c
%
% Analytical solution
  xi=x-c*t;
  if(xi<0)
    uanal=0;
  elseif((xi>=0)&(xi<=1))
    uanal=1;
  elseif(xi>1)
    uanal=0;
  end
```

LISTING 2.5: Function `ua_2.m` for IC (2.6).

The coding in `ua_2.m` follows from IC (2.6) except that the variable x is replaced by $\xi = x - ct$ (since eq. (2.3) indicates a traveling wave solution with independent variable $x - ct$). In other words, `ua_2.m` applies to all values of t and not just $t = 0$ of IC (2.6).

`ua_2.m` is now used to set IC (2.6) in `inital_2.m`.

```
  function u0=inital_2(t0)
%
% Function inital_2 sets the initial condition for the linear
% advection equation
%
  global xl xu x n
%
% Spatial domain and initial condition
  xl=-10;
  xu= 10;
  n=201;
  dx=(xu-xl)/(n-1);
%
% IC from analytical solution
  for i=1:n
    x(i)=xl+(i-1)*dx;
    u0(i)=ua_2(x(i),0.0);
  end
```

LISTING 2.6: Function `inital_2.m` for IC (2.6).

We can note the following points about `inital_2.m`:

- The interval in x has been increased from $-5 \leq x \leq 10$ in `inital_1.m` of Listing 2.3 to $-10 \leq x \leq 10$

  ```
  %
  % Spatial domain and initial condition
    xl=-10;
    xu= 10;
    n=201;
  ```

  The reason for this change is explained subsequently.

- The call to the analytical solution (at $t = 0$) is actually to ua_2 while the function in Listing 2.5 is defined as ua_2a.

```
u0(i)=ua_2(x(i),0.0);
```

  Thus, it would seem that the call to ua_2 would not execute. To explain why this is not so, Matlab calls a function by its **file name** (and not by the name in the first-line definition). Thus, as long as ua_2a.m of Listing 2.5 is saved as file ua_2.m, the call to ua_2.m in inital_2.m of Listing 2.6 will execute correctly.

  This raises the question of why this apparent complexity has been introduced. The answer, at least for our purpose, is that we can save a series of analytical solutions as files, ua_2a.m, ua_2b.m, ..., and to select any particular solution to be called by inital_2.m, we only have to create a file by the name used in the calling statement (ua_2.m). Thus, file ua_2a.m was also saved as file ua_2.m for this particular execution of the various routines. When ua_2b.m (a different analytical solution) is to be used later, it will also be saved as file ua_2.m.

The main program pde_2_main.m is the same as main program pde_1_main.m in Listing 2.1 except that the calls to ua_1 and pde_1 have been changed to ua_2 and pde_2. Files pde_2.m and jpattern_num_2.m are essentially the same as pde_1.m and jpattern_num_1.m. Execution of pde_2_main.m gives three plots as pde_1_main.m discussed previously. One of these plots is discussed next (the other two plots are not included in the discussion at this point).

We can note the following points about Fig. 2.4:

- The numerical solution is highly oscillatory and distorted. Clearly, the MOL solution for IC (2.6) is completely unsatisfactory.
- An error wave (oscillation around $x = 0$) propagates right to left (from the left face of the pulse). This error wave goes past $x = -5$, and when this occurs and the left boundary is $x = -5$, the distortions are even more severe. Thus, as mentioned previously, the left boundary was set at $x = -10$; this only reduced the distortions in the numerical solution but obviously did not eliminate them.
- We cannot achieve the instantaneous change in $f(x)$ from 0 to 1 at $x = 0$ and from 1 to 0 at $x = 1$ defined by eq. (2.6) using ua_2.m and the grid in x with a finite number of points. Rather, the best resolution in x we can achieve for these changes is over the grid spacing $(10-(-10))/(201-1) = 0.1$ for 201 points (since ua_2.m operates on this grid). In other words, $f(x)$ of eq. (2.6) defines two discontinuities at $x = 0, 1$ which cannot be represented exactly with computer arithmetic.

  We could do the plotting of the IC so that vertical lines are drawn at $x = 0, 1$ and $t = 0$, and the subsequent analytical solution having the vertical lines is moved to the right at velocity $c = 1$, but we did not add these details to the plotting (and therefore the plotting in pde_1_main and pde_2_main remains the same).

In conclusion, the use of the five-point FDs in dss004 did not work for IC (2.6). We, therefore, now consider some alternatives that might lead to a better numerical solution.

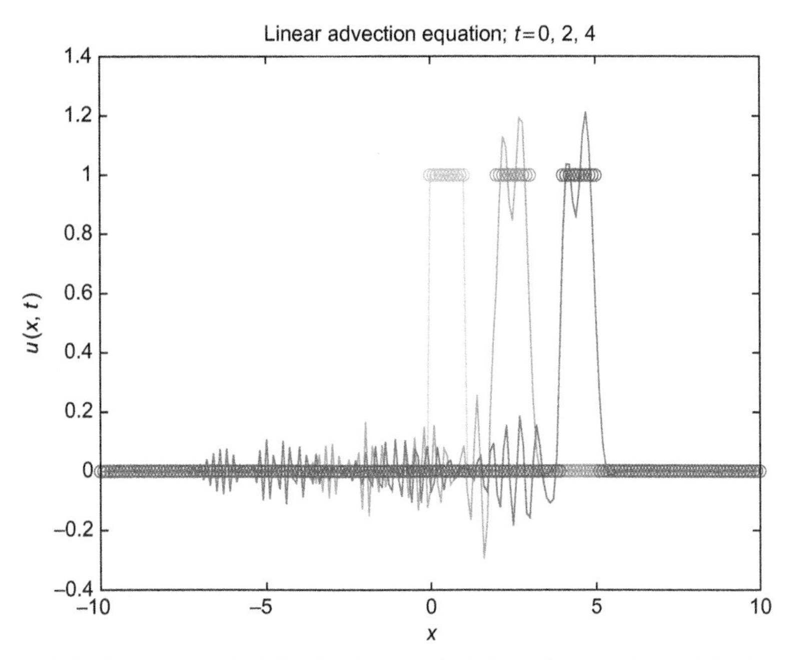

FIGURE 2.4: Numerical solution to eq. (2.1) (lines) with the analytical solution superimposed (circles) using five-point FD approximations in `dss004`.

To start, we consider another FD approximation in place of `dss004`. A FD for the derivative in x in eq. (2.1) can be constructed that uses just two values of the dependent variable.

$$\left(\frac{\partial u}{\partial x}\right)_i \approx \frac{u(i) - u(i-1)}{\Delta x} + O(\Delta x) \tag{2.7}$$

Note that the approximation of the x derivative at grid point i is based on $u(i)$ and $u(i-1)$, that is, at the point i and the point $i-1$ *upstream* or *upwind* of i with respect to the direction of flow (left to right for $c > 0$). Thus, the approximation of eq. (2.7) is termed a *two-point upwind FD approximation*. $O(\Delta x)$, read as *of order* Δx, indicates this approximation is *first order in* Δx or *of order one in* Δx (Δx to the first power).

Equation (2.7) can be built into `pde_2.m` very simply, i.e., the line that calls `dss004` is replaced with

```
%
% ux
  dx=(xu-xl)/(n-1);
  u(1)=0;
  for i=2:n
    ux(i)=(u(i)-u(i-1))/dx;
  end
```

(note that the BC $u(x = -10, t) =$ `u(1)=0` has been applied before the `for` loop so that the `for` loop starts at `i=2`).

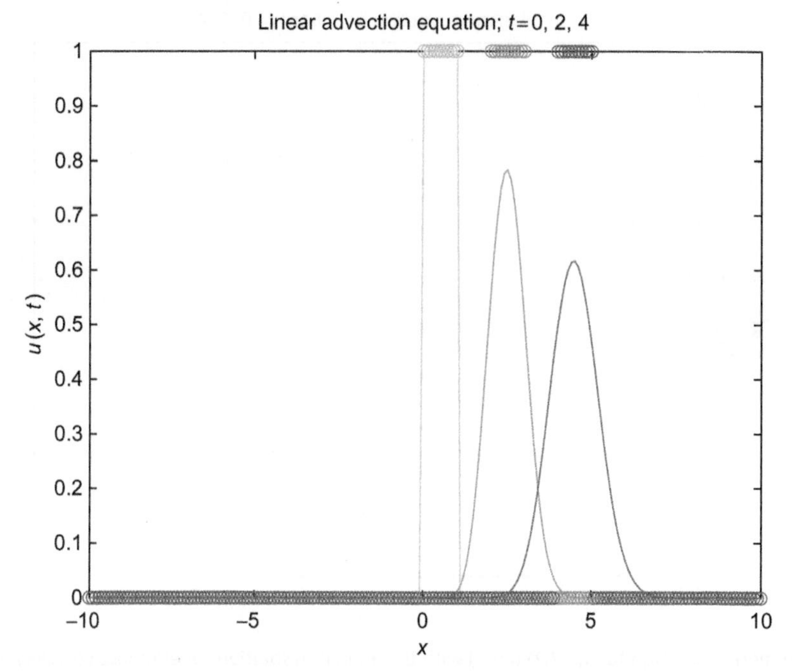

FIGURE 2.5: Numerical solution to eq. (2.1) (lines) with the analytical solution superimposed (circles) using a 2pu FD approximation.

If the preceding routines are executed with this change, the analytical and numerical solutions for $f(x)$ of eq. (2.6) are plotted in Fig. 2.5.

We can note the following points about these solutions:

- The numerical solution again does not match the analytical solution. Thus, the 2pu FD approximation of eq. (2.7) does not produce an acceptable numerical solution.
- The numerical solution does not have the oscillations of Fig. 2.4, but it is excessively smoothed. Figures 2.4 and 2.5 illustrated the two principal types of errors in numerical solutions, *numerical oscillation* (Fig. 2.4) and *numerical diffusion* (Fig. 2.5), which are generally observed when solving *strongly convective* or *strongly hyperbolic* PDEs such as eq. (2.1).
- The numerical distortions of Figs. 2.4 and 2.5 are a consequence of an important theorem, the *Godunov barrier theorem* [5] [6]. To explain,
  - Equations (2.1) and (2.6) are example of a *Riemann problem*, that is a *Cauchy* or *initial value problem* with a discontinuous initial condition, in this case, eq. (2.6).
  - Equation (2.7) is a *linear approximation* in the sense that u appears to the first power or degree.
  - Similarly, the five-point FD approximations in `dss004` are linear, since the dependent variable appears only to the first power. Also, these FD approximations are $O(\Delta x^4)$.

- Godunov's barrier theorem then states that *Thm: There is no linear approximation to the Riemann problem, higher than first order, that is nonoscillatory.* As an example, since the fourth-order FDs of `dss004` are linear (and higher than first order), the numerical solution from these approximations oscillates (as in Fig. 2.4).
- Also, the FD approximation of eq. (2.7) is first order and therefore does not oscillate, but it also produces excessive numerical diffusion.

- In other words, if we are to produce a numerical solution that does not oscillate using approximations of higher order than one (to achieve an acceptable accuracy in the numerical solution), the approximations cannot be linear (they must be *nonlinear*).

We now introduce a nonlinear numerical device, the so-called *flux limiter*, that generally reduces the level of oscillation in the solution. For many problems, flux limiters will either totally eliminate oscillations or, at least reduce them to an acceptable level. Some background to flux limiters is included in the appendix to this chapter.

To illustrate the idea, we repeat the solution of eqns. (2.1) and (2.6) with two representative *flux limiters*, the *van Leer*, and the *superbee* limiters - see Appendix at the end of this chapter for details on these plus 13 additional types. The van Leer and superbee limiters have been programmed in two routines, `vanl1.m` and `super.m`. To conserve space, we will not discuss these routines; they are discussed elsewhere ([26]) and are available with the downloads for this book.

The calls to these routines are easily included in `pde_1.m` (in place of the call to `dss004`).

```
%
% ux
  u(1)=0;
%
%   vanLeer, vl
    ux=vanl1(xl,xu,n,u,c);
%
%   superbee
%   ux=super(xl,xu,n,u,c);
```

We can note the following points about this coding:

- The two limiters have a fifth input argument, `c`, the velocity in eq. (2.1). This argument is required, since the limiters require the direction of flow or wave propagation (as the sign of the fifth argument; the numerical value of the argument is not used). In other words, the limiters are not centered or symmetric with respect to the point where the derivative `ux` is computed. In applications, the direction of flow is often known (from the physical situation such as flow through a pipe or reactor). If the direction of flow is not known, or changes due to flow reversals (e.g., due to reflection off a boundary or wall), flux limiters are available, which are *centered* and therefore work with flow in either direction without having to specify the direction; we will not consider these centered limiters here.

- The function for the van Leer limiter has the name `vanl1`. The `1` was added, since we also have available routines for two other limiters (in routines `vanl2.m` and `vanl3.m`) that are due to van Leer; we will not discuss these routines here.

  The plot of the numerical and analytical solutions for eq. (2.6) produced by `vanl1` appears in Fig. 2.6.

We can note the following points about these solutions:

- Although the numerical solution does not reproduce the analytical solution exactly, it is a major improvement over the numerical solutions of Figs. 2.4 and 2.5, particularly the absence of unrealistic oscillations, which occur for all linear approximations above first order (according to Godunov's theorem). However, there is a certain amount of dissipation or attenuation of the pulse from its initial shape. If the simulation is continued, eventually the dissipation will be considerable and the result will be totally unrealistic.
- In a sense, the IC function of eq. (2.6) presents an impossible problem, since at the points of discontinuity, $\frac{\partial u}{\partial x}$ is undefined (or infinite). Thus, at these points of discontinuity, we are asking the routine to calculate a derivative `ux` that is not defined. Viewed in this way, the fact that the numerical solution in Fig. 2.6 is as good as it is perhaps unexpected and surprising.

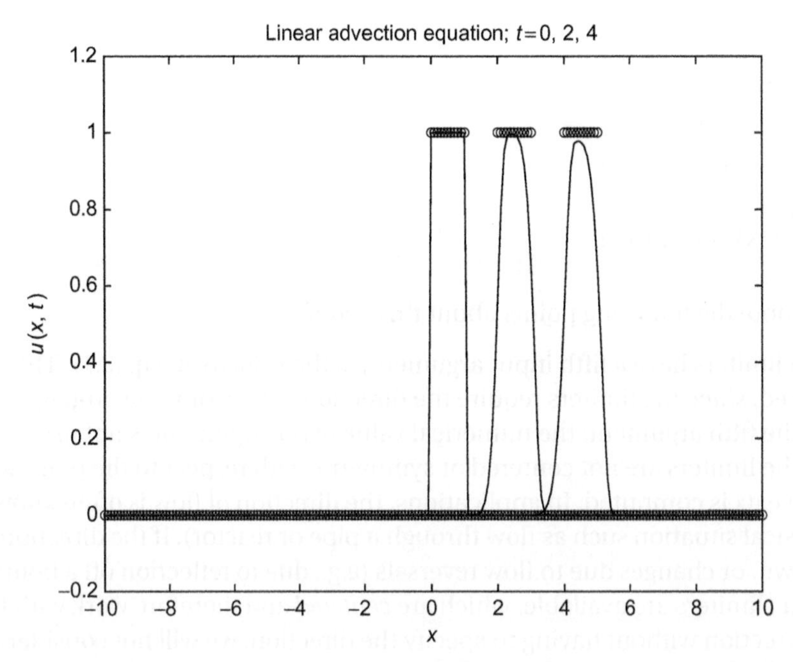

FIGURE 2.6: Numerical solution to eq. (2.1) (lines) with the analytical solution superimposed (circles) using a van Leer flux limiter.

As we might expect, since many flux limiters have been proposed, some will perform better than others. As an example, we consider also the *superbee* limiter for comparison with the van Leer limiter (and executed merely by activating the call to `super` in the preceding code and deactivating the call to `vanl1`). The plotted output follows in Fig. 2.7(A).

We can note the following points about this solution:

- The superbee limiter (Fig. 2.7(A)) performs considerably better than the van Leer limiter (Fig. 2.6). After the initial slight smearing of discontinuities, the superbee limiter performs well, as the pulse then propagates without further distortion or dissipation for a long time. This is highlighted by plot of Fig. 2.7(B) when the later pulses have been superimposed on earlier ones — in fact, apart from a small amount of dissipation, they are almost indistinguishable.
- However, this improved performance requires a somewhat greater calculational effort, `ncall = 1349` for van Leer, `ncall = 4033` for superbee.

We mention briefly that the extended duration numerical solution shown in Fig. 2.7(B) was obtained by using *periodic boundary conditions.* This technique simply connects together the two ends of a 1D linear spatial domain by the addition of an additional cell, thus effectively transforming it into a *ring* [1] see Fig. 2.7(C). The grid points 1 and n, therefore, become neighbors, which results in the system having the same number of grid points as cells, i.e., n. A simple method of implementing periodic BCs is to add *ghost cells* at either end of the spatial domain [14]. Note: for a *finite volume scheme* cell, values are used but for a *finite difference scheme,* grid point values are used. The number of ghost cells depends upon the size of the numerical stencil. For our application with $c > 0$, we use a finite difference scheme based on a *four-point upwind stencil* consisting of grid points numbered $i-2$, $i-1$, i, and $i+1$. This requires a minimum of two ghost cells to be added at the start of the spatial domain and one at the end. Similarly, for our application with $c < 0$, we use a finite difference scheme based on a four-point upwind stencil consisting of grid points numbered $i-1$, i, $i+1$, and $i+2$. But now, we require a minimum of one ghost cell to be added at the start of the spatial domain and two at the end. For ease of programming, it is usual to add the same number of ghost cells to each end. Therefore, assuming that the spatial domain consists of grid points numbered from 1 to n, which are populated with variables u_1 to u_n, respectively, we insert two ghost cells at each end and set the grid values as follows:

$$u_{-1} = u_{n-1}, \quad u_0 = u_n \qquad \text{Cells inserted at the start of spatial domain}$$

$$u_{n+1} = u_1, \quad u_{n+2} = u_2 \qquad \text{Cells inserted at the end of spatial domain}$$

The limiter routines provided along with the downloads for this book include a flag `periodic_BC` that can be set to `0` for nonperiodic BCs or `1` for periodic BCs. This facility can be used by the reader to explore the performance of flux limiters over extended time durations

[1]For a 2D *Cartesian coordinate* problem, periodic BCs effectively transform the spatial domain into a *torus.*

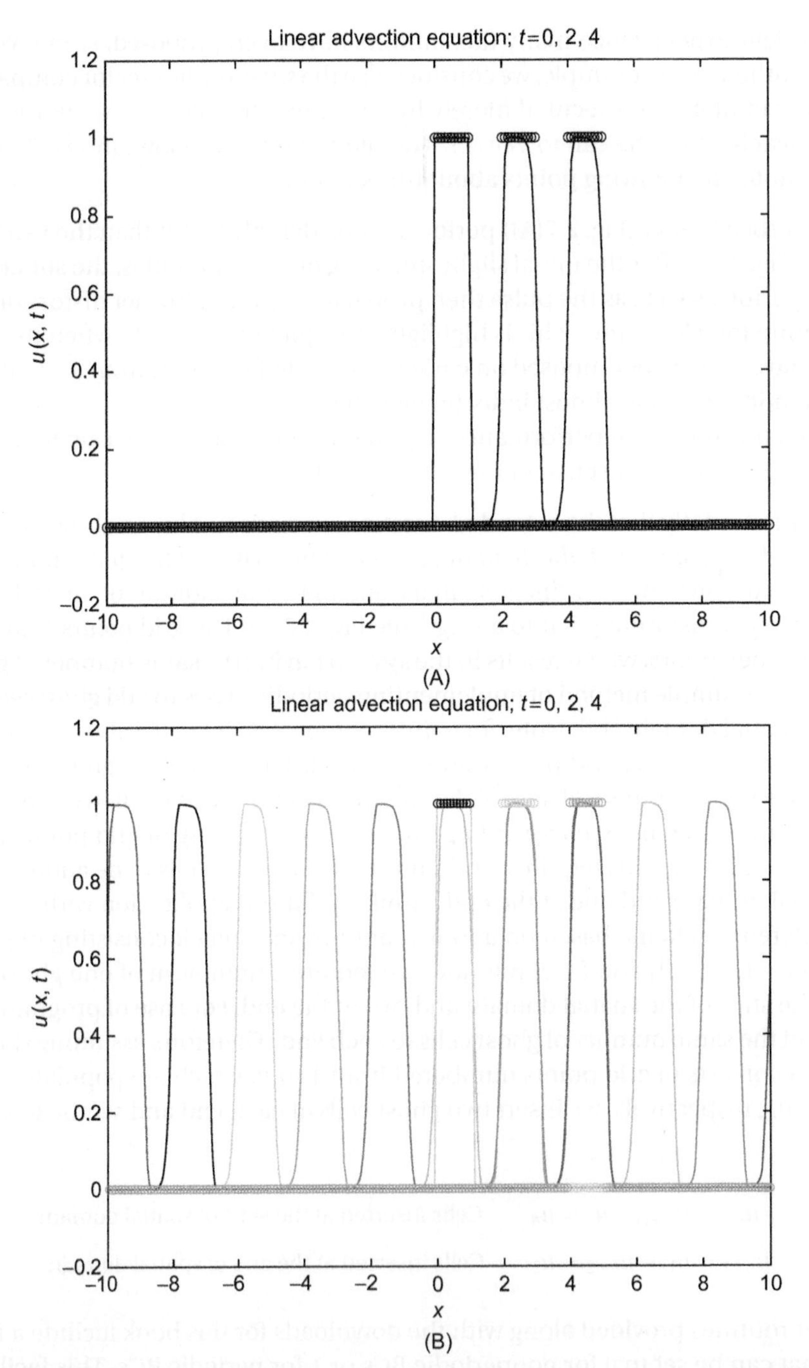

FIGURE 2.7: (A) Numerical solution to eq. (2.1) (lines) with the analytical solution superimposed (circles) using the superbee flux limiter. (B) Extended duration (`tf=24`) numerical solution to eq. (2.1) (lines) with the analytical solution superimposed (circles) using the superbee flux limiter. (C) (a) *Linear spatial domain* with n grid points and $n-1$ cells. (b) *Periodic spatial domain* defined with the linear domain repeated indefinitely left and right. (c) *Equivalent periodic spatial domain* defined as a *ring* with n cells and n grid points. Image (C) is shown on the following page.

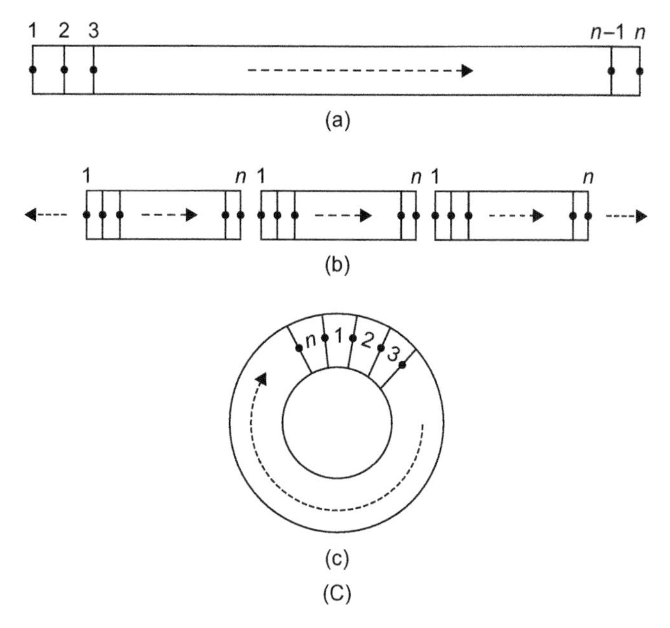

FIGURE 2.7: *(Continued)*

where it would be impractical to plot the entire spatial domain. It will be clear, therefore, that long simulation times (where $c \times t > x_{\max}$) result in the solution propagating around the spatial domain. Note: Periodic BCs can occur naturally in problems described in cylindrical coordinates.

Generally, we observe that the computational effort can vary widely for different limiters without a major improvement in the accuracy of the numerical solution; thus, we have found superbee to be a good compromise between accuracy and computational efficiency. The reader could certainly come to a different conclusion when comparing the performance of the many limiters that have been proposed (and which can be coded through minor variations in `van11.m` or `super.m`). It should also be noted that for some applications the superbee limiter can result in unrealistic sharpening of the solution.

We now consider a second IC function with a discontinuity, the *Heaviside* (unit) step function, defined in eq. (2.8)

$$f(x) = \begin{cases} 0, & x < 0 \\ 1, & x > 1 \end{cases} \tag{2.8}$$

Equations (2.1) and (2.8) again define a Riemann problem (because of the discontinuous IC of eq. (2.8)). A step as defined in eq. (2.8) occurs more frequently in applications than the pulse of eq. (2.6). As eq. (2.3) indicates, the analytical solution is a unit step that propagates from left to right with velocity $c > 0$. This type of solution is often termed a *moving front*, and an extensive literature is available for computing solutions to moving front problems. Here, we just repeat three methods we considered previously, five-point centered FDs (in `dss004`), 2pu, and the superbee flux limiter.

The analytical solution is easily programmed in ua_2.m

```
  function uanal=ua_2b(x,t)
%
% Function uanal computes the exact solution of the linear advection
% equation for comparison with the numerical solution
  global c
%
% Analytical solution
  xi=x-c*t;
  if(xi<0)
    uanal=0;
  elseif(xi>=0)
    uanal=1;
  end
```

LISTING 2.7: Function ua_2.m for IC (2.8).

Note again the use of the variable $\xi = x - ct$ so that we have a Lagrangian (moving coordinate) solution of eqns. (2.1) and (2.8) in function ua_2.m of Listing 2.7. All of the other preceding routines remained unchanged in using IC (2.8) rather than IC (2.6) (with ua_2b.m saved as file ua_2.m as explained previously).

The plotted output from the five-point centered FDs in dss004 appears in Fig. 2.8(A). We note that the numerical solution is highly oscillatory with an error wave propagating right to left from the left face of the step (as in Fig. 2.4)).

The plotted output with the 2pu approximation of the derivative in x in eq. (2.1) appears in Fig. 2.8(B). Again, as in Fig. 2.5, the numerical solution is smooth but excessively damped.

The plotted output with the superbee flux limiter appears in Fig. 2.9. Again, as in Fig. 2.6, the numerical solution is a substantial improvement over Figs. 2.8(A) and 2.8(B). The numerical solution is not a vertical line at the points of discontinuity as defined by eqns. (2.3) and (2.8), but there is no oscillation or *overshoot* as in Fig. 2.8(A) and the smoothing (numerical diffusion) is minimal compared with Fig. 2.8(B). Also, we again should appreciate that we are asking for the calculation of a derivative in x that is undefined at the points of discontinuity and viewed this way, the numerical solution of Fig. 2.9 is quite satisfactory. If the PDE problem has any smoothing, such as a second derivative in x, so that a discontinuity does not occur, the differences between the numerical and analytical solutions will probably be imperceptible when plotted. In other words, the superbee flux limiter provides a high-quality numerical solution with no additional programming (in pde_1.m).

To conclude the discussion of eqns. (2.1), (2.2), (2.6), and (2.8), because this strongly hyperbolic PDE propagates discontinuities, it requires special attention to compute a solution of reasonably good accuracy, i.e., the use of a nonlinear approximation for ux rather than a linear approximation. We can add some additional points:

- Generally for physical systems, true discontinuities do not occur. In other words, there is usually some smoothing from physical phenomena or processes, such as diffusion

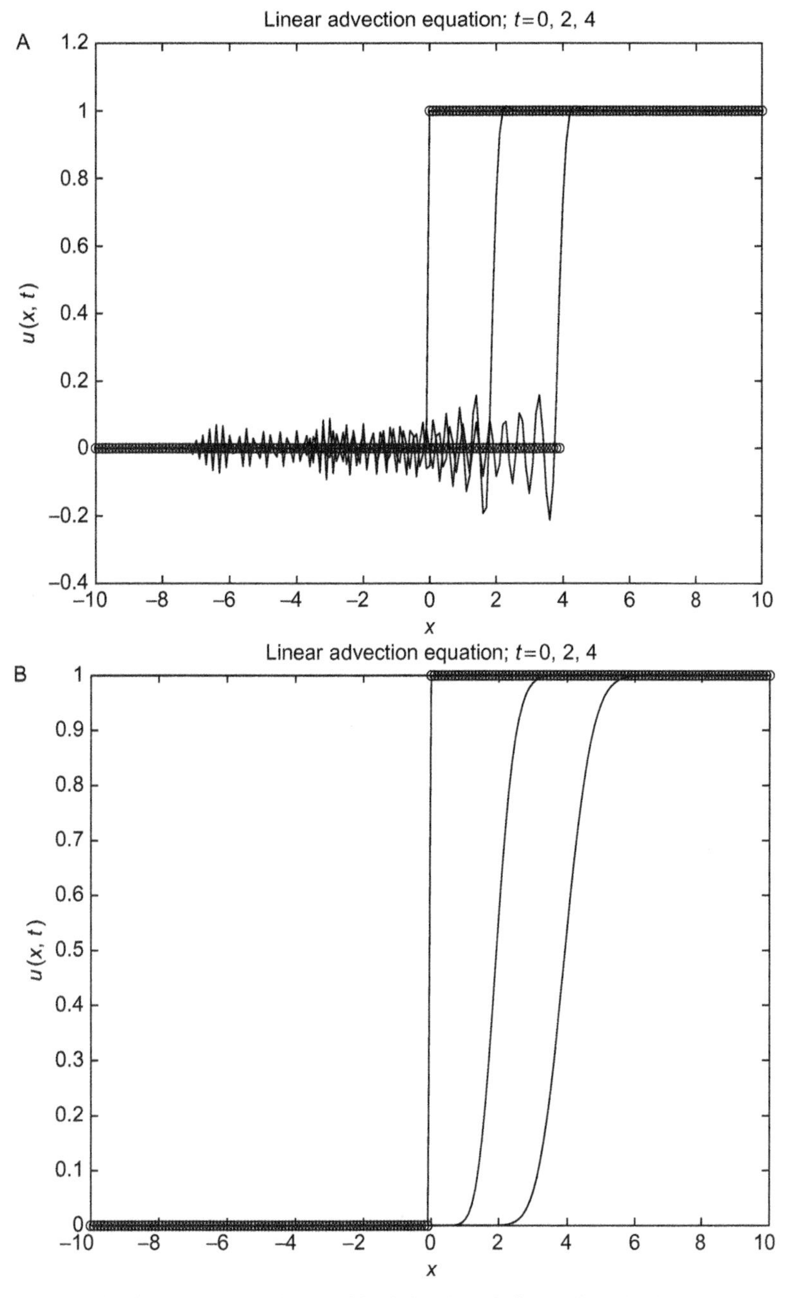

FIGURE 2.8: (A) Numerical solution to eqns. (2.1) and (2.8) (lines) with the analytical solution superimposed (circles) using five-point FDs in `dss004` (B) Numerical solution to eqns. (2.1) and (2.8) (lines) with the analytical solution superimposed (circles) using 2pu FDs.

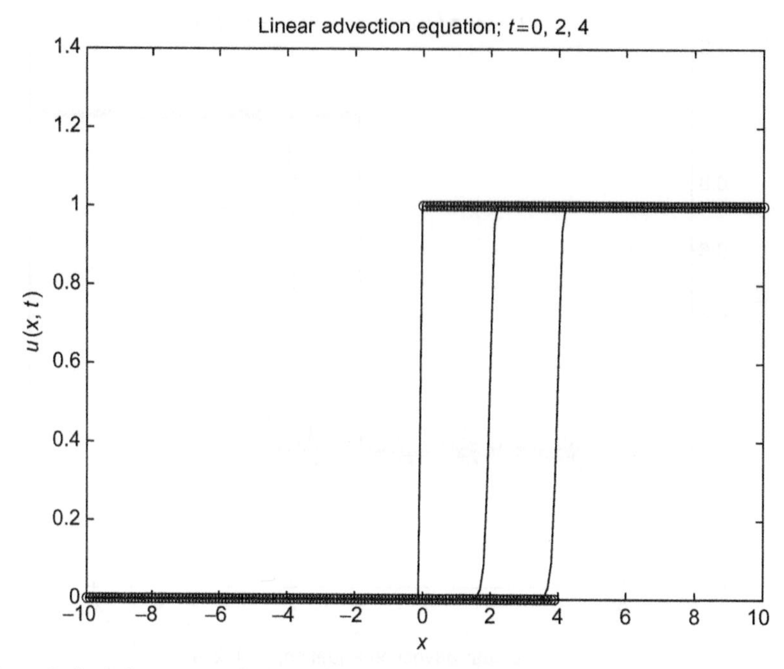

FIGURE 2.9: Numerical solution to eqns. (2.1) and (2.8) (lines) with the analytical solution superimposed (circles) using the superbee flux limiter.

or the effect of viscosity in fluid flow, that precludes a discontinuity. Thus, test problems with discontinuities, such as eqns. (2.6) and (2.8), are generally not completely realistic physically, or in other words, are overly stringent in testing the applicability of numerical methods to physical problems.

- When we used the FD approximations of `dss044` or the 2pu approximation of eq. (2.7), we might naturally think that they should produce an acceptable numerical solution, especially since the accuracy of these approximations improves with increasing numbers of grid points (smaller Δx). However, if we ran some additional cases for larger numbers of grid points (than $n = 201$), we would find the accuracy of the numerical solutions does not improve substantially (they still have excessive numerical diffusion and oscillation). Also, we might reason that using higher-order FDs (using more grid points than the five in `dss004`) would give better numerical solutions; in fact, if we used higher-order FDs, we would find that the numerical solutions oscillate even more than in Figs. 2.4 and 2.8. Thus, we conclude that using more grid points (generally termed *h refinement*) or higher-order approximations (generally termed *p refinement*) will not necessarily produce numerical solutions of acceptable accuracy, especially for strongly hyperbolic problems. Fortunately, for less difficult PDE problems, *h* and *p* refinement are effective.
- Somewhat ironically, although eq. (2.1) is perhaps the simplest PDE we can envision, it also is one of the most difficult to solve numerically, depending on the properties of

$f(x)$ in eq. (2.2). PDEs that have a small amount of smoothing or diffusion, typically through a second derivative in x, appear to be more complicated than eq. (2.1), but they are actually easier to integrate numerically than eq. (2.1). PDEs with first- and second-order derivatives in x are termed *hyperbolic–parabolic* (or *convective–diffusive*); we will consider some examples of these *mixed-type* PDEs subsequently.

- The resolution of steep moving fronts, such as those from ICs (2.6) and (2.8) can be achieved by numerical methods other than flux limiter schemes. They include *weighted essentially non-oscillatory* (WENO) [21], *level set* [16], and *discontinuous Galerkin* [8] methods. These recently developed schemes can be thought of as high-resolution extensions to the MOL approach. However, they are more advanced and are still the subject of ongoing research; as such, they will not be discussed further here. An additional possibility would be the *method of characteristics* (MOC), which would give exactly the correct numerical solution to eqns. (2.1), (2.6), and (2.8). It has been used successfully for many years in application areas, such as *adsorption, chromatography* [1, 17], and *water-hammer* [22]. However, our experience has indicated that more accurate numerical methods such as MOC are difficult to implement in a general-purpose format (they require an analysis for each particular PDE application, followed by some specialized coding), whereas the MOL approach is quite general and can be applied to a spectrum of PDE problems, both linear and nonlinear, with simultaneous PDEs. We demonstrate this generality in the subsequent applications. MOL and MOC have also been combined in the solution of the *Euler equations for fluid mechanics* [2].
- The accurate resolution of steep moving fronts is determined by the smoothness of the fronts (with discontinuities being the most difficult). Mathematically, the functions of eqns. (2.6) and (2.8) are discontinuous in the solution ($u(x,t)$ of eq. (2.1)). If these ICs are continuous in $u(x,t=0)$ but discontinuous in the first x derivative, the calculation of the numerical solution is much easier. To demonstrate this point, we consider a *triangular pulse* IC that is continuous in $u(x,t=0)$ but discontinuous in $\dfrac{\partial u(x,t=0)}{\partial x}$.

The triangular pulse is defined as

$$f(x) = \begin{cases} 0, & x < 0 \\ x/2, & 0 \le x < 2 \\ 2 - x/2, & 2 \le x < 4 \\ 0, & x \ge 4 \end{cases} \tag{2.9}$$

We can note the following points about this $f(x)$ in eq. (2.2):

- Equation (2.3) indicates that this pulse moves left to right with velocity $c = 1$, which is subsequently used to evaluate the numerical solution.
- The function consists of four sections:
  - For $x < 0$, the function of eq. (2.9) is zero.

- For $0 \leq x < 2$, the function increases linearly from $f(0) = 0$ to $f(2) = 1$.
- For $2 \leq x < 4$, the function decreases linearly from $f(2) = 1$ to $f(4) = 0$.
- For $x \geq 4$, the function is zero.

Thus, $f(x)$ is continuous in x, but its first derivative is discontinuous at $x = 0, 2, 4$. The first derivative undergoes its largest discontinuity at $x = 2$ so that it provides the most stringent test of the numerical solution at $x = 2$, as will be observed in the subsequent comparison of the analytical and numerical solutions.

Since the triangular pulse of eq. (2.9) is a significant departure from the previous test problems, the associated Matlab files will be given the designation 3. For example, the main programs are numbered:

- `pde_1_main.m` - Gaussian pulse of eq. (2.4)
- `pde_2_main.m` - square pulse of eq. (2.6) and step of eq. (2.8)
- `pde_3_main.m` - triangular pulse of eq. (2.9)

Thus, for the triangular pulse of eq. (2.9), the Matlab routines are `pde_3_main.m`, `pde_3.m`, `inital_3.m`, and `ua_3.m`.

The routine that implements IC (2.9) is as follows:

```
  function uanal=ua_3(x,t)
%
% Function uanal computes the exact solution of the linear advection
% equation for comparison with the numerical solution
  global c
%
% Analytical solution
  xi=x-c*t;
  if(xi<0)
    uanal=0;
  elseif((xi>=0)&(xi<=2))
    uanal=(x-c*t)/2;
  elseif((xi>=1)&(xi<=4))
    uanal=2-(x-c*t)/2;
  elseif(xi>1)
    uanal=0;
  end
```

LISTING 2.8: Function `ua_3.m` for IC (2.9).

The coding in Listing 2.8 is a straightforward implementation of eq. (2.9). `ua_3.m` is then called in `pde_3_main.m`, `pde_3.m`, and `inital_3.m`, `dss004` is called in `pde_3.m` to give a five-point FD numerical solution to eqns. (2.1), (2.2), and (2.9). The plotted output is in Fig. 2.10.

The agreement between the numerical and analytical solutions is generally quite good. The only points of significant difference are at the peak where the two linear segments of the solutions are connected ($x = 2, 4, 6$ corresponding to $t = 0, 2, 4$ for $c = 1$). The tabulated numerical output near these three points is listed in Table 2.2.

As expected, the IC ($t = 0$) values of the peak agree (since the numerical and analytical solutions always start from the same IC). At $t = 2$, the numerical peak value is `0.970590`,

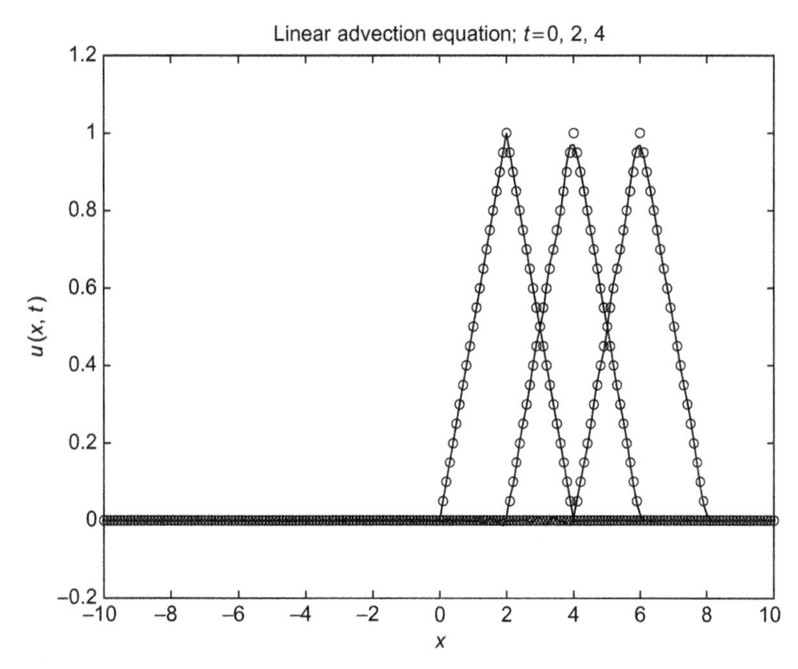

FIGURE 2.10: Numerical solution to eqns. (2.1) and (2.9) (lines) with the analytical solution superimposed (circles) using five-point FDs in `dss004`.

whereas at $t = 4$, it is `0.968542`. Although the errors are significant, they do not represent a major distortion of the numerical solution (as indicated in Fig. 2.10).

In an effort to improve these peak values, two other numerical solutions were computed:

- Nine-point, eighth-order FDs in routine `dss008` were used in place of `dss004`. The peak values are summarized in Table 2.3. Although the peak values improved marginally (closer to `1.000000`), unexpectedly, an error wave developed near the left boundary $x = -10$ as indicated in Fig. 2.11.
- The superbee flux limiter was called in `pde_3.m`. The peak values are summarized in Table 2.4. The peak values are actually less accurate than for `dss004`, and the number of calls to `pde_3.m` increased from `ncall = 515` (`dss004`) to `ncall = 3055` (`super`). Thus, both efforts to improve the peak values were unsuccessful.

In summary, the smoothness of the triangular pulse IC of eq. (2.9) relative to the square pulse IC of eq. (2.8) provided a generally satisfactory numerical solution. This conclusion also indicates why the numerical solution for the Gaussian IC of eq. (2.4) was relatively easy to compute, since this function is continuous and has continuous derivatives of all orders. Also, the numerical solution for the triangular pulse of eq. (2.9) deteriorates as the pulse is narrowed (below the width of 4) as might be expected; presumably, this could be countered by adding more grid points (in all of the previous cases, $n = 201$).

Table 2.2: Abbreviated and tabular numerical and analytical solutions for eqns. (2.1), (2.2), and (2.9) from dss004

t	x	u(it,i)	u_anal(it,i)	err(it,i)
0.00	0.000	0.000000	0.000000	0.000000
0.00	1.000	0.500000	0.500000	0.000000
0.00	2.000	1.000000	1.000000	0.000000
0.00	3.000	0.500000	0.500000	0.000000
0.00	4.000	0.000000	0.000000	0.000000
t	x	u(it,i)	u_anal(it,i)	err(it,i)
2.00	2.000	0.013343	0.000000	0.013343
2.00	3.000	0.495102	0.500000	−0.004898
2.00	4.000	0.970590	1.000000	−0.029410
2.00	5.000	0.503054	0.500000	0.003054
2.00	6.000	0.014590	0.000000	0.014590
t	x	u(it,i)	u_anal(it,i)	err(it,i)
4.00	4.000	0.012795	0.000000	0.012795
4.00	5.000	0.493201	0.500000	−0.006799
4.00	6.000	0.968542	1.000000	−0.031458
4.00	7.000	0.502513	0.500000	0.002513
4.00	8.000	0.017230	0.000000	0.017230
ncall=459				

As a concluding point, we mention that the various traveling wave functions (Gaussian, square pulse, and triangular pulse) did not reach the boundaries in x; thus, the boundary values remained at zero. If the traveling wave functions had reached either boundary in x, the solution would have become more complicated and would be determined by what is assumed and computed at the boundaries; that is, the BC in x would have become more consequential. We will not consider this added complication (however, in physical applications, it can be important). Additional details concerning MOL analysis for traveling waves including boundary effects can be found in [19].

This rather extended discussion of eq. (2.1) was provided at this point to demonstrate that PDE applications are not necessarily straightforward to solve and each PDE problem *is usually a new problem* with no guarantee of success in advance for computing an acceptable numerical solution; to a minimum extent, some trial and error may be required to produce a numerical solution of acceptable accuracy.

In subsequent chapters, we analyze a series of PDE applications for which the computation of acceptable numerical solutions was achieved (generally confirmed through comparisons with analytical solutions). We would not, however, wish to leave the impression that the programming and testing of the computer code led directly to an acceptable solution. Generally, we had to experiment to arrive at a workable numerical procedure, and we wish to convey to the reader that this is a normal part of this type of experimental investigation.

Table 2.3: Abbreviated and tabular numerical and analytical solutions for eqns. (2.1), (2.2), and (2.9) from dss008

t	x	u(it,i)	u_anal(it,i)	err(it,i)
0.00	0.000	0.000000	0.000000	0.000000
0.00	1.000	0.500000	0.500000	0.000000
0.00	2.000	1.000000	1.000000	0.000000
0.00	3.000	0.500000	0.500000	0.000000
0.00	4.000	0.000000	0.000000	0.000000
t	x	u(it,i)	u_anal(it,i)	err(it,i)
2.00	2.000	0.007239	0.000000	0.007239
2.00	3.000	0.497509	0.500000	−0.002491
2.00	4.000	0.983813	1.000000	−0.016187
2.00	5.000	0.501166	0.500000	0.001166
2.00	6.000	0.008799	0.000000	0.008799
t	x	u(it,i)	u_anal(it,i)	err(it,i)
4.00	4.000	0.009567	0.000000	0.009567
4.00	5.000	0.505752	0.500000	0.005752
4.00	6.000	0.979748	1.000000	−0.020252
4.00	7.000	0.497179	0.500000	−0.002821
4.00	8.000	0.009959	0.000000	0.009959
ncall=515				

Appendix[2]

Flux limiters are used in numerical schemes to solve problems in science and engineering, particularly fluid dynamics, that are described by *partial differential equations* (PDEs). Their main purpose is to avoid the spurious oscillations (wiggles) that would otherwise occur with high-order spatial discretization (as predicted by Godunov [5, 6]) due to shocks, discontinuities, or steep gradients in the solution domain. They can be used directly on *finite difference schemes* for simple applications, such as the 1D advection PDE of eq. (2.1). However, for more complex systems involving conservation laws, *finite volume high-resolution schemes* are employed that are also designed to avoid entropy violations.

A popular high-resolution scheme is the *MUSCL scheme*—MUSCL stands for *Monotone Upstream-centered Schemes for Conservation Laws* [3]. The term was introduced in a seminal paper by Bram van Leer [30]. In this paper, he constructed the first high-order, *total variation diminishing* (TVD) scheme, where he obtained second-order spatial accuracy. The concept of TVD was introduced by Ami Harten [7] and relates to a

[2]This appendix is based on the Wikipedia article, http://en.wikipedia.org/wiki/Flux_limiters, written by one of the authors.

[3]A readable overview of the MUSCL scheme is available on-line at: http://en.wikipedia.org/wiki/MUSCL_scheme.

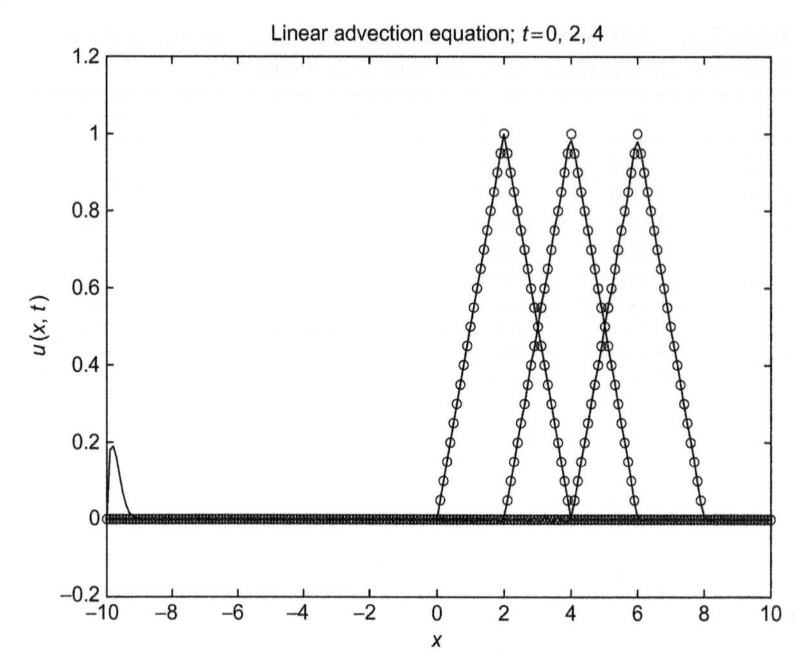

FIGURE 2.11: Numerical solution to eqns. (2.1) and (2.9) (lines) with the analytical solution superimposed (circles) using nine-point FDs in `dss008`.

monotone numerical scheme. A system is said to be *monotonicity preserving* if the following properties are maintained over time:

- *No new local extrema are created* within the solution spatial domain,
- The value of a *local minimum is nondecreasing*, and the value of a *local maximum is nonincreasing.*

Use of flux limiters, together with an appropriate high-resolution scheme, make the solutions *total variation diminishing.*

Flux limiters are also referred to as *slope limiters* because they both have the same mathematical form and both have the effect of limiting the solution gradient near shocks or discontinuities. In general, the term *flux limiter* is used when the limiter acts on system *fluxes,* and *slope limiter* is used when the limiter acts on system *states.*

How They Work

The main idea behind the construction of flux limiter schemes is to limit the spatial derivatives to realistic values—for scientific and engineering problems, this usually means physically realizable values. They are used in high-resolution schemes for solving problems described by PDE's and only come into operation when sharp wave fronts are present. For smoothly changing waves, the flux limiters do not operate and the spatial derivatives can

Table 2.4: Abbreviated and tabular numerical and analytical solutions for eqns. (2.1), (2.2), and (2.9) from `super.m`.

t	x	u(it,i)	u_anal(it,i)	err(it,i)
0.00	0.000	0.000000	0.000000	0.000000
0.00	1.000	0.500000	0.500000	0.000000
0.00	2.000	1.000000	1.000000	0.000000
0.00	3.000	0.500000	0.500000	0.000000
0.00	4.000	0.000000	0.000000	0.000000
t	x	u(it,i)	u_anal(it,i)	err(it,i)
2.00	2.000	0.012115	0.000000	0.012115
2.00	3.000	0.500211	0.500000	0.000211
2.00	4.000	0.950851	1.000000	−0.049149
2.00	5.000	0.498828	0.500000	−0.001172
2.00	6.000	0.002511	0.000000	0.002511
t	x	u(it,i)	u_anal(it,i)	err(it,i)
4.00	4.000	0.012292	0.000000	0.012292
4.00	5.000	0.500114	0.500000	0.000114
4.00	6.000	0.943321	1.000000	−0.056679
4.00	7.000	0.493475	0.500000	−0.006525
4.00	8.000	0.002608	0.000000	0.002608
	ncall=3055			

be represented by higher-order approximations without introducing nonreal oscillations. Consider the 1D *semidiscrete scheme* below,

$$\frac{du_i}{dt}+\frac{1}{\Delta x_i}\left[F\left(u_{i+\frac{1}{2}}\right)-F\left(u_{i-\frac{1}{2}}\right)\right]=0$$

where, for a *finite volume scheme*, $F(u_{i+\frac{1}{2}})$ and $F(u_{i-\frac{1}{2}})$ represent edge fluxes for the *ith* cell. Similarly, for a *finite difference scheme*, they represent flux values on the grid at point $x=x_{i+\frac{1}{2}}$ and point $x=x_{i-\frac{1}{2}}$. If these fluxes can be represented by *low-* and *high-*resolution schemes, then a flux limiter can switch between these schemes depending upon the gradients close to the particular cell as follows:

$$F\left(u_{i+\frac{1}{2}}\right)=f^{\text{low}}_{i+\frac{1}{2}}-\phi(r_i)\left(f^{\text{low}}_{i+\frac{1}{2}}-f^{\text{high}}_{i+\frac{1}{2}}\right)$$

$$F\left(u_{i-\frac{1}{2}}\right)=f^{\text{low}}_{i-\frac{1}{2}}-\phi(r_{i-1})\left(f^{\text{low}}_{i-\frac{1}{2}}-f^{\text{high}}_{i-\frac{1}{2}}\right)$$

where, $f^{low} =$ low-resolution flux, $f^{high} =$ high-resolution flux, $\phi(r) =$ flux limiter function, and r represents the ratio of successive gradients on the solution mesh, i.e.,

$$r_i = \frac{u_i - u_{i-1}}{u_{i+1} - u_i}$$

The limiter function is constrained to be greater than or equal to zero, i.e., $r \geq 0$. Therefore, when the limiter is equal to zero (sharp gradient, opposite slopes, or zero gradient), the flux is represented by a *low-resolution scheme*. Similarly, when the limiter is equal to 1 (smooth solution), it is represented by a *high-resolution scheme*. The various limiters listed below have differing switching characteristics and are selected to suit the particular problem and numerical solution scheme. No particular limiter has been found to work well for all problems, and a particular choice is usually made on a trial-and-error basis.

Limiter Functions

The following are common forms of flux/slope limiter function, $\phi(r)$:

*CHARM** [33]

$$\phi_{cm}(r) = \begin{cases} \dfrac{r(3r+1)}{(r+2)^2}, & r > 0, \quad \lim_{r\to\infty} \phi_{cm}(r) = 3 \\ 0, & r \leq 0 \end{cases}$$

*HCUS** [31]

$$\phi_{hc}(r) = \frac{1.5(r+|r|)}{(r+2)}; \quad \lim_{r\to\infty} \phi_{hc}(r) = 3$$

*HQUICK** [13]

$$\phi_{hq}(r) = \frac{2(r+|r|)}{(r+3)}; \quad \lim_{r\to\infty} \phi_{hq}(r) = 4$$

Koren [11]

$$\phi_{kn}(r) = \max[0, \min(2r, (1+2r)/3, 2)]; \quad \lim_{r\to\infty} \phi_{kn}(r) = 2$$

minmod# [18]

$$\phi_{mm}(r) = \max[0, \min(1, r)]; \quad \lim_{r\to\infty} \phi_{mm}(r) = 1$$

monotonized central (MC)# [29]

$$\phi_{mc}(r) = \max[0, \min(2r, 0.5(1+r), 2)]; \quad \lim_{r\to\infty} \phi_{mc}(r) = 2$$

Osher [3]

$$\phi_{os}(r) = \max\left[0, \min\left(r, \beta\right)\right], \quad (1 \le \beta \le 2); \quad \lim_{r\to\infty} \phi_{os}(r) = \beta$$

ospre# [31]

$$\phi_{op}(r) = \frac{1.5\left(r^2 + r\right)}{\left(r^2 + r + 1\right)}; \quad \lim_{r\to\infty} \phi_{op}(r) = 1.5$$

*smart** [4]

$$\phi_{sm}(r) = \max\left[0, \min\left(2r, (0.25 + 0.75r), 4\right)\right]; \quad \lim_{r\to\infty} \phi_{sm}(r) = 4$$

superbee# [18]

$$\phi_{sb}(r) = \max\left[0, \min\left(2r, 1\right), \min\left(r, 2\right)\right]; \quad \lim_{r\to\infty} \phi_{sb}(r) = 2$$

Sweby# [23]

$$\phi_{sw}(r) = \max\left[0, \min\left(\beta r, 1\right), \left(r, \beta\right)\right], \quad \min\left(1 \le \beta \le 2\right); \quad \lim_{r\to\infty} \phi_{sw}(r) = \beta$$

UMIST [15]

$$\phi_{um}(r) = \max\left[0, \min\left(2r, (0.25 + 0.75r), (0.75 + 0.25r), 2\right)\right]; \quad \lim_{r\to\infty} \phi_{um}(r) = 2$$

van Albada 1# [27]

$$\phi_{va1}(r) = \frac{r^2 + r}{r^2 + 1}; \quad \lim_{r\to\infty} \phi_{va1}(r) = 1$$

*van Albada 2** Alternative form used on high spatial order schemes [10]

$$\phi_{va2}(r) = \frac{2r}{r^2 + 1}; \quad \lim_{r\to\infty} \phi_{va2}(r) = 0$$

van Leer# [28]

$$\phi_{vl}(r) = \frac{r+|r|}{1+r}; \quad \lim_{r\to\infty} \phi_{vl}(r) = 2$$

* Limiter is *not second-order TVD*!
# Limiter is *symmetric* and exhibits the following symmetry property,

$$\frac{\phi(r)}{r} = \phi\left(\frac{1}{r}\right)$$

This is a desirable property, as it ensures that the limiting actions for forward and backward gradients operate in the same way.

Unless indicated to the contrary, the above limiter functions are second-order *total variation diminishing* (TVD). This means that they are designed such that they pass through a certain region of the solution, known as the TVD region, in order to guarantee stability of the scheme. Second-order, TVD limiters satisfy at least the following criteria:

- $r \le \phi(r) \le 2r, (0 \le r \le 1)$,
- $1 \le \phi(r) \le r, (1 \le r \le 2)$,
- $1 \le \phi(r) \le 2, (r > 2)$,
- $\phi(1) = 1$.

The admissible limiter region for second-order TVD schemes is shown in the *Sweby Diagram* Fig. 2.12 [23] and plots showing limiter functions overlaid onto the TVD region are shown in Fig. 2.13. In this image, plots for the Osher and Sweby limiters have been generated using $\beta = 1.5$.

For further information relating to the theory and application of *flux limiters*, the reader is referred to [9] [12] [25] [24], and [32].

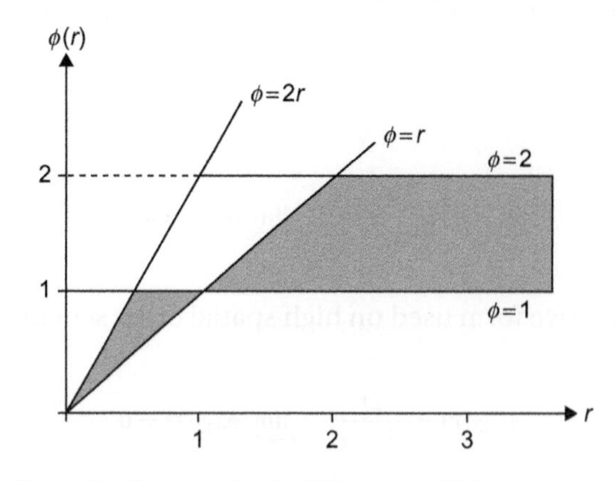

FIGURE 2.12: Admissible limiter region for second-order TVD schemes [23].

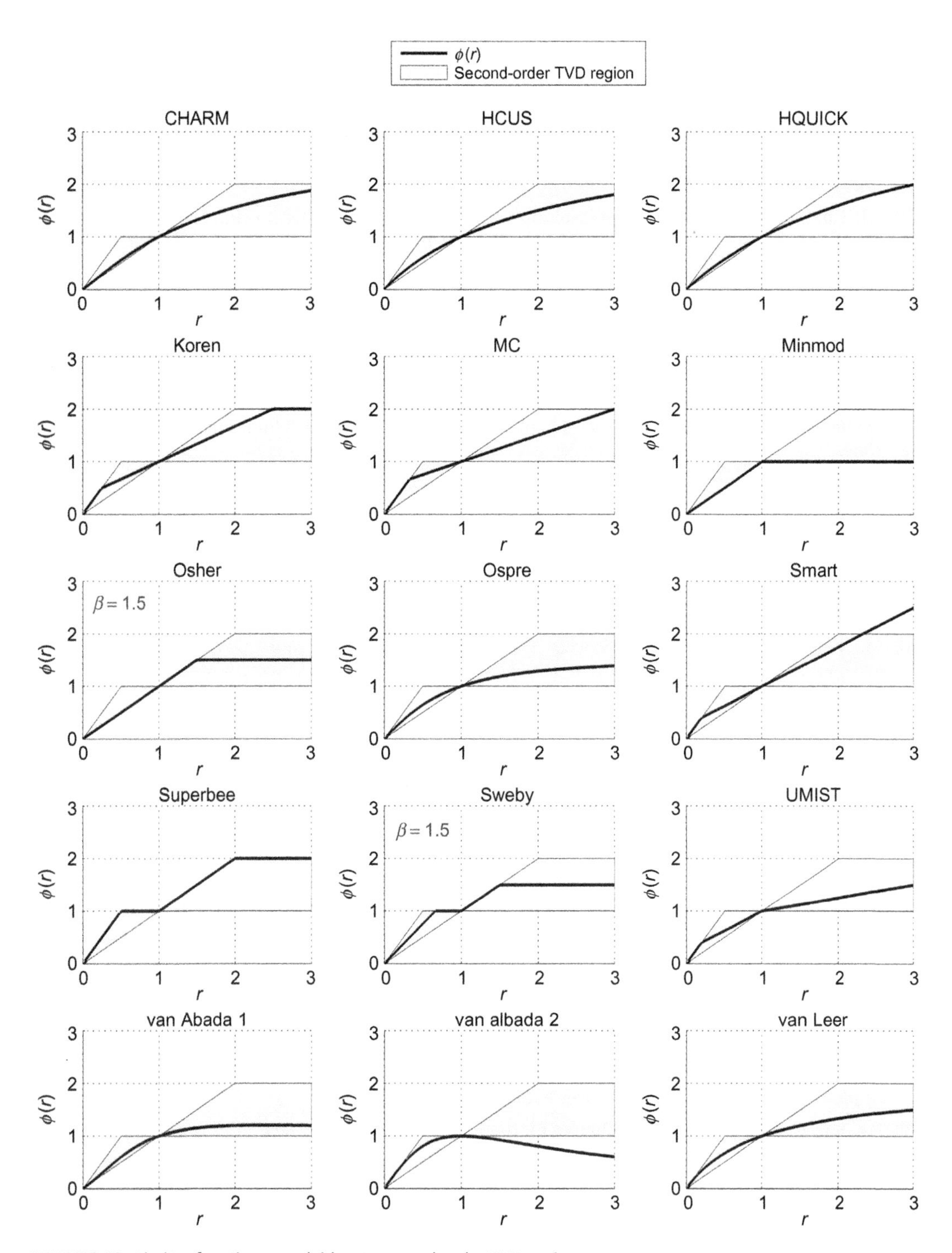

FIGURE 2.13: Limiter functions overlaid onto second-order TVD region.

References

[1] A. Acrivos, Method of characteristics technique: application to heat and mass transfer, *Ind. Eng. Chem.* 48 (4) (1956) 703–710.

[2] M.B. Carver, Pseudo characteristic method of lines solution of the conservation equations, *J. Computat. Phys.* 35 (1) (1980) 57–76.

[3] S.R. Chatkravathy, S. Osher, High resolution applications of the Osher upwind scheme for the Euler equations, in: *AIAA Paper 83-1943, Proceedings of the AIAA Sixth Comutational Fluid Dynamics Conference*, (1983) 363–373.

[4] P.H. Gaskell, A.K.C. Lau, Curvature-compensated convective transport: SMART, a new boundedness-preserving transport algorithm, *Int. J. Num. Meth. Fluids*, 8 (1988) 617.

[5] Godunov, K. Sergei, *Different Methods for Shock Waves*, Ph.D. Thesis, Moscow State University, Moscow, 1954.

[6] Godunov, K. Sergei, A Difference Scheme for Numerical Solution of Discontinuous Solution of Hydrodynamic Equations, *Math. Sbornik.*, 47 (1959) 271–306. Translated US Joint Publ. Res. Service, 7226, 1969.

[7] A. Harten, High resolution schemes for hyperbolic conservation laws, *J. Comput. Phys.* 49 (1983) 357–393.

[8] J.S. Hesthaven, T. Warburton, *Nodal Discontinuous Galerkin Methods: Algorithms, Analysis and Applications*, Springer, New York, 2008.

[9] C. Hirsch, *Numerical Computation of Internal and External Flows, volume 2: Computational Methods for Inviscid and Viscous Flows*, Wiley, Chichester, 1990.

[10] M.J. Kermani, A.G. Gerber, J.M. Stockie, Thermodynamically based moisture prediction using Roe's scheme, in: *4th Conference of Iranian AeroSpace Society, Amir Kabir University of Technology, Tehran, Iran*, January 27–29, 2003.

[11] B. Koren, A robust upwind discretisation method for advection, diffusion and source terms, In: C.B. Vreugdenhil & B. Koren (Eds), *Numerical Methods for Advection-Diffusion Problems*, Vieweg, Braunschweig, 1993.

[12] C.B. Laney, *Computational Gas Dynamics*, Cambridge University Press, UK, 1998.

[13] B.P. Leonard, M.A. Leschziner, J. McGuirk, The QUICK algorithm: a uniformly 3rd-order finite-difference method for highly convective flows, in: *Proceedings of the First Conference on Numerical Methods in Laminar & Turbulent Flow, Swansea*, 1978, pp. 807.

[14] R.J. Leveque, *Finite Volume Methods for Hyperbolic Problems*, Cambridge Universty Press, UK, 2002, pp. 129–131.

[15] F.S. Lien, M.A. Leschziner, Upstream monotonic interpolation for scalar transport with application to complex turbulent flows, *Int. J. Num. Meth. Fluids.*, 19 (1994) 527.

[16] S. Osher, R. Fedkiw, *Level Set Methods and Dynamic Implicit Surfaces*, Springer-Verlag, New York, 2003.

[17] H.-K. Rhee, R. Aris and N.R. Amundson, *First-order Partial Differential Equations, Theory and Application of Single Equations*, Dover Publications, NY, 1986.

[18] P. L. Roe, Characteristic-based schemes for the Euler equations, *Ann. Rev. Fluid. Mech.* 18 (1986) 337.

[19] P. Saucez, W.E. Schiesser, A. vande Wouwer, Upwinding in the method of lines, *Math. Comp. Simulation*, 56 (2001) 171–185.

[20] W.E. Schiesser, G.W. Griffiths, *A Compendium of Partial Differential Equation Models*, Cambridge University Press, UK, 2009.

[21] C. Shu, High Order Weighted Essentially Nonoscillatory Schemes for Convection Dominated Problems, *SIAM Review*, 51 (1) (2009) 82–126.

[22] V. Streeter, K.W. Bedford, E.B. Wylie, *Fluid Mechanics*, ninth ed., McGraw-Hill, New York, 1997.

[23] P.K. Sweby, High resolution schemes using flux-limiters for hyperbolic conservation laws, *SIAM J. Num. Anal.* 21 (1984) 995–1011.

[24] J.C. Tannehill, D.A. Anderson, R.H. Pletcher, *Computational Fluid mechanics and Heat Transfer*, second ed., Taylor and Francis, UK, 1997.

[25] E.F. Toro, *Riemann Solvers and Numerical Methods for Fluid Dynamics: A Practical Introduction*, Springer-Verlag, New York, 1999.

[26] A. vande Wouwer, Ph. Saucez, W.E. Schiesser (Eds.), *Adaptive Method of Lines*, Chapter 1, Chapman and Hall/CRC, Boca Raton, FL, 2001.

[27] G.D. van Albada, B. Van Leer, W.W. Roberts, A comparative study of computational methods in cosmic gas dynamics, *Astron. Astrophys.* 108 (1982) 76.

[28] B. van Leer, Towards the ultimate conservative difference scheme II. Monotonicity and conservation combined in a second order scheme, *J. Comp. Phys.* 14 (1974) 361–70.

[29] B. van Leer, Towards the ultimate conservative difference scheme III. Upstream-centered finite-difference schemes for ideal compressible flow, *J. Comp. Phys.* 23 (1977) 263–75.

[30] B. van Leer, Towards the ultimate conservative difference scheme V, *J. Comp. Phys.* 32 (1979) 101.

[31] N.P. Waterson, H. Deconinck, A unified approach to the design and application of bounded higher-order convection schemes, *VKI Preprint* 21, (1995).

[32] P. Wesseling, *Principles of Computational Fluid Dynamics*, Springer-Verlag, New York, 2001.

[33] G. Zhou, *Numerical simulations of physical discontinuities in single and multi-fluid flows for arbitrary Mach numbers*, PhD Thesis, Chalmers Univ. of Tech., Goteborg, Sweden, 1995.

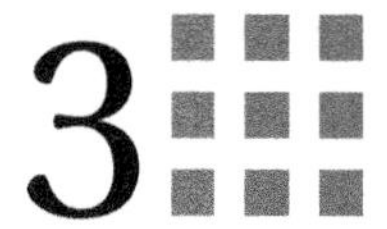

3 Linear Diffusion Equation

PDEs that model diffusion, technically classified as *parabolic* PDEs, can admit traveling wave solutions as we demonstrate in the following analysis. The linear diffusion equation is

$$\frac{\partial u}{\partial t} = D\frac{\partial^2 u}{\partial x^2} \tag{3.1}$$

with the initial condition (IC)

$$u\,(x, t = 0) = f\,(x) \tag{3.2}$$

For a traveling wave solution, we consider

$$U(\xi) = u(k(x - Dt));\; \xi = k(x - Dt) \tag{3.3}$$

Equation (3.3), when substituted into eq. (3.1) gives

$$\frac{dU}{d\xi}\frac{\partial \xi}{\partial t} = D\frac{d}{d\xi}\left(\frac{dU}{d\xi}\frac{\partial \xi}{\partial x}\right)\frac{\partial \xi}{\partial x}$$

$$\frac{dU}{d\xi}(-kD) = D\frac{d^2U}{d\xi^2}k^2 \tag{3.4}$$

Equation (3.4) is a second-order ODE that can be integrated once to give (after cancellation of D)

$$k\frac{dU}{d\xi} + U = C_1 \tag{3.5}$$

If we impose the conditions $U(\xi) = \dfrac{dU}{d\xi} = 0, \xi \to \infty$, the integration constant is $C_1 = 0$. A second integration gives

$$U(\xi) = Ce^{-\xi/k}$$

which satisfies the condition $U(\xi) = 0, \xi \to \infty$ (with $k = 1$). Thus,

$$u(x, t) = Ce^{-(x-Dt)} \tag{3.6}$$

For the IC for eq. (3.1) and eq. (3.2), we take

$$u(x,t=0)=f(x)=\mathrm{e}^{-x} \tag{3.7}$$

so $C=1$ in eq. (3.6).

Since eq. (3.1) is second order in x, it requires two boundary conditions (BCs), which we take as

$$u(x=0,t)=\mathrm{e}^{Dt}; \tag{3.8}$$

$$u(x\to\infty)=0 \tag{3.9}$$

Equation (3.6) (with $C=1$) is the analytical solution that will be used to evaluate the numerical solution.

Before proceeding to the Matlab routines, we confirm that eq. (3.6) is the analytical solution to eq. (3.1) with IC (3.7) and BCs (3.8) and (3.9). Substitution of eq. (3.6) into eq. (3.1) gives (with $C=1$)

Terms in PDE eq. (3.1)	Terms from eq. (3.6)
$\frac{\partial u}{\partial t}$	$D\mathrm{e}^{-(x-Dt)}$
$-D\frac{\partial^2 u}{\partial x^2}$	$-D\mathrm{e}^{-(x-Dt)}$
Sum of terms	Sum of terms
0	0

(3.10)

By inspection, eq. (3.6) satisfies IC (3.7) and BCs (3.8) and (3.9). Thus, we are assured that eq. (3.6) is a valid test of the numerical solution.

The Matlab routines closely resemble those of Chapter 2. Here, we list a few details pertaining to eqns. (3.1) and (3.6)–(3.9). First, the ODE routine `pde_1.m` is

```
  function ut=pde_1(t,u)
%
% Function pde_1 computes the t derivative vector for the linear
% diffusion equation
%
  global xl xu x n ncall
%
  global D
%
% uxx
  nl=1; nu=1;
  u(1)=exp(D*t);
  u(n)=0;
  ux(1)=0;
  uxx=dss044(xl,xu,n,u,ux,nl,nu);
%
% PDE
  for i=1:n
    ut(i)=D*uxx(i);
  end
```

```
  ut(1)=0;
  ut(n)=0;
  ut=ut';
%
% Increment calls to pde_1
  ncall=ncall+1;
```

LISTING 3.1: Function pde_1.m for eq. (3.1).

We can note the following points about pde_1.m:

- The function and some global parameters are first defined.

```
  function ut=pde_1(t,u)
%
% Function pde_1 computes the t derivative vector for the linear
% diffusion equation
%
  global xl xu x n ncall
%
  global D
```

- The second derivative in eq. (3.1), uxx, is then computed using the function dss044 with *Dirichlet BCs* (3.8) and (3.9) specified (nl=1, nu=1) at grid points 1 corresponding to $x = 0$ and n (= 51) corresponding to $x = \infty$ (subsequently set in function inital_1.m).

```
%
% uxx
  nl=1; nu=1;
  u(1)=exp(D*t);
  u(n)=0;
  ux(1)=0;
  uxx=dss044(xl,xu,n,u,ux,nl,nu);
```

  Note that ux(1)=0 is used only to satisfy the calling requirements of dss044 (in Matlab, all input arguments must have a value); ux is not actually used in dss044 for Dirichlet BCs.

- Equation (3.1) is then programmed

```
%
% PDE
  for i=1:n
    ut(i)=D*uxx(i);
  end
  ut(1)=0;
  ut(n)=0;
  ut=ut';
%
% Increment calls to pde_1
  ncall=ncall+1;
```

  Since Dirichlet BCs are used, the derivatives in t are set to zero at the boundaries (so that the ODE integrator does not move the boundary values away from their prescribed values). A transpose is included to meet the requirements of the ODE integrator ode15s. Finally, the counter for the number of calls to pde_1.m is incremented.

The IC of eq. (3.7) is programmed in inital_1.m listed next.

```
  function u0=inital_1(t0)
%
% Function inital_1 sets the initial condition for the linear
% advection equation
%
  global xl xu x n
%
% Spatial domain and initial condition
  xl= 0;
  xu=10;
  n=51;
  dx=(xu-xl)/(n-1);
%
% IC from analytical solution
  for i=1:n
    x(i)=xl+(i-1)*dx;
    u0(i)=ua_1(x(i),0.0);
  end
```

LISTING 3.2: Function inital_1.m for IC (3.7).

We can note the following points about inital_1.m:

- The function and some global parameters are first defined.

```
  function u0=inital_1(t0)
%
% Function inital_1 sets the initial condition for the linear
% advection equation
%
  global xl xu x n
```

- The grid in x is then defined over the interval $0 \le x \le 10$ for 51 points.

```
%
% Spatial domain and initial condition
  xl= 0;
  xu=10;
  n=51;
  dx=(xu-xl)/(n-1);
%
% IC from analytical solution
  for i=1:n
    x(i)=xl+(i-1)*dx;
    u0(i)=ua_1(x(i),0.0);
  end
```

  As the grid in x is defined in the for loop, function ua_1 (listed next) is called (for $t = 0$) to define IC (3.7).

- $x = 10$ effectively defines a boundary at $x = \infty$. This can be inferred from the analytical solution, eq. (3.6), since with $C = D = 1$,

  $$u(x = 10, t) = \mathrm{e}^{-(10-Dt)}$$

  which for the present example is small for any value of t we consider ($t \leq 4$). In other words, we have essentially implemented BC (3.9).

Function `ua_1.m` is a straightforward implementation of the analytical solution, eq. (3.6).

```
  function uanal=ua_1(x,t)
%
% Function uanal computes the exact solution of the linear diffusion
% equation for comparison with the numerical solution
%
  global D
%
% Analytical solution
  uanal=exp(-(x-D*t));
```

LISTING 3.3: Function `ua_1.m` for analytical solution (3.6).

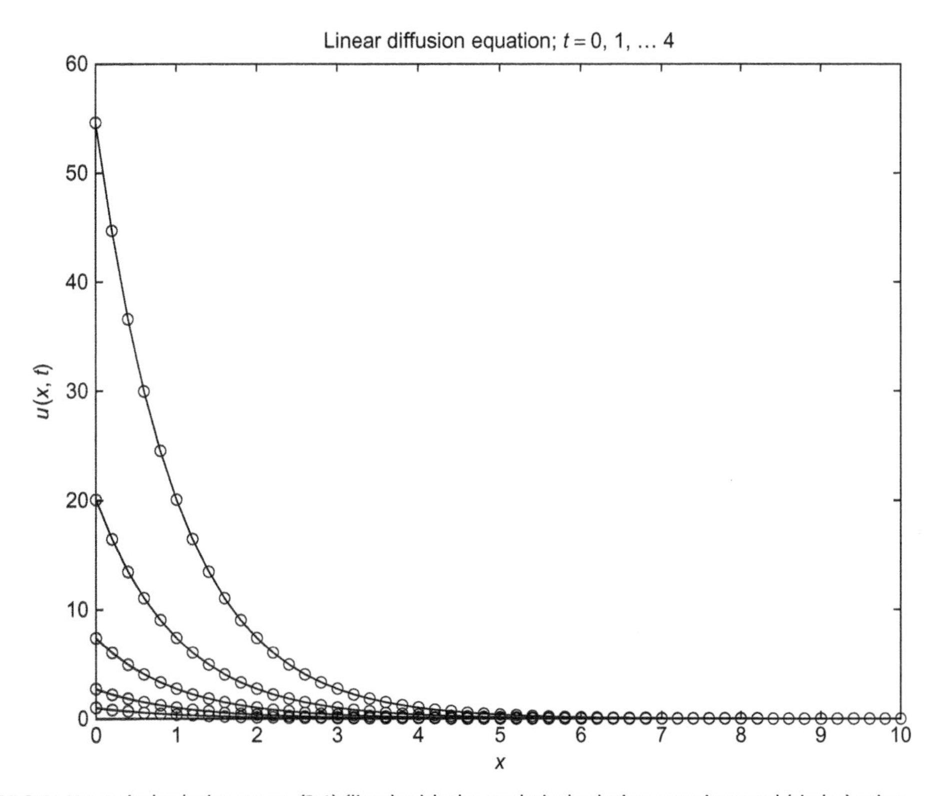

FIGURE 3.1: Numerical solution to eq. (3.1) (lines) with the analytical solution superimposed (circles) using five-point FD approximations in `dss044` [1].

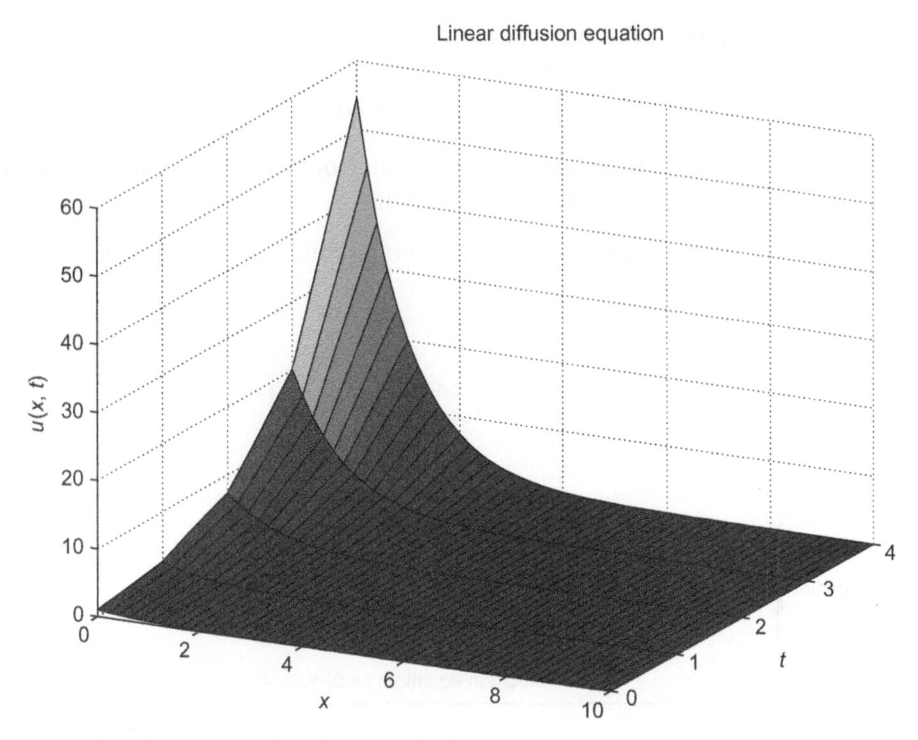

FIGURE 3.2: 3D plot of the numerical solution to eq. (3.1).

The main program, `pde_1_main`, is essentially the same as `pde_1_main` of Listing 2.1 and therefore it is not listed here. The problem parameter is set in the statements.

```
global D
D=1;
```

The main program produces the same three figures and tabulated output as in Chapter 2, which are now reviewed. Also, the Jacobian matrix routine `jpattern_num_1.m` is the same as `jpattern_num_1.m` in Chapter 2 and is therefore not reproduced here. Figure 3.1 indicates good agreement between the analytical and numerical solutions. Figure 3.2 is a 3D plot of the numerical solution. The map of the ODE Jacobian matrix, Fig. 3.3, reflects the banded structure of the ODEs produced by `dss044`.

The tabular analytical and numerical solutions given in Table 3.1 also reflect the good agreement between these two solutions. The computational effort reflected in `ncall = 150` is quite modest.

In summary, the solution of eq. (3.1) subject to IC (3.2) or (3.7) and BCs (3.8) and (3.9) is straightforward. This is to be expected, since parabolic problems tend to produce smooth solutions; that is, the propagation of steep fronts as in the case of the linear advection eq. (2.1) is generally not a problem.

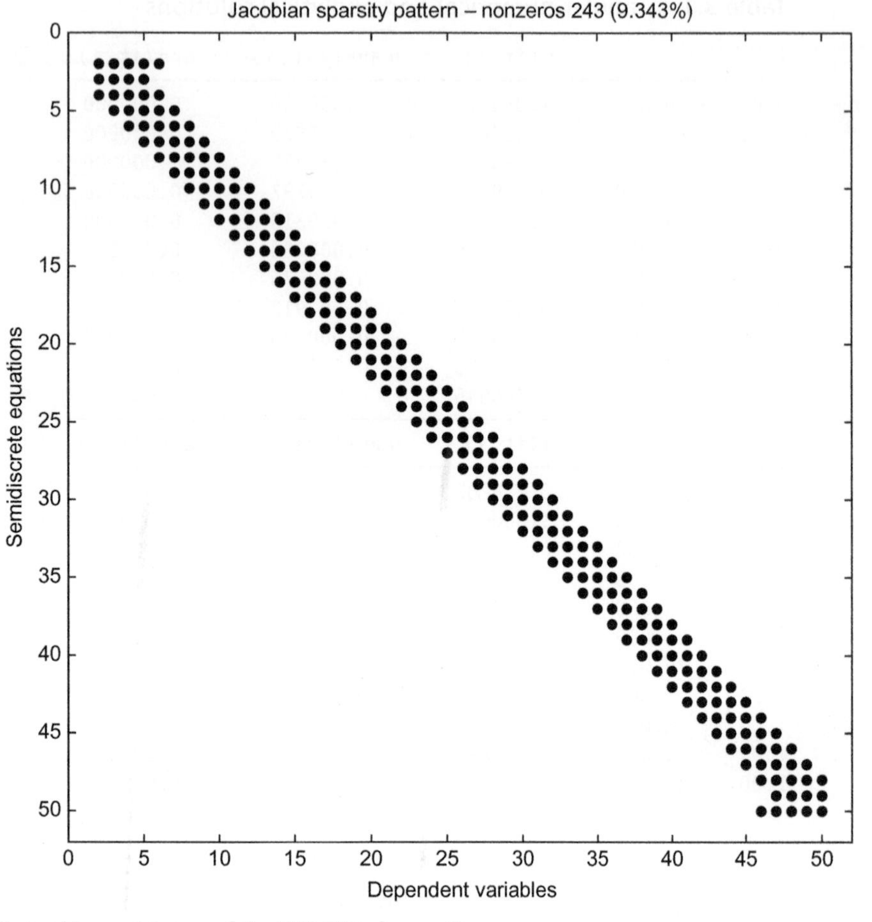

FIGURE 3.3: Jacobian matrix map of the MOL ODEs for $n = 51$.

However, because we have followed a numerical approach, extensions of the problem are straightforward. For example, if the problem included a second-order chemical reaction so that eq. (3.1) becomes

$$\frac{\partial u}{\partial t} = D\frac{\partial^2 u}{\partial x^2} + u^2 \tag{3.11}$$

the preceding coding of `pde_1` in Listing 3.1 would be simply changed to

```
ut(i)=D*uxx(i)+u(i)^2;
```

Although the change in the coding is trivial, the comparison of the numerical and analytical solutions becomes much more difficult because the analytical solution is not readily available. In other words, although we generally discuss PDE problems in the subsequent chapters for which analytical solutions are available to evaluate the numerical solutions,

Table 3.1: Tabular numerical and analytical solutions

t	x	u(it,i)	u_anal(it,i)	err(it,i)
0.00	0.000	1.000000	1.000000	0.000000
0.00	1.000	0.367879	0.367879	0.000000
0.00	2.000	0.135335	0.135335	0.000000
0.00	3.000	0.049787	0.049787	0.000000
0.00	4.000	0.018316	0.018316	0.000000
0.00	5.000	0.006738	0.006738	0.000000
0.00	6.000	0.002479	0.002479	0.000000
0.00	7.000	0.000912	0.000912	0.000000
0.00	8.000	0.000335	0.000335	0.000000
0.00	9.000	0.000123	0.000123	0.000000
0.00	10.000	0.000045	0.000045	0.000000
t	x	u(it,i)	u_anal(it,i)	err(it,i)
1.00	0.000	2.718282	2.718282	0.000000
1.00	1.000	1.000000	1.000000	-0.000000
1.00	2.000	0.367878	0.367879	-0.000001
1.00	3.000	0.135335	0.135335	-0.000001
1.00	4.000	0.049787	0.049787	-0.000000
1.00	5.000	0.018315	0.018316	-0.000000
1.00	6.000	0.006738	0.006738	-0.000000
1.00	7.000	0.002477	0.002479	-0.000002
1.00	8.000	0.000901	0.000912	-0.000010
1.00	9.000	0.000295	0.000335	-0.000040
1.00	10.000	0.000123	0.000123	0.000000
	.		.	
	.		.	
	.		.	
	output for t=2, 3 removed			
	.		.	
	.		.	
	.		.	
t	x	u(it,i)	u_anal(it,i)	err(it,i)
4.00	0.000	54.598150	54.598150	0.000000
4.00	1.000	20.085485	20.085537	-0.000052
4.00	2.000	7.388982	7.389056	-0.000074
4.00	3.000	2.718233	2.718282	-0.000048
4.00	4.000	0.999971	1.000000	-0.000029
4.00	5.000	0.367853	0.367879	-0.000026
4.00	6.000	0.135287	0.135335	-0.000048
4.00	7.000	0.049664	0.049787	-0.000123
4.00	8.000	0.017981	0.018316	-0.000334
4.00	9.000	0.005827	0.006738	-0.000911
4.00	10.000	0.002479	0.002479	0.000000

ncall=150

the intent is to demonstrate numerical procedures that can readily be extended to cases for which analytical solutions are not available. In fact, the possibility of solving numerically (at least in principle) a PDE problem of essentially any complexity is the principal reason for studying and using numerical methods.

Reference

[1] W.E. Schiesser, G.W. Griffiths, *A Compendium of Partial Differential Equation Models*, Cambridge University Press, UK, 2009.

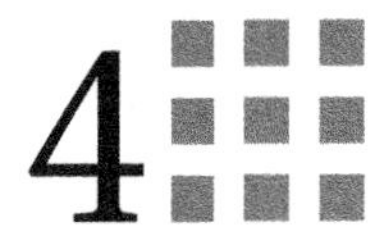

4 A Linear Convection Diffusion Reaction Equation

We considered previously in Chapter 2 a PDE with just a first-order (convective, hyperbolic) derivative in x, and in Chapter 3 a PDE with just a second-order (diffusive, parabolic) derivative in x. Now, we combine these two types of PDEs plus reaction to give a linear convection–diffusion-reaction (CDR) PDE.

$$\frac{\partial u}{\partial t} = -c\frac{\partial u}{\partial x} + D\frac{\partial^2 u}{\partial x^2} - r_c u \tag{4.1}$$

Equation (4.1) can also be classified as a *hyperbolic–parabolic* (or *convective–diffusive*) PDE. The IC is

$$u(x, t = 0) = f(x) \tag{4.2}$$

For a traveling wave solution to eqns. (4.1) and (4.2), we consider

$$U(\xi) = u[k(x - x_0 - ct)];\ \xi = k(x - x_0 - ct) \tag{4.3}$$

where $k, c,$ and x_0 are the wavenumber, velocity, and initial displacement, respectively. Equation (4.3), when substituted into eq. (4.1), gives

$$\frac{dU}{d\xi}\frac{\partial \xi}{\partial t} = -c\frac{dU}{d\xi}\frac{\partial \xi}{\partial x} + D\frac{d}{d\xi}\left(\frac{dU}{d\xi}\frac{\partial \xi}{\partial x}\right)\frac{\partial \xi}{\partial x} - r_c U$$

$$\frac{dU}{d\xi}(-kc) = -ck\frac{dU}{d\xi} + Dk^2\frac{d^2U}{d\xi^2} - r_c U \tag{4.4}$$

Equation (4.4) is a second-order ODE

$$Dk^2\frac{d^2U}{d\xi^2} - r_c U = 0 \tag{4.5}$$

Equation (4.5) can be integrated by assuming an exponential solution

$$U(\xi) = C\exp(B\xi)$$

Substitution in eq. (4.5) gives

$$Dk^2B^2 - r_c = 0$$

Thus, $B = \pm\sqrt{\frac{r_c}{Dk^2}}$ so that the solution to eq. (4.5) is

$$U(\xi) = C_+ \exp\left(\sqrt{\frac{r_c}{Dk^2}}\xi\right) + C_- \exp\left(-\sqrt{\frac{r_c}{Dk^2}}\xi\right)$$

where C_+ and C_- are constants to be determined.

If we impose the conditions $U(\xi) = \frac{dU}{d\xi} = 0, \xi \to \infty$ and $C_+ = 0$. For the IC for eq. (4.1), we take

$$u(x, t = 0) = \exp\left(-\sqrt{\frac{r_c}{Dk^2}}k(x - x_0)\right) \tag{4.6}$$

so $C_- = 1$, and the BCs are

$$u(x = 0, t) = \exp\left(-\sqrt{\frac{r_c}{Dk^2}}k(-x_0 - ct)\right) \tag{4.7}$$

$$u(x \to \infty, t) = 0 \tag{4.8}$$

The solution to eq. (4.1) is therefore

$$u(x, t) = \exp\left(-\sqrt{\frac{r_c}{Dk^2}}k(x - x_0 - ct)\right) \tag{4.9}$$

To verify this solution, we have:

$$\begin{array}{cc}
\text{Terms in PDE eq. (4.1)} & \text{Terms from eq. (4.9)} \\
\frac{\partial u}{\partial t} & -\sqrt{\frac{r_c}{Dk^2}}(-kc)\exp\left(-\sqrt{\frac{r_c}{Dk^2}}k(x - x_0 - ct)\right) \\
+c\frac{\partial u}{\partial x} & -c\sqrt{\frac{r_c}{Dk^2}}(k)\exp\left(-\sqrt{\frac{r_c}{Dk^2}}k(x - x_0 - ct)\right) \\
-D\frac{\partial^2 u}{\partial x^2} & -D\frac{r_c}{Dk^2}(k^2)\exp\left(-\sqrt{\frac{r_c}{Dk^2}}k(x - x_0 - ct)\right) \\
+r_c u & r_c\exp\left(-\sqrt{\frac{r_c}{Dk^2}}k(x - x_0 - ct)\right) \\
\text{Sum of terms} & \text{Sum of terms} \\
0 & 0
\end{array} \tag{4.10}$$

The IC, eq. (4.6), and BCs, eqns. (4.7) and (4.8), follow directly from eq. (4.9).

The Matlab routines closely resemble those of Chapter 3. Here, we list a few details pertaining to eqns. (4.1), (4.6), (4.7), (4.8), and (4.9). First, the ODE routine pde_1.m is

```
  function ut=pde_1(t,u)
%
% Function pde_1 computes the t derivative vector for the linear
% convection diffusion reaction equation
%
  global xl xu x n ncall
%
  global D rc k x0 c sr
%
% ux
  u(1)=exp(-sr*k*(-x0-c*t));
  u(n)=0;
  ux=dss004(xl,xu,n,u);
%
% uxx
  nl=1; nu=1;
  ux(1)=0;
  uxx=dss044(xl,xu,n,u,ux,nl,nu);
%
% PDE
  for i=1:n
    ut(i)=-c*ux(i)+D*uxx(i)-rc*u(i);
  end
  ut(1)=0;
  ut(n)=0;
  ut=ut';
%
% Increment calls to pde_1
  ncall=ncall+1;
```

LISTING 4.1: Function pde_1.m for eq. (4.1).

We can note the following points about pde_1.m:

- The function and some global parameters are first defined.

```
  function ut=pde_1(t,u)
%
% Function pde_1 computes the t derivative vector for the linear
% convection diffusion reaction equation
%
  global xl xu x n ncall
%
  global D rc k x0 c sr
```

- The first derivative in eq. (4.1), ux, is computed using the function dss004 with BCs (4.7) and (4.8) at grid point 1 corresponding to $x = 0$ and n $(= 51)$ corresponding to $x = \infty$ (n is set in function inital_1.m, discussed subsequently).

```
%
% ux
```

```
  u(1)=exp(-sr*k*(-x0-c*t));
  u(n)=0;
  ux=dss004(xl,xu,n,u);
```

- The second derivative in eq. (4.1), uxx, is computed using the function dss044 with *Dirichlet BCs* (4.7) and (4.8) specified (nl=1,nu=1) again at grid points 1 corresponding to $x = 0$ and n (= 51) corresponding to $x = \infty$, (subsequently set in function to inital_1.m).

```
%
% uxx
  nl=1; nu=1;
  ux(1)=0;
  uxx=dss044(xl,xu,n,u,ux,nl,nu);
```

Note that ux(1)=0 is used only to satisfy the calling requirements of dss044 (in Matlab, all input arguments must have a value); ux is not actually used in dss044 for Dirichlet BCs.

- Equation (4.1) is then programmed:

```
%
% PDE
  for i=1:n
    ut(i)=-c*ux(i)+D*uxx(i)-rc*u(i);
  end
  ut(1)=0;
  ut(n)=0;
  ut=ut';
%
% Increment calls to pde_1
  ncall=ncall+1;
```

Since Dirichlet BCs are used, the derivatives in t are set to zero at the boundaries (so that the ODE integrator does not move the boundary values away from their prescribed values). A transpose is included to meet the requirements of the ODE integrator ode15s. Finally, the counter for the number of calls to pde_1.m is incremented.

The IC of eq. (4.6) is programmed in inital_1.m.

```
  function u0=inital_1(t0)
%
% Function inital_1 sets the initial condition for the linear
% convection-diffusion reaction equation
%
  global xl xu x n
%
% Spatial domain and initial condition
  xl= 0;
  xu=10;
  n=51;
```

```
  dx=(xu-xl)/(n-1);
%
% IC from analytical solution
  for i=1:n
    x(i)=xl+(i-1)*dx;
    u0(i)=ua_1(x(i),0.0);
  end
```

LISTING 4.2: Function inital_1.m for IC (4.6).

We can note the following points about inital_1.m:

- The function and some global parameters are first defined.

```
  function u0=inital_1(t0)
%
% Function inital_1 sets the initial condition for the linear
% convection diffusion reaction equation
%
  global xl xu x n
```

- The grid in x is then defined over the interval $0 \leq x \leq 10$ for 51 points.

```
%
% Spatial domain and initial condition
  xl= 0;
  xu=10;
  n=51;
  dx=(xu-xl)/(n-1);
%
% IC from analytical solution
  for i=1:n
    x(i)=xl+(i-1)*dx;
    u0(i)=ua_1(x(i),0.0);
  end
```

  As the grid in x is defined in the for loop, function ua_1 (listed next) is called (for $t=0$) to define IC (4.6).

- $x = 10$ effectively defines a boundary at $x = \infty$. This can be inferred from the analytical solution, eq. (4.9) , since with $\sqrt{\frac{r_c}{DK^2}}k = 1,\ x_0 = 0$,

$$u(x = 10, t) = \mathrm{e}^{-(10-ct)}$$

  which for the present example is small for any value of t we consider ($t \leq 3$ and $c = 1$). In other words, we have essentially implemented BC (4.8).

Function ua_1.m is a straightforward implementation of the analytical solution, eq. (4.9).

```
  function uanal=ua_1(x,t)
%
```

```
% Function uanal computes the exact solution of the linear convection
% diffusion reaction equation for comparison with the numerical solution
%
  global D rc k x0 c sr
%
% Analytical solution
  uanal=exp(-sr*k*(x-x0-c*t));
```

LISTING 4.3: Function ua_1.m for analytical solution eq. (4.6).

The main program, pde_1_main, is essentially the same as pde_1_main of Listing 2.1 and therefore is not listed here. The problem parameters are set in the statements.

```
global D rc k x0 c sr
D=1; rc=1; k=1; x0=0; c=1;
sr=(rc/(D*k^2))^0.5;
```

The main program produces the same three figures and tabulated output as in Chapters 2 and 3, which are now reviewed. Also, the Jacobian matrix routine jpattern_num_1.m is the same as jpattern_num_1.m in Chapters 2 and 3 and is therefore not reproduced here.

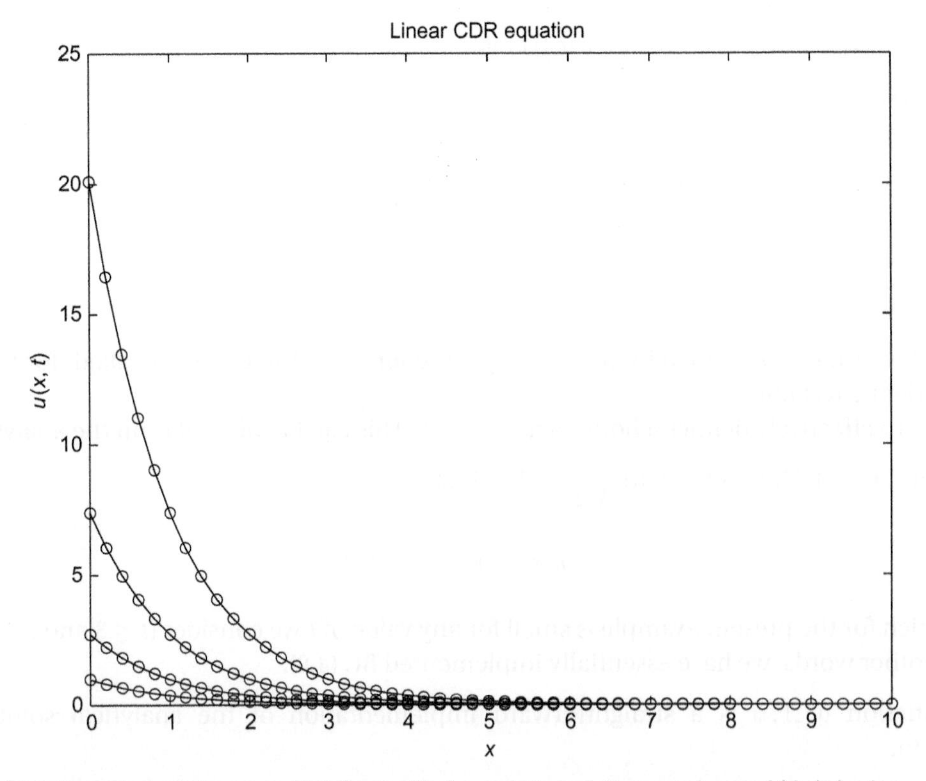

FIGURE 4.1: Numerical solution to eq. (4.1) (lines) with the analytical solution superimposed (circles) using five-point FD approximations in dss004 and dss044 [1] for $t = 0, 1, 2, 3$ (bottom to top).

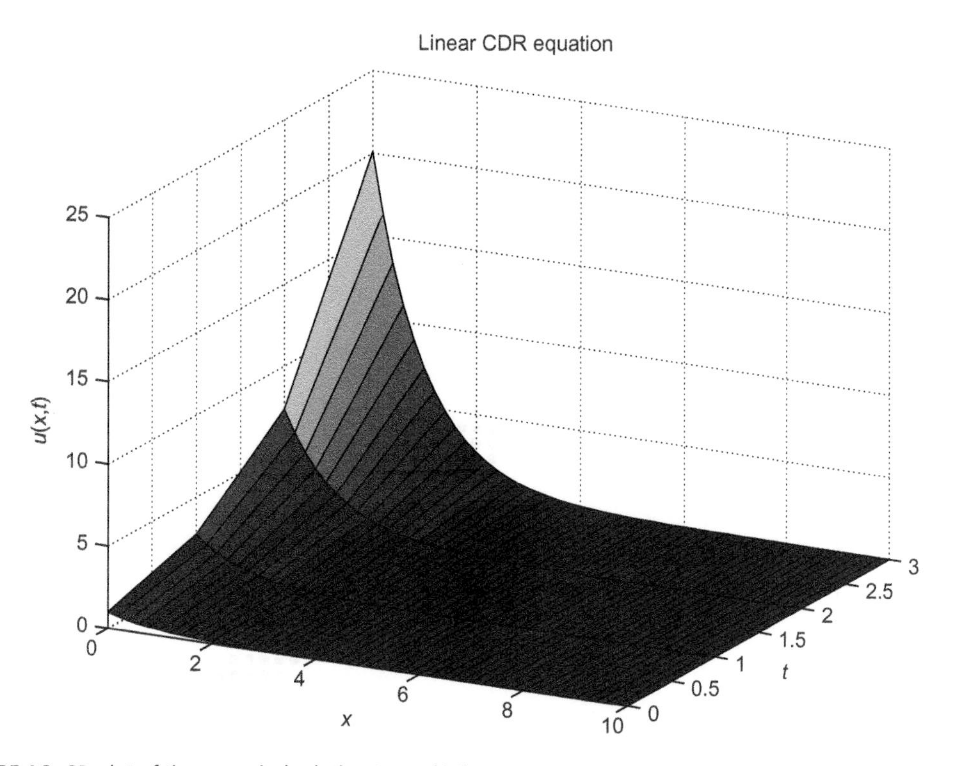

FIGURE 4.2: 3D plot of the numerical solution to eq. (4.1).

Figure 4.1 indicates good agreement between the analytical and numerical solutions. Figure 4.2 is a 3D plot of the numerical solution. The map of the ODE Jacobian matrix, Fig. 4.3, reflects the banded structure of the ODEs produced by `dss044` and `dss044`. The tabular analytical and numerical solutions given in Table 4.1 also reflect the good agreement between these two solutions. The computational effort reflected in `ncall = 140` is quite modest.

In summary, the solution of eq. (4.1) subject to IC (4.6) and BCs (4.7) and (4.8) is straightforward. This is to be expected, since this hyperbolic–parabolic problem has a smooth solution due to the diffusive second derivative in x; that is, the problem is strongly parabolic, so the propagation of a steep front such as in the case of the linear advection eq. (2.1) is not a problem.

However, because we have followed a numerical approach, extensions of the problem are straightforward. For example, if the problem included a nonlinear convective term, $u\dfrac{\partial u}{\partial x}$ and an (effective) 1.5-order chemical reaction so that eq. (4.1) becomes

$$\frac{\partial u}{\partial t} = -u\frac{\partial u}{\partial x} + D\frac{\partial^2 u}{\partial x^2} - r_c u^{1.5} \tag{4.11}$$

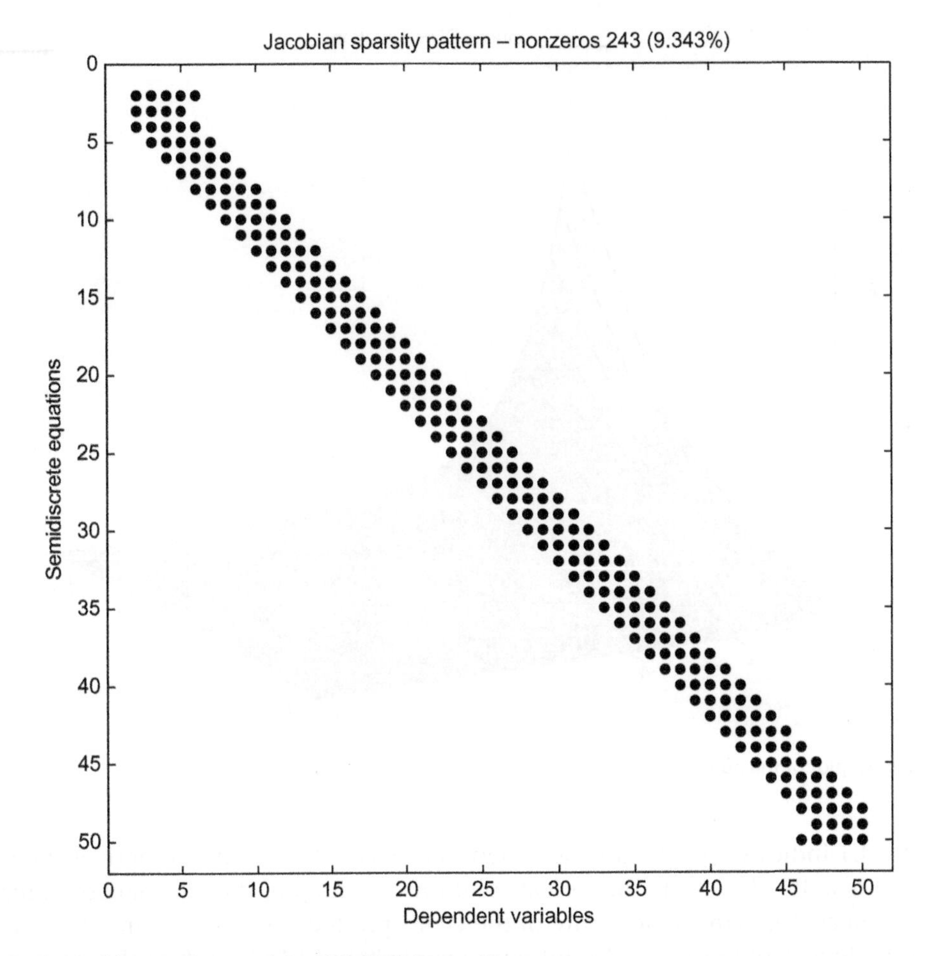

FIGURE 4.3: Jacobian matrix map of the MOL ODEs for $n = 51$.

the preceding coding of pde_1 in Listing 4.1 would be simply changed to

```
ut(i)=-u(i)*ux(i)+D*uxx(i)+u(i)^(1.5);
```

Although the change in the coding is trivial, the comparison of the numerical and analytical solutions becomes much more difficult because the analytical solution is not readily available. As we commented previously, although we generally discuss PDE problems in the subsequent chapters for which analytical solutions are available to evaluate the numerical solutions, the intent is to demonstrate numerical procedures that can readily be extended to cases for which analytical solutions are not available. In fact, the possibility of solving numerically (at least in principle) a PDE problem of essentially any complexity is the principal reason for studying and using numerical methods.

Table 4.1: Tabular numerical and analytical solutions

t	x	u(it,i)	u_anal(it,i)	err(it,i)
0.00	0.000	1.000000	1.000000	0.000000
0.00	1.000	0.367879	0.367879	0.000000
0.00	2.000	0.135335	0.135335	0.000000
0.00	3.000	0.049787	0.049787	0.000000
0.00	4.000	0.018316	0.018316	0.000000
0.00	5.000	0.006738	0.006738	0.000000
0.00	6.000	0.002479	0.002479	0.000000
0.00	7.000	0.000912	0.000912	0.000000
0.00	8.000	0.000335	0.000335	0.000000
0.00	9.000	0.000123	0.000123	0.000000
0.00	10.000	0.000045	0.000045	0.000000

.
.
.
output for t=1, 2 removed
.
.
.

t	x	u(it,i)	u_anal(it,i)	err(it,i)
3.00	0.000	20.085537	20.085537	0.000000
3.00	1.000	7.388959	7.389056	-0.000098
3.00	2.000	2.718184	2.718282	-0.000098
3.00	3.000	0.999941	1.000000	-0.000059
3.00	4.000	0.367849	0.367879	-0.000030
3.00	5.000	0.135321	0.135335	-0.000014
3.00	6.000	0.049781	0.049787	-0.000007
3.00	7.000	0.018311	0.018316	-0.000005
3.00	8.000	0.006720	0.006738	-0.000018
3.00	9.000	0.002355	0.002479	-0.000124
3.00	10.000	0.000912	0.000912	0.000000

ncall=140

Reference

[1] W.E. Schiesser, G.W. Griffiths, *A Compendium of Partial Differential Equation Models*, Cambridge University Press, UK, 2009.

5 Diffusion Equation with Nonlinear Source Terms

The diffusion equation with nonlinear source terms we will consider is ([3], p3)

$$\frac{\partial u}{\partial t} = a\frac{\partial^2 u}{\partial x^2} - bu^3 - cu^2 \tag{5.1}$$

where a, b, and c are arbitrary. An analytical solution is ([3], p3)

$$u(x,t) = \left(ct + \sqrt{\frac{b}{2a}}x + 1\right)^{-1} = \left(\sqrt{\frac{b}{2a}}\left(x + \sqrt{\frac{2a}{b}}ct\right) + 1\right)^{-1} \tag{5.2}$$

The second form of the analytical solution of eq. (5.2) indicates a *traveling wave solution* with a *wavenumber* equal to $\sqrt{\frac{b}{2a}}$ and a *wave velocity* equal to $-\sqrt{\frac{2a}{b}}c$.

We study eq. (5.1) to investigate the use of (1) *Dirichlet*, (2) *Neumann*, (3) *third-type*, (4) *nonlinear third-type, and* (5) *analytical Neumann* boundary conditions (BCs), both for the linear case $b = c = 0$ for which eq. (5.1) reduces to the *one-dimensional diffusion equation* of Chapter 3 and for the nonlinear case $b \neq 0, c \neq 0$.

For the linear case, eq. (5.2) reduces to the trivial solution $u(x,t) = 1$, so we look for another solution. Fortunately, since eq. (5.1) becomes the linear diffusion equation, finding an alternative to eq. (5.2) is quite straightforward. We will use the analytical solution

$$u(x,t) = \sin(\pi x/x_u)e^{-a(\pi/x_u)^2 t} \tag{5.3}$$

which provides the IC with $t = 0$. x_u is the value of x at the right boundary; the left boundary value of x will be denoted as x_l. That is, the interval in x is $x_l \leq x \leq x_u$.

To implement the 10 cases ((linear and nonlinear PDE)(5 BCs) = 10 cases), the Matlab routines still resemble those of previous chapters, but some branching is required for the 10 cases.

Case 1: Linear PDE, Dirichlet BCs (ncase=1, nbc=1, a=0.1, b=c=0)

The ODE routine, pde_1.m, is considered first (with ncase, nbc set in the main program).

```
  function ut=pde_1(t,u)
%
% Function pde_1 computes the t derivative vector for the diffusion
```

```
% equation with nonlinear source terms
%
  global xl xu x n ncall ncase nbc
%
% Model parameters
  global a b c
%
% Linear case
  if(ncase==1)
%
%   ux
    ux=dss004(xl,xu,n,u);
%
%   Dirichlet BCs
    if(nbc==1)
      u(1)=0; u(n)=0;
      ux=dss004(xl,xu,n,u);
    end
%
%   Neumann BCs
    if(nbc==2)
      ux(1)=0; ux(n)=0;
    end
%
%   Third type BCs
    if(nbc==3)
      ux(1)=-(1/a)*(1-u(1)); ux(n)=(1/a)*(1-u(n));
    end
%
%   Nonlinear Neumann BCs
    if(nbc==4)
      ux(1)=-(1/a)*(1-u(1)^4); ux(n)=(1/a)*(1-u(n)^4);
    end
%
%   Analytical Neumann BCs
    if(nbc==5)
      ux(1)=(pi/xu)*cos(pi*x(1)/xu)*exp(-a*(pi/xu)^2*t);
      ux(n)=(pi/xu)*cos(pi*x(n)/xu)*exp(-a*(pi/xu)^2*t);
    end
%
%   uxx
    uxx=dss004(xl,xu,n,ux);
  end

%
% Nonlinear case
  if(ncase==2)
%
%   ux
    ux=dss004(xl,xu,n,u);
%
%   Dirichlet BCs
    if(nbc==1)
      u(1)=ua_1(x(1),t); u(n)=ua_1(x(n),t);
```

```
      ux=dss004(xl,xu,n,u);
    end
%
%   Neumann BCs
    if(nbc==2)
      ux(1)=0; ux(n)=0;
    end
%
%   Third type BCs
    if(nbc==3)
      ux(1)=-(1/a)*(1-u(1)); ux(n)=(1/a)*(1-u(n));
    end
%
%   Nonlinear Neumann BCs
    if(nbc==4)
      ux(1)=-(1/a)*(1-u(1)^4); ux(n)=(1/a)*(1-u(n)^4);
    end
%
%   Analytical Neumann BCs
    if(nbc==5)
      ux(1)=-1/(c*t+(b/(2*a))^0.5*x(1)+1)^2*(b/(2*a))^0.5;
      ux(n)=-1/(c*t+(b/(2*a))^0.5*x(n)+1)^2*(b/(2*a))^0.5;
    end
%
%   uxx
    uxx=dss004(xl,xu,n,ux);
  end
%
% PDE
  for i=1:n
    sq=u(i)^2;
    cu=u(i)*sq;
    ut(i)=a*uxx(i)-b*cu-c*sq;
  end
  if(nbc==1)
    ut(1)=0; ut(n)=0;
  end
  ut=ut';
%
% Increment calls to pde_1
  ncall=ncall+1;
```

LISTING 5.1: Function pde_1.m for eq. (5.1).

Because of the branching for the 10 cases, we consider each case one at a time.

- The function and some global parameters are first defined.

```
  function ut=pde_1(t,u)
%
% Function pde_1 computes the t derivative vector for the diffusion
% equation with nonlinear source terms
%
  global xl xu x n ncall ncase nbc
%
```

```
% Model parameters
  global a b c
```

- Since eq. (5.1) is second order in x, two BCs are required. To implement these BCs, the routine starts with a call to `dss004` to compute `ux` (first for the linear case, $b = c = 0$ with `ncase=1` as defined numerically in the main program `pde_1_main.m` and received by `pde_1.m` as global variables).

```
%
% Linear case
  if(ncase==1)
%
%   ux
    ux=dss004(xl,xu,n,u);
```

- The routine then goes into the first BC type, `nbc=1` for Dirichlet BCs.

```
%
%   Dirichlet BCs
    if(nbc==1)
      u(1)=0; u(n)=0;
      ux=dss004(xl,xu,n,u);
    end
```

  Note that $u(x = xl = 0, t) = u(x = xu = 1, t) = 0$ are used as the BCs (as `u` at grid points `1,n`). Then, `dss004` is called again to use these boundary values. The zero BC values are generally termed *homogeneous*. However, in principle, any values could be used, and they could even be functions of t. Thus, this case illustrates an important advantage of the numerical approach to PDEs, namely the *ease with which BCs of essentially any form can be specified* (as illustrated by the 10 cases). This contrasts with the analytical approach in which changes in the BCs generally require major changes in the corresponding analytical solutions; in the case of nonlinear BCs, analytical solutions may not even be available, but as we will observe, nonlinear BCs are easily implemented numerically (for `nbc=4`).
- The second derivative in eq. (5.1), `uxx`, is computed by a second call to `dss004` (further down `pde_1.m`).

```
%
%   uxx
    uxx=dss004(xl,xu,n,ux);
```

  This completes the derivatives in x with the BCs for `ncase=1, nbc=1` (the first of the 10 cases). We have used so-called *stagewise differentiation* in which `uxx` is computed by two successive calls to `dss004`. An alternative would be to use `dss044` to directly compute `uxx` from `u`, as discussed in Chapter 4.
- Equation (5.1) is then programmed (at the end of `pde_1.m`).

```
%
% PDE
  for i=1:n
    sq=u(i)^2;
```

```
      cu=u(i)*sq;
      ut(i)=a*uxx(i)-b*cu-c*sq;
    end
    if(nbc==1)
      ut(1)=0; ut(n)=0;
    end
    ut=ut';
  %
  % Increment calls to pde_1
    ncall=ncall+1;
```

Recall again that $b = c = 0$ (for `ncase=1`) so that the nonlinear terms `sq` and `cu` have no effect, and we have only the linear diffusion equation `ut(i)=a*uxx(i)`. Since Dirichlet BCs are used (`nbc=1`), the derivatives in t are set to zero at the boundaries (so that the ODE integrator does not move the boundary values away from their prescribed values). A transpose is included to meet the requirements of the ODE integrator `ode15s`. Finally, the counter for the number of calls to `pde_1.m` is incremented.

This completes the coding for the first of the 10 cases (`ncase=nbc=1`) in `pde_1.m`. At this point, we complete the discussion of this first case, then return to the other nine cases. The IC from eq. (5.2) with $t = 0$ is programmed in `inital_1.m`.

```
  function u0=inital_1(t0)
%
% Function inital_1 sets the initial condition for the diffusion
% equation with nonlinear source terms
%
% Parameters shared with other routines
  global xl xu x n ncall
%
% Model parameters
  global a b c
%
% Spatial domain and initial condition
  xl=0;
  xu=1;
  n=26;
  dx=(xu-xl)/(n-1);
%
% IC from analytical solution
  for i=1:n
    x(i)=xl+(i-1)*dx;
    u0(i)=ua_1(x(i),0.0);
  end
```

LISTING 5.2: Function `inital_1.m` for IC from eq. (5.2) or eq. (5.3) with $t = 0$.

We can note the following points about `inital_1.m`:

- The function and some global parameters are first defined.

```
  function u0=inital_1(t0)
%
% Function inital_1 sets the initial condition for the diffusion equation
```

```
% with nonlinear source terms
%
% Parameters shared with other routines
  global xl xu x n ncall
%
% Model parameters
  global a b c
```

- The grid in x is then defined over the interval $0 \le x \le 1$ for 26 points.

```
%
% Spatial domain and initial condition
  xl=0;
  xu=1;
  n=26;
  dx=(xu-xl)/(n-1);
%
% IC from analytical solution
  for i=1:n
    x(i)=xl+(i-1)*dx;
    u0(i)=ua_1(x(i),0.0);
  end
```

As the grid in x is defined in the `for` loop, function `ua_1` (listed next) is called (for $t = 0$) to define the IC from eq. (5.3) with $t = 0$ for `ncase=1` or the IC from eq. (5.2) for `ncase=2`. `inital_1.m` serves both cases, `ncase=1,2`, since `ncase` is a global variable that is passed to the routine for the analytical solutions, `ua_1.m`, for eqns. (5.2) and (5.3).

Function `ua_1.m` is a straightforward implementation of the analytical solutions, eqns. (5.2) and (5.3).

```
  function uanal=ua_1(x,t)
%
% Function uanal computes the exact solution of the diffusion equation
% with nonlinear source terms for comparison with the numerical solution
%
% Model parameters
  global xl xu n ncall ncase nbc
  global a b c
%
% Analytical solution
%
%   Linear case
    if(ncase==1)
      uanal=sin(pi*x/xu)*exp(-a*(pi/xu)^2*t);
    end
%
%   Nonlinear case
    if(ncase==2)
      uanal=1/(c*t+(b/(2*a))^0.5*x+1);
    end
```

LISTING 5.3: Function `ua_1.m` for analytical solution (5.2) and (5.3).

The main program, pde_1_main, is essentially the same as pde_1_main of previous chapters, but since it contains the coding for the 10 cases, it is listed here along with some discussion.

```
%
% Diffusion equation with nonlinear source terms
%
% Clear previous files
  clear all
  clc
%
% Parameters shared with other routines
  global xl xu x n ncall ncase nbc
%
% Model parameters
  global a b c
%
% Select case
%
% 1 - linear; 2 - nonlinear
  ncase=1;
%
% Linear case
  if(ncase==1)
    a=0.1; b=0; c=0;
  end
%
% Nonlinear case
  if(ncase==2)
    a=1; b=1; c=1;
  end
%
% Select BCs
%
% 1 - Dirichlet; 2 - Neumann; 3 - Third type; 4 - Nonlinear third type,
% 5 - Analytical Neumann
  nbc=1;
%
% Initial condition
  t0=0.0;
  u0=inital_1(t0);
%
% Independent variable for ODE integration
  tf=0.9;
  tout=[t0:0.3:tf]';
  nout=4;
  ncall=0;
%
% ODE integration
  mf=2;
  reltol=1.0e-06; abstol=1.0e-06;
  options=odeset('RelTol',reltol,'AbsTol',abstol);
%
% Explicit (nonstiff) integration
```

```
  if(mf==1)[t,u]=ode45(@pde_1,tout,u0,options); end
%
% Implicit (sparse stiff) integration
  if(mf==2)
    S=jpattern_num_1;
    options=odeset(options,'JPattern',S)
    [t,u]=ode15s(@pde_1,tout,u0,options);
  end
%
% Store analytical solution, errors in numerical solution
  if(nbc==1 | nbc==5)
    for it=1:nout
      u(it,1)=ua_1(x(1),t(it));
      u(it,n)=ua_1(x(n),t(it));
      for i=1:n
        u_anal(it,i)=ua_1(x(i),t(it));
        err(it,i)=u(it,i)-u_anal(it,i);
      end
    end
    for it=1:nout
      fprintf('\n     t       x         u(it,i)   u_anal(it,i)
              err(it,i)\n');
      for i=1:n
        fprintf('%6.2f%8.3f%15.6f%15.6f%15.6f\n',...
              t(it),x(i),u(it,i),u_anal(it,i),err(it,i));
      end
    end
  end
  if(nbc==2 | nbc==3 | nbc==4)
    for it=1:nout
      fprintf('\n     t       x         u(it,i)\n');
      for i=1:n
        fprintf('%6.2f%8.3f%15.6f\n',t(it),x(i),u(it,i));
      end
    end
  end
    fprintf('    ncall = %4d\n\n',ncall);
%
%   Plot numerical and analytical solutions
    figure(2)
    if(nbc==1 | nbc==5)
      plot(x,u,'-',x,u_anal,'o')
      title('Diffusion equation with nonlinear source terms;
            t = 0, 0.3, 0.6, 0.9; solid - numerical; o - analytical')
    elseif(nbc==2 | nbc==3 | nbc==4)
      plot(x,u,'-')
      title('Diffusion equation with nonlinear source terms;
            t = 0, 0.3, 0.6, 0.9; solid - numerical')
    end
    xlabel('x')
    ylabel('u(x,t)')
    figure(3)
    surf(x,t,u)
    az=50; el=40;
```

```
  view(az,el)
  xlabel('x'); ylabel('t'); zlabel('u(x,t)');
  title('Diffusion equation with nonlinear source terms');
```

LISTING 5.4: Main program `pde_1_main.m` for eq. (5.1).

We can note the following points about `pde_1_main.m`:

- Previous files are cleared and the problem parameters are declared as global and defined numerically.

```
%
% Diffusion equation with nonlinear source terms
%
% Clear previous files
  clear all
  clc
%
% Parameters shared with other routines
  global xl xu x n ncall ncase nbc
%
% Model parameters
  global a b c
```

- The case is selected, which includes setting the values of a, b, c in eq. (5.1).

```
%
% Select case
%
% 1 - linear; 2 - nonlinear
  ncase=1;
%
% Linear case
  if(ncase==1)
    a=0.1; b=0; c=0;
  end
%
% Nonlinear case
  if(ncase==2)
    a=1; b=1; c=1;
  end
```

- One of the five BCs is then selected.

```
%
% Select BCs
%
% 1 - Dirichlet; 2 - Neumann; 3 - Third type; 4 - Nonlinear third type,
% 5 - Analytical Neumann
  nbc=1;
```

  Here, we specify `nbc=1` corresponding to the homogeneous Dirichlet BCs used in `pde_1.m` of Listing 5.1.

- The IC from `inital_1.m` is set, and the interval in t is defined as $0 \le t \le 0.9$ with four outputs at $t = 0, 0.3, 0.6, 0.9$.

```
%
% Initial condition
  t0=0.0;
  u0=inital_1(t0);
%
% Independent variable for ODE integration
  tf=0.9;
  tout=[t0:0.3:tf]';
  nout=4;
  ncall=0;
```

- The ODE integration is by `ode45` (for `mf=1`, nonstiff) or by `ode15s` (for `mf=2`, stiff). Note that these integrators call `pde_1.m` as the first argument.

```
%
% ODE integration
  mf=2;
  reltol=1.0e-06; abstol=1.0e-06;
  options=odeset('RelTol',reltol,'AbsTol',abstol);
%
% Explicit (nonstiff) integration
  if(mf==1)[t,u]=ode45(@pde_1,tout,u0,options); end
%
% Implicit (sparse stiff) integration
  if(mf==2)
    S=jpattern_num_1;
    options=odeset(options,'JPattern',S)
    [t,u]=ode15s(@pde_1,tout,u0,options);
  end
```

- The solution is displayed numerically. For `nbc=1,5`, analytical solutions (5.2) and (5.3) can be used for comparison with the numerical solution, and the difference between the two solutions (`err`) can be included in the output. For `nbc=2,3,4`, an analytical solution is not readily available and therefore only the numerical solution is displayed. These same considerations (availability of an analytical solution) apply to the plotted output as well.

```
%
% Store analytical solution, errors in numerical solution
  if(nbc==1 | nbc==5)
    for it=1:nout
      u(it,1)=ua_1(x(1),t(it));
      u(it,n)=ua_1(x(n),t(it));
      for i=1:n
        u_anal(it,i)=ua_1(x(i),t(it));
        err(it,i)=u(it,i)-u_anal(it,i);
      end
    end
    for it=1:nout
      fprintf('\n      t         x           u(it,i)   u_anal(it,i)
                err(it,i)\n');
      for i=1:n
        fprintf('%6.2f%8.3f%15.6f%15.6f%15.6f\n',...
                t(it),x(i),u(it,i),u_anal(it,i),err(it,i));
      end
```

```
    end
  end
  if(nbc==2 | nbc==3 | nbc==4)
    for it=1:nout
      fprintf('\n     t       x        u(it,i)\n');
      for i=1:n
        fprintf('%6.2f%8.3f%15.6f\n',t(it),x(i),u(it,i));
      end
    end
  end
    fprintf('    ncall = %4d\n\n',ncall);
```

The number of calls to `pde_1.m`, `ncall`, is displayed at the end to give an indication of the total computational effort required to compute the numerical solution.

- Finally, the solutions are displayed graphically in 2D (by `plot`) and 3D (by `surf`).

```
%
%   Plot numerical and analytical solutions
    figure(2)
    if(nbc==1 | nbc==5)
      plot(x,u,'-',x,u_anal,'o')
      title('Diffusion equation with nonlinear source terms;
            t = 0, 0.3, 0.6, 0.9; solid - numerical; o - analytical')
    elseif(nbc==2 | nbc==3 | nbc==4)
      plot(x,u,'-')
      title('Diffusion equation with nonlinear source terms;
            t = 0, 0.3, 0.6, 0.9; solid - numerical')
    end
    xlabel('x')
    ylabel('u(x,t)')
    figure(3)
    surf(x,t,u)
    az=50; el=40;
    view(az,el)
    xlabel('x'); ylabel('t'); zlabel('u(x,t)');
    title('Diffusion equation with nonlinear source terms');
```

The main program produces the same three figures and tabulated output as in previous chapters, which are now reviewed. The Jacobian matrix routine `jpattern_num_1.m` is the same as `jpattern_num_1.m` in previous chapters and is therefore not reproduced here.

Figure 5.1 indicates good agreement between the analytical and numerical solutions. Figure 5.2 is the 3D plot of the numerical solution.

The map of the ODE Jacobian matrix, Fig. 5.3, reflects the banded structure of the ODEs produced by `dss004`. In particular, since the number of grid points, $n = 26$, is relatively small, the individual elements of the Jacobian matrix are distinct. Also, note that the bandwidth is 9 and not 5 as might be expected from the five-point FDs in `dss004`. This greater bandwidth is due to the repeated use of `dss004` in `pde_1.m` to compute `uxx` from `u` by stagewise differentiation. This example illustrates one of the disadvantages of stagewise differentiation, that is, the increase in the bandwidth of the ODE Jacobian matrix through successive calls of the spatial differentiator such as `dss004`.

The tabular analytical and numerical solutions given in Table 5.1 also reflect the good agreement between these two solutions.

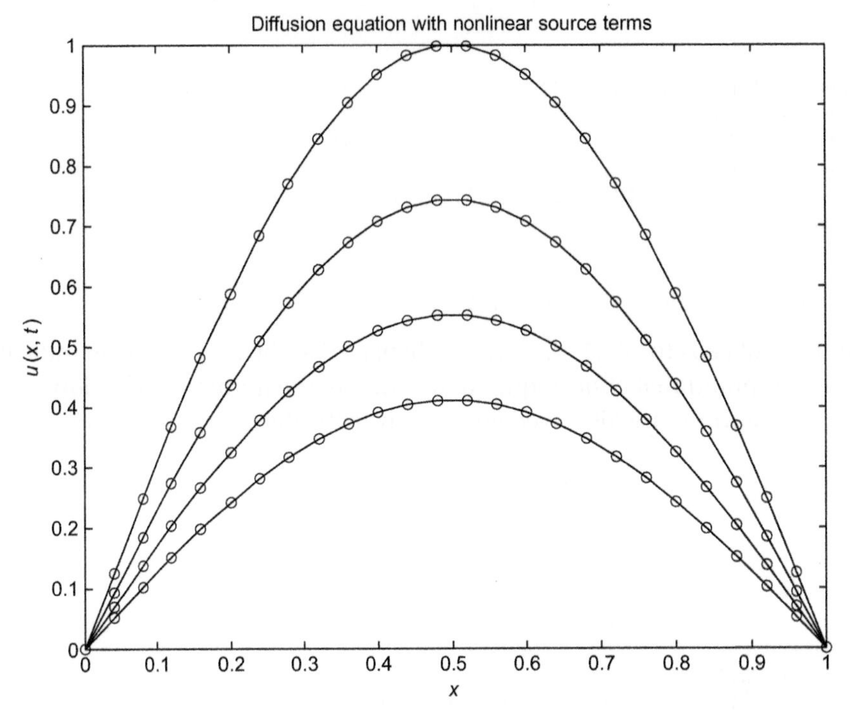

FIGURE 5.1: Numerical solution to eq. (5.1) (lines) with the analytical solution, eq. (5.3), superimposed (circles) using five-point FD approximations in dss004 for $t = 0, 0.3, 0.6, 0.9$ (top to bottom).

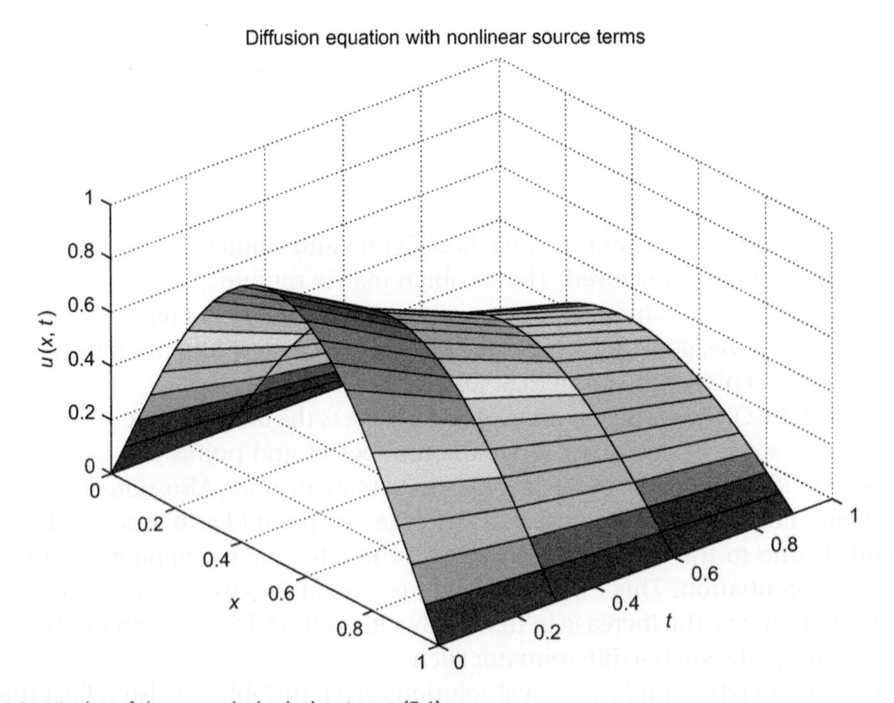

FIGURE 5.2: 3D plot of the numerical solution to eq. (5.1).

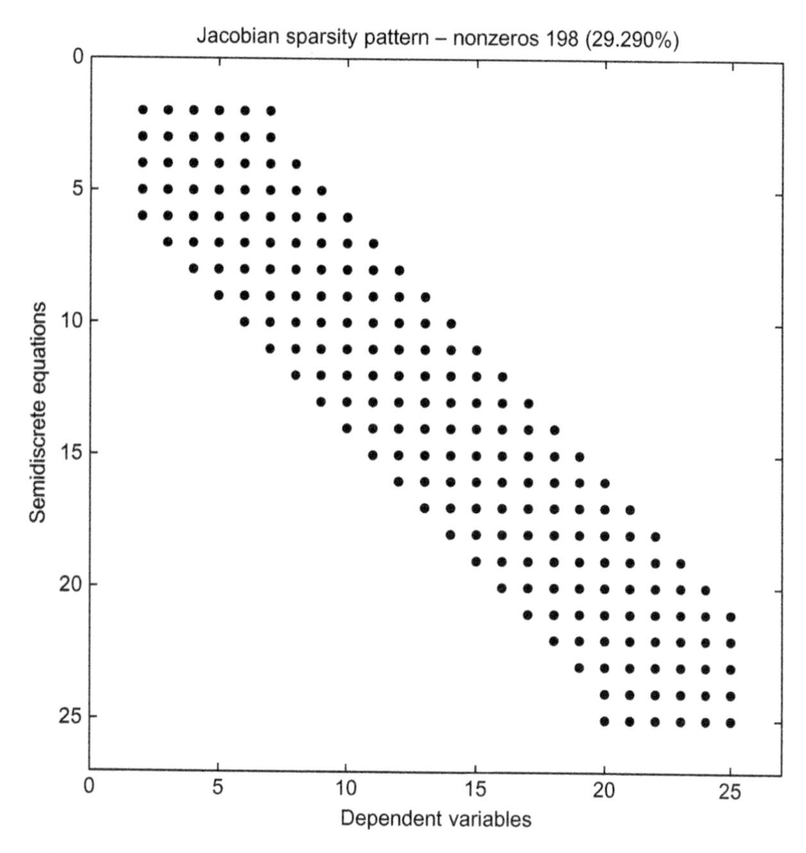

FIGURE 5.3: Jacobian matrix map of the MOL ODEs for $n = 26$.

The computational effort reflected in `ncall = 95` is quite modest. The agreement between the analytical and numerical solution of approximately six figures is due in part to the smoothness of the solution as reflected in Fig. 5.1, even when using only 26 grid points. Also, as a result of the homogeneous Dirichlet BCs, the solution has the limiting value $u(x, t \to \infty) = 0$ (although the interval in t would have to be expanded beyond $t = 0.9$ to demonstrate this limiting value, which is also apparent from eq. (5.3)).

In summary, the solution of eq. (5.1) subject to the IC from eq. (5.3) (with $t = 0$) and two homogeneous Dirichlet BCs (with $x_l = 0, x_u = 1$) is straightforward, particularly, since the solution is quite smooth. We now consider the other nine cases.

Case 2: Linear PDE, Neumann BCs (`ncase=1, nbc=2, a=0.1, b=c=0`)

For `ncase=1,nbc=2` (linear case, homogeneous Neumann BCs), the BCs are

$$\frac{\partial u(x = x_l, t)}{\partial x} = \frac{\partial u(x = x_u, t)}{\partial x} = 0 \tag{5.4}$$

The relevant coding in `pde_1.m` of Listing 5.1 is

Table 5.1: Abbreviated tabular numerical and analytical solutions for ncase=1,nbc=1

t	x	u(it,i)	u_anal(it,i)	err(it,i)
0.00	0.000	0.000000	0.000000	0.000000
0.00	0.040	0.125333	0.125333	0.000000
0.00	0.080	0.248690	0.248690	0.000000
0.00	0.120	0.368125	0.368125	0.000000
0.00	0.160	0.481754	0.481754	0.000000
	.		.	
	.		.	
	.		.	
0.00	0.840	0.481754	0.481754	0.000000
0.00	0.880	0.368125	0.368125	0.000000
0.00	0.920	0.248690	0.248690	0.000000
0.00	0.960	0.125333	0.125333	0.000000
0.00	1.000	0.000000	0.000000	0.000000

```
                .                              .
                .                              .
                .                              .
          Output for t=0.3, 0.6 removed
                .                              .
                .                              .
                .                              .
```

t	x	u(it,i)	u_anal(it,i)	err(it,i)
0.90	0.000	0.000000	0.000000	0.000000
0.90	0.040	0.051559	0.051558	0.000000
0.90	0.080	0.102304	0.102303	0.000000
0.90	0.120	0.151436	0.151435	0.000001
0.90	0.160	0.198180	0.198179	0.000001
0.90	0.200	0.241799	0.241797	0.000002
	.		.	
	.		.	
	.		.	
0.90	0.800	0.241799	0.241797	0.000002
0.90	0.840	0.198180	0.198179	0.000001
0.90	0.880	0.151436	0.151435	0.000001
0.90	0.920	0.102304	0.102303	0.000000
0.90	0.960	0.051559	0.051558	0.000000
0.90	1.000	0.000000	0.000000	0.000000

ncall=95

```
      .
      .
      .
%
% Linear case
  if(ncase==1)
```

```
%
%     ux
      ux=dss004(xl,xu,n,u);
          .
          .
          .
%
%     Neumann BCs
      if(nbc==2)
        ux(1)=0; ux(n)=0;
      end
          .
          .
          .
%
%     uxx
      uxx=dss004(xl,xu,n,ux);
    end
```

LISTING 5.5: Programming from `pde_1.m` for eq. (5.1) with `ncase=1, nbc=2`.

The coding for homogeneous Neumann BCs (`nbc=2`, eq. (5.4)) resets the boundary values of the first derivative, `ux(1),ux(n)`. The plotted numerical solution is in Fig. 5.4.

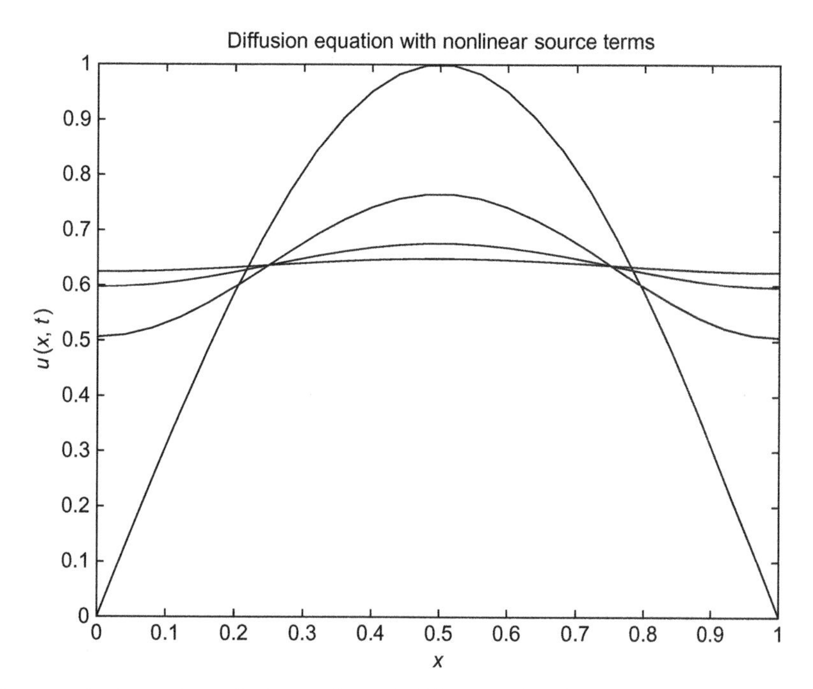

FIGURE 5.4: Numerical solution to eq. (5.1) using five-point FD approximations in `dss004` for $t = 0, 0.3, 0.6, 0.9$ for `ncase=1,nbc=2`.

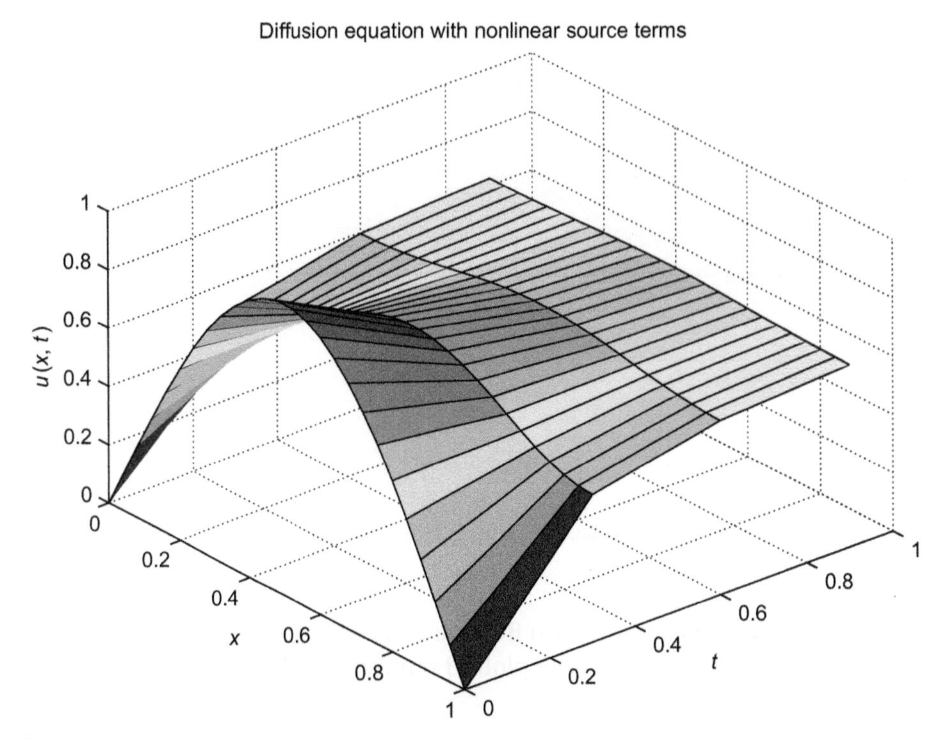

FIGURE 5.5: 3D plot of the numerical solution to eq. (5.1).

We can note, in particular, the zero slope (homogeneous Neumann BCs eq. (5.4)) at the boundaries. Also, $u(x,t \to \infty)$ approaches a constant value $2/\pi = 0.636620$, as explained in Appendix 1 at the end of this chapter.

Figure 5.5 is the 3D plot of the numerical solution. The limiting value $u(x,t \to \infty) = 2/\pi = 0.636620$ is apparent.

The map of the ODE Jacobian matrix is basically the same as in Fig. 5.3 and is therefore not reproduced here (since the PDE is the same and the numerical approximations are nearly the same with differences only at the boundaries for `nbc=1,2`). Also, an analytical solution is not presented for this case (that could be used to evaluate the numerical solution). However, another test of the numerical solution based on *conservation of mass (or energy)* is possible; this idea is explained in Appendix 1 at the end of this chapter.

Case 3: Linear PDE, Third-Type BCs (`ncase=1, nbc=3, a=0.1, b=c=0`)

For `ncase=1,nbc=3` (linear case, third-type BCs), the BCs are

$$\frac{\partial u(x=x_l,t)}{\partial x} = -a(1-u(x=x_l,t)), \quad \frac{\partial u(x=x_u,t)}{\partial x} = a(1-u(x=x_u,t)) \tag{5.5}$$

The relevant coding in `pde_1.m` of Listing 5.1 is

```
      .
      .
      .
%
% Linear case
  if(ncase==1)
%
%   ux
    ux=dss004(xl,xu,n,u);
      .
      .
      .
%
%   Third type BCs
    if(nbc==3)
      ux(1)=-(1/a)*(1-u(1)); ux(n)=(1/a)*(1-u(n));
    end
      .
      .
      .
%
%   uxx
    uxx=dss004(xl,xu,n,ux);
  end
```

LISTING 5.6: Programming from `pde_1.m` for eq. (5.1) with `ncase=1, nbc=3`.

The coding for the third-type BCs (`nbc=3`) follows directly from eq. (5.5). The plotted numerical solution is in Fig. 5.6.

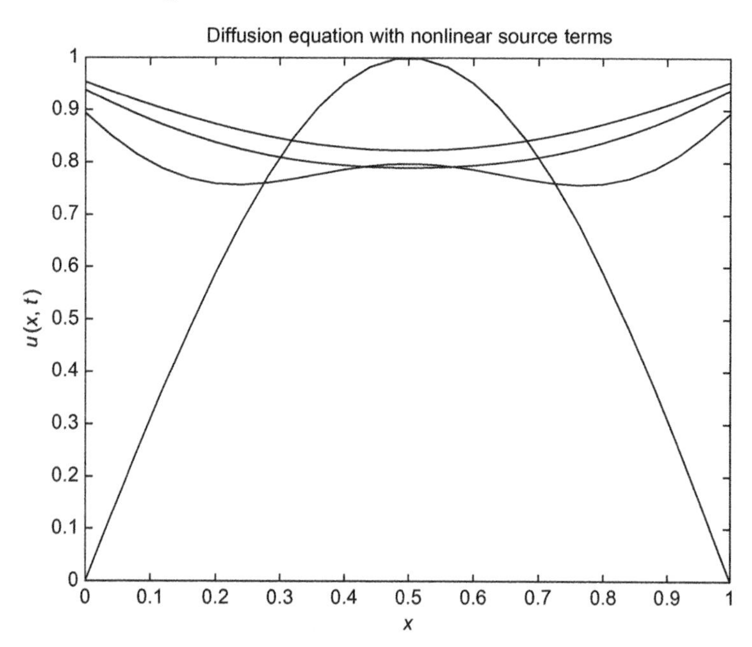

FIGURE 5.6: Numerical solution to eq. (5.1) using five-point FD approximations in `dss004` for $t = 0, 0.3, 0.6, 0.9$ for `ncase=1,nbc=3`.

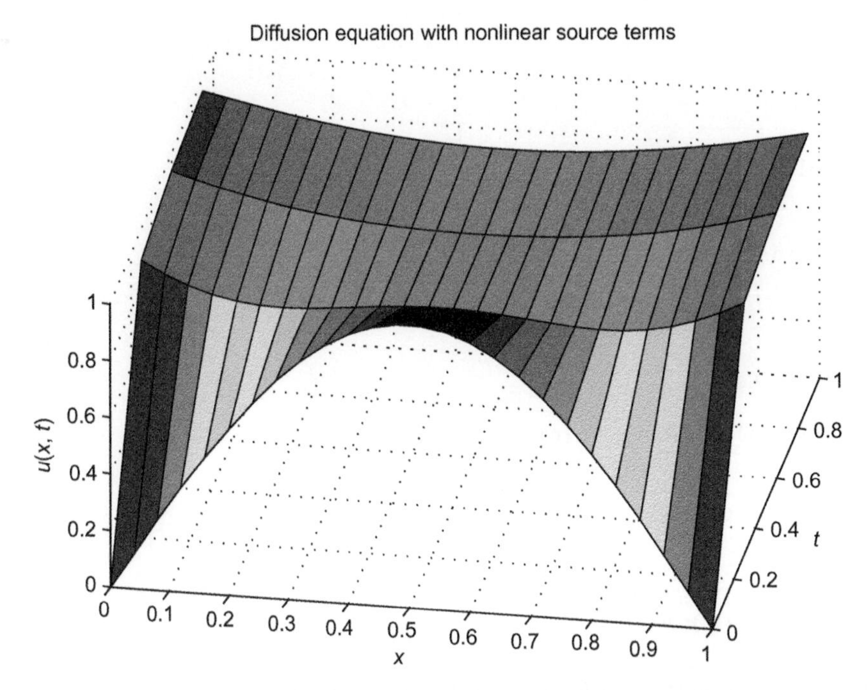

FIGURE 5.7: 3D plot of the numerical solution to eq. (5.1).

We can note in particular that the limiting value $u(x,t \to \infty) = 1$ follows from BCs (5.5); this would be more apparent numerically if the final value of t was increased from $t = 0.9$ (the same interval in t was used in all 10 cases to minimize the coding).

Figure 5.7 is the 3D plot of the numerical solution.

The limiting value $u(x,t \to \infty) = 1$ is apparent (particularly if t is extended beyond $t = 0.9$).

The map of the ODE Jacobian matrix is basically the same as in Fig. 5.3 and is therefore not reproduced here (since the PDE is the same and the numerical approximations are nearly the same with differences only at the boundaries for `nbc=1,3`). Also, an analytical solution is not presented for this case (that could be used to evaluate the numerical solution).

Case 4: Linear PDE, Nonlinear Third-Type BCs (`ncase=1, nbc=4, a=0.1, b=c=0`)

For `ncase=1,nbc=4` (linear case, nonlinear third-type BCs), the BCs are

$$\frac{\partial u(x=x_l,t)}{\partial x} = -a[1-u(x=x_l,t)^4], \quad \frac{\partial u(x=x_u,t)}{\partial x} = a[1-u(x=x_u,t)^4] \tag{5.6}$$

Equation (5.6) may appear to be a trivial extension of eq. (5.5). However, the nonlinear terms $u(x = x_l, t)^4, u(x = x_u, t)^4$ are significant for the following reasons:

- The fourth power has an important application in radiation heat transfer (according to the *Stefan–Boltzmann law*).
- Although the solution for the nonlinear case (nbc=4) is a minor extension of the linear case (nbc=3) numerically (as discussed subsequently), analytically, the nonlinear case is much more difficult, and in fact, analytical solutions for the nonlinear case may not be available. In other words, this example illustrates a major advantage of numerical methods, that is, the accommodation of nonlinear BCs.
- To extend this idea further, nonlinear BCs of the general form (for a second-order PDE)

$$f_b(u, u_x, x, t) = 0 \tag{5.7}$$

  can be accommodated numerically by using a root finder applied to f_b to compute u_x at the boundary, followed by application of the method for nonlinear third-type BCs discussed next.

The relevant coding in pde_1.m of Listing 5.1 is

```
        .
        .
        .
%
% Linear case
  if(ncase==1)
%
%   ux
    ux=dss004(xl,xu,n,u);
        .
        .
        .
%
%   Nonlinear Neumann BCs
    if(nbc==4)
      ux(1)=-(1/a)*(1-u(1)^4); ux(n)=(1/a)*(1-u(n)^4);
    end
        .
        .
        .
%
%   uxx
    uxx=dss004(xl,xu,n,ux);
  end
```

LISTING 5.7: Programming from pde_1.m for eq. (5.1) with ncase=1, nbc=4.

The coding for the nonlinear third-type BCs (nbc=4) follows directly from eq. (5.6). The plotted numerical solution is in Fig. 5.8.

We can note in particular that the limiting value $u(x, t \to \infty) = 1$ follows from BCs (5.6); this would be more apparent numerically if the final value of t was increased from $t = 0.9$

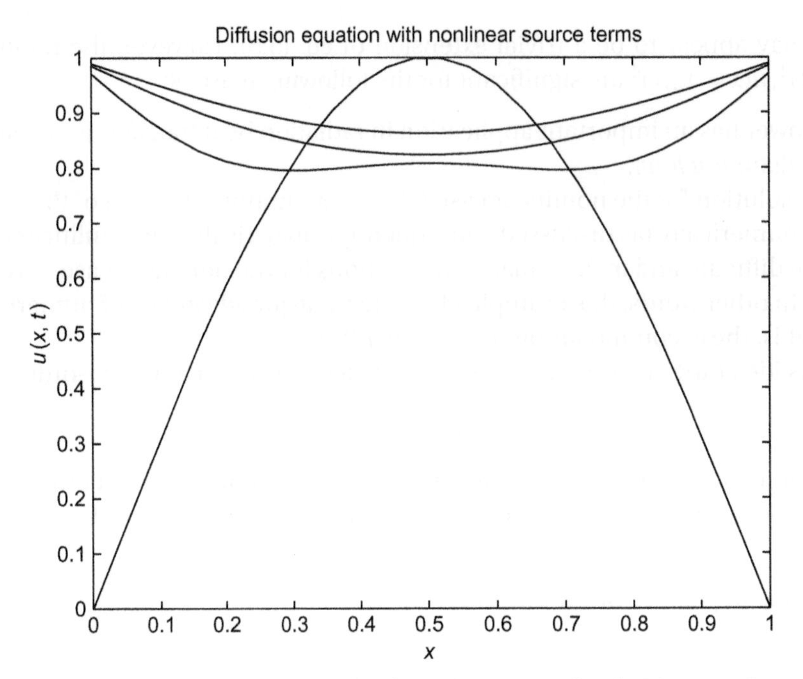

FIGURE 5.8: Numerical solution to eq. (5.1) using five-point FD approximations in `dss004` for $t = 0, 0.3, 0.6, 0.9$ for `ncase=1,nbc=4`.

(again, the same interval in t was used in all 10 cases to minimize the coding). Figure 5.9 is the 3D plot of the numerical solution.

The limiting value $u(x, t \to \infty) = 1$ is apparent (particularly if t is extended beyond $t = 0.9$).

The map of the ODE Jacobian matrix is basically the same as in Fig. 5.3 and is therefore not reproduced here (since the PDE is the same and the numerical approximations are nearly the same with differences only at the boundaries for `nbc=1,4`). Also, an analytical solution is not presented for this case (that could be used to evaluate the numerical solution); this is primarily due to the unavailability of an analytical solution for this nonlinear case.

Case 5: Linear PDE, Analytical Neumann BCs (`ncase=1, nbc=5, a=0.1, b=c=0`)

As a final linear case, we consider analytical Neumann BCs (`ncase=1, nbc=5`)

$$\frac{\partial u(x = x_l, t)}{\partial x} = f_{b1}(t), \quad \frac{\partial u(x = x_u, t)}{\partial x} = f_{b2}(t) \tag{5.8}$$

where the BC functions, f_{b1}, f_{b2}, are obtained by differentiating eq. (5.3) with $x = x_l, x_u$.

$$\frac{\partial u(x, t)}{\partial x} = (\pi/x_u)\cos(\pi x/x_u)e^{-a(\pi/x_u)^2 t} \tag{5.9}$$

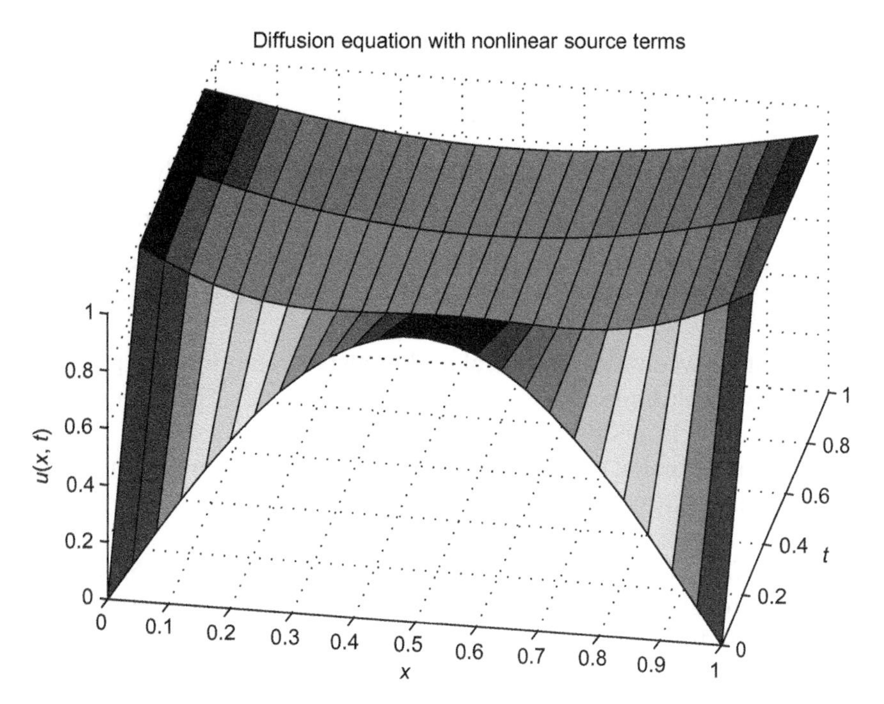

FIGURE 5.9: 3D plot of the numerical solution to eq. (5.1).

The relevant coding in `pde_1.m` of Listing 5.1 is

```
        .
        .
        .
%
% Linear case
  if(ncase==1)
%
%   ux
    ux=dss004(xl,xu,n,u);
        .
        .
        .
%
%   Analytical Neumann BCs
    if(nbc==5)
      ux(1)=(pi/xu)*cos(pi*x(1)/xu)*exp(-a*(pi/xu)^2*t);
      ux(n)=(pi/xu)*cos(pi*x(n)/xu)*exp(-a*(pi/xu)^2*t);
    end
        .
        .
        .
%
%   uxx
```

```
    uxx=dss004(xl,xu,n,ux);
  end
```

LISTING 5.8: Programming from `pde_1.m` for eq. (5.1) with `ncase=1, nbc=5`.

The coding for the analytical Neumann BCs (`nbc=5`) follows directly from eq. (5.8) with $\frac{\partial u(x_l,t)}{\partial x}$= `ux(1)` and $\frac{\partial u(x_u,t)}{\partial x}$= `ux(n)`. The plotted numerical solution is in Fig. 5.10, which includes the numerical (lines) and analytical (circles) solutions; the latter, again eq. (5.3), applies since Neumann BCs based on this solution were used (in eqns. (5.7) and (5.8)).

The agreement between the numerical and analytical solutions is quite satisfactory; a portion of the two solutions at $t = 0.9$ is listed in Table 5.2. The computational effort reflected in `ncall = 71` is modest.

Figure 5.11 is the 3D plot of the numerical solution. Recall again that the numerical and plotted output from `pde_1_main` in Listing 5.4 (Figs. 5.10, 5.11, Table 5.2) is produced by the code for `nbc=1,5`.

The map of the ODE Jacobian matrix is basically the same as in Fig. 5.3 and is therefore not reproduced here (since the PDE is the same and the numerical approximations are nearly the same with differences only at the boundaries for `nbc=1,5`).

This completes the discussion of the five BC cases for the linear PDE ($b = c = 0$, `ncase=1`). We now essentially repeat the preceding discussion for the five BC cases

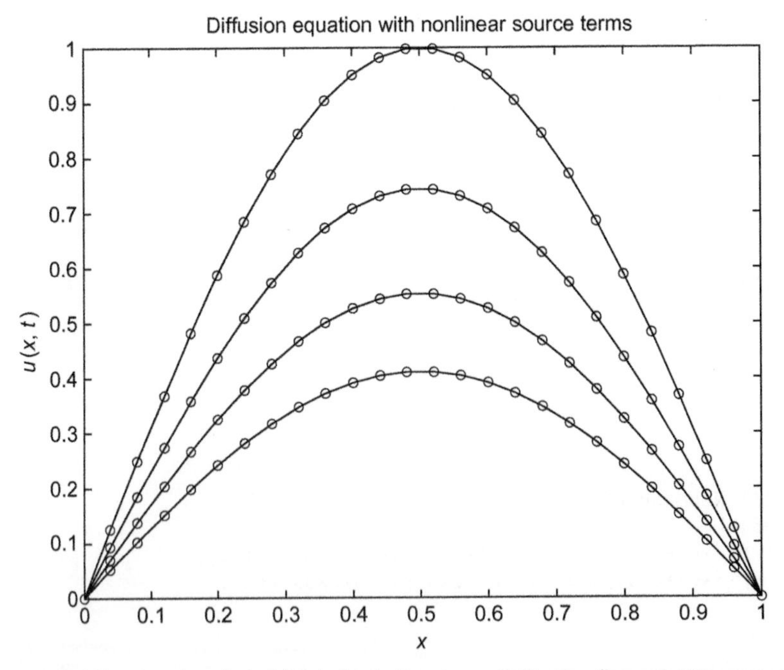

FIGURE 5.10: Numerical (lines) and analytical (circles) solutions to eq. (5.1) using five-point FD approximations in `dss004` $t = 0, 0.3, 0.6, 0.9$ for `ncase=1,nbc=5` (top to bottom).

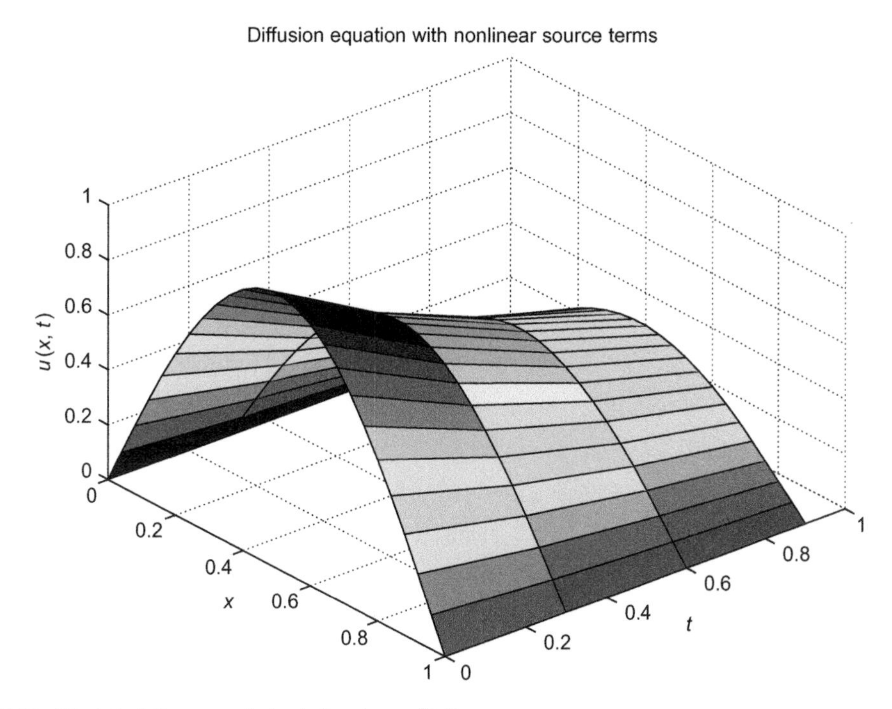

FIGURE 5.11: 3D plot of the numerical solution to eq. (5.1).

Table 5.2: Abbreviated tabular numerical and analytical solutions for `ncase=1,nbc=5`

t	x	u(it,i)	u_anal(it,i)	err(it,i)
	.		.	
	.		.	
	.		.	
0.90	0.400	0.391239	0.391235	0.000003
0.90	0.440	0.404086	0.404083	0.000004
0.90	0.480	0.410561	0.410557	0.000004
0.90	0.520	0.410561	0.410557	0.000004
0.90	0.560	0.404086	0.404083	0.000004
0.90	0.600	0.391239	0.391235	0.000003
	.		.	
	.		.	
	.		.	
ncall=71				

(`nbc=1,2,3,4,5`) for the nonlinear PDE ($b \neq 0, c \neq 0$, `ncase=2`). The purpose of this discussion is to demonstrate, at least in principle, that a nonlinear PDE numerical solution is as readily computed as for a linear PDE. Since the following discussion is basically a repetition of the previous discussion, we present only the details for the few differences.

Case 6: Nonlinear PDE, Dirichlet BCs (ncase=2, nbc=1, a=b=c=1)

For Dirichlet BCs (ncase=2, nbc=1), the relevant coding in pde_1.m of Listing 5.1 is

```
          .
          .
          .
%
% Nonlinear case
  if(ncase==2)
%
%   ux
    ux=dss004(xl,xu,n,u);
%
%   Dirichlet BCs
    if(nbc==1)
      u(1)=ua_1(x(1),t); u(n)=ua_1(x(n),t);
      ux=dss004(xl,xu,n,u);
    end
          .
          .
          .
%
%   uxx
    uxx=dss004(xl,xu,n,ux);
  end
%
% PDE
  for i=1:n
    sq=u(i)^2;
    cu=u(i)*sq;
    ut(i)=a*uxx(i)-b*cu-c*sq;
  end
  if(nbc==1)
    ut(1)=0; ut(n)=0;
  end
  ut=ut';
%
% Increment calls to pde_1
  ncall=ncall+1;
  end
```

LISTING 5.9: Programming from pde_1.m for eq. (5.1) with ncase=2, nbc=1.

ua_1.m can again be used for the Dirichlet BCs through the use of ncase as a global variable (refer to Listing 5.3). The programming of eq. (5.1) for the nonlinear case (ncase=2) is the same as for the linear case (ncase=1); only the values of a, b, c (programmed in pde_1_main.m of Listing 5.4) are changed.

inital_1.m and pde_1_main.m of Listings 5.2 and 5.4 remain the same for ncase=2, and the map of the Jacobian matrix does not change significantly from the preceding five linear cases. The plotted solutions follow.

The agreement between the numerical and analytical solutions is to at least six figures; a portion of the two solutions at $t = 0.9$ is listed in Table 5.3. The computational effort reflected in ncall=87 is quite modest.

Table 5.3: Abbreviated tabular numerical and analytical solutions for `ncase=2,nbc=1`

t	x	u(it,i)	u_anal(it,i)	err(it,i)
	.		.	
	.		.	
	.		.	
0.90	0.400	0.458118	0.458118	0.000000
0.90	0.440	0.452258	0.452258	0.000000
0.90	0.480	0.446546	0.446546	0.000000
0.90	0.520	0.440976	0.440976	0.000000
0.90	0.560	0.435544	0.435544	0.000000
0.90	0.600	0.430244	0.430244	0.000000
	.		.	
	.		.	
	.		.	
ncall=87				

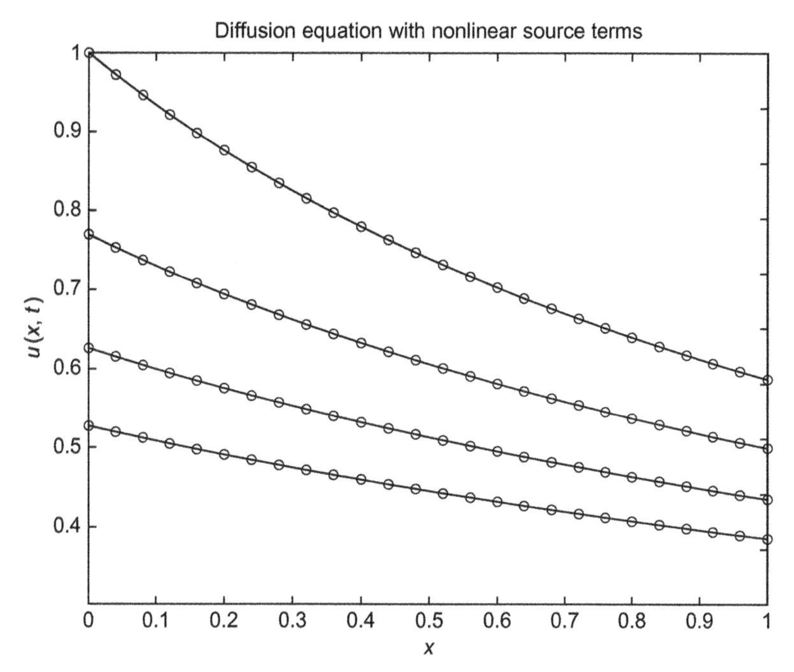

FIGURE 5.12: Numerical (lines) and analytical (circles) solutions to eq. (5.1) using five-point FD approximations in `dss004` $t = 0, 0.3, 0.6, 0.9$ for `ncase=2,nbc=1` (top to bottom).

Figure 5.13 is the 3D plot of the numerical solution. Recall again that the numerical and plotted output from `pde_1_main` in Listing 5.4 (Figs. 5.12, 5.13, Table 5.3) is produced by the code for `ncase=2, nbc=1,5`.

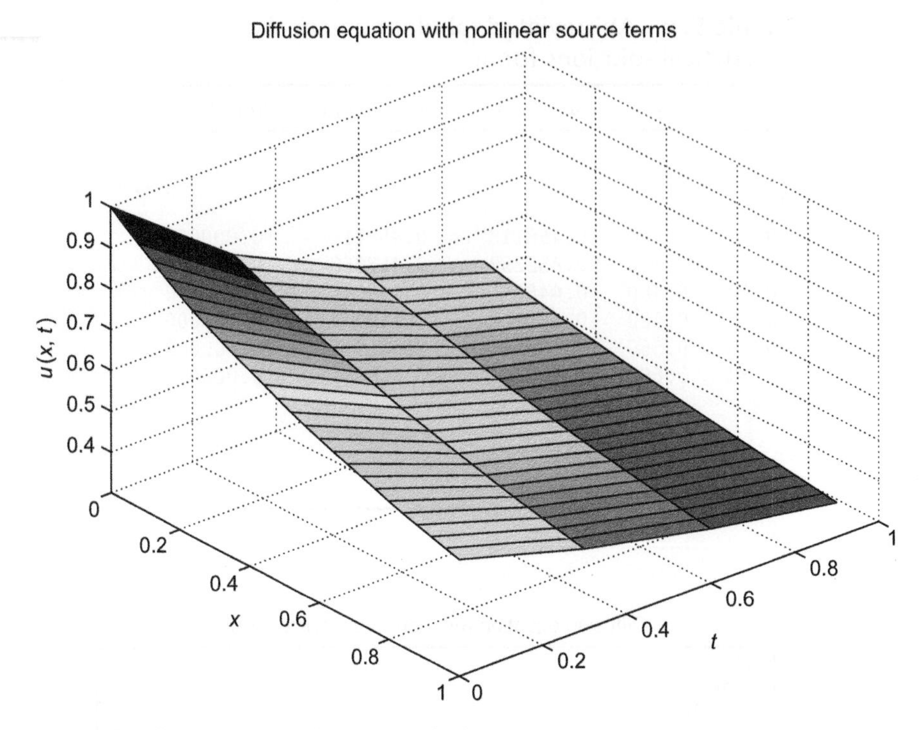

FIGURE 5.13: 3D plot of the numerical solution to eq. (5.1).

Case 7: Nonlinear PDE, Neumann BCs (`ncase=2, nbc=2, a=b=c=1`)

For the Neumann BCs of eq. (5.4) (`ncase=2, nbc=2`), the relevant coding in `pde_1.m` of Listing 5.1 is

```
          .
          .
          .
%
% Nonlinear case
  if(ncase==2)
%
%   ux
    ux=dss004(xl,xu,n,u);
          .
          .
          .
%
%   Neumann BCs
    if(nbc==2)
      ux(1)=0; ux(n)=0;
    end
          .
          .
          .
```

```
%
%    uxx
     uxx=dss004(xl,xu,n,ux);
   end
%
% PDE
   for i=1:n
     sq=u(i)^2;
     cu=u(i)*sq;
     ut(i)=a*uxx(i)-b*cu-c*sq;
   end
   if(nbc==1)
     ut(1)=0; ut(n)=0;
   end
   ut=ut';
%
% Increment calls to pde_1
   ncall=ncall+1;
   end
```

LISTING 5.10: Programming from `pde_1.m` for eq. (5.1) with `ncase=2, nbc=2`.

The programming of eq. (5.1) for the nonlinear case (`ncase=2`) is the same as for the linear case (`ncase=1`); only the values of a, b, c (programmed in `pde_1_main.m` of Listing 5.4) are changed.

`inital_1.m` and `pde_1_main.m` of Listings 5.2 and 5.4 remain the same for `ncase=2`, and the map of the Jacobian matrix does not change significantly from the preceding cases. The plotted solutions follow.

A portion of the solution at $t = 0.9$ is listed in Table 5.4.

The computational effort reflected in `ncall = 188` is modest, and the approach to a constant value $u(x, t \to \infty) = 0.3722$ is clear. This value could also be confirmed by

Table 5.4: Abbreviated tabular numerical and analytical solutions for `ncase=2,nbc=2`

t	x	u(it,i)
	.	
	.	
	.	
0.90	0.400	0.372289
0.90	0.440	0.372288
0.90	0.480	0.372288
0.90	0.520	0.372287
0.90	0.560	0.372286
0.90	0.600	0.372286
	.	
	.	
	.	
ncall=188		

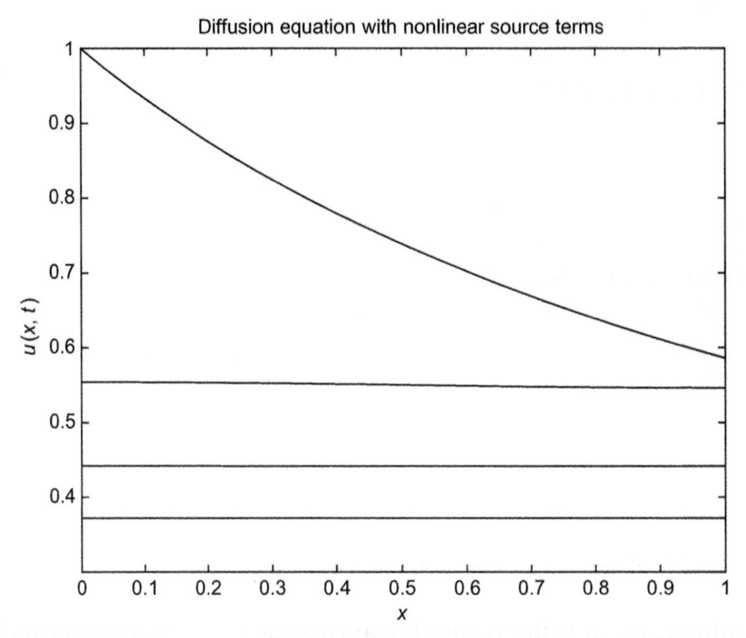

FIGURE 5.14: Numerical solution to eq. (5.1) using five-point FD approximations in `dss004` $t = 0, 0.3, 0.6, 0.9$ for `ncase=2,nbc=2` (top to bottom).

application of mass conservation as explained in Appendix 1 at the end of this chapter; note that it is different than the value $2/\pi = 0.636620$ for the linear case.

Figure 5.15 is the 3D plot of the numerical solution. Recall again that the numerical and plotted output from `pde_1_main` in Listing 5.4 (Figs. 5.14, 5.15, Table 5.4) is produced by the code for `ncase=2, nbc=2,3,4`.

Case 8: Nonlinear PDE, Third-type BCs (`ncase=2, nbc=3, a=b=c=1`)

For the third-type BCs of eq. (5.5) (`ncase=2, nbc=3`), the relevant coding in `pde_1.m` of Listing 5.1 is

```
        .
        .
        .
%
% Nonlinear case
  if(ncase==2)
%
%    ux
     ux=dss004(xl,xu,n,u);
        .
        .
        .
%
%    Third type BCs
     if(nbc==3)
```

```
      ux(1)=-(1/a)*(1-u(1)); ux(n)=(1/a)*(1-u(n));
    end
        .
        .
        .
%
%   uxx
    uxx=dss004(xl,xu,n,ux);
  end
%
% PDE
  for i=1:n
    sq=u(i)^2;
    cu=u(i)*sq;
    ut(i)=a*uxx(i)-b*cu-c*sq;
  end
  if(nbc==1)
    ut(1)=0; ut(n)=0;
  end
  ut=ut';
%
% Increment calls to pde_1
  ncall=ncall+1;
  end
```

LISTING 5.11: Programming from `pde_1.m` for eq. (5.1) with `ncase=2, nbc=3`.

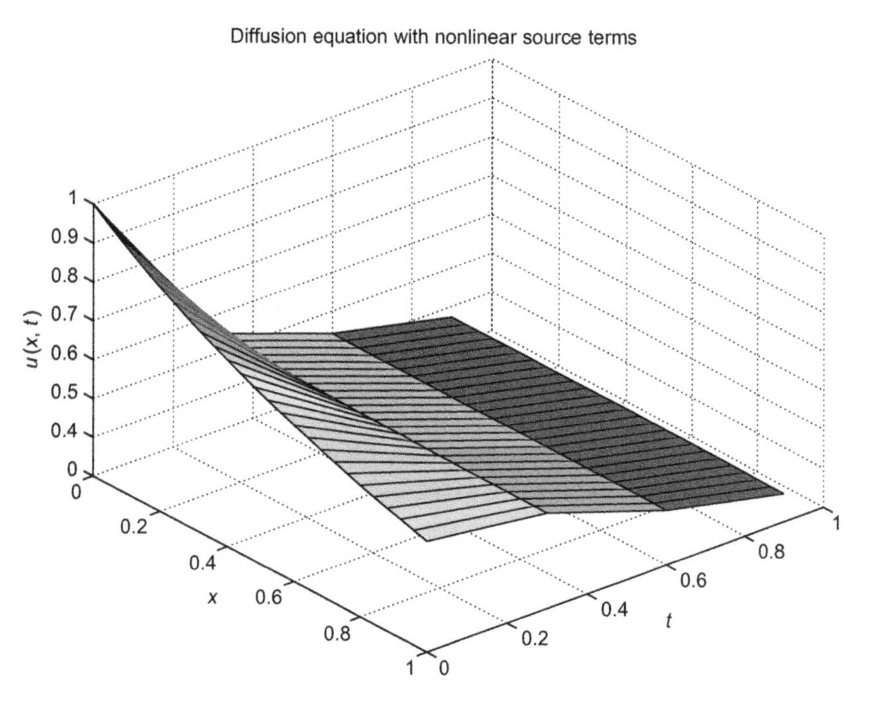

FIGURE 5.15: 3D plot of the numerical solution to eq. (5.1).

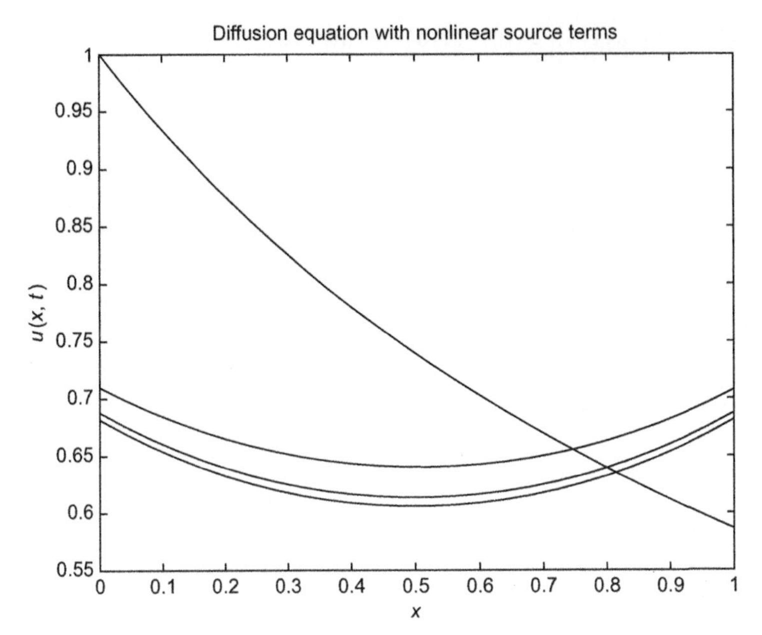

FIGURE 5.16: Numerical solution to eq. (5.1) using five-point FD approximations in `dss004` $t = 0, 0.3, 0.6, 0.9$ for `ncase=2,nbc=3` (top to bottom).

The programming of eq. (5.1) for the nonlinear case (`ncase=2`) is the same as for the linear case (`ncase=1`); only the values of a, b, c (programmed in `pde_1_main.m` of Listing 5.4) are changed.

`inital_1.m` and `pde_1_main.m` of Listings 5.2 and 5.4 remain the same for `ncase=2`, and the map of the Jacobian matrix does not change significantly from the preceding cases. The plotted solutions follow. A portion of the solution at $t = 0.9$ is listed in Table 5.5. The computational effort reflected in `ncall = 183` is modest. The approach to a solution invariant in t, $u(x, t \to \infty)$, is not clear, but it could be studied further by extending the solution beyond $t = 0.9$. Whatever the limiting solution might be, it will be the solution to eq. (5.1) for $\frac{\partial u}{\partial t} \to 0$.

This is an important point, since for this case, eq. (5.1) reduces to an ODE (with only a second derivative in x). In other words, the preceding numerical solution to eq. (5.1) (for large t) is also a solution to a *two-point nonlinear boundary value ODE (BVODE)*. To emphasize this point, if we start with a BVODE, we could in principle compute a numerical solution by appending a derivative in an initial value variable, such as $\frac{\partial u}{\partial t}$ in eq. (5.1), then integrate numerically in t until this derivative effectively vanishes; the resulting solution is then for the BVODE.

Figure 5.17 is the 3D plot of the numerical solution. Recall again that the numerical and plotted output from `pde_1_main` in Listing 5.4 (Figs. 5.16, 5.17, Table 5.5) is produced by the code for `ncase=2, nbc=2,3,4`.

Table 5.5: Abbreviated tabular numerical and analytical solutions for `ncase=2,nbc=3`

t	x	u(it,i)
	.	
	.	
	.	
0.90	0.400	0.608456
0.90	0.440	0.606608
0.90	0.480	0.605685
0.90	0.520	0.605685
0.90	0.560	0.606608
0.90	0.600	0.608456
	.	
	.	
	.	
ncall=183		

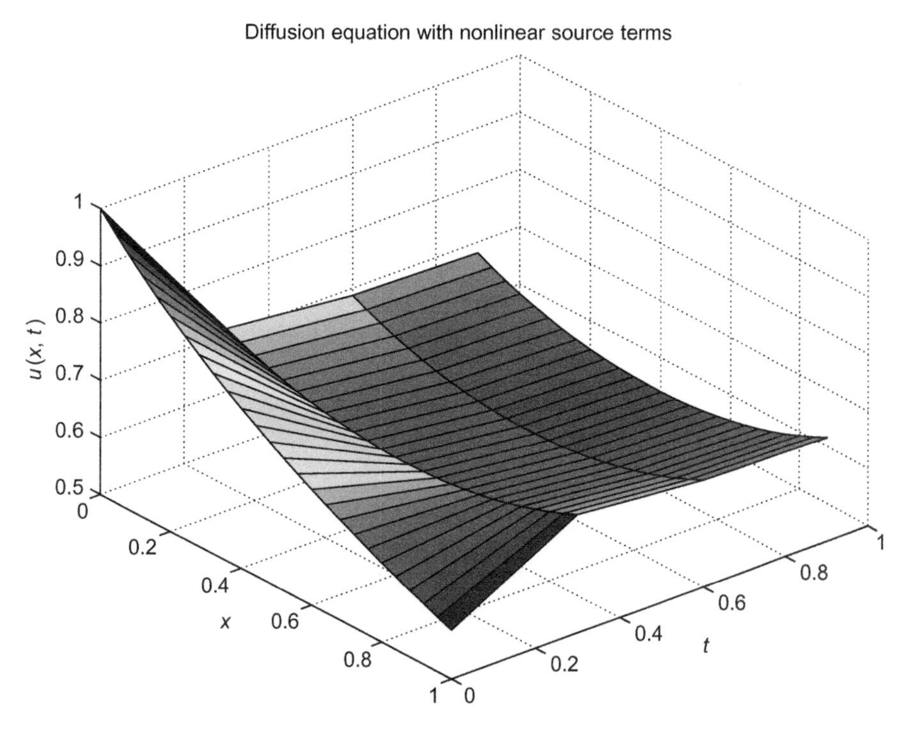

FIGURE 5.17: 3D plot of the numerical solution to eq. (5.1).

Case 9: Nonlinear PDE, Nonlinear Third-Type BCs (ncase=2, nbc=4, a=b=c=1)

For the nonlinear third-type BCs of eq. (5.6) (ncase=2, nbc=4), the relevant coding in pde_1.m of Listing 5.1 is

```
        .
        .
        .
%
% Nonlinear case
  if(ncase==2)
%
%   ux
    ux=dss004(xl,xu,n,u);
        .
        .
        .
%
%   Nonlinear Neumann BCs
    if(nbc==4)
      ux(1)=-(1/a)*(1-u(1)^4); ux(n)=(1/a)*(1-u(n)^4);
    end
        .
        .
        .
%
%   uxx
    uxx=dss004(xl,xu,n,ux);
  end
%
% PDE
  for i=1:n
    sq=u(i)^2;
    cu=u(i)*sq;
    ut(i)=a*uxx(i)-b*cu-c*sq;
  end
  if(nbc==1)
    ut(1)=0; ut(n)=0;
  end
  ut=ut';
%
% Increment calls to pde_1
  ncall=ncall+1;
  end
```

LISTING 5.12: Programming from pde_1.m for eq. (5.1) with ncase=2, nbc=4.

The programming of eq. (5.1) for the nonlinear case (ncase=2) is the same as for the linear case (ncase=1); only the values of a, b, c (programmed in pde_1_main.m of Listing 5.4) are changed.

Table 5.6: Abbreviated tabular numerical and analytical solutions for `ncase=2,nbc=4`

t	x	u(it,i)
	.	
	.	
	.	
0.90	0.400	0.723781
0.90	0.440	0.720924
0.90	0.480	0.719499
0.90	0.520	0.719499
0.90	0.560	0.720924
0.90	0.600	0.723781
	.	
	.	
	.	
ncall=228		

`inital_1.m` and `pde_1_main.m` of Listings 5.2 and 5.4 remain the same for `ncase=2`, and the map of the Jacobian matrix does not change significantly from the preceding cases. The plotted solutions follow.

A portion of the solution at $t = 0.9$ is listed in Table 5.6. The computational effort reflected in `ncall = 228` is modest.

In contrast with `nbc=3`, the approach to a solution invariant in t, $u(x, t \rightarrow \infty)$ is clear, since the curves for $t = 0.6$ and $t = 0.9$ are nearly the same; this could be confirmed by extending the solution beyond $t = 0.9$. Whatever the limiting solution might be, it will be the solution to eq. (5.1) for $\frac{\partial u}{\partial t} \rightarrow 0$. Note also that the solutions of Table 5.5 (linear BCs) and Table 5.6 (nonlinear BCs) are substantially different, which reflects the difference in eqns. (5.5) and (5.6); that is, the BCs have a significant effect in the two cases (even though they are similar).

Again, as with `nbc=3`, this approach to a solution invariant in t is an important point, since for this case, eq. (5.1) reduces to an ODE (with only a second derivative in x). In other words, the preceding numerical solution to eq. (5.1) (for large t) is also a solution to a *two-point nonlinear boundary value ODE (BVODE)*. There is also an important distinction from the preceding `ncase=2, nbc=3` case, since now the ODE and the *boundary conditions are nonlinear.* In other words, if we start with a BVODE that is nonlinear in both the ODE and its boundary conditions, we could in principle compute a numerical solution by appending a derivative in an initial value variable, such as $\frac{\partial u}{\partial t}$ in eq. (5.1), then integrate numerically in t until this derivative effectively vanishes; the resulting solution is then for the nonlinear BVODE.

Figure 5.19 is the 3D plot of the numerical solution. Recall again that the numerical and plotted output from `pde_1_main` in Listing 5.4 (Figs. 5.18, 5.19, Table 5.6) is produced by the code for `ncase=2, nbc=2,3,4`.

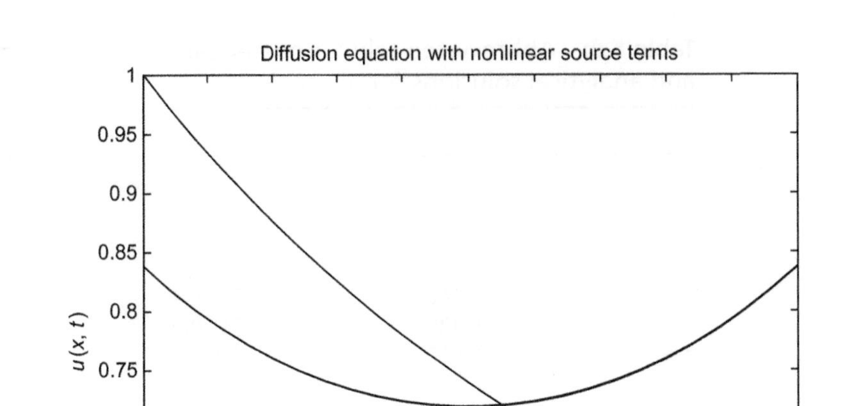

FIGURE 5.18: Numerical solution to eq. (5.1) using five-point FD approximations in `dss004` $t = 0, 0.3, 0.6, 0.9$ for `ncase=2,nbc=4` (top to bottom).

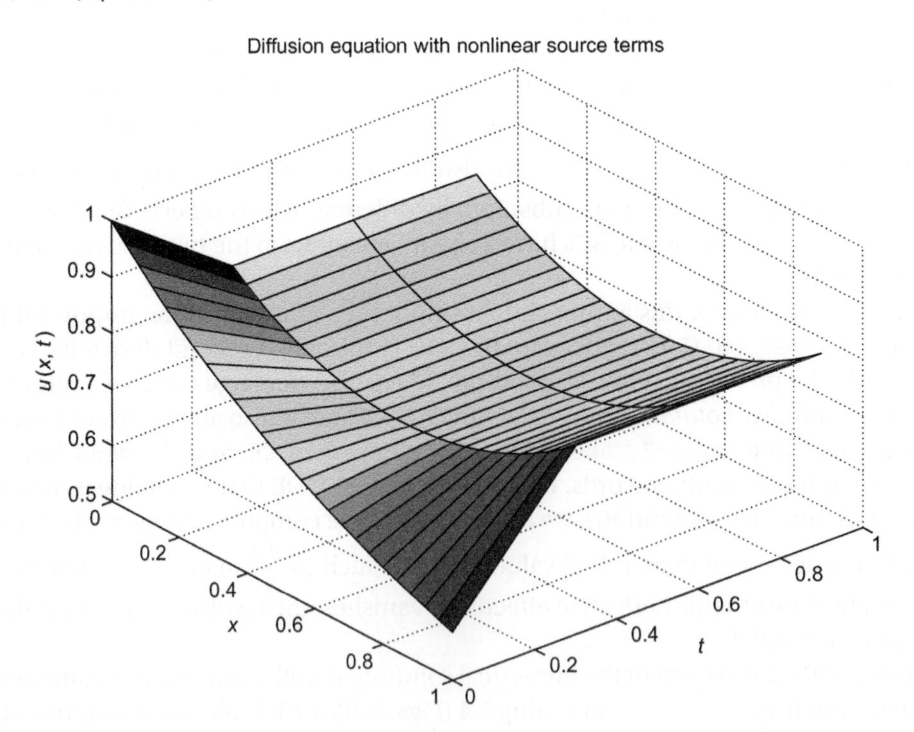

FIGURE 5.19: 3D plot of the numerical solution to eq. (5.1).

Case 10: Nonlinear PDE, Analytical Neumann BCs (`ncase=2, nbc=5, a=b=c=1`)

Finally, we conclude (with the tenth case) using the analytical Neumann BCs obtained by differentiation eq. (5.2) with respect to x.

$$\frac{\partial u(x,t)}{\partial x} = -\left(\sqrt{\frac{b}{2a}}(x+\sqrt{\frac{2a}{b}}ct)+1\right)^{-2}\sqrt{\frac{b}{2a}} \tag{5.10}$$

For `ncase=2, nbc=5`, the relevant coding in `pde_1.m` of Listing 5.1 is

```
        .
        .
        .
%
% Nonlinear case
  if(ncase==2)
%
%   ux
    ux=dss004(xl,xu,n,u);
        .
        .
        .
%
%   Analytical Neumann BCs
    if(nbc==5)
      ux(1)=-1/(c*t+(b/(2*a))^0.5*x(1)+1)^2*(b/(2*a))^0.5;
      ux(n)=-1/(c*t+(b/(2*a))^0.5*x(n)+1)^2*(b/(2*a))^0.5;
    end
%
%   uxx
    uxx=dss004(xl,xu,n,ux);
  end
%
% PDE
  for i=1:n
    sq=u(i)^2;
    cu=u(i)*sq;
    ut(i)=a*uxx(i)-b*cu-c*sq;
  end
  if(nbc==1)
    ut(1)=0; ut(n)=0;
  end
  ut=ut';
%
% Increment calls to pde_1
  ncall=ncall+1;
  end
```

LISTING 5.13: Programming from `pde_1.m` for eq. (5.1) with `ncase=2, nbc=5`.

Table 5.7: Abbreviated tabular numerical and analytical solutions for `ncase=2,nbc=5`

t	x	u(it,i)	u_anal(it,i)	err(it,i)
	.		.	
	.		.	
	.		.	
0.90	0.400	0.458118	0.458118	0.000000
0.90	0.440	0.452258	0.452258	0.000000
0.90	0.480	0.446546	0.446546	0.000000
0.90	0.520	0.440976	0.440976	0.000000
0.90	0.560	0.435544	0.435544	0.000000
0.90	0.600	0.430244	0.430244	0.000000
	.		.	
	.		.	
	.		.	
ncall=84				

The programming of eq. (5.1) for the nonlinear case (`ncase=2`) is the same as for the linear case (`ncase=1`); only the values of a, b, c (programmed in `pde_1_main.m` of Listing 5.4) are changed.

`inital_1.m` and `pde_1_main.m` of Listings 5.2 and 5.4 remain the same for `ncase=2`, and the map of the Jacobian matrix does not change significantly from the preceding cases. The plotted solutions follow.

A portion of the solution at $t = 0.9$ is listed in Table 5.7. The computational effort reflected in `ncall = 84` is quite modest, and the numerical and analytical solutions agree to at least six figures. Figure 5.21 is the 3D plot of the numerical solution.

Recall again that the numerical and plotted output from `pde_1_main` in Listing 5.4 (Figs. 5.20, 5.21, Table 5.7) is produced by the code for `ncase=2, nbc=1,5`.

In conclusion, we have considered 10 cases of the solution of eq. (5.1) as a linear PDE ($b = c = 0$) and a nonlinear PDE ($b \neq 0, c \neq 0$) with linear and nonlinear BCs. The intent is to demonstrate the generality and flexibility of the numerical approach to PDE solutions. The obvious limitation to the preceding cases is that they are all 1D. We are developing analogous procedures for 2D and 3D PDEs, although these additional cases are also discussed to a limited extent in [5].

Appendix 1

In the previous chapters, we have evaluated numerical PDE solutions by comparison with analytical solutions. This is a valuable procedure, since agreement between numerical and analytical solutions gives some assurance that the numerical procedures are sound and

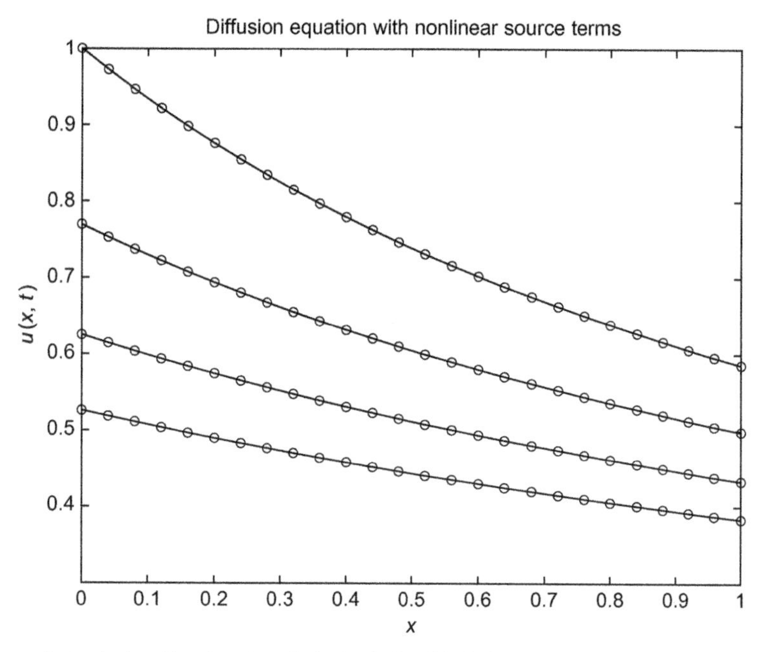

FIGURE 5.20: Numerical solution (lines) and analytical solution (circles) to eq. (5.1) using five-point FD approximations in `dss004` $t = 0, 0.3, 0.6, 0.9$ for `ncase=2,nbc=5`.

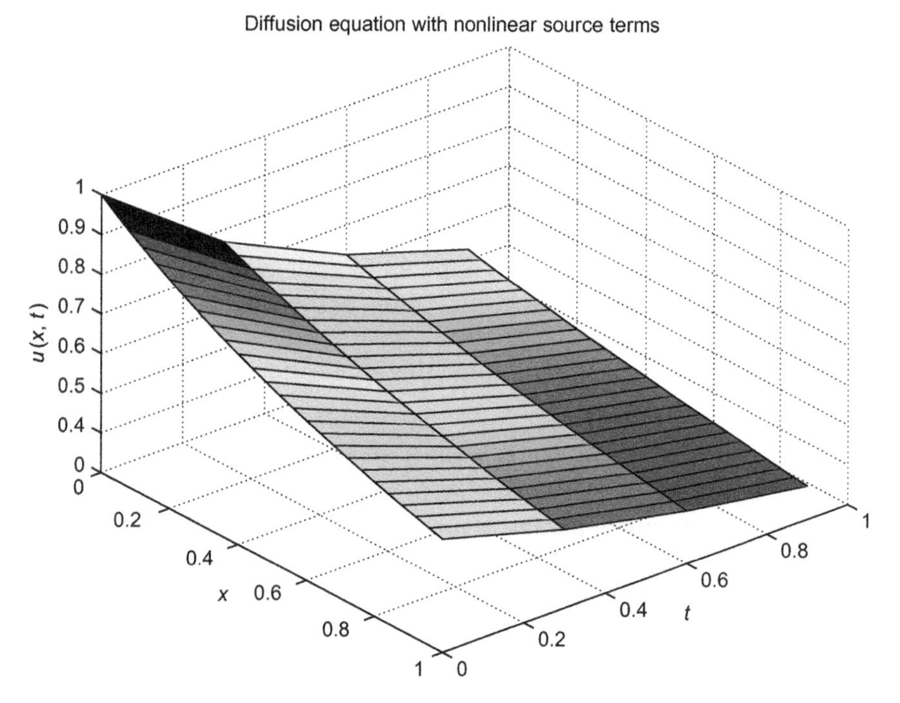

FIGURE 5.21: 3D plot of the numerical solution to eq. (5.1).

can therefore be used in the solution of PDE systems for which an analytical solution is not available.

A second widely used method for the validation of numerical methods is to apply a *conservation principle* such as *conservation of mass, momentum, or energy*. Here, we demonstrate how the conservation of mass (or energy) can be applied to the numerical solution of eq. (5.1) for the linear case (`ncase = 1`) and homogeneous Neumann BCs (`nbc=2`). Physically, the BCs specify *zero flux* or *insulated boundaries* so that the initial mass or energy of the system should remain constant with increasing t.

For the linear case (`ncase=1`), the IC is eq. (5.3) (with $t = 0$), and for $t \to \infty$, the solution approaches a constant value given by the *integral average of the IC*.

$$\frac{\int_{x_l}^{x_u} \sin(\pi x/x_u)dx}{x_u - x_l} = -\frac{\cos(\pi x/x_u)(x_u/\pi)|_{x_l}^{x_u}}{x_u - x_l}$$

$$= \frac{(x_u/\pi)[\cos(\pi x_l/x_u) - \cos(\pi x_u/x_u)]}{x_u - x_l} = \frac{(x_u/\pi)[1 + \cos(\pi x_l/x_u)]}{x_u - x_l} \tag{5.11}$$

For $x_u = 1, x_l = 0$, eq. (5.11) becomes

$$\frac{\int_{x_l}^{x_u} \sin(\pi x/x_u)dx}{x_u - x_l} = 2/\pi = 0.636620 \tag{5.12}$$

We can then compare the numerical value ($2/\pi = 0.636620$) with the value that the numerical solution approaches for $t \to \infty$. To accomplish this, we replace the relatively short time interval $0 \le t \le 0.9$ in `pde_1_main.m` with the expanded interval $0 \le t \le 2.7$ (with the four output points $t = 0, 0.9, 1.8, 2.7$); $t = 2.7$ is large enough that the solution reaches a constant value as will be observed in the following numerical output. All of the other Matlab code remain unchanged.

The output for `ncase=1, nbc=2` is listed in Table 5.8. The agreement with the integral average value $2/\pi = 0.636620$ at $t = 2.7$ is apparent. Recall again that this agreement was accomplished with only $n = 26$ points, which is due to the smoothness of the solution of eq. (5.3).

We can make three additional points about this application of the conservation of mass (energy):

- This test of conservation could not be applied to the nonlinear case `ncase=2` ($b \ne 0, c \ne 0$ in eq. (5.1)). Physically, the two nonlinear terms $-bu^3, -cu^2$ can be considered as representing depletion of mass due to a third- and second-order reaction, respectively. Thus, mass does not remain constant for $t > 0$.
- Equation (5.11) can be considered as the application of an *integral constraint* or *invariant*. However, to use this method, it is not necessary to evaluate the integral analytically as we have done for eq. (5.11). Rather, the integral can be evaluated numerically, for example, by using *Simpson's rule*. In other words, for a more complicated IC than eq. (5.3) (with $t = 0$), numerical integration may be simpler than

Table 5.8: Abbreviated tabular numerical and analytical solutions for $0 \le t \le 2.7$

t	x	u(it,i)
0.00	0.000	0.000000
0.00	0.020	0.062791
0.00	0.040	0.125333
0.00	0.060	0.187381
0.00	0.080	0.248690
0.00	0.100	0.309017
	.	.
	.	.
	.	.
0.00	0.460	0.992115
0.00	0.480	0.998027
0.00	0.500	1.000000
0.00	0.520	0.998027
0.00	0.540	0.992115
	.	.
	.	.
	.	.
0.00	0.900	0.309017
0.00	0.920	0.248690
0.00	0.940	0.187381
0.00	0.960	0.125333
0.00	0.980	0.062791
0.00	1.000	0.000000
	.	.
	.	.
	.	.
output for t=0.9, 1.8 removed		
	.	.
	.	.
	.	.
t	x	u(it,i)
2.70	0.000	0.636610
2.70	0.020	0.636610
2.70	0.040	0.636610
2.70	0.060	0.636611
2.70	0.080	0.636611
2.70	0.100	0.636612
	.	.
	.	.
	.	.
2.70	0.460	0.636630
2.70	0.480	0.636630
2.70	0.500	0.636630

(Continued)

Table 5.8: *(Continued)*

t	x	u(it,i)
2.70	0.520	0.636630
2.70	0.540	0.636630
	.	.
	.	.
	.	.
2.70	0.900	0.636612
2.70	0.920	0.636611
2.70	0.940	0.636611
2.70	0.960	0.636610
2.70	0.980	0.636610
2.70	1.000	0.636610
ncall=226		

analytical integration. In fact, the use of numerical integration in applying an integral constraint is commonplace.

- With either method (analytical or numerical), the integral constraint can be applied at any $t > 0$ (it does not apply at just $t \to \infty$). Although the shape of the solution will change with t, its total mass or energy will remain the same and should therefore be independent (invariant) with respect to t. In other words, a check that the integral does not change with t confirms the conservation principle for any $t > 0$.

Appendix 2

We conclude this chapter by solving the diffusion equation with nonlinear source terms using the *factorization method* as outlined in the main Appendix (after Chapter 19). We repeat the problem eq. (5.1) below for convenience, as we will be using different constant names in our analysis

$$\frac{\partial u}{\partial t} - \alpha \frac{\partial^2 u}{\partial x^2} + \beta u^3 + \delta u^2 = 0, \quad u = u(x,t),\ t > 0 \tag{5.13}$$

If we assume a traveling wave solution of the form $u(x,t) = U(\xi)$, where $\xi = k(x - ct)$, $c =$ velocity, and $k =$ wavenumber, eq. (5.13) reduces to the traveling wave ODE

$$\frac{d^2U}{dx^2} + g\frac{dU}{dx} + F(U) = 0 \tag{5.14}$$

where $g = \dfrac{c}{\alpha k}$ and $F(U) = \dfrac{-1}{\alpha k^2} U^2(\delta + \beta U)$.

We factor the polynomial function of eq. (5.14) (the third term) as

$$\frac{F(U)}{U} = f_1 f_2 = -\frac{1}{\alpha k^2} U(\delta + \beta U)$$

and choose

$$f_1 = -\frac{a}{k} U, \quad f_2 = \frac{1}{a\alpha k}(\delta + b\beta U), \quad \alpha \neq 0 \tag{5.15}$$

where we have also introduced the constant a.

From eqns. (A.41) and (5.15), we obtain the following ODE

$$\frac{df_1}{dU} U + f_1 + f_2 = -\frac{2\alpha}{k} U + \frac{1}{a\alpha k}(\delta + \beta U) = -g = -\frac{c}{\alpha k} \tag{5.16}$$

Collecting terms and equating the coefficients of U to zero (as the left-hand side of eq. (5.16) is equal to a constant) gives $a = \pm\sqrt{\frac{\beta}{2\alpha}}$. Also, as g is a constant and independent of the value of U, we find that $c = -\frac{\delta}{a}$ on setting $U = 0$. Now, adopting the grouping of Cornejo-Perez [2], i.e., eq. (A.40b),

$$\frac{d^2U}{d\xi^2} - \left(\frac{df_1}{dU} U + f_1 + f_2\right)\frac{dU}{d\xi} + f_1 f_2 U = 0 \tag{5.17}$$

the corresponding factorization eq. (A.39), i.e., $[D - f_2(U)][D - f_1(U)]U = 0$, becomes

$$\left[D \pm \frac{1}{a\alpha k}(\delta + \beta U)\right]\left[D \mp \left(-\frac{a}{k} U\right)\right] U = 0 \tag{5.18}$$

where $D = \frac{d}{d\xi}$. Therefore, it follows that eq. (5.17) is compatible with the first-order ODE

$$\frac{dU}{d\xi} \pm \frac{1}{k}\sqrt{\frac{\beta}{2\alpha}} U^2 = 0 \tag{5.19}$$

Integrating eq. (5.18) either manually or using Maple yields

$$U = \pm\left(\frac{1}{k}\sqrt{\frac{\beta}{2\alpha}}\xi + K\right)^{-1} \tag{5.20}$$

where K is an arbitrary constant of integration. We take the positive value of eq. (5.20) and set $K = 1$; then, transforming back to $u(x,t)$ leads to the final solution

$$u(x,t) = \left(\sqrt{\frac{\beta}{2\alpha}}(x - ct) + 1\right)^{-1} \tag{5.21}$$

From eq. (5.21), we deduce that the wavenumber $k = \sqrt{\frac{\beta}{2\alpha}}$. Also, from the above definitions, we see that the wave velocity is given by $c = -\delta\sqrt{\frac{2\alpha}{\beta}}$. This solution is the same as the analytical solution of eq. (5.2) (but with different constant names) used in the numerical simulation discussed in the main body of this chapter. However, if we swap the equation for f_1 with that for f_2, and similarly the equation for f_2 with that for f_1, the same calculation procedure arrives at the following entirely different form of traveling wave solution

$$u(x,t) = -\frac{\delta}{\beta}\left[1 - K\frac{\delta}{\beta}\exp\left(-\frac{\delta(x-ct)}{\sqrt{2\alpha\beta}}\right)\right]^{-1} \tag{5.22}$$

where K is an arbitrary constant. The Maple script of Listing 5.14 will perform the above calculations.

Readers are referred to the papers by Berkovich [1] and Cornejo-Perez and Rosu [2, 4] for more information on this method and additional examples of its use.

```
># Diffusion equation with nonlinear source terms.
 # Some calculations to confirm the results of a
 # factorization solution.
 # Ref: Cornejo-Perez, O. and H. C. Rosu (2005).
 #       Nonlinear second order ODE: Factorizations and
 #       particular solutions, Progress of Theoretical
 #       Physics, vol 114, No 3, pp 533-538
 restart; with(DEtools): with(PDEtools): with(plots):
 alias(u=u(x,t)): alias(U=U(xi)):
>NumericEventHandler(division_by_zero = proc (operator, operands, defVal)
   if operator = ln then return -infinity else return defVal end if end proc);
 division_by_zero = proc (operator, operands, defVal) defVal end proc;
># Define PZ-1 equation
 pde1:=diff(u,t)-alpha*diff(u,x,x)+beta*u^3+delta*u^2=0;
># Convert PDE to ODE
 tr1:={x=(xi/k+c*tau),t=tau,u=U};
 ode1:=dchange(tr1,pde1,[xi,tau,U]);
># Define F(U), g
 F:=-(1/alpha/k^2)*U^2*(delta+beta*U);
 g:=c/(alpha*k);
># Factor F(U)/U - Note: 'a' introduced
 f[1]:=-a*U/k;
 f[2]:=(delta+beta*U)/(alpha*k*a);
># Check that factorization is correct
 F_chk:=simplify(eval(f[1]*f[2]*U));
># Use the C-P grouping
 alias(U=U):f[1]:=subs(U(xi)=U,f[1]):f[2]:=subs(U(xi)=U,f[2]):
 eqn1:=diff(f[1],U)*U+f[1]+f[2]=-g;
 sol1:=simplify(subs(U=0,eqn1),size);
 c:=solve(sol1,c);
># Collect terms in U
 eqn2:=collect(simplify(lhs(eqn1)-rhs(eqn1),size),U,'recursive');
># Equate coeff of U to zero
 eqn3:=coeff(eqn2,U)=0;
 sol2:=solve(eqn3,a);
```

```
 a:=sol2[1];
># Fomulate 1st order ode from [D-f[1]]U=0
 f[1]:=subs(U=U(xi),f[1]);ode2:=diff(U(xi),xi)-f[1]*U(xi);
># Obtain solution to ode2
 sol4:=dsolve(ode2);
># Check solution U(xi) satisfies ode1
 odeCHK:=simplify(eval(subs({U(xi)=rhs(sol4)},lhs(ode1))),symbolic);
 if odeCHK = 0 then
   print('solution PASSES');
 else
   print('solution FAILS!');
 end if;
># Obtain solution to pde1
 sol5:=u=simplify(eval(subs({xi=k*(x-c*t)},rhs(sol4))),size);
># Check solution u(x,t) satisfies pde1:
 pdeCHK:=simplify(pdetest(sol5,pde1),symbolic);
 if pdeCHK = 0 then
   print('solution PASSES');
 else
   print('solution FAILS!');
 end if;
># Plot results
 # ============
 x0:=0; alpha:=1; beta:=1; delta:=1; _C1:=1;
 zz:=rhs(sol5);
 animate(zz,x=0..10,t=0..10,axes=framed,
   labels=["x","u"],
   thickness=3,frames=100,numpoints=300,
   title="Diffusion equation with nonlinear source terms",
   labelfont=[TIMES, ROMAN, 16],axesfont=[TIMES, ROMAN, 16],
   titlefont=[TIMES, ROMAN, 16]);
>plot3d(zz,x=0..10,t=0..10,axes='framed',
   labels=["x","t","u(x,t)"],
   labeldirections=[HORIZONTAL,HORIZONTAL,VERTICAL],
   orientation=[-34,67],grid=[100,100],
   style=patchnogrid,shading=Z,
   title="Diffusion equation with nonlinear source terms",
   labelfont=[TIMES, ROMAN, 16],axesfont=[TIMES, ROMAN, 16],
   titlefont=[TIMES, ROMAN, 16]);
```

LISTING 5.14: Maple code to derive a solution to the diffusion equation with nonlinear source terms using the *factorization method.*

Finally, additional Maple scripts are included with the downloads for this book which obtain other solutions to the diffusion equation with nonlinear terms using *tanh-*, *exp-*, and *Riccati*-based methods.

References

[1] L.M. Berkovich, Factorization as a method of finding exact invariant solutions of the Kolmogorov-Petrovskii-Piskunov equation and the related Semenov and Zeldovich equations, *Sov. Math. Dokl.* 45, (1992) 162–167.

[2] O. Cornejo-Perez, H.C. Rosu, Nonlinear second order ODE: factorizations and particular solutions, *Progr. Theor. Phys.* 114 (3) (2005) 533–538.

[3] A.D. Polyanin, V.F. Zaitsev, *Handbook of Nonlinear Partial Differential Equations*, Chapman & Hall/ CRC, Boca Raton, FL, 2004.

[4] H.C. Rosu, O. Cornejo-Perez, Supersymmetric pairing of kinks for polynomial nonlinearities, http://arxiv.org/PS_cahe/math-ph/pdf/0401/0401040v3.pdf, last accessed 23 April, 2010.

[5] W.E. Schiesser, G.W. Griffiths, *A Compendium of Partial Differential Equation Models*, Cambridge University Press, Cambridge, UK, 2009.

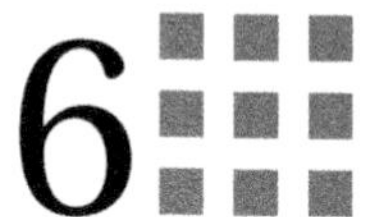

6 Burgers–Huxley Equation

We now consider the *Burgers–Huxley equation*, which has a spectrum of applications in nonlinear physics and physiology [5]

$$\frac{\partial u}{\partial t}+u^2\frac{\partial u}{\partial x}-\frac{\partial^2 u}{\partial x^2}=\frac{2}{3}u^3(1-u^2), \quad t>0, 0\leq x\leq 1 \tag{6.1}$$

with initial condition (IC)

$$u(x,0)=\left[\frac{1}{2}+\frac{1}{2}\tanh\left(\frac{1}{3}x\right)\right]^{\frac{1}{2}} \tag{6.2}$$

and BCs at $x=-15,10$ given by the analytical solution [3]

$$u(x,t)=\left[\frac{1}{2}+\frac{1}{2}\tanh\left(\frac{1}{9}(3x+t)\right)\right]^{\frac{1}{2}} \tag{6.3}$$

The Matlab routines closely resemble those of Chapters 2–5. Here, we list a few details pertaining to eqns. (6.1), (6.2), and (6.3). First, the ODE routine `pde_1.m` is

```
  function ut=pde_1(t,u)
%
% Function pde_1 computes the t derivative vector for the Burgers-Huxley
% equation
%
  global xl xu x n ncall
%
% BCs at x = -15,10
  u(1)=ua_1(x(1),t);
  u(n)=ua_1(x(n),t);
%
% ux
  ux=dss004(xl,xu,n,u);
%
% uxx
  uxx=dss004(xl,xu,n,ux);
%
% PDE
  for i=2:n-1
    ut(i)=-(u(i)^2)*ux(i)+uxx(i)+(2/3)*u(i)^3*(1-u(i)^2);
  end
  ut(1)=0;
  ut(n)=0;
  ut=ut';
```

```
%
% Increment calls to pde_1
  ncall=ncall+1;
```

LISTING 6.1: Function pde_1.m for eq. (6.1).

We can note the following points about pde_1.m:

- The function and some global parameters are first defined.

```
  function ut=pde_1(t,u)
%
% Function pde_1 computes the t derivative vector for the Burgers-Huxley
% equation
%
  global xl xu x n ncall
```

- The first derivative in eq. (6.1), ux, is computed using the function dss004. Since eq. (6.1) is second order in x, the two required BCs are taken from eq. (6.2) with $x = -15, 10$ at grid points i=1,n, respectively (with n=51 subsequently set in function inital_1.m).

```
%
% BCs at x = -15,10
  u(1)=ua_1(x(1),t);
  u(n)=ua_1(x(n),t);
%
% ux
  ux=dss004(xl,xu,n,u);
```

- The second derivative in eq. (6.1), uxx, is computed with dss004 by differentiating ux, so-called *stagewise differentiation*. The alternative would be to use dss044 to directly compute uxx from u, as discussed in Chapter 4.

```
%
% uxx
  uxx=dss004(xl,xu,n,ux);
```

- Equation 6.1 is then programmed.

```
%
% PDE
  for i=2:n-1
    ut(i)=-(u(i)^2)*ux(i)+uxx(i)+(2/3)*u(i)^3*(1-u(i)^2);
  end
  ut(1)=0;
  ut(n)=0;
  ut=ut';
%
% Increment calls to pde_1
  ncall=ncall+1;
```

Since Dirichlet BCs are used, the derivatives in t are set to zero at the boundaries (so that the ODE integrator does not move the boundary values away from their prescribed values). A transpose is included to meet the requirements of the ODE integrator `ode15s`. Finally, the counter for the number of calls to `pde_1.m` is incremented.

The IC of eq. (6.2) is programmed in `inital_1.m`.

```
  function u0=inital_1(t0)
%
% Function inital_1 sets the initial condition for the Burgers-
% Huxley equation
%
  global xl xu x n
%
% Spatial domain and initial condition
  xl=-15;
  xu= 10;
  n=51;
  dx=(xu-xl)/(n-1);
%
% IC from analytical solution
  for i=1:n
    x(i)=xl+(i-1)*dx;
    u0(i)=ua_1(x(i),0.0);
  end
```

LISTING 6.2: Function `inital_1.m` for IC from eq. (6.2).

We can note the following points about `inital_1.m`:

- The function and some global parameters are first defined.

```
  function u0=inital_1(t0)
%
% Function inital_1 sets the initial condition for the Burgers-
% Huxley equation
%
  global xl xu x n
```

- The grid in x is then defined over the interval $-15 \le x \le 10$ for 51 points (these grid parameters were selected by trial and error to produce a numerical solution with acceptable accuracy).

```
%
% Spatial domain and initial condition
  xl=-15;
  xu= 10;
  n=51;
  dx=(xu-xl)/(n-1);
%
% IC from analytical solution
  for i=1:n
    x(i)=xl+(i-1)*dx;
    u0(i)=ua_1(x(i),0.0);
  end
```

  As the grid in x is defined in the `for` loop, function `ua_1` (listed next) is called (for $t = 0$) to define the IC from eq. (6.2).

Function ua_1.m is a straightforward implementation of the analytical solution, eq. (6.3).

```
  function uanal=ua_1(x,t)
%
% Function uanal computes the exact solution of the Burgers-Huxley
% equation for comparison with the numerical solution
%
% Analytical solution
  expp=exp( (1/3)*x+(1/9)*t);
  expm=exp(-(1/3)*x-(1/9)*t);
  uanal=((1/2)*(1+(expp-expm)/(expp+expm)))^0.5;
```

LISTING 6.3: Function ua_1.m for analytical solution (6.2).

The main program, pde_1_main, is similar to pde_1_main Listing 2.1 and therefore is not listed here. The main program produces the same three figures and tabulated output as in Chapters 2–5 which are now reviewed. The Jacobian matrix routine jpattern_num_1.m is the same as jpattern_num_1.m in Chapters 2–4 and is therefore not reproduced here.

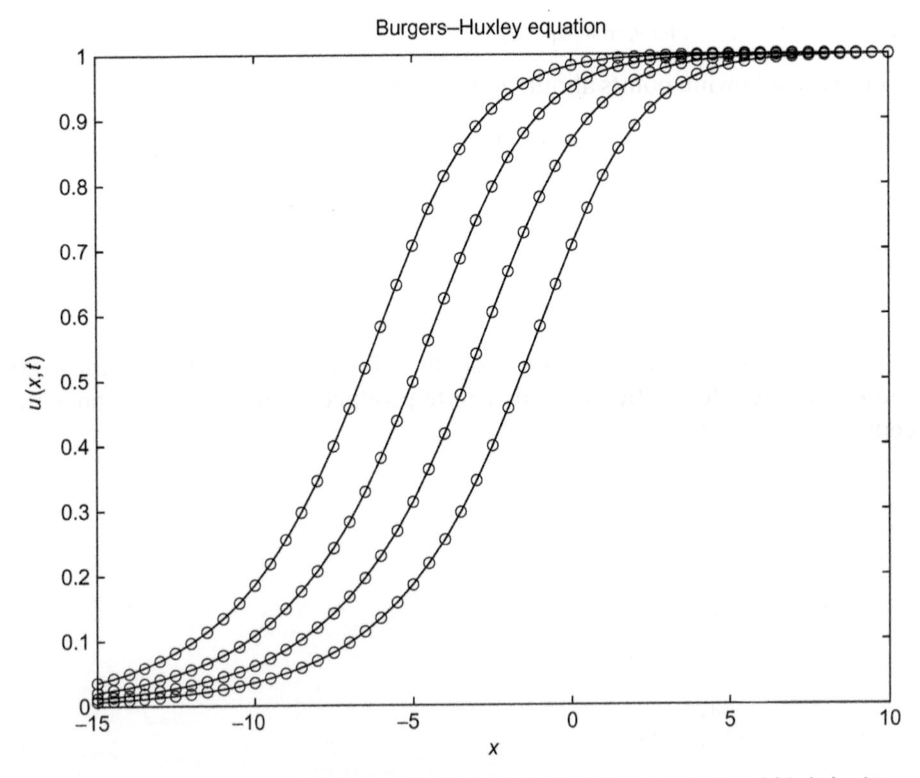

FIGURE 6.1: Numerical solution to eq. (6.1) (lines) with the analytical solution superimposed (circles) using five-point FD approximations in dss004 for $t = 0, 5, 10, 15$ (right to left).

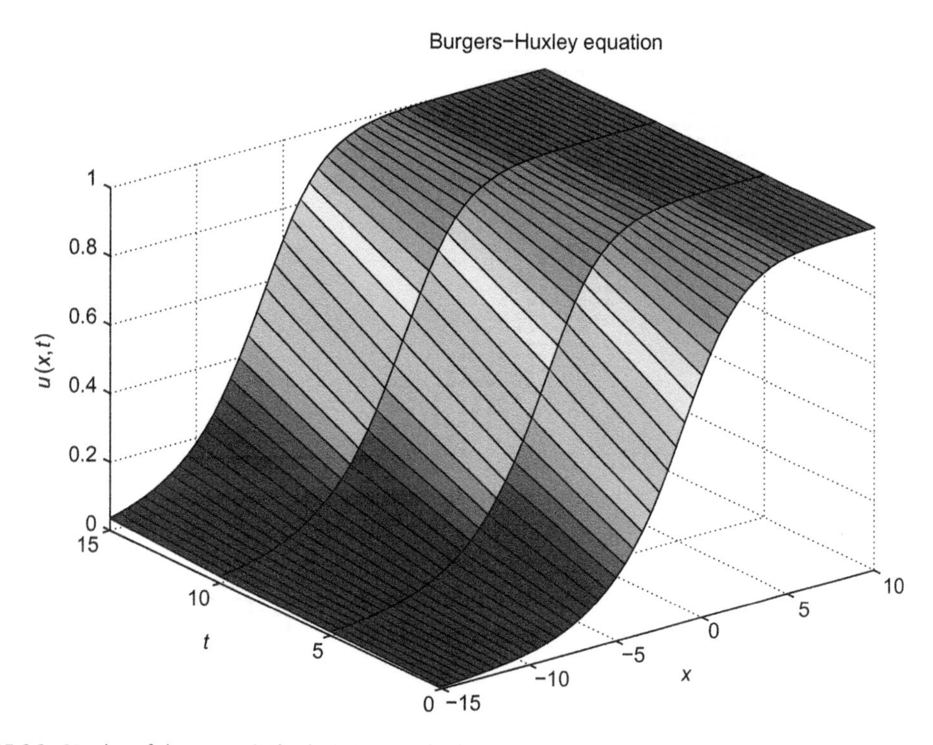

FIGURE 6.2: 3D plot of the numerical solution to eq. (6.1).

Figure 6.1 indicates good agreement between the analytical and numerical solutions. Also, the solution appears to be a traveling wave as specified by eq. (6.3) in the sense that the successive curves are displaced by a constant distance in t. Note that the curves move right to left, since from eq. (6.2), the velocity is $c = -3/9$ (the *Lagrangian* variable is $\xi = k(x - ct) = (1/3)(x + 3/9t))$. Figure 6.2 is a 3D plot of the numerical solution.

The map of the ODE Jacobian matrix, Fig. 6.3, reflects the banded structure of the ODEs produced by `dss004`. In particular, since the number of grid points, $n = 51$, is relatively small, the individual elements of the Jacobian matrix are distinct. Also, note that the bandwidth is 9 and not 5 as might be expected from the five-point FDs in `dss004`. This greater bandwidth is due to the repeated use of `dss004` in `pde_1.m` to compute `uxx` from `u` by stagewise differentiation. This example illustrates a disadvantage of stagewise differentiation, that is, the increase in the bandwidth of the ODE Jacobian matrix through successive calls of the spatial differentiator such as `dss004`.

The tabular analytical and numerical solutions also reflect the good agreement between these two solutions. The computational effort reflected in `ncall = 129` is quite modest.

As required, the analytical and numerical solutions agree for $t = 0$ (since both solutions are from eq. (6.2)). For $t > 0$, the agreement between the analytical and numerical solutions of approximately five figures is quite acceptable, even with only 51 grid points.

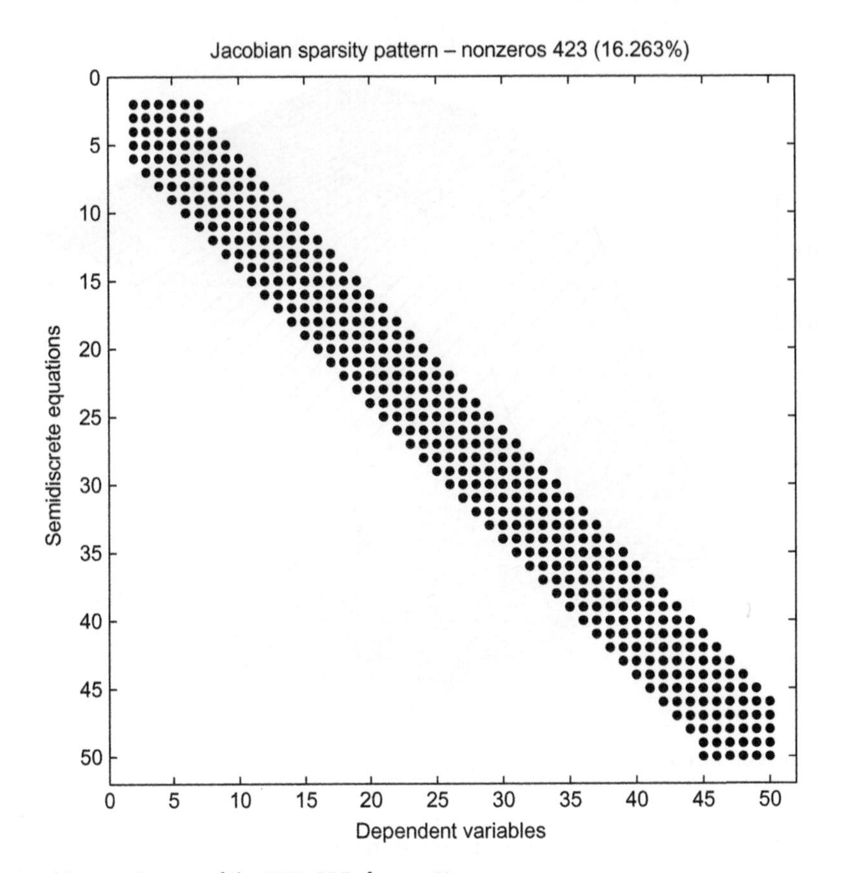

FIGURE 6.3: Jacobian matrix map of the MOL ODEs for $n = 51$.

The numerical solution of Table 6.1 can be used to estimate the velocity of the traveling wave of Fig. 6.1. For example, we can determine the distance (in x) between the two solution values $u(x,t) = 0.5$ for $t = 0$ and $t = 5$. For $u(x,t = 0) = 0.5$, the value of x can be estimated by linear interpolation within the solution values of Table 6.1 at $t = 0$.

$$x = -2.0 + (-1.5 + 2.0)(0.5 - 0.456737)/(0.518596 - 0.456737) = -1.6503$$

Similarly, for $t = 5$, linear interpolation to determine x for $u(x,t = 5) = 0.5$ gives

$$x = -3.5 + (-3.0 + 3.5)(0.5 - 0.477017)/(0.539758 - 0.477017) = -3.3168$$

Thus, the estimated velocity (of the value $u(x,t) = 0.5$) is

$$c = \frac{\Delta x}{\Delta t} = \frac{-3.3168 - (-1.6503)}{5 - 0} = -0.3333$$

which compares with the analytical value of $c = -3/9 = -0.3333$ (as discussed previously).

Table 6.1: Tabular numerical and analytical solutions

t	x	u(it,i)	u_anal(it,i)	err(it,i)
0.00	−15.000	0.006738	0.006738	0.000000
0.00	−14.500	0.007960	0.007960	0.000000
0.00	−14.000	0.009403	0.009403	0.000000
0.00	−13.500	0.011108	0.011108	0.000000
0.00	−13.000	0.013123	0.013123	0.000000
	⋮		⋮	
0.00	−2.500	0.398584	0.398584	0.000000
0.00	−2.000	0.456737	0.456737	0.000000
0.00	−1.500	0.518596	0.518596	0.000000
0.00	−1.000	0.582446	0.582446	0.000000
0.00	−0.500	0.646088	0.646088	0.000000
0.00	0.000	0.707107	0.707107	0.000000
	⋮		⋮	
0.00	8.000	0.997595	0.997595	0.000000
0.00	8.500	0.998275	0.998275	0.000000
0.00	9.000	0.998763	0.998763	0.000000
0.00	9.500	0.999113	0.999113	0.000000
0.00	10.000	0.999364	0.999364	0.000000
t	**x**	**u(it,i)**	**u_anal(it,i)**	**err(it,i)**
5.00	−15.000	0.011743	0.011743	0.000000
5.00	−14.500	0.013872	0.013872	0.000000
5.00	−14.000	0.016387	0.016387	−0.000000
5.00	−13.500	0.019358	0.019358	0.000000
5.00	−13.000	0.022868	0.022867	0.000000
	⋮		⋮	
5.00	−4.000	0.417432	0.417475	−0.000043
5.00	−3.500	0.476965	0.477017	−0.000053
5.00	−3.000	0.539705	0.539758	−0.000053
5.00	−2.500	0.603761	0.603802	−0.000041
5.00	−2.000	0.666818	0.666837	−0.000019
	⋮		⋮	
5.00	8.000	0.999207	0.999206	0.000000
5.00	8.500	0.999431	0.999431	0.000000
5.00	9.000	0.999593	0.999592	0.000000
5.00	9.500	0.999708	0.999708	0.000000
5.00	10.000	0.999791	0.999791	0.000000
	⋮		⋮	

```
output for t=10, 15 removed
```

⋮

```
ncall=129
```

In summary, the solution of eq. (6.1) subject to the IC from eq. (6.2) (with $t = 0$) and two Dirichlet BCs from eq. (6.2) (with $x = -15, 10$) is straightforward. Also, eq. (6.1) is nonlinear, yet the programming in `pde_1.m` (Listing 6.1) is straightforward. Consequently, variations in the PDE can easily be made for cases for which an analytical solution might not be available.

Appendix

We conclude this chapter with an analysis of the *generalized Burgers–Huxley* equation which has the following form [5]

$$\frac{\partial u}{\partial t} + \alpha u^{\delta}\frac{\partial u}{\partial x} - \frac{\partial^2 u}{\partial x^2} = \beta u\left(1-u^{\delta}\right)\left(u^{\delta}-\gamma\right), \quad u = u(x,t),\ t \geq 0 \tag{6.4}$$

When $\alpha = 1$, $\beta = 2/3$, $\delta = 2$, and $\gamma = 0$, eq. (6.4) reduces to eq. (6.1).

We will apply the *factorization method* to eq. (6.4), as outlined in the main Appendix, and start by applying the traveling wave transformation $u(x,t) = U(\xi)$, $\xi = k(x - ct)$. This converts the PDE to the following ODE.

$$\frac{d^2U}{d\xi^2} + g(U)\frac{dU}{d\xi} + F(U) = 0 \tag{6.5}$$

where $g(U) = \frac{1}{k}\left(c - \alpha U^{\delta}\right)$ and $F(U) = \frac{\beta}{k^2}U\left(1-U^{\delta}\right)\left(U^{\delta}-\gamma\right)$.

In order to proceed with this method, we factorize $F(U)$ as

$$\frac{F(U)}{U} = f_1 f_2 = \frac{\beta}{k^2}\left(1-U^{\delta}\right)\left(U^{\delta}-\gamma\right)$$

and choose

$$f_1(U) = a\frac{\sqrt{\beta}}{k}\left(1-U^{\delta}\right), \quad f_2(U) = a^{-1}\frac{\sqrt{\beta}}{k}\left(U^{\delta}-\gamma\right) \tag{6.6}$$

where we have also introduced the constant a to be determined later.

From eqns. (A.41) and (6.6), we obtain the following ODE

$$\frac{df_1}{dU}U + f_1 + f_2 = -a\delta\frac{\sqrt{\beta}}{k}U^{\delta} + a\frac{\sqrt{\beta}}{k}\left(1-U^{\delta}\right) + a^{-1}\frac{\sqrt{\beta}}{k}\left(U^{\delta}-\gamma\right) = -g(U) = -\frac{1}{k}\left(c-\alpha U^{\delta}\right) \tag{6.7}$$

Rearranging and collecting terms, we have

$$\left[\sqrt{\beta}\left(-a + a^{-1} - a\delta\right) - \alpha\right]U^{\delta} + \left[\sqrt{\beta}\left(a - \gamma a^{-1}\right) + c\right] = 0 \tag{6.8}$$

Now, because U is a variable, for consistency, we are able to equate each of the terms within square brackets to zero, yielding

$$c = -\sqrt{\beta}\left(a - \gamma a^{-1}\right), \quad a = \frac{-\alpha \pm \sqrt{\alpha^2 + 4\beta(1+\delta)}}{2\sqrt{\beta}(1+\delta)}$$

We now adopt the grouping of Cornejo-Perez [2, 4], i.e., eq. (A.40b), from which it follows that

$$\frac{d^2U}{d\xi^2} + \frac{1}{k}\sqrt{\beta}\left[\left(a - \gamma a^{-1}\right) - \left(a(1+\delta) - a^{-1}\right)U^\delta\right]\frac{dU}{d\xi} + f_1 f_2 U = 0 \tag{6.9}$$

Thus, the corresponding factorization of eq. (6.5), i.e., $[D - f_2(U)][D - f_1(U)]U = 0$ (see eq. (A.39) of main Appendix), becomes

$$\left[D - a^{-1}\frac{\sqrt{\beta}}{k}\left(U^\delta - \gamma\right)\right]\left[D - a\frac{\sqrt{\beta}}{k}\left(1 - U^\delta\right)\right]U = 0 \tag{6.10}$$

where $D = \dfrac{d}{d\xi}$. Therefore, it follows that eq. (6.9) is compatible with the first-order ODE

$$\frac{dU}{d\xi} \mp a\frac{\sqrt{\beta}}{k}\left(1 - U^\delta\right)U = 0 \tag{6.11}$$

Integrating eq. (6.11) yields a general solution of the form

$$\begin{aligned} U &= \left(1 \pm K\exp\left[-\left(a\sqrt{\beta}\delta/k\right)\xi\right]\right)^{-1/\delta} \\ &= \left\{1 \pm K\exp\left[-\left(\frac{-\alpha \pm \sqrt{\alpha^2 + 4\beta(1+\delta)}}{2(1+\delta)k}\right)\delta\xi\right]\right\}^{-1/\delta} \end{aligned} \tag{6.12}$$

where K is an arbitrary constant.

Using the same values for constants as in the numerical simulation, i.e., $\alpha = 1$, $\beta = 2/3$, $\delta = 2$, and $\gamma = 0$, we obtain

$$U = \left\{1 \pm K\exp\left[-\left(\frac{-1 \pm 3}{3k}\right)\xi\right]\right\}^{-1/2} \tag{6.13}$$

Finally, setting $K = +1$ and applying the inverse transformation $U(\xi) = u(x,t)$, $\xi = k(x - ct)$ yields the solutions

$$u_1(x,t) = \left\{1 + \exp\left[-\left(\frac{2}{3}x + \frac{2}{9}t\right)\right]\right\}^{-1/2} = \left[\frac{1}{2} + \frac{1}{2}\tanh\left(\frac{1}{3}x + \frac{1}{9}t\right)\right]^{1/2} \tag{6.14}$$

$$u_2(x,t) = \left\{1 + \exp\left[+\left(\frac{4}{3}x + \frac{4}{9}t\right)\right]\right\}^{-1/2} = \left[\frac{1}{2} - \frac{1}{2}\tanh\left(\frac{2}{3}x + \frac{2}{9}t\right)\right]^{1/2} \tag{6.15}$$

Alternatively, setting $K = -1$ yields the solutions

$$u_3(x,t) = \left\{1 - \exp\left[-\left(\frac{2}{3}x + \frac{2}{9}t\right)\right]\right\}^{-1/2} = \left[\frac{1}{2} + \frac{1}{2}\coth\left(\frac{1}{3}x + \frac{1}{9}t\right)\right]^{1/2} \quad (6.16)$$

$$u_4(x,t) = \left\{1 - \exp\left[+\left(\frac{4}{3}x + \frac{4}{9}t\right)\right]\right\}^{-1/2} = \left[\frac{1}{2} - \frac{1}{2}\coth\left(\frac{2}{3}x + \frac{2}{9}t\right)\right]^{1/2} \quad (6.17)$$

Solution u_1 is the same solution, eq. (6.3), that is used in the main body of this chapter for the numerical simulation. By inspection, we see that for solutions u_1 and u_3, the wavenumber k is equal to $1/3$ and for solutions u_2 and u_4, it is equal to $2/3$. For all solutions, the wave velocity c is equal to $-1/3$. Note: solutions u_3 and u_4 have singularities at $(x = 0, t = 0)$. A Maple script that derives these solutions is given in Listing 6.4.

Additional solutions can be found by choosing different factorizations of eq. (6.5), for example, eq. (6.10) could be changed to

$$\left[D - a^{-1}\frac{\sqrt{\beta}}{k}\left(1 - U^{\delta}\right)\right]\left[D - a\frac{\sqrt{\beta}}{k}\left(U^{\delta} - \gamma\right)\right]U = 0$$

and the same procedure followed. This is left as an exercise for the reader.

Readers are referred to the papers by Berkovich [1] and Cornejo-Perez and Rosu [2, 4] for more information on this method and additional examples of its use.

```
# Some calculations to confirm the results of a
# factorization solution to the Burgers-Huxley Equation
# Ref: Cornejo-Perez, O. and H. C. Rosu (2005).
#      Nonlinear second order ODE: Factorizations and
#      particular solutions, Progress of Theoretical
#      Physics, vol 114, No 3, pp 533-538
>restart; with(DEtools): with(PDEtools): with(plots):
 unprotect(gamma):
 alias(u=u(x,t)): alias(U=U(xi)):
># Define PDE equation
 pde1:=diff(u,t)+alpha*u^delta*diff(u,x)-diff(u,x,x)
        -beta*u*(1-u^delta)*(u^delta-gamma)=0;
># Convert PDE to ODE
 tr1:={x=(xi/k+c*tau),t=tau,u=U};
 ode1:=dchange(tr1,pde1,[xi,tau,U]);
># Define F(U), g
 F:=(beta/k^2)*U*(1-U^delta)*(U^delta-gamma);
 g:=(1/k)*(c-alpha*U^delta);
># Factor F(U)/U - Note: 'a' introduced
 f[1]:=a*sqrt(beta)*(1-U^delta)/k;
 f[2]:=(1/a)*sqrt(beta)*(U^delta-gamma)/k;
># Check that factorization is correct
 F_chk:=simplify(eval(f[1]*f[2]*U));
># Use the C-P grouping
 alias(U=U):f[1]:=subs(U(xi)=U,f[1]):f[2]:=subs(U(xi)=U,f[2]):
 eqn1:=diff(f[1],U)*U+f[1]+f[2]=-subs(U(xi)=U,g);
 sol1:=simplify(subs(U=0,eqn1),size);
 c:=solve(sol1,c);
```

```
># Collect terms in U
 eqn2:=collect(simplify(lhs(eqn1)-rhs(eqn1),size),U,'recursive');
># Equate coeff of U to zero
 eqn3:=coeff(eqn2,U^delta)=0;
 sol2:=solve(eqn3,a);
 a:=sol2[1];
># Fomulate 1st order ode from [D-f[1]]U=0
 f[1]:=subs(U=U(xi),f[1]);ode2:=diff(U(xi),xi)-f[1]*U(xi);
># Obtain solution to ode2
 sol4:=dsolve(ode2);
># Check solution U(xi) satisfies ode1
 odeCHK:=simplify(eval(subs({U(xi)=rhs(sol4)},lhs(ode1))),symbolic);
 if odeCHK = 0 then
   print('solution PASSES');
 else
   print('solution FAILS!');
 end if;
># Obtain solution to pde1
 sol5:=u=simplify(eval(subs({xi=k*(x+x0-c*t)},rhs(sol4))),size);
># Check solution u(x,t) satisfies pde1:
 pdeCHK:=simplify(pdetest(sol5,pde1),symbolic);
 if pdeCHK = 0 then
   print('solution PASSES');
 else
   print('solution FAILS!');
 end if;
># Plot results
 # ============
 x0:=0; _C1:=1;alpha:=1;beta:=2/3;delta:=2;gamma:=0;
 zz:=simplify(eval(rhs(sol5)));
   animate(zz,x=-20..10,t=0..30,axes=framed,
   thickness=3,frames=50,numpoints=100,
   title="Burgers-Huxley Equation");
>plot3d(zz,x=-20..10,t=0..30,axes='framed',
   labels=["x","t","u(x,t)"],
   labeldirections=[HORIZONTAL,HORIZONTAL,VERTICAL],
   labelfont=[TIMES, ROMAN, 16],
   orientation=[-122,68],grid=[100,100],
   style=patchnogrid,axesfont=[TIMES, ROMAN, 16],
   shading=Z,title="Burgers-Huxley Equation",
   titlefont=[TIMES, ROMAN, 16]);
```

LISTING 6.4: Maple code to derive a solution to the *Burgers–Huxley equation* using the *factorization method.*

Finally, additional Maple scripts are included with the downloads for this book that solve the Burgers–Huxley equation using *tanh-*, *exp-*, and *Riccati*-based methods.

References

[1] L.M. Berkovich, Factorization as a method of finding exact invariant solutions of the Kolmogorov-Petrovskii-Piskunov equation and the related Semenov and Zeldovich equations, *Sov. Math. Dokl.* 45 (1992) 162–167.

[2] O. Cornejo-Perez, H.C. Rosu, Nonlinear second order ODE: Factorizations and particular solutions, *Prog. Theor. Phys.*, 114 (3) (2005) 533–538.

[3] A. Molabahrami, F. Khani, The homotopy analysis method to solve the Burgers-Huxley equation, *Nonlinear Anal. R. World Appl.* 10 (2) (2009) 589–600.

[4] H.C. Rosu, O. Cornejo-Perez, Supersymmetric pairing of kinks for polynomial nonlinearities, `http://arxiv.org/PS_cache/math-ph/pdf/0401/0401040v3.pdf`, last accessed 23 April, 2010.

[5] X.Y. Wang, Z.S. Zhu, Y.K. Lu, Solitary wave solutions of the generalized Burgers-Huxley equation, *J. Phys. A: Math. Gen.* 23 (1990) 271–274.

7 Burgers–Fisher Equation

We consider the *Burgers–Fisher equation*, which has applications principally in biology [1]

$$\frac{\partial u}{\partial t} + u^2 \frac{\partial u}{\partial x} - \frac{\partial^2 u}{\partial x^2} = u(1-u^2), \quad t > 0 \tag{7.1}$$

with an initial condition (IC) and two boundary conditions (BCs) at $x = -10, 10$ given by the analytical solution [1]

$$u(x,t) = \frac{1}{2}\left[1 - \tanh\left(\frac{1}{3}x - \frac{10}{9}t\right)\right]^{\frac{1}{2}} \tag{7.2}$$

The Matlab routines closely resemble those of Chapters 3–6. Here, we list a few details pertaining to eqns. (7.1) and (7.2). First, the ODE routine pde_1.m, is

```
  function ut=pde_1(t,u)
%
% Function pde_1 computes the t derivative vector for the Burgers-Fisher
% equation
%
  global xl xu x n ncall
%
% BCs at x = -10,10
  u(1)=ua_1(x(1),t);
  u(n)=ua_1(x(n),t);
%
% ux
  ux=dss004(xl,xu,n,u);
%
% uxx
  uxx=dss004(xl,xu,n,ux);
%
% PDE
  for i=2:n-1
    ut(i)=-(u(i)^2)*ux(i)+uxx(i)+u(i)*(1-u(i)^2);
  end
  ut(1)=0;
  ut(n)=0;
  ut=ut';
%
% Increment calls to pde_1
  ncall=ncall+1;
```

LISTING 7.1: Function pde_1.m for eq. (7.1).

We can note the following points about pde_1.m:

- The function and some global parameters are first defined.

```
%
% Function pde_1 computes the t derivative vector for the Burgers-Fisher
% equation
%
  global xl xu x n ncall
```

- The first derivative in eq. (7.1), ux, is computed using the function dss004. Since eq. (7.1) is second order in x, the two required BCs are taken from eq. (7.2) with $x = -10, 10$ at grid points i=1,n, respectively (with n=51 subsequently set in function inital_1.m).

```
%
% BCs at x = -10,10
  u(1)=ua_1(x(1),t);
  u(n)=ua_1(x(n),t);
%
% ux
  ux=dss004(xl,xu,n,u);
```

- The second derivative in eq. (7.1), uxx, is computed with dss004 by differentiating ux, so-called *stagewise differentiation*. The alternative would be to use dss044 to directly compute uxx from u, as discussed in Chapter 4.

```
  ux=dss004(xl,xu,n,u);
%
% uxx
  uxx=dss004(xl,xu,n,ux);
```

- Equation (7.1) is then programmed.

```
%
% PDE
  for i=2:n-1
    ut(i)=-(u(i)^2)*ux(i)+uxx(i)+u(i)*(1-u(i)^2);
  end
  ut(1)=0;
  ut(n)=0;
  ut=ut';
%
% Increment calls to pde_1
  ncall=ncall+1;
```

Since Dirichlet BCs are used, the derivatives in t are set to zero at the boundaries (so that the ODE integrator does not move the boundary values away from their prescribed values). A transpose is included to meet the requirements of the ODE integrator ode15s. Finally, the counter for the number of calls to pde_1.m is incremented.

The IC from eq. (7.2) with $t = 0$ is programmed in inital_1.m.

```
  function u0=inital_1(t0)
%
% Function inital_1 sets the initial condition for the Burgers-
% Fisher equation
%
  global xl xu x n
%
% Spatial domain and initial condition
  xl=-10;
  xu= 10;
  n=51;
  dx=(xu-xl)/(n-1);
%
% IC from analytical solution
  for i=1:n
    x(i)=xl+(i-1)*dx;
    u0(i)=ua_1(x(i),0.0);
  end
```

LISTING 7.2: Function `inital_1.m` from (7.2) with $t = 0$.

We can note the following points about `inital_1.m`:

- The function and some global parameters are first defined.

```
  function u0=inital_1(t0)
%
% Function inital_1 sets the initial condition for the Burgers-
% Fisher equation
%
  global xl xu x n
```

- The grid in x is then defined over the interval $-10 \le x \le 10$ for 51 points.

```
%
% Spatial domain and initial condition
  xl=-10;
  xu= 10;
  n=51;
  dx=(xu-xl)/(n-1);
%
% IC from analytical solution
  for i=1:n
    x(i)=xl+(i-1)*dx;
    u0(i)=ua_1(x(i),0.0);
  end
```

As the grid in x is defined in the `for` loop, function `ua_1` (listed next) is called (for $t = 0$) to define the IC from eq. (7.2)

Function `ua_1.m` is a straightforward implementation of the analytical solution, eq. (7.2).

```
  function uanal=ua_1(x,t)
%
% Function uanal computes the exact solution of the Burgers-Fisher
```

```
% equation for comparison with the numerical solution
%
% Analytical solution
  expp=exp( (1/3)*x-(10/9)*t);
  expm=exp(-(1/3)*x+(10/9)*t);
  uanal=((1/2)*(1-(expp-expm)/(expp+expm)))^0.5;
```

LISTING 7.3: Function ua_1.m for analytical solution (7.2).

The main program, pde_1_main, is essentially the same as pde_1_main of Listing 2.1 and therefore is not listed here. The main program produces the same three figures and tabulated output as in Chapters 2–6, which are now reviewed. Also, the Jacobian matrix routine jpattern_num_1.m is the same as jpattern_num_1.m in Chapters 2–6 and is therefore not reproduced here.

Figure 7.1 indicates good agreement between the analytical and numerical solutions. Also, the solution does not appear to be exactly a traveling wave as specified by eq. (7.2) in the sense that the successive curves should be displaced by a constant distance in t. However, this apparent discrepancy is due to the termination of the plot on the right at

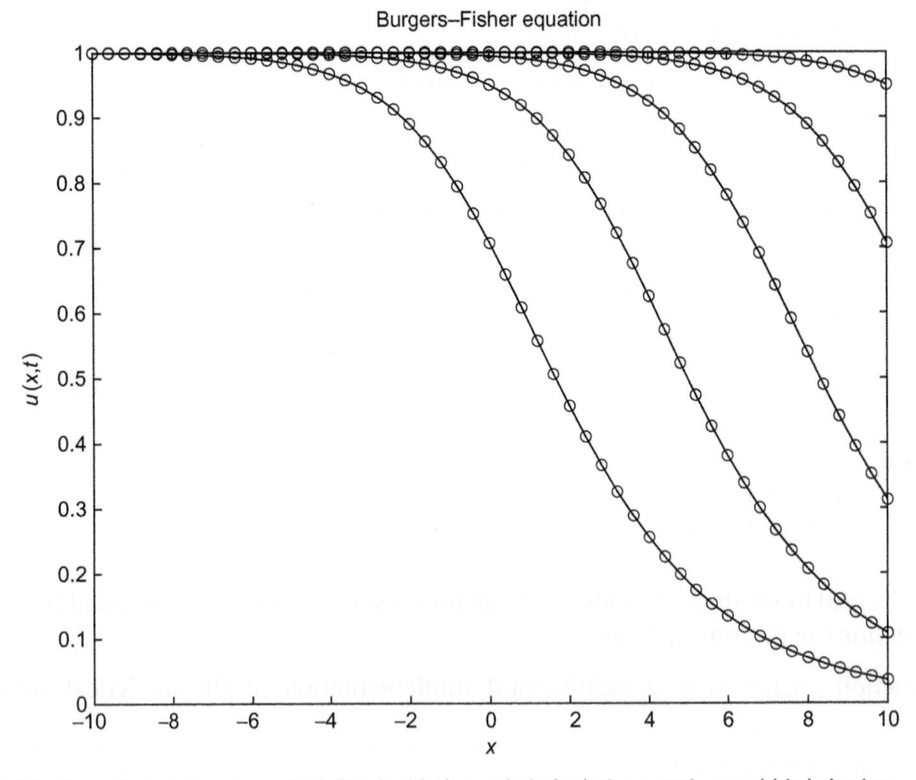

FIGURE 7.1: Numerical solution to eq. (7.1) (lines) with the analytical solution superimposed (circles) using five-point FD approximations in dss004 for $t = 0, 1, 2, 3, 4$ (left to right).

$x = 10$ (if the plot was extended beyond $x = 10$, the traveling wave characteristic, that is, displacement along the x axis without changing shape, would be apparent). Figure 7.2 is a 3D plot of the numerical solution.

The map of the ODE Jacobian matrix, Fig. 7.3, reflects the banded structure of the ODEs produced by `dss004`. In particular, since the number of grid points, $n = 51$, is relatively small, the individual elements of the Jacobian matrix are distinct. Also, note that the bandwidth is 9 and not 5 as might be expected from the five-point FDs in `dss004`. This greater bandwidth is due to the repeated use of `dss004` in `pde_1.m` to compute `uxx` from `u` by stagewise differentiation. This example illustrates a disadvantage of stagewise differentiation, that is, the increase in the bandwidth of the ODE Jacobian matrix through successive calls of the spatial differentiator such as `dss004`.

The tabular analytical and numerical solutions of Table 7.1 also reflect the good agreement between these two solutions (see Table 7.1). The computational effort reflected in `ncall = 190` is quite modest. As required, the analytical and numerical solutions agree for $t = 0$ (since both solutions are from eq. (7.2) with $t = 0$). For $t > 0$, the agreement between the analytical and numerical solutions of approximately five figures is quite acceptable, even with only 51 grid points.

In summary, the solution of eq. (7.1) subject to the IC and two BCs from eq. (7.2) with $t = 0$ for the IC and $x = -10, 10$ for the BCs is straightforward, particularly since the

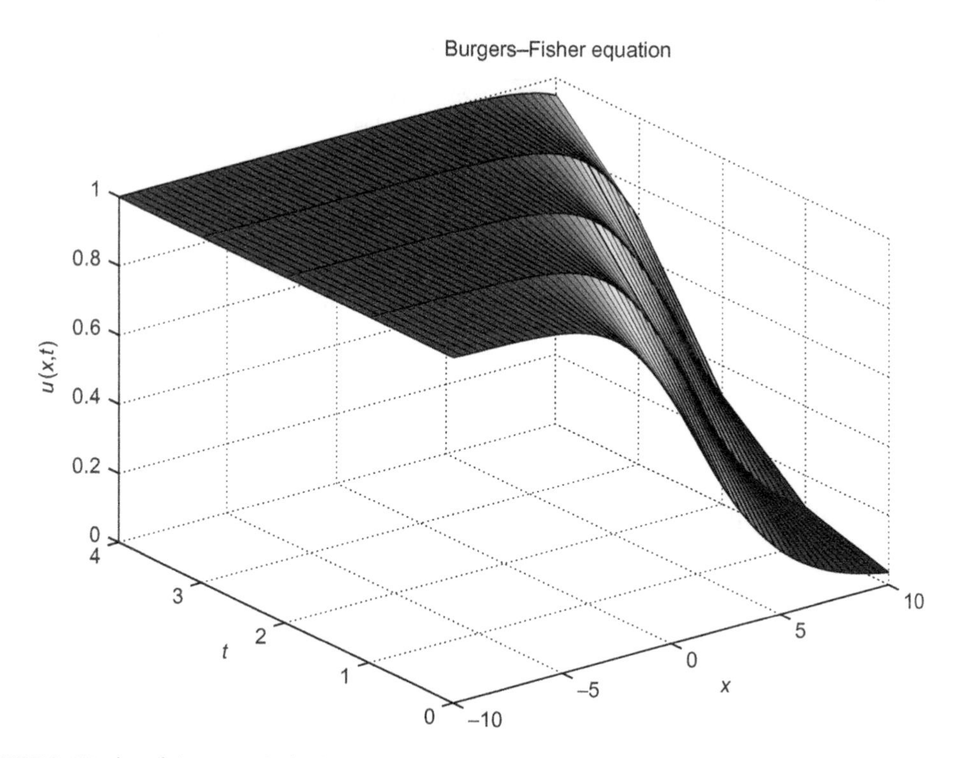

FIGURE 7.2: 3D plot of the numerical solution to eq. (7.1).

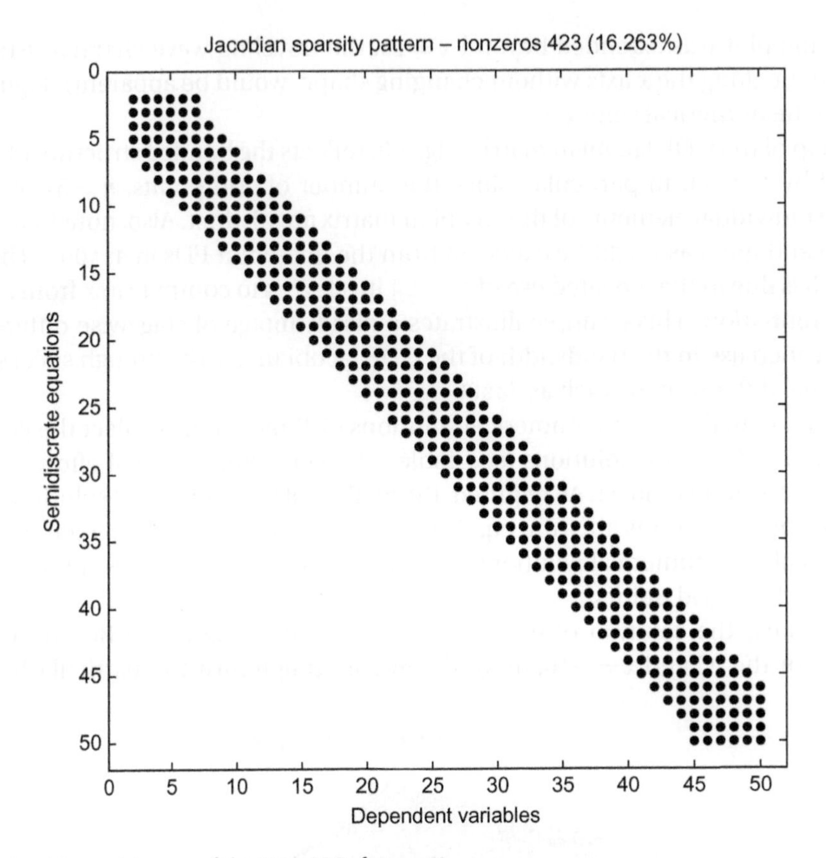

FIGURE 7.3: Jacobian matrix map of the MOL ODEs for $n = 51$.

solution is quite smooth. Also, eq. (7.1) is nonlinear, yet the programming in `pde_1.m` is straightforward. Consequently, variations in the PDE can easily be made for cases for which an analytical solution might not be available.

Appendix

We conclude this chapter by outlining an approach to solving the *generalized Burgers–Fisher* equation

$$\frac{\partial}{\partial t}v(x,t) + pv(x,t)^s \frac{\partial}{\partial x}v(x,t) - \frac{\partial^2}{\partial x^2}v(x,t) - qv(x,t)\left(1 - v(x,t)^s\right) = 0 \tag{7.3}$$

On first sight, this appears to be a rather daunting equation for which we have to find an analytical traveling wave solution. In fact, the *tanh, exp,* and *Riccati* methods all fail on this equation in its current form. However, we shall show that by applying certain transformations, it will yield a rather simple solution.

Table 7.1: Tabular numerical and analytical solutions

t	x	u(it,i)	u_anal(it,i)	err(it,i)
0.00	−10.000	0.999364	0.999364	0.000000
0.00	−9.600	0.999170	0.999170	0.000000
0.00	−9.200	0.998917	0.998917	0.000000
0.00	−8.800	0.998587	0.998587	0.000000
0.00	−8.400	0.998156	0.998156	0.000000
0.00	−8.000	0.997595	0.997595	0.000000
	⋮		⋮	
0.00	8.000	0.069316	0.069316	0.000000
0.00	8.400	0.060698	0.060698	0.000000
0.00	8.800	0.053144	0.053144	0.000000
0.00	9.200	0.046526	0.046526	0.000000
0.00	9.600	0.040728	0.040728	0.000000
0.00	10.000	0.035651	0.035651	0.000000
t	x	u(it,i)	u_anal(it,i)	err(it,i)
1.00	−10.000	0.999931	0.999931	0.000000
1.00	−9.600	0.999910	0.999910	0.000000
1.00	−9.200	0.999882	0.999882	0.000000
1.00	−8.800	0.999847	0.999847	0.000000
1.00	−8.400	0.999800	0.999800	0.000000
1.00	−8.000	0.999739	0.999739	0.000000
	⋮		⋮	
1.00	−2.800	0.991724	0.991725	−0.000001
1.00	−2.400	0.989236	0.989237	−0.000001
1.00	−2.000	0.986015	0.986016	−0.000001
1.00	−1.600	0.981857	0.981858	−0.000001
1.00	−1.200	0.976507	0.976508	−0.000001
1.00	−0.800	0.969652	0.969653	−0.000001
1.00	−0.400	0.960916	0.960916	−0.000000
1.00	0.000	0.949857	0.949857	0.000000
1.00	0.400	0.935977	0.935976	0.000001
1.00	0.800	0.918740	0.918737	0.000002
1.00	1.200	0.897604	0.897600	0.000003
1.00	1.600	0.872084	0.872080	0.000004
1.00	2.000	0.841824	0.841820	0.000004
1.00	2.400	0.806677	0.806675	0.000002
1.00	2.800	0.766788	0.766788	−0.000001
1.00	3.200	0.722636	0.722639	−0.000004
1.00	3.600	0.675029	0.675035	−0.000006
1.00	4.000	0.625038	0.625046	−0.000007
1.00	4.400	0.573886	0.573893	−0.000007
1.00	4.800	0.522808	0.522813	−0.000004

(Continued)

Table 7.1: *(Continued)*

t	x	u(it,i)	u_anal(it,i)	err(it,i)
1.00	5.200	0.472929	0.472931	-0.000001
1.00	5.600	0.425177	0.425175	0.000002
1.00	6.000	0.380238	0.380234	0.000004
1.00	6.400	0.338553	0.338548	0.000005
1.00	6.800	0.300350	0.300345	0.000006
1.00	7.200	0.265679	0.265673	0.000005
1.00	7.600	0.234460	0.234455	0.000005
1.00	8.000	0.206526	0.206522	0.000004
1.00	8.400	0.181655	0.181651	0.000003
1.00	8.800	0.159596	0.159594	0.000002
1.00	9.200	0.140092	0.140091	0.000002
1.00	9.600	0.122887	0.122886	0.000001
1.00	10.000	0.107737	0.107737	0.000000

```
            .                     .
            .                     .
            .                     .
  output for t = 2, 3, 4 removed
            .                     .
            .                     .
            .                     .
  ncall=190
```

We start by introducing the following new variable

$$v(x,t) = u(x,t)^{1/s}$$

from which we obtain

$$\frac{u^{1/s}}{su}\frac{\partial}{\partial t}u + \frac{p\left(u^{1/s}\right)^{s}u^{1/s}}{su}\frac{\partial}{\partial x}u - \frac{u^{1/s}}{s^{2}u^{2}}\left(\frac{\partial}{\partial x}u\right)^{2} - \frac{u^{1/s}}{su}\frac{\partial^{2}}{\partial x^{2}}u + \frac{u^{1/s}}{su^{2}}\left(\frac{\partial}{\partial x}u\right)^{2} - qu^{1/s}\left[1-\left(u^{1/s}\right)^{s}\right] = 0$$

where, for brevity, we have written u for $u(x,t)$. Eliminating the denominators by cross-multiplication, dividing by the common factor $u^{1/s}$, and collecting like terms yields the much simpler form

$$su\frac{\partial}{\partial t} - su\frac{\partial^{2}}{\partial x^{2}}u + (s-1)\left(\frac{\partial}{\partial x}u\right)^{2} + psu^{2}\frac{\partial}{\partial x}u - qs^{2}u^{2}(1-u) = 0$$

We are now able to apply the tanh method, which yields the following traveling wave solution,

$$u = \frac{1}{2} + \frac{1}{2}\tanh\left[\frac{s\left(-xp - xps + tp^{2} + tqs^{2} + tq + 2tqs\right)}{2\left(1+s\right)^{2}}\right]$$

Then, by applying the inverse transformation $v = \sqrt{u}$, we obtain

$$v = \frac{1}{2}\sqrt{1 + \tanh\left[\frac{s(-xp - xps + tp^2 + tqs^2 + tq + 2tqs)}{2(1+s)^2}\right]} \tag{7.4}$$

which is a solution to the generalized Burgers–Fisher equation (7.3). Using the Maple procedure `pdetest()`, we confirm that eq. (7.4) does indeed satisfy eq. (7.3). Finally, substituting the values $p = 1$, $q = 1$, and $s = 2$ into eq. (7.4), we arrive at the solution given in eq. (7.2), i.e.,

$$v = \frac{1}{2}\sqrt{1 - \tanh\left[\frac{1}{3}\left(x - \frac{10}{3}t\right)\right]} \tag{7.5}$$

From eq. (7.5), we see that the wavenumber is equal to $k = 1/3$ and the wave velocity is equal to $c = 10/3$. 2D and 3D plots of this solution are given in Figs. 7.4 and 7.5. Figure 7.4 is the initial condition (at $t = 0$), which then moves left to right when the animation (see

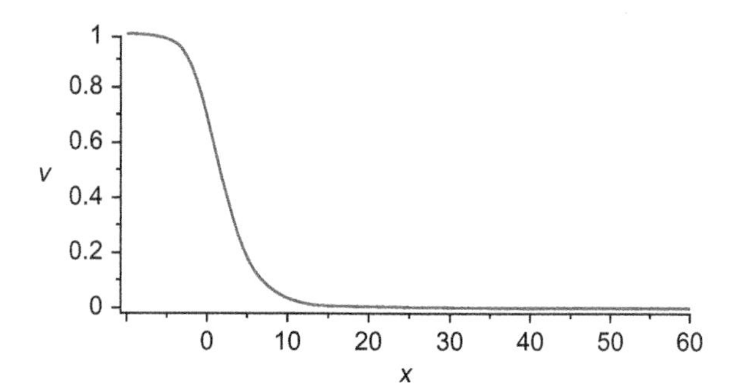

FIGURE 7.4: 2D plot of the solution to Burgers–Fisher equation at $t = 0$.

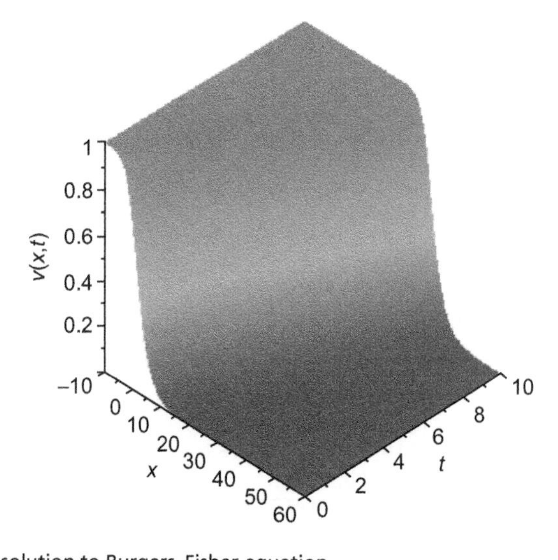

FIGURE 7.5: 3D plot of the solution to Burgers–Fisher equation.

Listing 7.4) is activated. Figure 7.5 is a 3D plot that demonstrates the movement of the solution through x and t.

A Maple script that performs the calculations for Figs. 7.3 and 7.4 is given in Listing 7.4.

```
># Burgers-Fisher Equation
 # Attempt at Malfliet's tanh solution
 restart; with(PDEtools): with(PolynomialTools):
 with(plots):
># Set up alias
 alias(u=u(x,t)): alias(v=v(x,t)):
># Define generalized B-F pde
 pde0:=diff(v,t)+p*v^s*diff(v,x)-diff(v,x,x)-q*v*(1-v^s)=0;
># Apply a transformation
 tr0:={v=u^(1/s)};
 pde1:=dchange(tr0,pde0,[u]);
 pde1 := numer(lhs(pde1))*denom(rhs(pde1)) = numer(rhs(pde1))*denom(lhs(pde1));
># Divide through by u^1/s
 pde1:=simplify(pde1/(u^(1/s)),symbolic):
 pde1:=collect(pde1,{diff(u,t),diff(u,x),diff(u,x,x)});
>read("tanhMethod.txt");
 # Solve transformed equation
 intFlg:=0: # No integration of U(xi) needed !
 M:=1; # Set order of approximation
 infoLevOut:=0;
 tanhMethod(M,pde1,intFlg,infoLevOut);
># Test Solution (May take a long time!)
 s:=2;p:=1;q:=1;x0:=0;
 soln:=v=sqrt(rhs(sol[4]));
 testSol:=pdetest(soln,pde0);
 if testSol=0 then
   print("pdetest(): PASSED");
 else
   print("pdetest(): FAILED");
 end if;
># Set parameter values
 #p:=1;q:=1;s:=2;x0:=0;
># Animate the solution
 zz:=rhs(soln);
 animate(zz,x=-10..60, t=0..10, axes=framed,
   title="Burgers-Fisher Equation",
   labels=["x","v"],
   thickness=3,frames=50,numpoints=100,
   title="Burgers-Fisher Equation",
   labelfont=[TIMES, ROMAN, 16],axesfont=[TIMES, ROMAN, 16],
   titlefont=[TIMES, ROMAN, 16]);
># Generate a 3D surface plot of standard solution
 plot3d(zz,x=-10..60,t=0..10,axes=framed,
   labels=["x","t","v(x,t)"],
   labeldirections=[HORIZONTAL,HORIZONTAL,VERTICAL],
   orientation=[-45,53],grid=[100,100],
   style=patchnogrid,axesfont=[TIMES, ROMAN, 16],
   shading=Z,title="Burgers-Fisher Equation",
   labelfont=[TIMES, ROMAN, 16],axesfont=[TIMES, ROMAN, 16],
   titlefont=[TIMES, ROMAN, 16]);
```

LISTING 7.4: Maple code used to confirm correctness of analytical solutions to the generalized Burgers–Fisher equation.

Additional Maple scripts that find traveling wave solutions to the Burgers–Fisher equation using *exp*- and *Riccati*-based methods are included with the downloads for this book.

Reference

[1] A. Molabahrami, F. Khani, *The homotopy analysis method to solve the Burgers-Huxley equation,* Nonlinear Anal. R. World Appl. 10 (2) (2009) 589–600.

8

Fisher–Kolmogorov Equation

The *Fisher–Kolmogorov equation*, with applications in biology [3], is

$$\frac{\partial u}{\partial t} = \frac{\partial^2 u}{\partial x^2} + u(1 - u^q), \quad q > 0 \tag{8.1}$$

with the analytical solution

$$u(x,t) = \frac{1}{(1 + ae^{b\xi})^s}, \quad \xi = x - ct \tag{8.2}$$

where

$$s = \frac{2}{q} \tag{8.3}$$

$$b = \frac{q}{[2(q+2)]^{1/2}} \tag{8.4}$$

$$c = \frac{q+4}{[2(q+2)]^{1/2}} \tag{8.5}$$

We take

$$a = \sqrt{2} - 1 \tag{8.6}$$

as suggested in [3], although a is arbitrary.

The Matlab routines closely resemble those of Chapters 3 and 7. Here, we list a few details pertaining to eqns. (8.1)–(8.6). First, the ODE routine `pde_1.m` is

```
  function ut=pde_1(t,u)
%
% Function pde_1 computes the t derivative vector for the Fisher-
% Kolmogorov equation
%
  global xl xu x n ncall
%
% Model parameters
  global a b c s q
%
% BCs at x = -5,10
  u(1)=ua_1(x(1),t);
  u(n)=ua_1(x(n),t);
%
% ux
  ux=dss004(xl,xu,n,u);
%
```

```
% uxx
  uxx=dss004(xl,xu,n,ux);
%
% PDE
  for i=2:n-1
    ut(i)=uxx(i)+u(i)*(1-u(i)^q);
  end
  ut(1)=0;
  ut(n)=0;
  ut=ut';
%
% Increment calls to pde_1
  ncall=ncall+1;
```

LISTING 8.1: Function pde_1.m for eq. (8.1).

We can note the following points about pde_1.m:

- The function and some global parameters are first defined.

```
  function ut=pde_1(t,u)
%
% Function pde_1 computes the t derivative vector for the Fisher-
% Kolmogorov equation
%
  global xl xu x n ncall
%
% Model parameters
  global a b c s q
```

- The first derivative in eq. (8.1), ux, is computed using the function dss004. Since eq. (8.1) is second order in x, the two required BCs are taken from eq. (8.2) with $x = -5, 10$ at grid points i=1,n, respectively (with n=51 subsequently set in function inital_1.m).

```
%
% BCs at x = -5,10
  u(1)=ua_1(x(1),t);
  u(n)=ua_1(x(n),t);
%
% ux
  ux=dss004(xl,xu,n,u);
```

- The second derivative in eq. (8.1), uxx, is computed with dss004 by differentiating ux, so-called *stagewise differentiation*. The alternative would be to use dss044 to directly compute uxx from u, as discussed in Chapter 4.

```
%
% uxx
  uxx=dss004(xl,xu,n,ux);
```

- Equation (8.1) is then programmed.

```
%
% PDE
  for i=2:n-1
```

```
      ut(i)=uxx(i)+u(i)*(1-u(i)^q);
    end
    ut(1)=0;
    ut(n)=0;
    ut=ut';
  %
  % Increment calls to pde_1
    ncall=ncall+1;
```

Since Dirichlet BCs are used, the derivatives in t are set to zero at the boundaries (so that the ODE integrator does not move the boundary values away from their prescribed values). A transpose is included to meet the requirements of the ODE integrator `ode15s`. Finally, the counter for the number of calls to `pde_1.m` is incremented.

The IC from eq. (8.2) is programmed in `inital_1.m`.

```
  function u0=inital_1(t0)
%
% Function inital_1 sets the initial condition for the Fisher-
% Kolmogorov equation
%
% Parameters shared with other routines
  global xl xu x n ncall
%
% Model parameters
  global a b c s q
%
% Spatial domain and initial condition
  xl=-5;
  xu=10;
  n=51;
  dx=(xu-xl)/(n-1);
%
% IC from analytical solution
  for i=1:n
    x(i)=xl+(i-1)*dx;
    u0(i)=ua_1(x(i),0.0);
  end
```

LISTING 8.2: Function `inital_1.m` for IC from eq. (8.2).

We can note the following points about `inital_1.m`:

- The function and some global parameters are first defined.

```
    function u0=inital_1(t0)
  %
  % Function inital_1 sets the initial condition for the Fisher-
  % Kolmogorov equation
  %
  % Parameters shared with other routines
    global xl xu x n ncall
  %
  % Model parameters
    global a b c s q
```

- The grid in x is then defined over the interval $-5 \leq x \leq 10$ for 51 points (these grid parameters were selected by trial and error to produce a numerical solution with acceptable accuracy).

```
%
% Spatial domain and initial condition
  xl=-5;
  xu=10;
  n=51;
  dx=(xu-xl)/(n-1);
%
% IC from analytical solution
  for i=1:n
    x(i)=xl+(i-1)*dx;
    u0(i)=ua_1(x(i),0.0);
  end
```

As the grid in x is defined in the `for` loop, function `ua_1` (listed next) is called (for $t = 0$) to define the IC from eq. (8.2).

Function `ua_1.m` is a straightforward implementation of the analytical solution, eq. (8.2).

```
  function uanal=ua_1(x,t)
%
% Function uanal computes the exact solution of the Fisher-Kolmogorov
% equation for comparison with the numerical solution
%
% Model parameters
  global a b c s q
%
% Analytical solution
  z=x-c*t;
  uanal=1/(1+a*exp(b*z))^s;
```

LISTING 8.3: Function `ua_1.m` for analytical solution (8.2).

The main program, `pde_1_main`, is similar to `pde_1_main` of Listing 2.1 and therefore we list only a few selected parts of it.

```
        .
        .
        .
%
% Model parameters
  global a b c s q
%
  q=1;
  a=2^(1/2)-1;
  b=q/(2*(q+2))^(1/2);
  c=(q+4)/(2*(q+2))^(1/2);
  s=2/q;
        .
        .
        .
```

```
%
%   Display selected output
    fprintf('\n q = %4.2f,    a = %4.2f,    b = %4.2f,    c = %4.2f
            s = 4.2f\n',q,a,b,c,s);
    for it=1:nout
      fprintf('\n     t        x         u(it,i)    u_anal(it,i)
                err(it,i)\n');
      for i=1:n
        fprintf('%6.2f%8.3f%15.6f%15.6f%15.6f\n',...
                t(it),x(i),u(it,i),u_anal(it,i),err(it,i));
      end
    end
    fprintf('    ncall = %4d\n\n',ncall);
          .
          .
          .
```

LISTING 8.4: A selected portion of main program pde_1_main.m.

We can note the following points about this code:

- The parameter q in eq. (8.1) is first defined, and the parameters a, b, c, s of eqns. (8.3)–(8.6) are then computed. In particular, the wave velocity c of eq. (8.2) is computed for subsequent use in analyzing the numerical solution (note the *Lagrangian* or *traveling wave* or *moving* coordinate of eq. (8.2), $\xi = x - ct$).
- These parameters are then displayed, and the numerical and analytical solutions (u(it,i), u_anal(it,i)) and their difference (err) are displayed at the output points ($t = 0, 0.5, 1, \ldots, 5$ or 11 values) through the for it=1:nout loop (with nout=11).
- These constants and parameters are then passed as global variables (not shown) so that they can be used in pde_1.m of Listing 8.1 and ua_1.m of Listing 8.3.

The main program produces the same three figures and tabulated output as in Chapters 3–7, which are now reviewed. The Jacobian matrix routine jpattern_num_1.m is the same as jpattern_num_1.m in Chapters 3–7 and is therefore not reproduced here.

Figure 8.1 indicates good agreement between the analytical and numerical solutions. Also, the solution does not appear to be exactly a traveling wave as specified by eq. (8.2) in the sense that the successive curves should be displaced by a constant distance in t. However, this apparent discrepancy is due to the termination of the plot on the right at $x = 10$ (if the plot was extended beyond $x = 10$, the traveling wave characteristic, that is, displacement along the x axis without changing shape, would be apparent). Figure 8.2 is a 3D plot of the numerical solution.

The map of the ODE Jacobian matrix, Fig. 8.3, reflects the banded structure of the ODEs produced by dss004. In particular, since the number of grid points, $n = 51$, is relatively small, the individual elements of the Jacobian matrix are distinct. Also, note that the bandwidth is 9 and not 5 as might be expected from the five-point FDs in dss004. This greater bandwidth is due to the repeated use of dss004 in pde_1.m to compute uxx from u by stagewise differentiation. This example illustrates a disadvantage of stagewise differentiation, that is, the increase in the bandwidth of the ODE Jacobian matrix through successive calls of the spatial differentiator such as dss004.

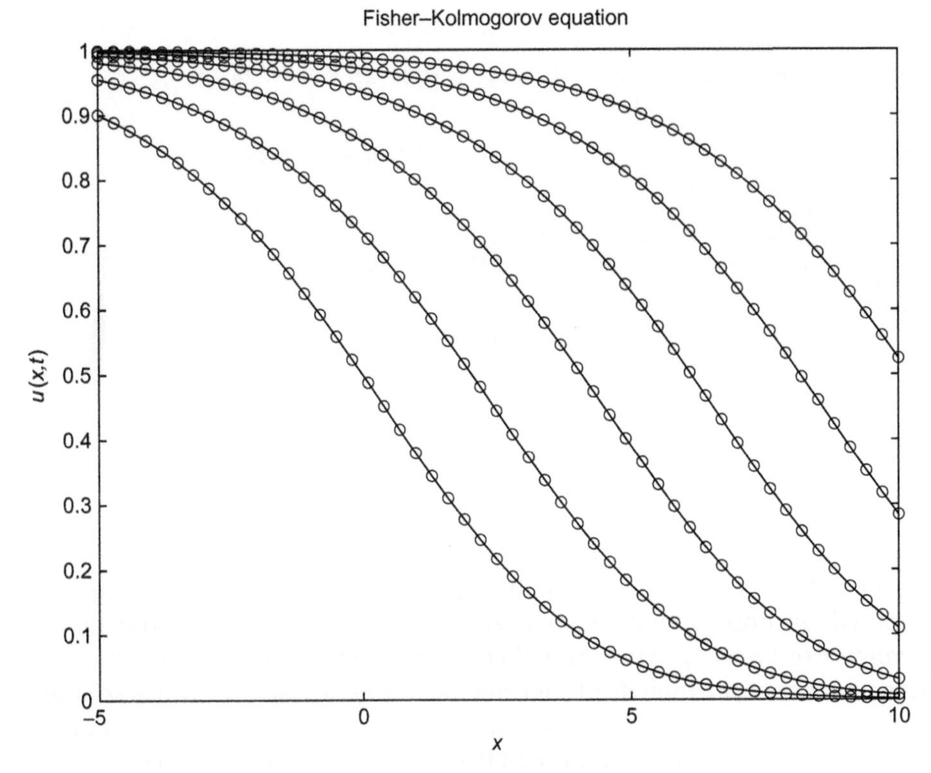

FIGURE 8.1: Numerical solution to eq. (8.1) (lines) with the analytical solution superimposed (circles) using five-point FD approximations in `dss004` for $t = 0, 0.5, 1, \ldots, 5$ (left to right).

The tabular analytical and numerical solutions also reflect the good agreement between these two solutions; see Table 8.1. The computational effort reflected in `ncall = 150` is quite modest. As required, the analytical and numerical solutions agree for $t = 0$ (since both solutions are from eq. (8.2) with $t = 0$). For $t > 0$, the agreement between the analytical and numerical solutions of approximately six figures is quite acceptable, even with only 51 grid points. In this case, the low number of grid points worked in our favor, but, of course, this is not always the case.

The numerical solution of Table 8.1 can be used to estimate the velocity of the traveling wave of Fig. 8.1. For $u(x, t = 0) = 0.5$, $x = 0$ (this follows from the values of q and a set in `pde_1_main.m` and used in `ua_1.m` with $x = t = 0$). By linear interpolation, the point on the x axis where $u = 0.5$ at time $t = 1$ (from Table 8.1) is

$$x = 1.9 + (2.2 - 1.9)(0.5 - 0.516824)/(0.480950 - 0.516824) = 2.0407$$

Thus, the estimated velocity is

$$c = \frac{\Delta x}{\Delta t} = \frac{2.0407 - 0}{1 - 0} = 2.0407$$

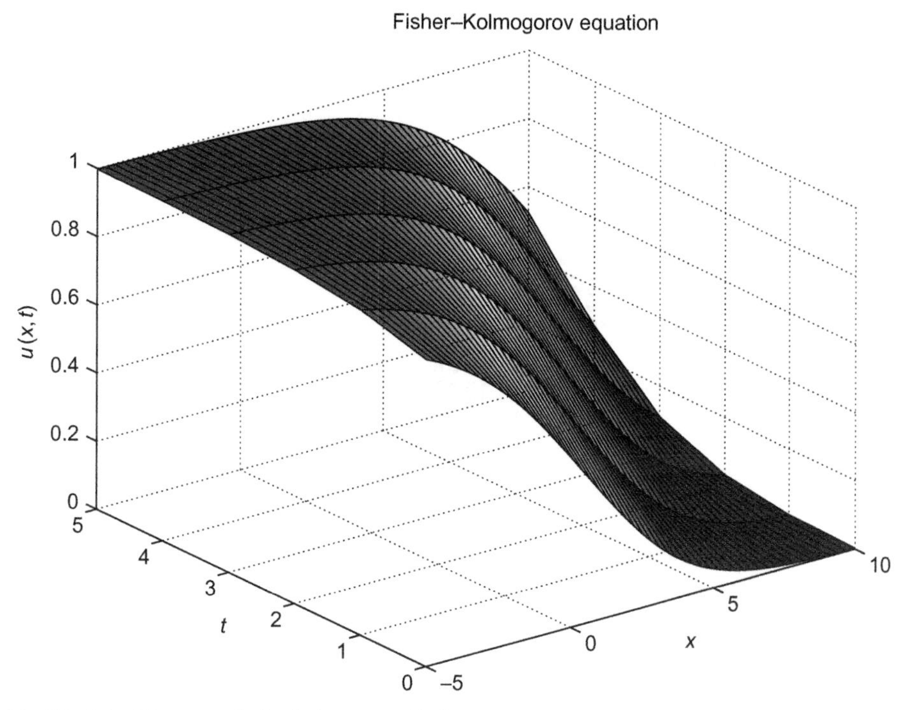

FIGURE 8.2: 3D plot of the numerical solution to eq. (8.1).

which compares with the computed value of $c = 2.04$ (see the parameter values in Table 8.1).

In summary, the solution of eq. (8.1) subject to the IC from eq. (8.2) (with $t = 0$) and two Dirichlet BCs from eq. (8.2) (with $x = -5, 10$) is straightforward. Also, eq. (8.1) is nonlinear, yet the programming in `pde_1.m` (Listing 8.1) is straightforward. Consequently, variations in the PDE can easily be made for cases for which an analytical solution might not be available.

Appendix

We conclude this chapter by solving the *Fisher–Kolmogorov* (F–K) equation using the *factorization method* [1, 4, 5], as outlined in the main Appendix. The F–K equation is a simplified version of the KPP equation, eq. (10.6), where for this problem, the KPP constants would become: $a = 1$, $b = -1$, $\delta = 1$, and $m = q + 1$. We follow the general approach used in the appendix to Chapter 10, and we repeat the problem equation below for convenience

$$\frac{\partial u}{\partial t} - \frac{\partial^2 u}{\partial x^2} - u(1 - u^q) = 0 \tag{8.1}$$

$$u = u(x, t),\ t > 0,\ q > 0 \tag{8.7}$$

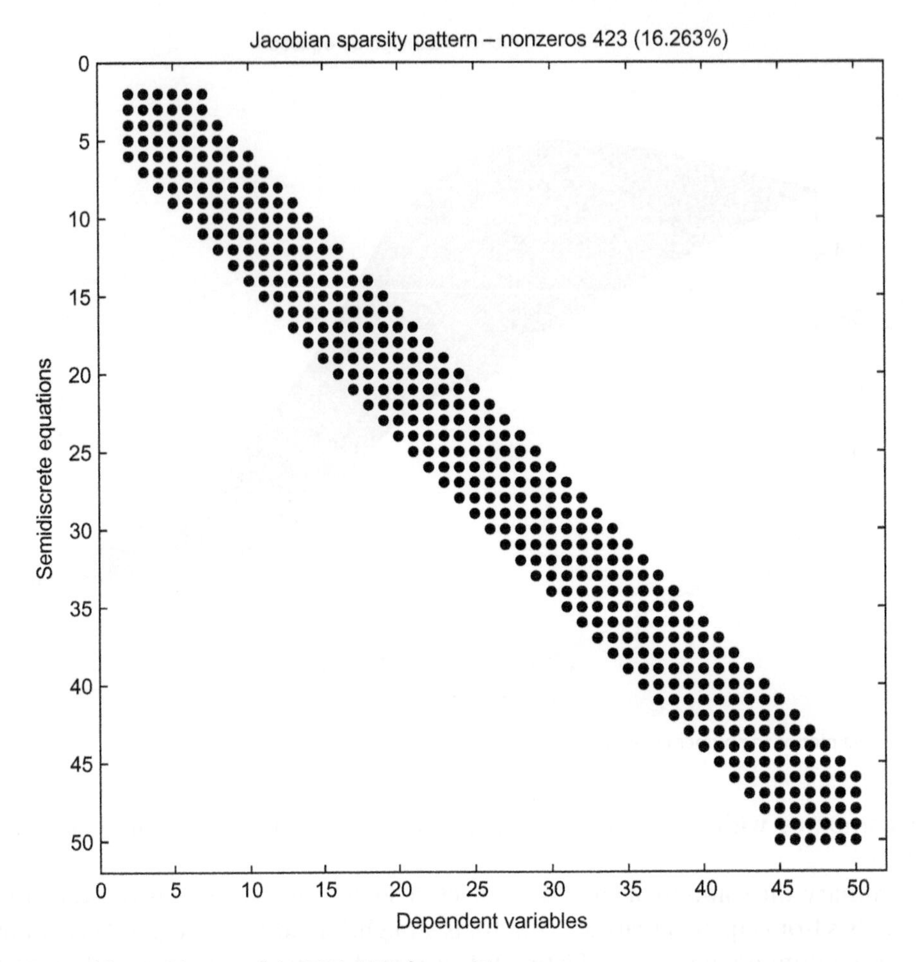

FIGURE 8.3: Jacobian matrix map of the MOL ODEs for $n = 51$.

If we assume a traveling wave solution of the form $u(x,t) = U(\xi)$, where $\xi = k(x - ct)$, $c =$ velocity, and $k =$ wavenumber, eq. (8.7) reduces to the traveling wave ODE

$$\frac{d^2U}{dx^2} + \gamma \frac{dU}{dx} + F(U) = 0 \tag{8.8}$$

where $\gamma = \dfrac{c}{k}$ and $F(U) = \dfrac{1}{k^2}U(1 - U^q)$.

We factor the polynomial function of eq. (8.8) (the third term) as

$$\frac{F(U)}{U} = f_1 f_2 = \frac{1}{k^2}(1 - U^{q/2})(1 + U^{q/2})$$

Table 8.1: Tabular numerical and analytical solutions

q=1.00,	a=0.41,	b=0.41,	c=2.04	s=2.00
t	x	u(it,i)	u_anal(it,i)	err(it,i)
0.00	-5.000	0.900512	0.900512	0.000000
0.00	-4.700	0.888652	0.888652	0.000000
0.00	-4.400	0.875527	0.875527	0.000000
0.00	-4.100	0.861038	0.861038	0.000000
0.00	-3.800	0.845091	0.845091	0.000000
	.		.	
	.		.	
	.		.	
0.00	-0.500	0.558806	0.558806	0.000000
0.00	-0.200	0.523780	0.523780	0.000000
0.00	0.100	0.488015	0.488015	0.000000
0.00	0.400	0.451830	0.451830	0.000000
0.00	0.700	0.415572	0.415572	0.000000
0.00	1.000	0.379608	0.379608	0.000000
	.		.	
	.		.	
	.		.	
0.00	8.800	0.003883	0.003883	0.000000
0.00	9.100	0.003084	0.003084	0.000000
0.00	9.400	0.002445	0.002445	0.000000
0.00	9.700	0.001936	0.001936	0.000000
0.00	10.000	0.001531	0.001531	0.000000
t	x	u(it,i)	u_anal(it,i)	err(it,i)
1.00	-5.000	0.954833	0.954833	0.000000
1.00	-4.700	0.949175	0.949175	-0.000000
1.00	-4.400	0.942839	0.942839	-0.000000
1.00	-4.100	0.935754	0.935754	-0.000000
1.00	-3.800	0.927842	0.927842	-0.000000
	.		.	
	.		.	
	.		.	
1.00	1.000	0.619243	0.619243	-0.000000
1.00	1.300	0.586240	0.586240	-0.000000
1.00	1.600	0.552017	0.552018	-0.000000
1.00	1.900	0.516824	0.516824	-0.000000
1.00	2.200	0.480950	0.480950	-0.000000
1.00	2.500	0.444722	0.444723	-0.000000
	.		.	
	.		.	
	.		.	
1.00	8.500	0.021720	0.021717	0.000003
1.00	8.800	0.017594	0.017591	0.000002
1.00	9.100	0.014202	0.014201	0.000002

(Continued)

Table 8.1: *(Continued)*

t	x	u(it,i)	u_anal(it,i)	err(it,i)
1.00	9.400	0.011428	0.011427	0.000001
1.00	9.700	0.009170	0.009169	0.000001
1.00	10.000	0.007338	0.007338	0.000000
	.		.	
	.		.	
	.		.	
	output for t = 2, 3, 4, 5 removed			
	.		.	
	.		.	
	.		.	
	ncall=150			

and choose

$$f_1 = (1 - U^{q/2})\frac{\alpha}{k}, \quad f_2 = (1 + U^{q/2})\frac{1}{\alpha k}, \quad \alpha \neq 0 \tag{8.9}$$

where we have also introduced the constant α.

From eqns. (A.41) and (8.9), we obtain the following ODE

$$\begin{aligned}\frac{df_1}{dU}U + f_1 + f_2 = -\frac{q}{2}U^{q/2}\frac{\alpha}{k} + (1 - U^{q/2})\frac{\alpha}{k} \\ + (1 + U^{q/2})\frac{1}{\alpha k} = -\gamma = -\frac{c}{k}\end{aligned} \tag{8.10}$$

Collecting terms and equating the coefficients of $U^{q/2}$ to zero (as the left-hand side of eq. (8.10) is equal to a constant) gives $\alpha = \pm\frac{2}{\sqrt{2(q+2)}}$ and $c = \left(\frac{q+4}{\sqrt{2(q+2)}}\right)$. Also, as γ is a constant and independent of the value of U, on setting $U = 0$, we find that $\gamma = -\frac{(\alpha+\alpha^{-1})}{k}$. Therefore, adopting the grouping of Cornejo-Perez [2], i.e., eq. (A.40b), it follows that

$$\frac{d^2U}{d\xi^2} \mp \frac{(\alpha + \alpha^{-1})}{k}\frac{dU}{d\xi^2} + f_1 f_2 U = 0 \tag{8.11}$$

Thus, the corresponding factorization eq. (A.39), i.e., $[D - f_2(U)][D - f_1(U)]U = 0$, becomes

$$\left[D \pm \frac{1}{\alpha k}\left(1 + U^{q/2}\right)\right]\left[D \mp \frac{\alpha}{k}\left(1 - U^{q/2}\right)\right]U = 0 \tag{8.12}$$

where $D = \dfrac{d}{d\xi}$. Therefore, it follows that eq. (8.11) is compatible with the first-order ODE

$$\frac{dU}{d\xi} \mp \frac{\alpha}{k}(1 - U^{q/2})U = 0 \tag{8.13}$$

Integrating eq. (8.13) either manually or using Maple yields

$$U = \left[1 + K\exp\left(-\frac{1}{2k}\alpha q\xi\right)\right]^{-2/q}$$

where K is an arbitrary constant of integration. Substituting back values for ξ, c, and α, we find that $k = \left(\frac{q}{\sqrt{2(q+2)}}\right)$, which leads to the final solution

$$U = \left\{1 + K\exp[k(x - ct)]\right\}^{-2/q} \tag{8.14}$$

where k and c are as defined above.

If we let $K = \pm\hat{K}\exp[kx_0]$, we arrive at the standard form of traveling wave solution

$$U^{\pm} = \left\{1 \pm \hat{K}\exp[k(x - x_0 - ct)]\right\}^{-2/q} \tag{8.15}$$

The above solution with $x_0 = 0$ is the same as the analytical solution, eqns. (8.2–8.5), used in the numerical simulation discussed in the main body of this chapter. The Maple code of Listing 8.5 will perform the above calculations.

```
># Some calculations to confirm the results of a
 # factorization solution to the FisherKolmogorov Equation
 # Ref: Rosu, H. C. and O. Cornejo-Perez (2008).
 #      Supersymmetric pairing of kinks for polynomial
 #      nonlinearities, [arXiv:math-ph/0401040v3 24 Dec 2004|
 #      |ArXiv: math-ph/0401040]
 restart; with(DEtools): with(PDEtools):
 alias(u=u(x,t)): alias(U=U(xi)):
># Define F-K equation
 pde1:=diff(u,t)-diff(u,x,x)-u*(1-u^q)=0;
># Convert PDE to ODE
 tr1:={x=(xi/k+c*tau),t=tau,u=U};
 ode1:=dchange(tr1,pde1,[xi,tau,U]);
># Define F(U), g
 F:=U*(1-U^(q))/(k^2);
 g:=c/k;
># Factor F(U)/U - Note: alpha introduced
 f[1]:=alpha*(1-U^(q/2))/k;
 f[2]:=(1+U^(q/2))/(alpha*k);
># Check that factorization is correct
 F_chk:=simplify(eval(f[1]*f[2]*U));
># Use the C-P grouping
 alias(U=U):f[1]:=subs(U(xi)=U,f[1]):f[2]:=subs(U(xi)=U,f[2]):
 eqn1:=diff(f[1],U)*U+f[1]+f[2]=-subs(U(xi)=U,g);
 sol1:=simplify(subs(U=0,eqn1),size);
```

```
 c:=solve(sol1,c);
># Collect terms in U^(q/2))
 eqn2:=collect(simplify(lhs(eqn1)-rhs(eqn1),size),U,'recursive');
># Equate coeff of U to zero
 eqn3:=coeff(eqn2,U^(q/2))=0;
 sol2:=solve(eqn3,alpha);
 alpha:=sol2[2];
># Fomulate 1st order ode from [D-f[1]]U=0
 f[1]:=subs(U=U(xi),f[1]);ode2:=diff(U(xi),xi)-f[1]*U(xi);
># Obtain solution to ode2
 sol4:=dsolve(ode2);
># Check solution U(xi) satisfies ode1
 odeCHK:=simplify(eval(subs({U(xi)=rhs(sol4)},rhs(ode1))),symbolic):
 if odeCHK = 0 then
   print('solution PASSES');
 else
   print('solution FAILS!');
 end if;
># Obtain solution to pde1
 sol5:=u=simplify(expand(subs({xi=k*(x-c*t)},rhs(sol4))),size);
># Check solution u satisfies pde1
 pdeCHK:=simplify(pdetest(sol5,pde1),symbolic);
 if pdeCHK = 0 then
   print('solution PASSES');
 else
   print('solution FAILS!');
 end if;
```

LISTING 8.5: Maple code to derive a solution to the Fisher–Kolmogorov equation using the *factorization* method.

Finally, additional Maple scripts that obtain different solutions to the Fisher-Kolmogorov equation using *tanh-*, *exp-*, and *Riccati*-based methods are included with the downloads for this book.

References

[1] L.M. Berkovich, Factorization as a method of finding exact invariant solutions of the Kolmogorov-Petrovskii-Piskunov equation and the related Semenov and Zeldovich equations, *Sov. Math. Dokl.* 45 (1992) 162–167.

[2] O. Cornejo-Perez, H.C. Rosu, Nonlinear second order ODE: Factorizations and particular solutions, *Prog. Theor. Phys.* 114 (3) (2005) 533–538.

[3] J.D. Murray, *Mathematical Biology, I: An Introduction*, third ed., Springer, New York, 2002, 450–451.

[4] A.D. Polyanin, V.F. Zaitsev, *Handbook of Nonlinear Partial Differential equations*, Chapman and Hall/CRC Press, Boca Raton, FL, 2004.

[5] H.C. Rosu, O. Cornejo-Perez, Supersymmetric pairing of kinks for polynomial nonlinearities, http://arxiv.org/PS_cache/math-ph/pdf/0401/0401040v3.pdf, Accessed 23 April, 2010.

9 Fitzhugh–Nagumo Equation

One of the most widely studied biological systems with excitable behavior is neural communication by nerve cells via electrical signaling. The *Fitzhugh–Nagumo equation* is a simplification of the Hodgin–Huxley model [6] for the membrane potential of a nerve axon. The first version was developed by Fitzhugh [3] and consisted of a 2×2 (two equations in two unknowns) system of ODEs.

Nagumo et al. [10] studied a related equation that added a diffusion term for the conduction process of action potentials along nerves. This system, where spatial diffusion in the transmembrane potential is allowed but without any applied external current, can be described by these equations [9]

$$\frac{\partial u}{\partial t} = D\frac{\partial^2 u}{\partial x^2} + f(u) - v, \quad f(u) = u(u-1)(a-u)$$
$$\frac{\partial v}{\partial t} = bu - \gamma v$$

Here, u is directly related to the membrane potential and v represents several variables associated with terms that contribute to the membrane current from sodium, potassium, and other ions. The diffusion constant D is associated with axial current in the axon. The parameters $0 < a < 1$, b, and γ are all positive.

It is analytically easier to see what is going on if we consider b and γ to be small so that $b = \epsilon L$, $\gamma = \epsilon M$, $0 < \epsilon \ll 1$, and the preceding equations become

$$\frac{\partial u}{\partial t} = D\frac{\partial^2 u}{\partial x^2} + f(u) - v, \quad f(u) = u(u-1)(a-u)$$
$$\frac{\partial v}{\partial t} = \epsilon(Lu - Mv)$$

In the limit when $\epsilon \to 0$, we have $v \simeq$ constant, and it turns out that this constant is equal to zero. Thus, under these conditions, the Fitzhugh–Nagumo system reduces to the nonlinear reaction–diffusion equation ([11], p4),

$$\frac{\partial u}{\partial t} = D\frac{\partial^2 u}{\partial x^2} - u(1-u)(a-u) \tag{9.1}$$

(a and D are arbitrary, with $0 \le a \le 1$ and $D > 0$); D has a marked effect on the solution of eq. (9.1), as we discuss later.

Equation (9.1) has the following analytical traveling wave solution ([11], p4)

$$u(x,t) = \frac{1}{1+\exp\left[\frac{x}{\sqrt{2D}}+\left(a-\frac{1}{2}\right)t\right]} \tag{9.2}$$

Equation (9.2) is the analytical solution to the reduced Fitzhugh–Nagumo equation (9.1) that we use to evaluate the numerical solution of eq. (9.1). For further details of the Fitzhugh–Nagumo system, the reader is referred to [4, 5, 9, 11].

The Matlab routines closely resemble those of Chapters 3–8. Here, we list a few details pertaining to eqns. (9.1) and (9.2). First, the ODE routine pde_1.m, is

```
  function ut=pde_1(t,u)
%
% Function pde_1 computes the t derivative vector for the Fitzhugh-Nagumo
% equation
%
  global xl xu x n ncall
%
  global a D
%
% BCs at x = 0,1
  u(1)=ua_1(x(1),t);
  u(n)=ua_1(x(n),t);
%
% ux
  ux=dss004(xl,xu,n,u);
%
% uxx
  uxx=dss004(xl,xu,n,ux);
%
% PDE
  for i=2:n-1
    ut(i)=D*uxx(i)-u(i)*(1-u(i))*(a-u(i));
  end
  ut(1)=0;
  ut(n)=0;
  ut=ut';
%
% Increment calls to pde_1
  ncall=ncall+1;
```

LISTING 9.1: Function pde_1.m for eq. (9.1).

We can note the following points about pde_1.m:

- The function and some global parameters are first defined.

```
  function ut=pde_1(t,u)
%
% Function pde_1 computes the t derivative vector for the Fitzhugh-Nagumo
% equation
%
```

```
  global xl xu x n ncall
%
  global a D
```

- The first derivative in eq. (9.1), `ux`, is computed using the function `dss004`. Since eq. (9.1) is second order in x, the two required BCs are taken from eq. (9.2) programmed in `ua_1.m` with $x = 0, 1$ at grid points `i=1,n`, respectively (with `n=26`, subsequently set in function `inital_1.m`).

```
%
% BCs at x = 0,1
  u(1)=ua_1(x(1),t);
  u(n)=ua_1(x(n),t);
%
% ux
  ux=dss004(xl,xu,n,u);
```

- The second derivative in eq. (9.1), `uxx`, is computed with `dss004` by differentiating `ux`, so-called *stagewise differentiation*. The alternative would be to use `dss044` to directly compute `uxx` from `u`, as discussed in Chapter 4.

```
%
% uxx
  uxx=dss004(xl,xu,n,ux);
%
% PDE
  for i=2:n-1
    ut(i)=D*uxx(i)-u(i)*(1-u(i))*(a-u(i));
  end
  ut(1)=0;
  ut(n)=0;
  ut=ut';
%
% Increment calls to pde_1
  ncall=ncall+1;
```

Equation (9.1) is programmed in a `for` loop over the interior points `i=2` to `i=n-1`. Since Dirichlet BCs are used, the derivatives in t are set to zero at the boundaries (so that the ODE integrator does not move the boundary values away from their prescribed values). A transpose is included to meet the requirements of the ODE integrator `ode15s`. Finally, the counter for the number of calls to `pde_1.m` is incremented.

The IC of eq. (9.2) (in `ua_1.m` with $t = 0$) is programmed in `inital_1.m`.

```
  function u0=inital_1(t0)
%
% Function inital_1 sets the initial condition for the Fitzhugh-Nagumo
% equation
%
  global xl xu x n
%
```

```
% Spatial domain and initial condition
  xl=-60; xu=20; n=101;
  dx=(xu-xl)/(n-1);
%
% IC from analytical solution
  for i=1:n
    x(i)=xl+(i-1)*dx;
    u0(i)=ua_1(x(i),0.0);
  end
```

LISTING 9.2: Function `inital_1.m` for IC from eq. (9.2) with $t = 0$.

We can note the following points about `inital_1.m`:

- The function and some global parameters are first defined.

```
  function u0=inital_1(t0)
%
% Function inital_1 sets the initial condition for the Fitzhugh-Nagumo
% equation
%
  global xl xu x n
```

- The grid in x is then defined over the interval $-60 \le x \le 20$ for 101 points.

```
%
% Spatial domain and initial condition
  xl=-60; xu=20; n=101;
  dx=(xu-xl)/(n-1);
%
% IC from analytical solution
  for i=1:n
    x(i)=xl+(i-1)*dx;
    u0(i)=ua_1(x(i),0.0);
  end
```

As the grid in x is defined in the `for` loop, function `ua_1` (listed next) is called (for $t = 0$) to define the IC from eq. (9.2).

Function `ua_1.m` is a straightforward implementation of the analytical solution, eq. (9.2).

```
  function uanal=ua_1(x,t)
%
% Function uanal computes the exact solution of the Fitzhugh-Nagumo
% equation for comparison with the numerical solution
%
  global a D
%
% Analytical solution
  uanal=1/(1+exp((1/2^0.5)*x/D^0.5+(a-1/2)*t));
```

LISTING 9.3: Function `ua_1.m` for analytical solution (9.2).

The main program, `pde_1_main`, is similar to `pde_1_main` of Chapters 2–8. However, we consider some essential differences in detail.

```
%
% Clear previous files
  clear all
  clc
%
% Parameters shared with other routines
  global xl xu x n ncall
%
% Model parameters
  global a D
%
% Select ncase
%
%   Smooth solution
%   ncase=1;
%
%   Moving front
%   ncase=2;
  ncase=1;
  if(ncase==1)a=1; D=1;   end
  if(ncase==2)a=1; D=0.1; end
%
% Independent variable for ODE integration
  t0=0;
  tf=60;
  tout=[t0:2:tf]';
  nout=31;
  ncall=0;
%
% Initial condition
  u0=inital_1(t0);
%
% ODE integration
  mf=2;
  reltol=1.0e-06; abstol=1.0e-06;
  options=odeset('RelTol',reltol,'AbsTol',abstol);
%
% Explicit (nonstiff) integration
  if(mf==1)[t,u]=ode45(@pde_1,tout,u0,options); end
%
% Implicit (sparse stiff) integration
  if(mf==2)
    S=jpattern_num_1;
%   pause
    options=odeset(options,'JPattern',S)
    [t,u]=ode15s(@pde_1,tout,u0,options);
  end
%
% Store analytical solution, errors in numerical solution
  iplot=0;
  for it=1:nout
    u(it,1)=ua_1(x(1),t(it));
    u(it,n)=ua_1(x(n),t(it));
    if((it-1)*(it-11)*(it-21)*(it-31)==0)
```

```
        iplot=iplot+1;
        fprintf('\n     t        x         u(it,i)    u_anal(it,i)
                  err(it,i)\n');
        for i=1:n
          tplot(iplot)=t(it);
          uplot(iplot,i)=u(it,i);
          u_anal(iplot,i)=ua_1(x(i),t(it));
          err(iplot,i)=uplot(iplot,i)-u_anal(iplot,i);
          fprintf('%6.2f%8.3f%15.6f%15.6f%15.6f\n',...
            tplot(iplot),x(i),uplot(iplot,i),u_anal(iplot,i),err(iplot,i));
        end
      end
    end
    fprintf('    ncall = %4d\n\n',ncall);
%
% Plot numerical and analytical solutions
  figure(2)
  plot(x,uplot,'-',x,u_anal,'o')
  xlabel('x')
  ylabel('u(x,t)')
  title('Fitzhugh-Nagumo equation; t = 0, 20, 40, 60; solid - numerical;
         o - analytical')
  figure(3)
  surf(x,t,u)
  xlabel('x'); ylabel('t'); zlabel('u(x,t)');
  title('Fitzhugh-Nagumo equation');
  view(47,20); axis tight
  shading interp
  colormap cool
```

LISTING 9.4: Main program pde_1_main.

We can note the following points about Listing 9.4:

- The model parameters are declared as global. Then, two cases are programmed (for $-60 \le x \le 20, n = 101$ from inital_1.m).

```
%
% Clear previous files
  clear all
  clc
%
% Parameters shared with other routines
  global xl xu x n ncall
%
% Model parameters
  global a D
%
% Select ncase
%
%   Smooth solution
%   ncase=1;
%
%   Moving front
%   ncase=2;
```

```
ncase=1;
if(ncase==1)a=1; D=1;   end
if(ncase==2)a=1; D=0.1; end
```

- For the first case, $D = 1$ produces a moderately steep moving front solution (as reflected in the subsequent output). For the second case, $D = 0.1$ produces a considerably steeper moving front solution as might be expected since the diffusion term in eq. (9.1) is smaller.

```
if(ncase==1)a=1; D=1;   end
if(ncase==2)a=1; D=0.1; end
```

- t varies over the interval $0 \leq t \leq 60$ with the solution from ODE integrator `ode45` and `ode15s` produced at intervals of 2. Thus, a total of 31 outputs (counting $t = 0$) result for $t = 0, 2, 4, \ldots 60$.

```
%
% Independent variable for ODE integration
  t0=0;
  tf=60;
  tout=[t0:2:tf]';
  nout=31;
  ncall=0;
%
% Initial condition
  u0=inital_1(t0);
%
% ODE integration
  mf=2;
  reltol=1.0e-06; abstol=1.0e-06;
  options=odeset('RelTol',reltol,'AbsTol',abstol);
%
% Explicit (nonstiff) integration
  if(mf==1)[t,u]=ode45(@pde_1,tout,u0,options); end
%
% Implicit (sparse stiff) integration
  if(mf==2)
    S=jpattern_num_1;
%   pause
    options=odeset(options,'JPattern',S)
    [t,u]=ode15s(@pde_1,tout,u0,options);
  end
```

The IC from eq. (9.2) with $t = 0$ is then defined by a call to function `inital_1.m`. The integration of the 101 ODEs by `ode45` (nonstiff) or `ode15s` (stiff) follows after the IC is defined.

- The 31 solution values from the ODE integration provide a smooth 3D plot from `surf`. However, they produce an excessive number of curves for the 2D plot from `plot`. Therefore, the solution for only $t = 0, 20, 40, 60$ is stored in `uplot` for the 2D plot; the corresponding four analytical solutions are put in array `u_anal`, and the difference between the numerical and analytical solutions is put in `err`.

```
%
% Store analytical solution, errors in numerical solution
  iplot=0;
  for it=1:nout
    u(it,1)=ua_1(x(1),t(it));
    u(it,n)=ua_1(x(n),t(it));
    if((it-1)*(it-11)*(it-21)*(it-31)==0)
      iplot=iplot+1;
      fprintf('\n      t          x          u(it,i)    u_anal(it,i)
                err(it,i)\n');
      for i=1:n
        tplot(iplot)=t(it);
        uplot(iplot,i)=u(it,i);
        u_anal(iplot,i)=ua_1(x(i),t(it));
        err(iplot,i)=uplot(iplot,i)-u_anal(iplot,i);
        fprintf('%6.2f%8.3f%15.6f%15.6f%15.6f\n',...
          tplot(iplot),x(i),uplot(iplot,i),u_anal(iplot,i),err(iplot,i));
      end
    end
  end
  fprintf('    ncall = %4d\n\n',ncall);
```

Note that the four output points from the total of 31 are selected by

```
if((it-1)*(it-11)*(it-21)*(it-31)==0)
```

The counter for the calls to `pde_1.m`, `ncall`, is displayed at the end ($t = 60$) to give an indication of the total computational effort.

- The 2D plot (with four solution curves) and the 3D plot (with 31 solution curves) are produced by calls to `plot` and `surf`, respectively.

```
%
% Plot numerical and analytical solutions
  figure(2)
  plot(x,uplot,'-',x,u_anal,'o')
  xlabel('x')
  ylabel('u(x,t)')
  title('Fitzhugh-Nagumo equation; t = 0, 20, 40, 60; solid - numerical;
         o - analytical')
  figure(3)
  surf(x,t,u)
  xlabel('x'); ylabel('t'); zlabel('u(x,t)');
  title('Fitzhugh-Nagumo equation');
  view(47,20); axis tight
  shading interp
  colormap cool
```

The appearance of the 3D plot is enhanced by calls to four Matlab routines (`view, axis, shading, colormap`).

The main program produces the same three figures and tabulated output as in Chapters 2–8, which are now reviewed. The Jacobian matrix routine `jpattern_num_1.m` is the same as `jpattern_num_1.m` in Chapters 2–8 and is therefore not reproduced here.

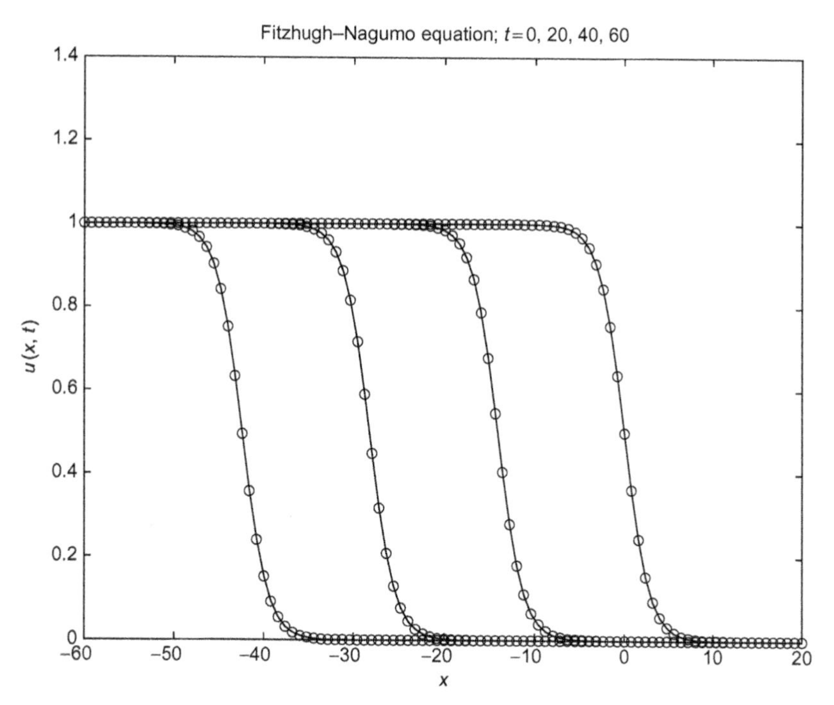

FIGURE 9.1: Numerical solution to eq. (9.1) (lines) with the analytical solution superimposed (circles) using five-point FD approximations in `dss004` for `ncase=1`.

Figure 9.1 indicates good agreement between the analytical and numerical solutions (the successive curves are right to left for $t = 0, 20, 40, 60$). Also, the traveling wave characteristic of eq. (9.2) is clear (the solution is merely displaced along the x axis). The boundary points $x = -60, 20$ are large enough that the traveling waves does not reach the boundaries for $t \leq 60$. As a cautionary note, if the wave did reach a boundary, some unexpected numerical effects might occur that would require additional analysis for an acceptable resolution of the solution ([12]).

Figure 9.2 is a 3D plot of the numerical solution. The map of the ODE Jacobian matrix, Fig. 9.3, reflects the banded structure of the ODEs produced by `dss004`. Also, the bandwidth is 9 and not 5 as might be expected from the five-point FDs in `dss004`. This greater bandwidth is due to the repeated use of `dss004` in `pde_1.m` to compute `uxx` from `u` by stagewise differentiation.

The tabular analytical and numerical solutions given in Table 9.1 also reflect the good agreement between these two solutions. The solution near the end points $x = -60, 20$ and around the front (from approximately $u = 0.01$ to $u = 0.99$) is retained in Table 9.1. The computational effort reflected in `ncall = 536` is modest.

The analytical and numerical solutions agree exactly at $t = 0$ as expected (from the use of `ua_1.m` in `inital_1.m`). The agreement between the analytical and numerical solutions of approximately four figures (at $t = 60$) is due in part to the smoothness of the solutions as analytical solution of eq. eq. (9.2) reflected in Fig. 9.1. This smoothness is perhaps unexpected considering the exponential in the analytical solution of eq. (9.2).

Table 9.1: Portion of the numerical and analytical solutions for `ncase=1`

t	x	u(it,i)	u_anal(it,i)	err(it,i)
0.00	-60.0	1.000000	1.000000	0.000000
0.00	-59.2	1.000000	1.000000	0.000000
0.00	-58.4	1.000000	1.000000	0.000000
0.00	-57.6	1.000000	1.000000	0.000000
0.00	-56.8	1.000000	1.000000	0.000000
	.		.	
	.		.	
	.		.	
0.00	-6.4	0.989287	0.989287	0.000000
0.00	-5.6	0.981290	0.981290	0.000000
0.00	-4.8	0.967520	0.967520	0.000000
0.00	-4.0	0.944193	0.944193	0.000000
0.00	-3.2	0.905744	0.905744	0.000000
0.00	-2.4	0.845150	0.845150	0.000000
0.00	-1.6	0.756092	0.756092	0.000000
0.00	-0.8	0.637767	0.637767	0.000000
0.00	0.0	0.500000	0.500000	0.000000
0.00	0.8	0.362233	0.362233	0.000000
0.00	1.6	0.243908	0.243908	0.000000
0.00	2.4	0.154850	0.154850	0.000000
0.00	3.2	0.094256	0.094256	0.000000
0.00	4.0	0.055807	0.055807	0.000000
0.00	4.8	0.032480	0.032480	0.000000
0.00	5.6	0.018710	0.018710	0.000000
0.00	6.4	0.010713	0.010713	0.000000
	.		.	
	.		.	
	.		.	
0.00	16.8	0.000007	0.000007	0.000000
0.00	17.6	0.000004	0.000004	0.000000
0.00	18.4	0.000002	0.000002	0.000000
0.00	19.2	0.000001	0.000001	0.000000
0.00	20.0	0.000001	0.000001	0.000000
t	x	u(it,i)	u_anal(it,i)	err(it,i)
20.00	-60.0	1.000000	1.000000	0.000000
20.00	-59.2	1.000000	1.000000	0.000000
20.00	-58.4	1.000000	1.000000	0.000000
20.00	-57.6	1.000000	1.000000	0.000000
20.00	-56.8	1.000000	1.000000	-0.000000
	.		.	
	.		.	
	.		.	
20.00	-20.0	0.984345	0.984360	-0.000014
20.00	-19.2	0.972730	0.972786	-0.000057

Table 9.1: (*Continued*)

t	x	u(it,i)	u_anal(it,i)	err(it,i)
20.00	−18.4	0.952968	0.953058	−0.000091
20.00	−17.6	0.920176	0.920201	−0.000025
20.00	−16.8	0.867859	0.867541	0.000317
20.00	−16.0	0.789200	0.788133	0.001067
20.00	−15.2	0.680686	0.678748	0.001938
20.00	−14.4	0.547600	0.545459	0.002142
20.00	−13.6	0.406753	0.405320	0.001434
20.00	−12.8	0.279836	0.279079	0.000756
20.00	−12.0	0.180804	0.180241	0.000564
20.00	−11.2	0.111505	0.111016	0.000489
20.00	−10.4	0.066618	0.066230	0.000388
20.00	−9.6	0.038988	0.038725	0.000263
20.00	−8.8	0.022527	0.022369	0.000158
20.00	−8.0	0.012916	0.012829	0.000087
	.		.	
	.		.	
	.		.	
20.00	16.8	0.000000	0.000000	−0.000000
20.00	17.6	0.000000	0.000000	−0.000000
20.00	18.4	0.000000	0.000000	−0.000000
20.00	19.2	0.000000	0.000000	−0.000000
20.00	20.0	0.000000	0.000000	0.000000
t	x	u(it,i)	u_anal(it,i)	err(it,i)
40.00	−60.0	1.000000	1.000000	0.000000
40.00	−59.2	1.000000	1.000000	0.000000
40.00	−58.4	1.000000	1.000000	0.000000
40.00	−57.6	1.000000	1.000000	0.000000
40.00	−56.8	1.000000	1.000000	0.000000
	.		.	
	.		.	
	.		.	
40.00	−34.4	0.986987	0.986932	0.000055
40.00	−33.6	0.977279	0.977219	0.000060
40.00	−32.8	0.960662	0.960574	0.000089
40.00	−32.0	0.932825	0.932605	0.000220
40.00	−31.2	0.887747	0.887128	0.000619
40.00	−30.4	0.818447	0.816984	0.001463
40.00	−29.6	0.719760	0.717148	0.002611
40.00	−28.8	0.593474	0.590172	0.003302
40.00	−28.0	0.452776	0.449916	0.002860
40.00	−27.2	0.319111	0.317194	0.001917
40.00	−26.4	0.210123	0.208766	0.001357
40.00	−25.6	0.131366	0.130327	0.001038
40.00	−24.8	0.079197	0.078439	0.000758
40.00	−24.0	0.046621	0.046114	0.000508

(*Continued*)

Table 9.1: *(Continued)*

t	x	u(it,i)	u_anal(it,i)	err(it,i)
40.00	−23.2	0.027032	0.026724	0.000309
40.00	−22.4	0.015532	0.015355	0.000177
	.		.	
	.		.	
	.		.	
40.00	16.8	0.000000	0.000000	−0.000000
40.00	17.6	0.000000	0.000000	−0.000000
40.00	18.4	0.000000	0.000000	−0.000000
40.00	19.2	0.000000	0.000000	−0.000000
40.00	20.0	0.000000	0.000000	0.000000

t	x	u(it,i)	u_anal(it,i)	err(it,i)
60.00	−60.0	0.999996	0.999996	0.000000
60.00	−59.2	0.999993	0.999993	0.000000
60.00	−58.4	0.999988	0.999988	0.000001
60.00	−57.6	0.999979	0.999978	0.000001
60.00	−56.8	0.999963	0.999961	0.000002
	.		.	
	.		.	
	.		.	
60.00	−48.8	0.989190	0.989087	0.000103
60.00	−48.0	0.981087	0.980944	0.000143
60.00	−47.2	0.967147	0.966928	0.000219
60.00	−46.4	0.943602	0.943201	0.000402
60.00	−45.6	0.904979	0.904138	0.000841
60.00	−44.8	0.844429	0.842690	0.001738
60.00	−44.0	0.755718	0.752632	0.003086
60.00	−43.2	0.637691	0.633442	0.004249
60.00	−42.4	0.499598	0.495332	0.004266
60.00	−41.6	0.361191	0.357930	0.003261
60.00	−40.8	0.242814	0.240481	0.002333
60.00	−40.0	0.154152	0.152422	0.001730
60.00	−39.2	0.093908	0.092674	0.001234
60.00	−38.4	0.055656	0.054831	0.000825
60.00	−37.6	0.032407	0.031898	0.000509
60.00	−36.8	0.018667	0.018370	0.000297
60.00	−36.0	0.010683	0.010517	0.000166
	.		.	
	.		.	
	.		.	
60.00	16.8	0.000000	0.000000	0.000000
60.00	17.6	0.000000	0.000000	−0.000000
60.00	18.4	0.000000	0.000000	−0.000000
60.00	19.2	0.000000	0.000000	−0.000000
60.00	20.0	0.000000	0.000000	0.000000

ncall=536

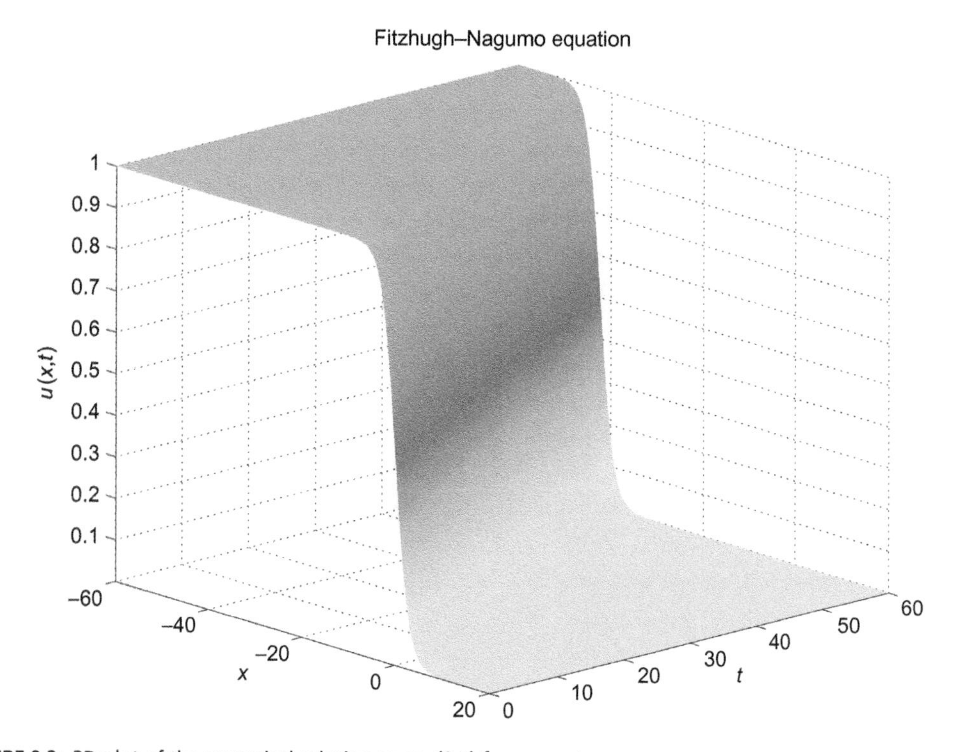

FIGURE 9.2: 3D plot of the numerical solution to eq. (9.1) for `ncase=1`.

For Case 2, D is reduced to 0.1 so that the variation of the solution with x is more pronounced, as reflected in the following plotted and tabular solutions (Figs. 9.4, 9.5, Table 9.2).

Figure 9.4 indicates good agreement between the analytical and numerical solutions (the successive curves are right to left for $t = 0, 20, 40, 60$), even with the considerably steeper moving front (compare Figs. 9.1 and 9.4). Figure 9.5 is a 3D plot of the numerical solution, which also reflects the considerably steeper moving front (compare Figs. 9.2 and 9.5).

The map of the ODE Jacobian matrix (not reproduced here) is the same as in Fig. 9.3 as expected, since the only change is in D from 1 to 0.1 (the ODE structure does not change).

The tabular analytical and numerical solutions given in Table 9.2 also reflect the good agreement between these two solutions. The computational effort reflected in `ncall = 727` is still modest, even though the error is rather large for certain points along the solution; for example, at $t = 60$ (from Table 9.2).

```
60.00   -13.6         0.736960        0.601214        0.135746
```

which is due to the relatively steep moving front as explained subsequently.

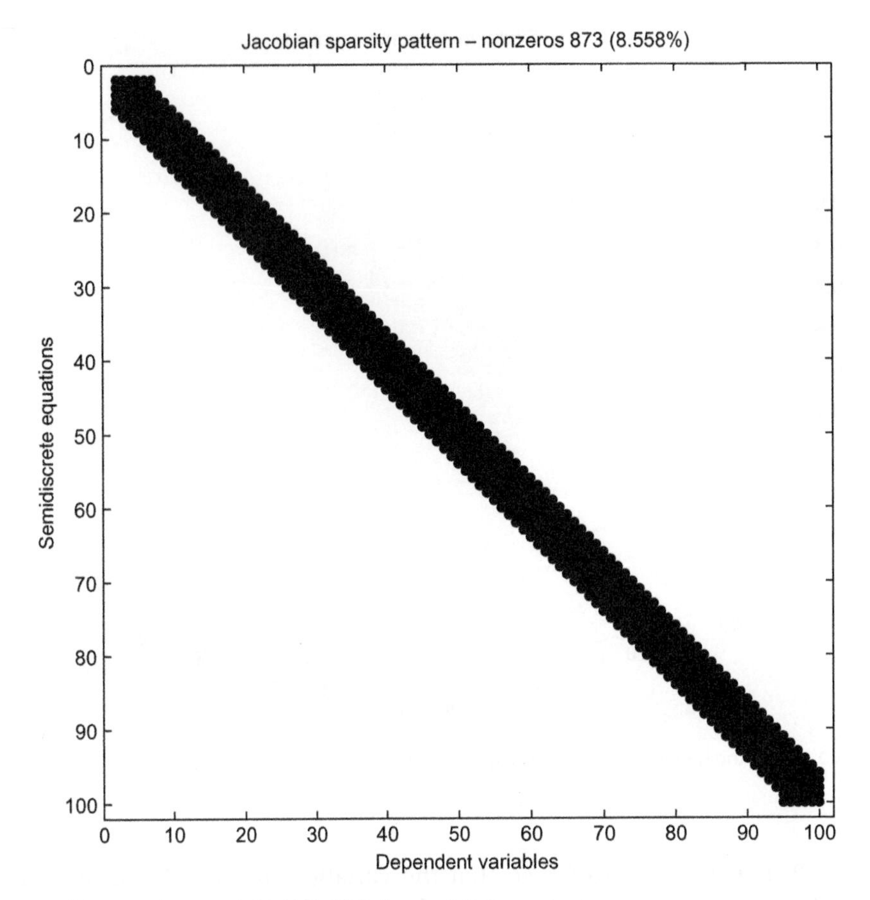

FIGURE 9.3: Jacobian matrix map of the MOL ODEs for $n = 101$ for `ncase=1`.

We can note the following points about Fig. 9.4:

- The solution is changing rapidly (is strongly vertical at the moving front) so that the occasional large errors do not distort the plot substantially.
- However, the maximum error along the moving front is increasing with t, as reflected in the three successive plots for $t = 20, 40, 60$, which implies that the numerical solution might fail for $t > 60$.
- The sharp front is marginally resolved with $n = 101$ so that for $D < 0.1$ (steeper front), more points in x might be required. In other words, the adequacy of a particular spatial grid is dependent on the particular PDE problem conditions (in this case, value of D).
- Thus, the selection of the number of spatial points is typically determined by a trial-and-error process. An alternative is to use an *adaptive grid*, termed *adaptive mesh refinement*, AMR, in which the number of points is adjusted by the numerical algorithm to adequately resolve the solution. Although AMR methods are well developed [14], they are inherently more complicated than fixed grid methods (such as

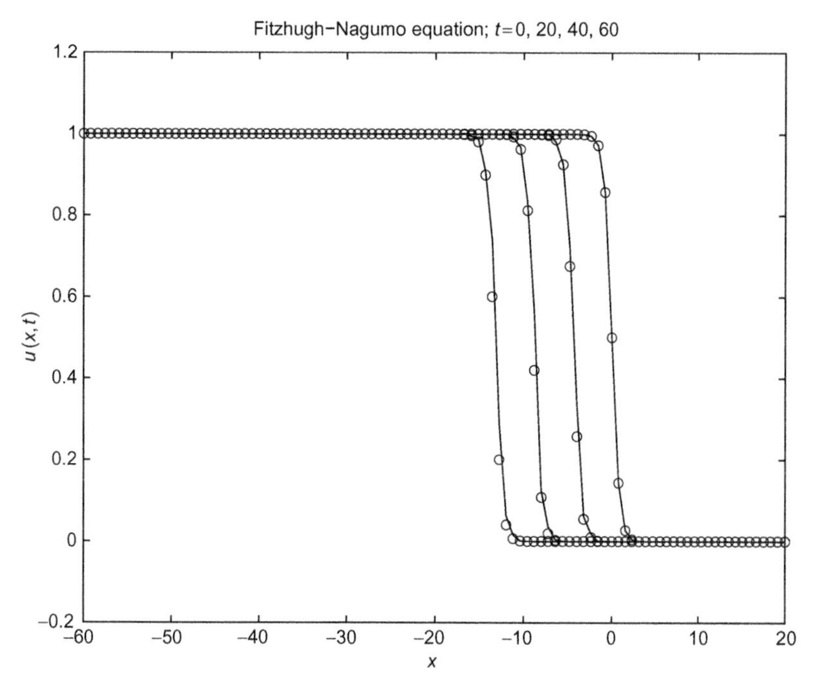

FIGURE 9.4: Numerical solution to eq. (9.1) (lines) with the analytical solution superimposed (circles) using five-point FD approximations in `dss004` for `ncase=2`.

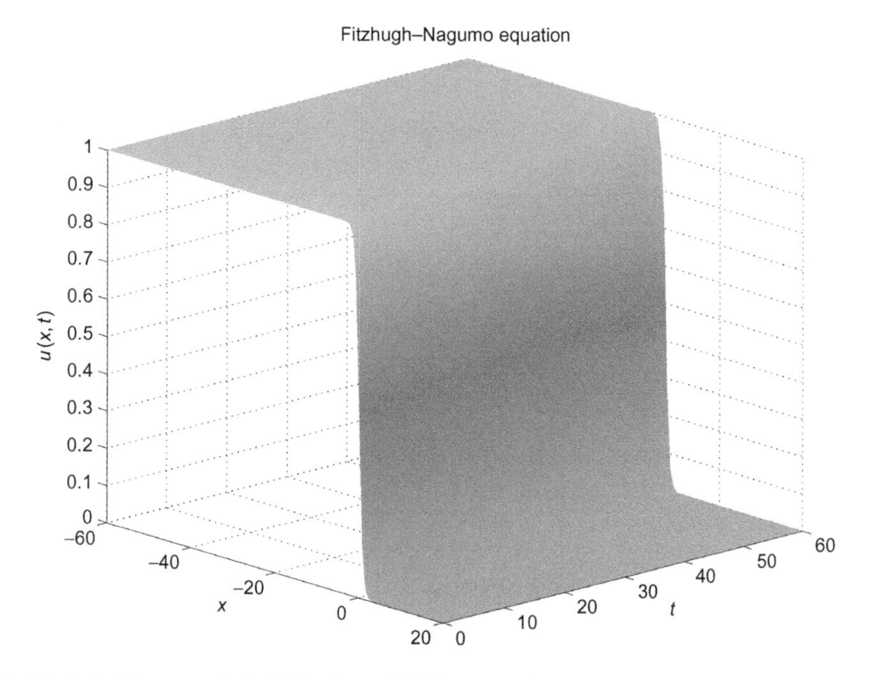

FIGURE 9.5: 3D plot of the numerical solution to eq. (9.1) for `ncase=2`.

Table 9.2: Abbreviated tabular numerical and analytical solutions for ncase=2

t	x	u(it,i)	u_anal(it,i)	err(it,i)
0.00	−60.0	1.000000	1.000000	0.000000
0.00	−59.2	1.000000	1.000000	0.000000
0.00	−58.4	1.000000	1.000000	0.000000
0.00	−57.6	1.000000	1.000000	0.000000
0.00	−56.8	1.000000	1.000000	0.000000
	.		.	
	.		.	
	.		.	
0.00	−2.4	0.995352	0.995352	0.000000
0.00	−1.6	0.972820	0.972820	0.000000
0.00	−0.8	0.856787	0.856787	0.000000
0.00	0.0	0.500000	0.500000	0.000000
0.00	0.8	0.143213	0.143213	0.000000
0.00	1.6	0.027180	0.027180	0.000000
0.00	2.4	0.004648	0.004648	0.000000
	.		.	
	.		.	
	.		.	
0.00	16.8	0.000000	0.000000	0.000000
0.00	17.6	0.000000	0.000000	0.000000
0.00	18.4	0.000000	0.000000	0.000000
0.00	19.2	0.000000	0.000000	0.000000
0.00	20.0	0.000000	0.000000	0.000000
t	x	u(it,i)	u_anal(it,i)	err(it,i)
20.00	−60.0	1.000000	1.000000	0.000000
20.00	−59.2	1.000000	1.000000	0.000000
20.00	−58.4	1.000000	1.000000	0.000000
20.00	−57.6	1.000000	1.000000	0.000000
20.00	−56.8	1.000000	1.000000	0.000000
	.		.	
	.		.	
	.		.	
20.00	−7.2	1.000052	0.997761	0.002290
20.00	−6.4	0.981269	0.986755	−0.005487
20.00	−5.6	0.928172	0.925668	0.002504
20.00	−4.8	0.722641	0.675491	0.047150
20.00	−4.0	0.335632	0.258127	0.077505
20.00	−3.2	0.060646	0.054962	0.005684
20.00	−2.4	0.019506	0.009628	0.009878
	.		.	
	.		.	
	.		.	
20.00	16.8	0.000000	0.000000	0.000000
20.00	17.6	−0.000000	0.000000	−0.000000
20.00	18.4	0.000000	0.000000	0.000000
20.00	19.2	−0.000000	0.000000	−0.000000
20.00	20.0	0.000000	0.000000	0.000000

Table 9.2: *(Continued)*

t	x	u(it,i)	u_anal(it,i)	err(it,i)
40.00	−60.0	1.000000	1.000000	0.000000
40.00	−59.2	1.000000	1.000000	0.000000
40.00	−58.4	1.000000	1.000000	0.000000
40.00	−57.6	1.000000	1.000000	0.000000
40.00	−56.8	1.000000	1.000000	0.000000
	.		.	
	.		.	
	.		.	
40.00	−11.2	0.990736	0.993593	−0.002857
40.00	−10.4	0.968699	0.962856	0.005843
40.00	−9.6	0.835951	0.812487	0.023463
40.00	−8.8	0.565642	0.420042	0.145600
40.00	−8.0	0.124694	0.107988	0.016706
40.00	−7.2	0.036314	0.019834	0.016479
40.00	−6.4	0.004377	0.003371	0.001006
	.		.	
	.		.	
	.		.	
40.00	16.8	0.000000	0.000000	0.000000
40.00	17.6	−0.000000	0.000000	−0.000000
40.00	18.4	0.000000	0.000000	0.000000
40.00	19.2	−0.000000	0.000000	−0.000000
40.00	20.0	0.000000	0.000000	0.000000

t	x	u(it,i)	u_anal(it,i)	err(it,i)
60.00	−60.0	1.000000	1.000000	0.000000
60.00	−59.2	1.000000	1.000000	0.000000
60.00	−58.4	1.000000	1.000000	0.000000
60.00	−57.6	1.000000	1.000000	0.000000
60.00	−56.8	1.000000	1.000000	0.000000
	.		.	
	.		.	
	.		.	
60.00	−16.0	0.995499	0.996912	−0.001413
60.00	−15.2	0.988144	0.981805	0.006339
60.00	−14.4	0.910405	0.900194	0.010212
60.00	−13.6	0.736960	0.601214	0.135746
60.00	−12.8	0.285822	0.201277	0.084545
60.00	−12.0	0.063044	0.040419	0.022625
60.00	−11.2	0.015916	0.006992	0.008924
	.		.	
	.		.	
	.		.	
60.00	16.8	0.000000	0.000000	0.000000
60.00	17.6	−0.000000	0.000000	−0.000000
60.00	18.4	0.000000	0.000000	0.000000
60.00	19.2	−0.000000	0.000000	−0.000000
60.00	20.0	0.000000	0.000000	0.000000

ncall=727

the use of $n = 101$ points in the present example), and they typically require tuning of the AMR algorithm parameters to arrive at an acceptable numerical solution; however, once an adaptive grid is operational, it typically will require fewer grid points than a fixed grid method, particularly in the resolution of steep moving fronts, and it can therefore result in a substantially reduced computational effort.

We note in eq. (9.2) that x and t appear in a linear combination, $\dfrac{x}{\sqrt{2D}} + \left(a - \dfrac{1}{2}\right)t$, so that eq. (9.2) represents a traveling wave solution. If we consider the *Lagrangian* variable to be $k(x - vt)$ where k and v are the *wavenumber* and *wave velocity*, respectively, then $k = \dfrac{1}{(2D)^{0.5}}$ and $v = -(2D)^{0.5}\left(a - \dfrac{1}{2}\right)$. For Case 2, with $a = 1, D = 0.1$, and $v = -((2)(0.1))^{0.5}\left(1 - \dfrac{1}{2}\right) = -0.2236$.

The velocity can also be estimated from the numerical output of Table 9.2. For example, if we estimate (by linear interpolation) the value of x at which the numerical solution is 0.5 at $t = 60$, then $x = -13.6 + (-12.8 - (-13.6))(0.5 - 0.736960)/(0.285822 - 0.736960) = -13.1798$. Thus, the velocity (for movement of the point $u(x,t) = 0.5$) is approximately (using also the value $u(x = 0, t = 0) = 0.5$ from Table 9.2) $v \approx \dfrac{\Delta x}{\Delta t} = (-13.1798 - 0)/(60 - 0) = -0.2197$, which agrees to $(0.2197 - 0.2236)/0.2236 \times 100 = -1.74\%$ with the value of -0.2236 from eq. (9.2) (recall again the second calculation is based on the part of the solution that is changing most rapidly, and it therefore has the greatest error).

In summary, the solution of eq. (9.1) subject to the IC from eq. (9.2) (with $t = 0$) and two Dirichlet BCs from eq. (9.2) (with $x = -60, 20$) is straightforward. Also, eq. (9.1) is nonlinear, yet the programming in `pde_1.m` is straightforward. Consequently, variations in the PDE can easily be made for cases for which an analytical solution might not be available. This example also demonstrates the possible sensitivity of a PDE solution to embedded parameters (such as D).

Analytical solutions, for example, eq. (9.2), can be derived by procedures such as the *exp* and *tanh* methods, particularly through the use of a *computer algebra system (CAS)* that also facilitates the verification of analytical solutions from the literature. This approach based on the CAS system Maple is demonstrated in the following Appendix.

Appendix

We conclude this chapter by first returning to the coupled two PDE systems at the beginning of the chapter, and then an example illustrating how to obtain the analytical traveling wave solution to the single Fitzhugh–Nagumo equation used in the numerical simulation.

The Coupled Fitzhugh–Nagumo Equations

The coupled diffusive Fitzhugh–Nagumo equations are classified as *reaction–diffusion equations* and are a simplified version of the more complex *Hodgkin–Huxley model* that describes the dynamics of the voltage $V(x,t)$ across a nerve cell [13]. They are derived from the Fitzhugh–Nagumo ordinary differential equations [7, 15]

$$\frac{dV}{d\tau} = 10\left(V - \frac{V^3}{3} - R\right)$$

$$\frac{dR}{d\tau} = 0.08\,(-R + 1.25V + 1.5)$$

where V represents the neural excitation voltage and R represents a recovery variable. By setting the time derivatives to zero, the steady-state values for the state variables are found to be $V = -(3/2)$ and $R = -(3/8)$. However, as it is more usual to work in *normalized units,* we choose to use the following form of transformed equations

$$\frac{dv}{dt} = v\,(v-1)\,(\mu - v) - r$$

$$\frac{dr}{dt} = \epsilon\,(v - \gamma r)$$

where μ represents *excitation threshold*, ϵ *excitability*, and γ a *model parameter*. Again, by setting the time derivatives to zero, the steady-state values for the state variables are found to be $v = 0$ and $r = 0$.

In order to model waves that travel along an excitable medium, we introduce the *spatially distributed* Fitzhugh–Nagumo equations, where a diffusion term has been incorporated into the above ODEs and which also include an external stimulus I, i.e.,

$$\frac{\partial v}{\partial t} = D\frac{\partial^2 v}{\partial x^2} + v\,(v-1)\,(a - v) - r + I \tag{9.3a}$$

$$\frac{\partial r}{\partial t} = \epsilon\,(v - \gamma r) \tag{9.3b}$$

The stimulus I represents an external current applied to an *axon* (or nerve fiber) and D represents electrical diffusivity of the axon. In mammals, this process is controlled chemically by, primarily, a combination of positively charged ions of *sodium* (Na^+) and *potassium* (K^+) and, to a lesser extent by positively charged ions of *calcium* (Ca^{2+}) and negatively charged ions of *chloride* (Cl^-), which pass through ion channels in the surrounding insulating *myelin membrane* of the axon. If the stimulus reaches a certain threshold level, the result is an impulse that propagates at speeds up to 100 m/s, which is considerably slower than the speed of electricity flowing along a wire. This is because, within a wire, electrical charge is carried by electrons that are very much smaller than the ions, which are the charge carriers within an axon. On reaching a synapse, the impulses can trigger the associated neuron to fire, resulting in an infusion of positive ions into its axon and another propagating impulse, and so on. For a full description of the associated biophysical processes, refer to [2] and for electrical analogs refer to [8].

For our subsequent analysis, we take the parameter values to be [1]

$$D = 0.03, \quad a = 0.139, \quad \gamma = 2.54, \quad \epsilon = 0.008, \quad I = 0.15$$

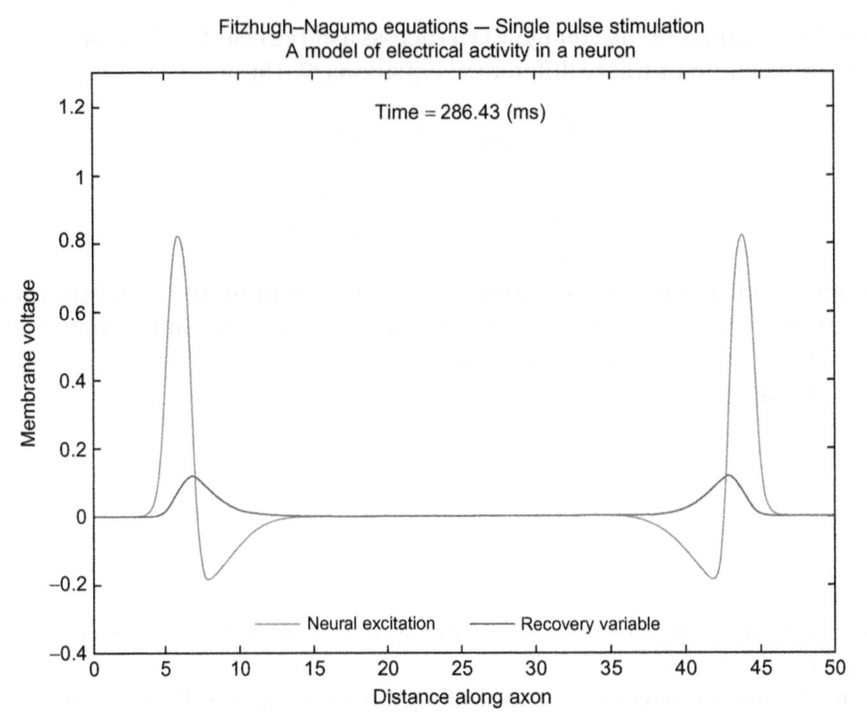

FIGURE 9.6: Solution resulting from a single-pulse stimulus applied at $x = 25$ (normalized units).

Equations (9.3a) and (9.3b) represent a nonlinear system for which an analytical solution is not available. However, we can easily calculate numerical solutions that demonstrates the general utility of the numerical approach. In this case, the solutions demonstrate rich dynamics with traveling pulses that propagate along the simulated axon. If the stimulation is applied as a single pulse, this results in two voltage spikes traveling in opposite directions that emanate from the point of stimulation (see Figs. 9.6 and 9.7).

However, if the stimulation is applied continuously, a stream of voltage spikes is generated (known as *bursting or firing*) that propagate in each direction away from the point of stimulation (see Figs. 9.8 and 9.9).

The Maple code for producing the above figures will not be discussed here, but it is included in the downloads for this chapter.

Analytical Solution of the Single Fitzhugh–Nagumo Equation

We can find the analytical solution to the single Fitzhugh–Nagumo equation, i.e.,

$$\frac{\partial u}{\partial t} = D\frac{\partial^2 u}{\partial x^2} - u(1-u)(a-u)$$

by application of the Maple procedures outlined in the main Appendix. Applying the `tanhMethod()` procedure as detailed in Listing 9.5

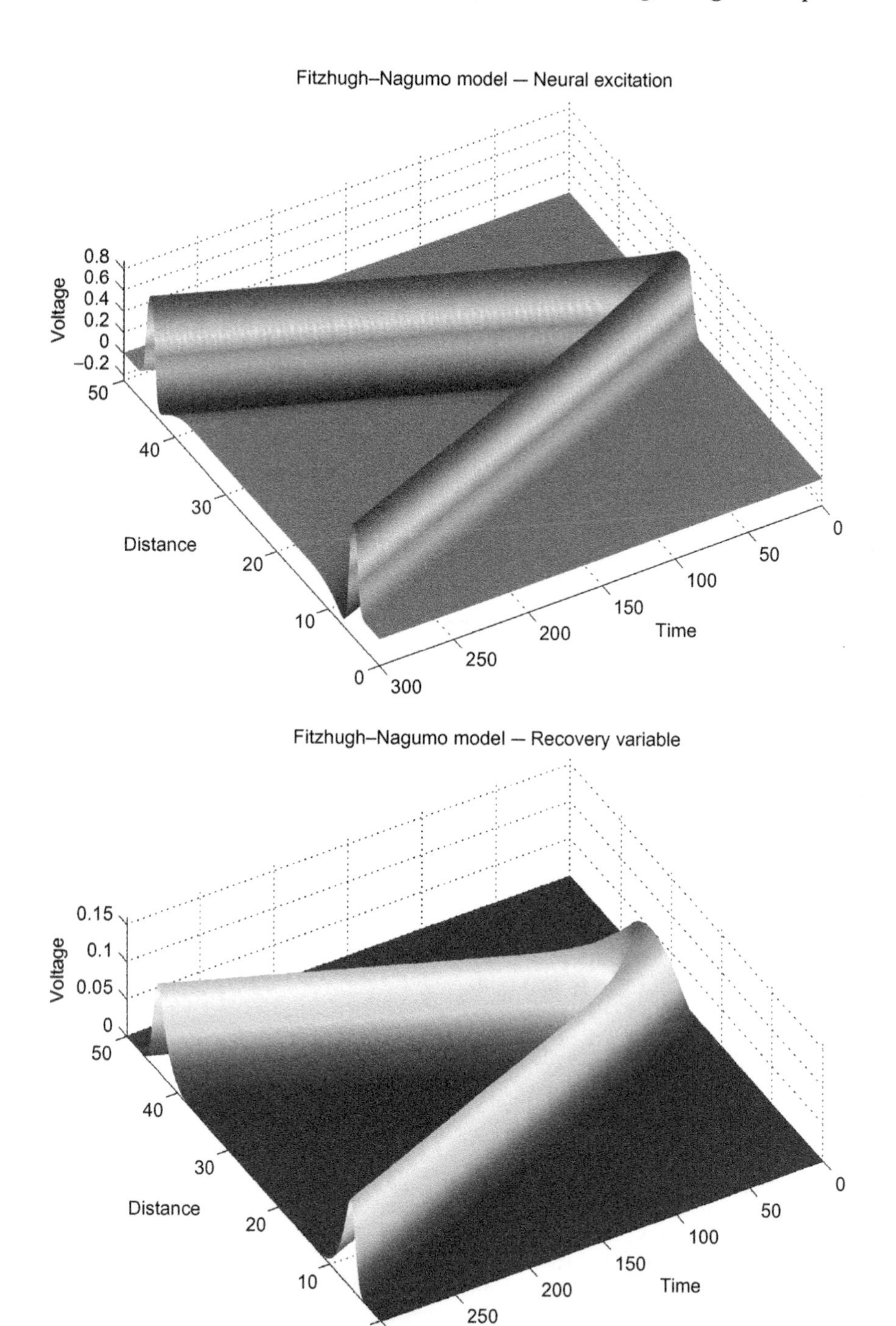

FIGURE 9.7: 3D plot resulting from a single pulse stimulus applied at $x = 25$. top: neural excitation, bottom: recovery variable (normalized units).

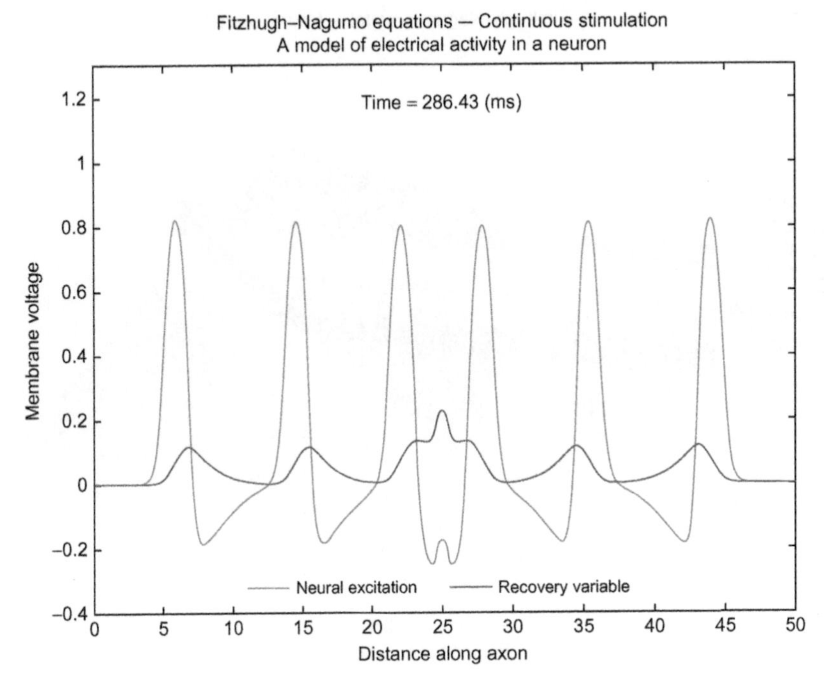

FIGURE 9.8: Solution resulting from a continuous stimulus applied at $x = 25$ (normalized units).

```
># Fitzhugh-Nagumo Equation
 # Attempt at Malfliet's tanh solution
 restart; with(PDEtools): with(PolynomialTools):
 with(plots): unprotect(D);
>alias(u=u(x,t)):
>pde1:=diff(u,t)-D*diff(u,x,x)+u*(1-u)*(a-u)=0;
>read("tanhMethod.txt");
>intFlg:=0: # No integration of U(xi) needed !
 M:=1; # Set order of approximation
 infoLevOut:=0;
 tanhMethod(M,pde1,intFlg,infoLevOut);
># Plot Case 1:
 # ===========
 D:=1.0; a:=1;x0:=0;n:=4;
 animate(rhs(sol[n]),x=-10..50,t=0..60,axes=framed,
   title="Fitzhugh-Nagumo Equation",labels=["x","u"],
   thickness=3,frames=50,numpoints=100,
   labelfont=[TIMES, ROMAN, 16],axesfont=[TIMES, ROMAN, 16],
   titlefont=[TIMES, ROMAN, 16]);
 plot3d(rhs(sol[n]),x=-10..50,t=0..60,axes='framed',
   title="Fitzhugh-Nagumo Equation",
   labels=["x","t","u(x,t)"],
   labeldirections=[HORIZONTAL,HORIZONTAL,VERTICAL],
   orientation=[116,51],grid=[100,100],
   style=patchnogrid,axesfont=[TIMES, ROMAN, 16],
   shading=Z,labelfont=[TIMES, ROMAN, 16],
   axesfont=[TIMES, ROMAN, 16], titlefont=[TIMES, ROMAN, 16]);
```

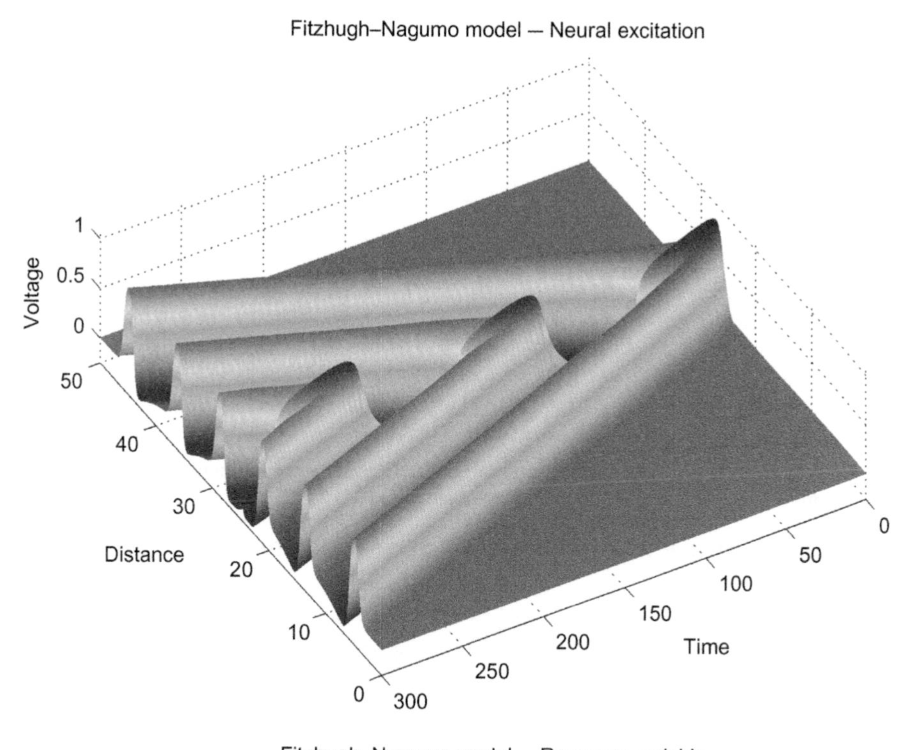

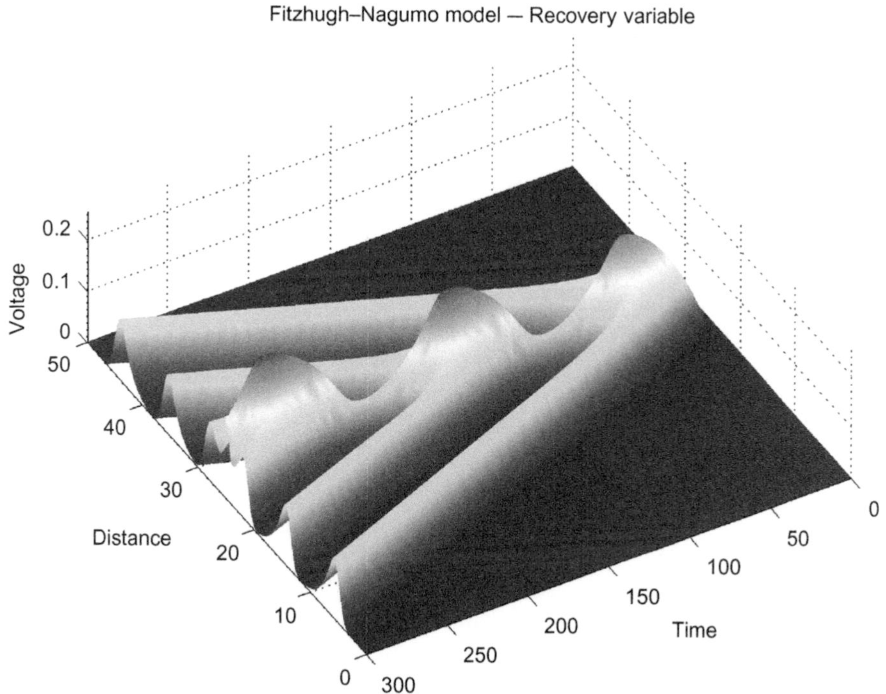

FIGURE 9.9: 3D plots resulting from a continuous stimulus applied at $x = 25$. top: neural excitation, bottom: recovery variable (normalized units).

```
># Plot Case 2:
 # ===========
 D:=0.1; a:=1;
 animate(rhs(sol[n]),x=-5..20,t=0..60,axes=framed,
   title="Fitzhugh-Nagumo Equation",labels=["x","u"],
   thickness=3,frames=50,numpoints=100,
   labelfont=[TIMES, ROMAN, 16],axesfont=[TIMES, ROMAN, 16],
   titlefont=[TIMES, ROMAN, 16]);
 plot3d(rhs(sol[n]),x=-5..20,t=0..60,axes='framed',
   title="Fitzhugh-Nagumo Equation",
   labels=["x","t","u(x,t)"],
   labeldirections=[HORIZONTAL,HORIZONTAL,VERTICAL],
   orientation=[116,51],grid=[100,100],
   style=patchnogrid,axesfont=[TIMES, ROMAN, 16],
   shading=Z,labelfont=[TIMES, ROMAN, 16],
   axesfont=[TIMES, ROMAN, 16], titlefont=[TIMES, ROMAN, 16]);
```

LISTING 9.5: Maple code to find traveling wave solutions of the Fitzhugh–Nagumo equation using the *tanh* method. Animations and 3D plots are also produced for cases 1 and 2.

produces 15 traveling wave solutions from which we select the following as an example

$$u := \frac{1}{2} - \frac{1}{2}\tanh\left(\frac{\sqrt{2}\left(2x + 2\sqrt{D}\sqrt{2}ta - \sqrt{D}\sqrt{2}t\right)}{8\sqrt{D}}\right)$$

which simplifies to

$$u = \frac{1}{2} - \frac{1}{2}\tanh\left(\frac{x}{2\sqrt{2D}} + \frac{1}{4}(2a-1)\,t\right) \tag{9.4}$$

This demonstrates the traveling wave property of the solution, where the wavenumber is given by $k = \dfrac{1}{2\sqrt{2D}}$ and the velocity by $c = \dfrac{\sqrt{2D}}{2}(2a-1)$.

Alternatively, we can apply the `expMethod()` procedure as detailed in Listing 9.6,

```
># Fitzhugh-Nagumo Equation
 # Attempt at Exp solution
 restart; with(PDEtools): with(PolynomialTools):
 with(plots):unprotect(D);
>alias(u=u(x,t)):
>pde1:=diff(u,t)-D*diff(u,x,x)+u*(1-u)*(a-u)=0;
>read("expMethod.txt");
>intFlg:=0: # No integration of U(xi) needed !
 Mn:=1; Md:=1; # Set order of approximation
 infoLevOut:=0;
 expMethod(Md,Mn,pde1,intFlg,infoLevOut);
># Plot Case 1:
 # ===========
 n:=3; x0:=0;
 zz:=rhs(sol[n]);
 D:=1.0; a:=1; _b[-1]:=1;
 animate(zz,x=-55..10,t=0..60,axes=framed,
   title="Fitzhugh-Nagumo Equation",labels=["x","u"],
```

```
    thickness=3,frames=50,numpoints=100,
    labelfont=[TIMES, ROMAN, 16],axesfont=[TIMES, ROMAN, 16],
    titlefont=[TIMES, ROMAN, 16]);
  plot3d(zz,x=-55..10,t=0..60,axes='framed',
    title="Fitzhugh-Nagumo Equation",
    labels=["x","t","u(x,t)"],
    labeldirections=[HORIZONTAL,HORIZONTAL,VERTICAL],
    orientation=[-31,33],grid=[100,100],
    style=patchnogrid,axesfont=[TIMES, ROMAN, 16],
    shading=Z,labelfont=[TIMES, ROMAN, 16],
    axesfont=[TIMES, ROMAN, 16], titlefont=[TIMES, ROMAN, 16]);
># Plot Case 2:
  # ===========
  D:=0.1; a:=1;
  animate(zz,x=-20..10,t=0..60,axes=framed,
    title="Fitzhugh-Nagumo Equation",labels=["x","u"],
    thickness=3,frames=50,numpoints=100,
    labelfont=[TIMES, ROMAN, 16],axesfont=[TIMES, ROMAN, 16],
    titlefont=[TIMES, ROMAN, 16]);
  plot3d(zz,x=-20..10,t=0..60,axes='framed',
    title="Fitzhugh-Nagumo Equation",
    labels=["x","t","u(x,t)"],
    labeldirections=[HORIZONTAL,HORIZONTAL,VERTICAL],
    orientation=[-31,33],grid=[100,100],
    style=patchnogrid,axesfont=[TIMES, ROMAN, 16],
    shading=Z,labelfont=[TIMES, ROMAN, 16],
    axesfont=[TIMES, ROMAN, 16], titlefont=[TIMES, ROMAN, 16]);
```

LISTING 9.6: Maple code to find traveling wave solutions of the Fitzhugh–Nagumo equation using the *exp* method. Animations and 3D plots are also produced for cases 1 and 2.

which produces 33 traveling wave solutions, including three trivial solutions and the following solution as outputted from Maple

$$u := \frac{\_b_{-1} e^{-\frac{\sqrt{2}\left(2x - \sqrt{d}\sqrt{2}t + 2\sqrt{d}\sqrt{2}t\alpha\right)}{8\sqrt{d}}}}{\_b_{-1} e^{-\frac{\sqrt{2}\left(2x - \sqrt{d}\sqrt{2}t + 2\sqrt{d}\sqrt{2}t\alpha\right)}{8\sqrt{d}}} + e^{\frac{\sqrt{2}\left(2x - \sqrt{d}\sqrt{2}t + 2\sqrt{d}\sqrt{2}t\alpha\right)}{8\sqrt{d}}}}$$

where $\_b_{-1}$ is an arbitrary constant. On letting $\_b_{-1} = 1$, we obtain, after simplification,

$$u = \frac{1}{1 + \exp\left[\frac{x}{\sqrt{2d}} + \left(\alpha - \frac{1}{2}\right)t\right]} \tag{9.5}$$

which is the traveling wave solution we used in the above numerical simulation (eq. (9.2)). Also, using the hyperbolic identity $\tanh(\theta) = 1 - 2/(1 + \exp(2\theta))$, we note that the *tanh* and *exp* solutions of eqns. (9.4) and (9.5) are, in fact, equivalent.

Finally, an additional Maple code that finds traveling wave solutions to the Fitzhugh–Nagumo equation using the Riccati-based method is included with the downloads for this book.

References

[1] J.G. Alford, G. Auchmuty, Rotating wave solutions of the Fitzhugh-Nagumo equations, *J. Math. Biol.* 53 (2006) 797–820.

[2] M.F. Bear, B.W. Conners, M.A. Paradiso, *Neuroscience: Exploring the Brain*, second ed., Lippincott, Williams & Wilkins, Philadelphia, 2001.

[3] R. Fitzhugh, Mathematical models of threshold phenomena in the nerve membrane, *Bull. Math. Biophysics*, 17 (1955) 257–269.

[4] J. Guckenheimer, C. Kuehn, Homoclinic orbits of the Fitzhugh-Nagumo equation: The singular-limit, discrete and continuous dynamical systems Series S, 2, no. 4 (2009) 851–872.

[5] S.P. Hastings, Some mathematical problems from neurobiology, *Am. Math. Mon.*, 82 (1975) 881–895.

[6] A.L. Hodgin, A.F. Huxley, A quantitative description of membrane current and its application to conduction and excitation in nerve, *J. Physiol.* 117 (1952) 500–505.

[7] J. Keener, J. Sneyd, *Mathematical Physiology I: Cellular Physiology*, second ed., Springer, New York, 2009.

[8] C. Koch, *Biophysics of Computation*, Oxford University Press, Oxford, UK, 1999.

[9] J.D. Murray, *Mathematical Biology II: Spatial Models and Biomedical Applications*, Third ed., Springer, New York, 2000.

[10] J. Nagumo, S. Arimoto, S. Yoshizawa, An active pulse transmission line simulating nerve axon, *Proc. IRE*, 50 (1962) 2061–2070.

[11] A.D. Polyanin, V.F. Zaitsev, *Handbook of Nonlinear Partial Differential Equations*, Chapman & Hall/CRC, Boca Raton, FL, 2004.

[12] P. Saucez, W.E. Schiesser, A. Vande Wouwer, Upwinding in the method of lines, *Math. Comp. Simul.* 56 (2001) 171–185; 59 (2002) 541.

[13] A. Scott, *Encyclopedia of Nonlinear Science*, Routledge, UK, 2005.

[14] A. Vande Wouwer, P. Saucez, W.E. Schiesser, *Adaptive Method of Lines*, Chapman & Hall/CRC, Boca Raton, FL, 2001.

[15] H.R. Wilson, *Spikes, Decisions, and Actions: The Dynamical Foundations of Neuroscience*, Oxford University Press, Oxford, UK, 1999; available on-line at `http://cvr.yorku.ca/webpages/wilson.htm`.

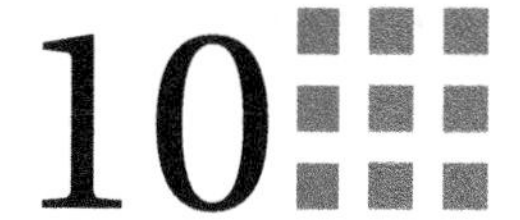

10 Kolmogorov–Petrovskii–Piskunov Equation

The *Kolmogorov–Petrovskii–Piskunov (KPP) equation* ([4], p7), eq. (10.1), also called the *Fisher–KPP equation* or just the *KPP equation*, is used in biological applications; for example, KPP is used to model tumor growth and invasion [3].

$$\frac{\partial u}{\partial t} = D\frac{\partial^2 u}{\partial x^2} + au + bu^m \tag{10.1}$$

where a, b, $m(\neq 1)$, and D are arbitrary.

An analytical solution is ([4], p7)

$$u(x,t) = \left[\beta + \exp(\lambda t + \mu x/D^{0.5})\right]^{2/(1-m)} \tag{10.2}$$

where

$$\lambda = \frac{a(1-m)(m+3)}{2(m+1)}, \tag{10.3}$$

$$\mu = \sqrt{\frac{a(1-m)^2}{2(m+1)}}, \tag{10.4}$$

$$\beta = \sqrt{-\frac{b}{a}} \tag{10.5}$$

We note that eq. (10.1) is similar to eq. (9.1), and it also clearly demonstrates a traveling wave solution as explained subsequently.

The Matlab routines closely resemble those of Chapter 9. Here, we list a few details pertaining to eqns. (10.1)–(10.5). First, the ODE routine, `pde_1.m`, is

```
  function ut=pde_1(t,u)
%
% Function pde_1 computes the t derivative vector for the Kolmogorov-
% Petrovskii-Piskunov equation
%
  global xl xu x n ncall
%
% Model parameters
  global a b m D
%
% BCs
```

```
  u(1)=ua_1(x(1),t);
  u(n)=ua_1(x(n),t);
%
% ux
  ux=dss004(xl,xu,n,u);
%
% uxx
  uxx=dss004(xl,xu,n,ux);
%
% PDE
  for i=2:n-1
    ut(i)=D*uxx(i)+a*u(i)+b*u(i)^m;
  end
  ut(1)=0;
  ut(n)=0;
  ut=ut';
%
% Increment calls to pde_1
  ncall=ncall+1;
```

LISTING 10.1: Function pde_1.m for eq. (10.1).

We can note the following points about pde_1.m:

- The function and some global parameters are first defined.

```
  function ut=pde_1(t,u)
%
% Function pde_1 computes the t derivative vector for the Kolmogorov-
% Petrovskii-Piskunov equation
%
  global xl xu x n ncall
%
% Model parameters
  global a b m D
```

- The first derivative in eq. (10.1), ux, is computed using the function dss004. Since eq. (10.1) is second order in x, the two required BCs are taken from eq. (10.2) with $xl = x(1) = -10, xu = x(n) = 10$ at grid points i=1,n, respectively (with n=101, subsequently set in the main program pde_1_main.m), and passed as global variables.

```
%
% BCs
  u(1)=ua_1(x(1),t);
  u(n)=ua_1(x(n),t);
%
% ux
  ux=dss004(xl,xu,n,u);
```

- The second derivative in eq. (10.1), uxx, is computed with dss004 by differentiating ux, so-called *stagewise differentiation*. The alternative would be to use dss044 to directly compute uxx from u, as discussed in Chapter 4.

```
%
% uxx
  uxx=dss004(xl,xu,n,ux);
```

- Equation (10.1) is then programmed.

```
%
% PDE
  for i=2:n-1
    ut(i)=D*uxx(i)+a*u(i)+b*u(i)^m;
  end
  ut(1)=0;
  ut(n)=0;
  ut=ut';
%
% Increment calls to pde_1
  ncall=ncall+1;
```

Since Dirichlet BCs are used, the derivatives in t are set to zero at the boundaries (so that the ODE integrator does not move the boundary values away from their prescribed values). A transpose is included to meet the requirements of the ODE integrator `ode15s`. Finally, the counter for the number of calls to `pde_1.m` is incremented.

The IC of eq. (10.2) (with $t = 0$) is programmed in `inital_1.m`.

```
  function u0=inital_1(t0)
%
% Function inital_1 sets the initial condition for the Kolmogorov-
% Petrovskii-Piskunov equation
%
% Parameters shared with other routines
  global xl xu x n ncall
%
% Spatial domain and initial condition
  dx=(xu-xl)/(n-1);
%
% IC from analytical solution
  for i=1:n
    x(i)=xl+(i-1)*dx;
    u0(i)=ua_1(x(i),0.0);
  end
```

LISTING 10.2: Function `inital_1.m` for IC from eq. (10.2) with $t = 0$.

We can note the following points about `inital_1.m`:

- The function and some global parameters are first defined.

```
  function u0=inital_1(t0)
%
% Function inital_1 sets the initial condition for the Kolmogorov-
% Petrovskii-Piskunov equation
%
% Parameters shared with other routines
  global xl xu x n ncall
```

- The grid in x is then defined over the interval $xl = -10 \leq x \leq xu = 10$ for 101 points (these grid parameters were selected by trial and error to produce a numerical solution with acceptable accuracy, and they are defined numerically in the main program, pde_1_main.m, discussed subsequently).

```
%
% Spatial domain and initial condition
  dx=(xu-xl)/(n-1);
%
% IC from analytical solution
  for i=1:n
    x(i)=xl+(i-1)*dx;
    u0(i)=ua_1(x(i),0.0);
  end
```

As the grid in x is defined in the for loop, function ua_1 (listed next) is called (for $t = 0$) to define the IC from eq. (10.2) (with $t = 0$).

Function ua_1.m is a straightforward implementation of the analytical solution, eq. (10.2).

```
  function uanal=ua_1(x,t)
%
% Function uanal computes the exact solution of the Kolmogorov-
% Petrovskii-Piskunov equation for comparison with the numerical
% solution
%
% Model parameters
  global a b m D lambda mu beta
%
% Analytical solution
  uanal=(beta+exp(lambda*t+mu*x/D^0.5))^(2/(1-m));
```

LISTING 10.3: Function ua_1.m for analytical solution (10.2).

The main program, pde_1_main, is similar to pde_1_main of Listing 2.1; therefore, we consider only a few specific lines of code pertaining to the parameters defined in eqns. (10.3), (10.4), and (10.5). Note that the parameters a, b, m, D, λ, μ, and β are declared global so that they can be shared with other routines.

```
                    .
                    .
                    .
%
% Parameters shared with other routines
  global xl xu x n ncall
%
% Model parameters
  global a b m D lambda mu beta
  a=1; b=-1; m=2; D=0.1; xl=-10; xu=10; n=101;
  lambda=a*(1-m)*(m+3)/(2*(m+1));
  mu=(a*(1-m)^2/(2*(m+1)))^0.5;
  beta=(-b/a)^0.5;
                    .
                    .
                    .
```

LISTING 10.4: Selected lines from the main program pde_1_main.m for eqns. (10.3), (10.4), and (10.5).

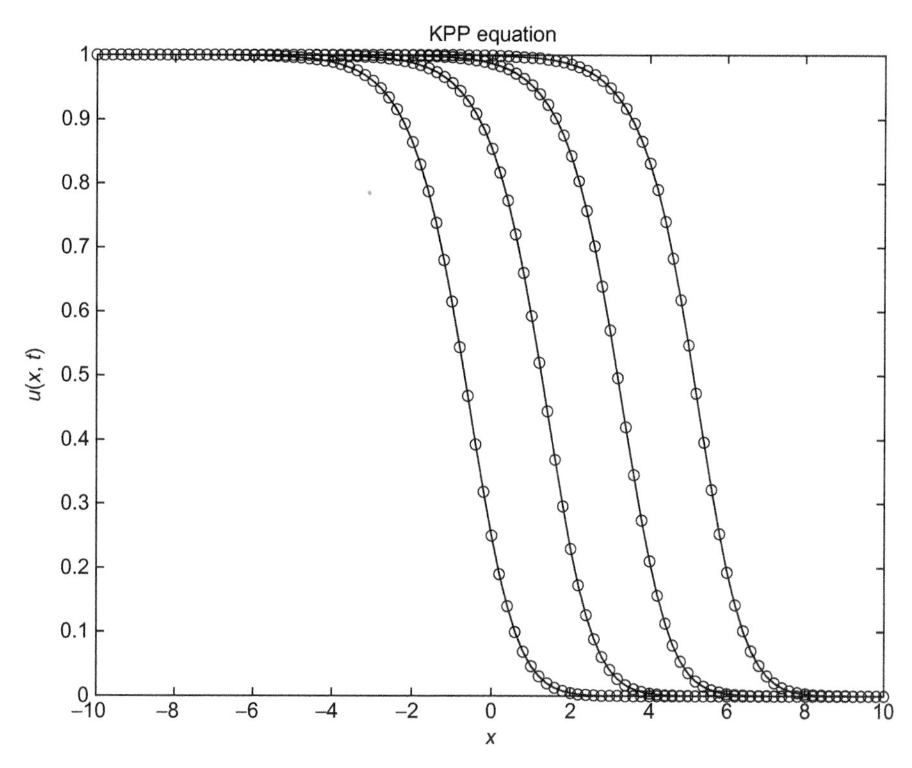

FIGURE 10.1: Numerical solution to eq. (10.1) (lines) with the analytical solution superimposed (circles) using five-point FD approximations in `dss004` for $t = 0, 3, 6, 9$ (left to right).

The main program produces the same three figures and tabulated output as in Chapters 3–9, which are now reviewed. The Jacobian matrix routine `jpattern_num_1.m` is the same as `jpattern_num_1.m` in Chapters 2 –9 and is therefore not reproduced here.

Figure 10.1 indicates good agreement between the analytical and numerical solutions. Also, the solution appears to be a traveling wave as specified by eq. (10.2) in the sense that the successive curves are displaced by a constant distance in t. Note that the curves move left to right, since from eq. (10.2), the *Lagrangian* variable is $\xi = \lambda t + \mu x/D^{0.5} = \mu/D^{0.5}(x + (D^{0.5}/\mu)\lambda t)$; in other words, the *wave velocity* is $c = -(D^{0.5}/\mu)\lambda$ (which will be confirmed in the subsequent discussion of the numerical output; note that $\lambda < 0$). Figure 10.2 is a 3D plot of the numerical solution.

The map of the ODE Jacobian matrix, Fig. 10.3, reflects the banded structure of the ODEs produced by `dss004`. The number of grid points, $n = 101$, is large enough that the individual elements of the Jacobian matrix are not distinct.

The tabular analytical and numerical solutions given in Table 10.1 also reflect the good agreement between these two solutions. The computational effort reflected in `ncall = 260` is quite modest. As required, the analytical and numerical solutions agree for $t = 0$ (since both solutions are from eq. (10.2) with $t = 0$). For $t > 0$, the agreement between the

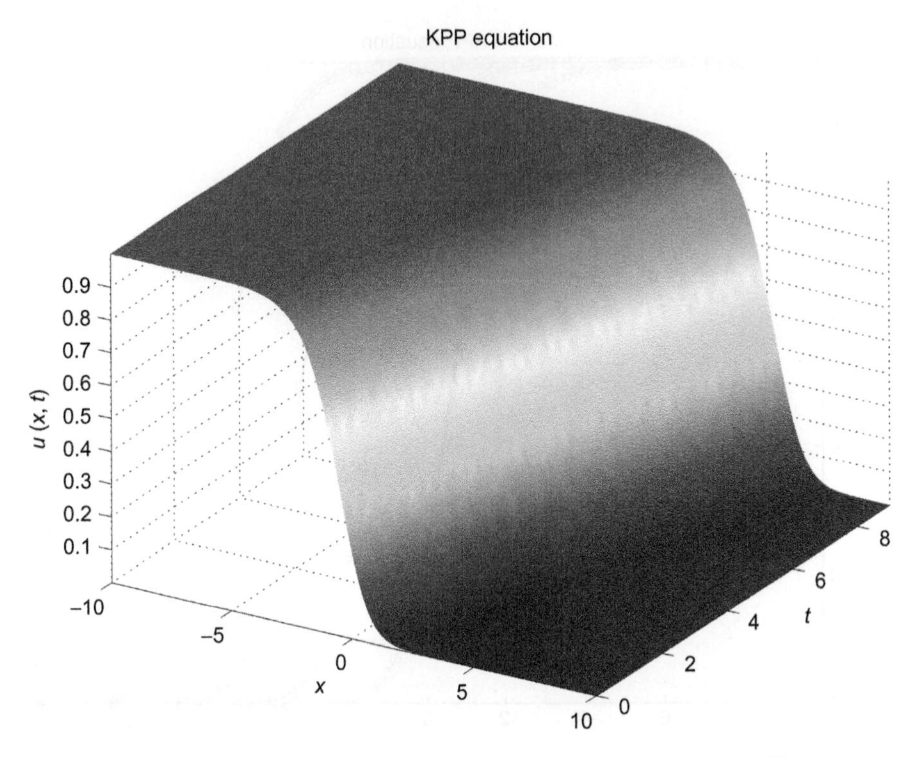

FIGURE 10.2: 3D plot of the numerical solution to eq. (10.1).

analytical and numerical solutions of approximately five figures is quite acceptable, even with only 101 grid points.

The numerical solution of Table 10.1 can be used to estimate the velocity of the traveling wave of Fig. 10.1. For example, we can determine the distance (in x) between the two solution values $u(x,t) = 0.5$ for $t = 0$ and $t = 3$. For $u(x,t = 0) = 0.5$, the value of x can be estimated by linear interpolation within the solution values of Table 10.1 at $t = 0$.

$$x = -0.80 + (-0.60 + 0.80)(0.5 - 0.543844)/(0.468560 - 0.543844) = -0.6835$$

Similarly, for $t = 3$, linear interpolation to determine x for $u(x,t = 3) = 0.5$ gives

$$x = 1.20 + (1.40 - 1.20)(0.5 - 0.520241)/(0.444285 - 0.520241) = 1.2533$$

Thus, the estimated velocity (of the value $u(x,t) = 0.5$) is

$$c = \frac{\Delta x}{\Delta t} = \frac{1.2533 - (-0.6835)}{3 - 0} = 0.6456$$

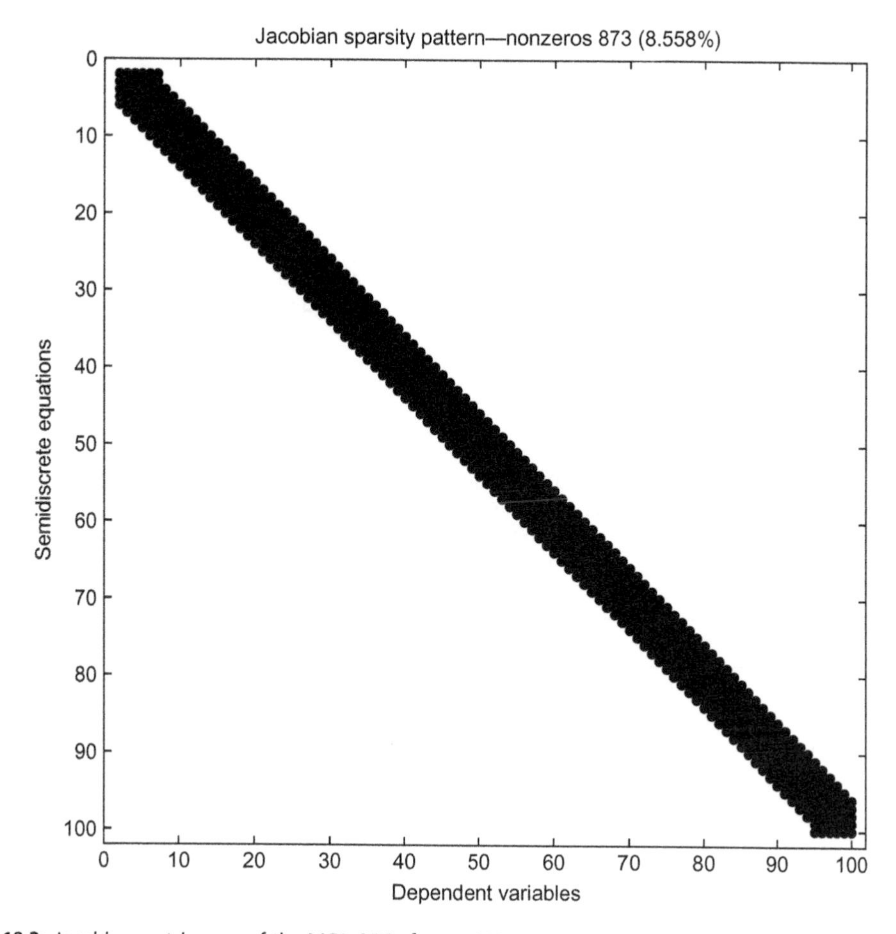

FIGURE 10.3: Jacobian matrix map of the MOL ODEs for $n = 101$.

which compares with the analytical value (with $\lambda = -0.8333, \mu = 0.4082, D = 0.1$) $c = -(D^{0.5}/\mu)\lambda = -(0.1^{0.5}/0.4082)(-0.8333) = 0.6455$ (λ and μ were computed according to eqns. (10.3) and (10.4), as programmed in Listing 10.4).

In summary, the solution of eq. (10.1) subject to the IC from eq. (10.2) (with $t = 0$) and two Dirichlet BCs from eq. (10.2) (with $xl = -10, xu = 10$) is straightforward. Also, eq. (10.1) is nonlinear, yet the programming in `pde_1.m` (Listing 10.1) is straightforward. Consequently, variations in the PDE can easily be made for cases for which an analytical solution might not be available.

Appendix

We conclude this chapter by solving the KPP equation using the *factorization method* as outlined in the main Appendix. The equation is repeated below for convenience, as we use

Table 10.1: Tabular numerical and analytical solutions

t	x	u(it,i)	u_anal(it,i)	err(it,i)
0.00	-10.000	0.999995	0.999995	0.000000
0.00	-9.800	0.999994	0.999994	0.000000
0.00	-9.600	0.999992	0.999992	0.000000
0.00	-9.400	0.999989	0.999989	0.000000
0.00	-9.200	0.999986	0.999986	0.000000
	.		.	
	.		.	
	.		.	
0.00	-1.000	0.615151	0.615151	0.000000
0.00	-0.800	0.543844	0.543844	0.000000
0.00	-0.600	0.468560	0.468560	0.000000
0.00	-0.400	0.392258	0.392258	0.000000
0.00	-0.200	0.318314	0.318314	0.000000
	.		.	
	.		.	
	.		.	
0.00	9.200	0.000000	0.000000	0.000000
0.00	9.400	0.000000	0.000000	0.000000
0.00	9.600	0.000000	0.000000	0.000000
0.00	9.800	0.000000	0.000000	0.000000
0.00	10.000	0.000000	0.000000	0.000000
t	x	u(it,i)	u_anal(it,i)	err(it,i)
3.00	-10.000	1.000000	1.000000	0.000000
3.00	-9.800	0.999999	0.999999	0.000000
3.00	-9.600	0.999999	0.999999	0.000000
3.00	-9.400	0.999999	0.999999	0.000000
3.00	-9.200	0.999999	0.999999	0.000000
	.		.	
	.		.	
	.		.	
3.00	0.800	0.660375	0.660371	0.000004
3.00	1.000	0.593090	0.593089	0.000001
3.00	1.200	0.520238	0.520241	-0.000003
3.00	1.400	0.444278	0.444285	-0.000007
3.00	1.600	0.368354	0.368359	-0.000005
	.		.	
	.		.	
	.		.	
3.00	9.200	0.000000	0.000000	-0.000000
3.00	9.400	0.000000	0.000000	-0.000000
3.00	9.600	0.000000	0.000000	-0.000000
3.00	9.800	0.000000	0.000000	-0.000000
3.00	10.000	0.000000	0.000000	0.000000

```
              .                           .
              .                           .
              .                           .
        output for t = 6, 9 removed
              .                           .
              .                           .
              .                           .
        ncall=260
```

slightly different notation than in eq. (10.1).

$$\frac{\partial u}{\partial t} - \delta \frac{\partial^2 u}{\partial x^2} - au - bu^m = 0 \tag{10.6}$$

If we assume a traveling wave solution of the form $u(x,t) = U(\xi)$, where $\xi = k(x - ct)$, $c =$ velocity, and $k =$ wavenumber, eq. (10.6) reduces to the traveling wave ODE

$$\frac{d^2 U}{dx^2} + \gamma \frac{dU}{dx} + F(U) = 0 \tag{10.7}$$

where $\gamma = \dfrac{c}{k\delta}$ and $F(U) = \dfrac{1}{k^2\delta}(aU + bU^m)$.

We factor the polynomial function of eq. (10.7) (the third-hand term) as

$$\frac{F(U)}{U} = f_1 f_2 = \frac{a}{\delta k^2}\left(1 - \sqrt{-\frac{b}{a}}U^{1/2m-1/2}\right)\left(1 + \sqrt{-\frac{b}{a}}U^{1/2m-1/2}\right)$$

and choose

$$f_1 = \sqrt{\frac{a}{\delta}}\left(1 - \sqrt{-\frac{b}{a}}U^{(m-1)/2}\right)\frac{\alpha}{k}, \quad f_2 = \sqrt{\frac{a}{\delta}}\left(1 + \sqrt{-\frac{b}{a}}U^{(m-1)/2}\right)\frac{1}{\alpha k}, \quad \alpha \neq 0 \tag{10.8}$$

where we have also introduced the constant α.

From eqns. (A.41) and (10.8), we obtain the following ODE

$$\begin{aligned}\frac{df_1}{dU}U + f_1 + f_2 = -\sqrt{-b/\delta}\frac{(m-1)}{2}U^{(m-1)/2}\frac{\alpha}{k} + \sqrt{a/\delta}\left(1 - \sqrt{-b/a}U^{(m-1)/2}\right)\frac{\alpha}{k} \\ + \sqrt{a/\delta}\left(1 + \sqrt{-b/a}U^{(m-1)/2}\right)\frac{1}{\alpha k} = -\gamma = -\frac{c}{k\delta}\end{aligned} \tag{10.9}$$

Collecting terms and equating the coefficients of $U^{(m-1)/2}$ to zero (as the left-hand side of eq. (10.9) is equal to a constant) gives $\alpha = \pm\dfrac{2}{\sqrt{2(m+1)}}$ and $c = -\sqrt{a\delta}\left(\dfrac{m+3}{\sqrt{2(m+1)}}\right)$. Also, as γ is a constant and is independent of the value of U, on setting $U = 0$, we find that $\gamma = -\dfrac{\sqrt{a/\delta}\left(\alpha + \alpha^{-1}\right)}{k}$. Therefore, adopting the grouping of Cornejo-Perez [2], i.e., eq. (A.40b), it follows that

$$\frac{d^2 U}{d\xi^2} \mp \frac{\sqrt{a/\delta}\left(\alpha + \alpha^{-1}\right)}{k}\frac{dU}{d\xi^2} + f_1 f_2 U = 0 \tag{10.10}$$

Thus, the corresponding factorization eq. (A.39), i.e., $[D - f_2(U)][D - f_1(U)]U = 0$, becomes

$$\left[D \pm \frac{1}{\alpha k}\sqrt{\frac{a}{\delta}}\left(1 + \sqrt{-\frac{b}{a}}U^{(m-1)/2}\right)\right]\left[D \mp \frac{\alpha}{k}\sqrt{\frac{a}{\delta}}\left(1 - \sqrt{-\frac{b}{a}}U^{(m-1)/2}\right)\right]U = 0 \tag{10.11}$$

where $D = \dfrac{d}{d\xi}$. Therefore, it follows that eq. (10.10) is compatible with the first-order ODE

$$\frac{dU}{d\xi} \mp \frac{\alpha}{k}\sqrt{\frac{a}{\delta}}\left(1 - \sqrt{-\frac{b}{a}}U^{(m-1)/2}\right)U = 0 \tag{10.12}$$

Integrating eq. (10.12) either manually or using Maple yields

$$U = \left(\sqrt{-\frac{b}{a}} + K\exp\left[-\frac{\sqrt{a/\delta}}{2k}\alpha(m-1)\xi\right]\right)^{-2/(m-1)}$$

where K is an arbitrary constant of integration. Substituting back values for ξ, c, and a_n, we find that $k = \sqrt{a/\delta}\left(\dfrac{m-1}{\sqrt{2(m+1)}}\right)$, which leads to the final solution

$$U = \left(\sqrt{-\frac{b}{a}} + K\exp\left[k(x-ct)\right]\right)^{-2/(m-1)} \tag{10.13}$$

where k and c are as defined above.

If we let $K = \pm\exp\left[kx_0\right]$, we arrive at the standard form of traveling wave solution given in eq. (10.2)

$$U^{\pm} = \left(\sqrt{-\frac{b}{a}} \pm \exp\left[k(x - x_0 - ct)\right]\right)^{-2/m-1} \tag{10.14}$$

The above solution with $x_0 = 0$ is the same as the analytical solution, eqns. (10.2)–(10.5), used in the numerical simulation discussed in the main body of this chapter. Maple code that performs the above calculations is included in Listing 10.5.

Readers are referred to the papers by Berkovich [1], Cornejo-Perez, and Rosu [2, 5] for more information on this method and additional examples of its use. Additional solutions to eq. (10.6) are given by Polyanin and Zaitsev [4, chapter 1].

```
># Some calculations to confirm the results of a
 # factorization solution to the KPP Equation
 # Ref: Rosu, H. C. and O. Cornejo-Perez (2008).
 #       Supersymmetric pairing of kinks for polynomial
 #       nonlinearities, [arXiv:math-ph/0401040v3 24 Dec 2004|
 #       |ArXiv: math-ph/0401040]
 restart; with(DEtools): with(PDEtools):
>alias(u=u(x,t)): alias(U=U(xi)):
># Define KPP equation
 pde1:=diff(u,t)-d*diff(u,x,x)-a*u-b*u^m=0;
># Convert PDE to ODE
 tr1:={x=(xi/k+c*tau),t=tau,u=U};
 ode1:=dchange(tr1,pde1,[xi,tau,U]);
># Define F(U), g
```

```
 F:=U*(a+b*U^(m-1))/(d*k^2);
 g:=c/(d*k);
># Factor F(U)/U - Note: alpha introduced
 f[1]:=alpha*sqrt(a/(d))*(1-sqrt(-b/a)*U^((m-1)/2))/k;
 f[2]:=sqrt(a/(d))/(alpha)*(1+sqrt(-b/a)*U^((m-1)/2))/k;
># Check that factorization is correct
 F_chk:=simplify(eval(f[1]*f[2]*U));
># Use the C-P grouping
 alias(U=U):f[1]:=subs(U(xi)=U,f[1]):f[2]:=subs(U(xi)=U,f[2]):
 eqn1:=diff(f[1],U)*U+f[1]+f[2]=-subs(U(xi)=U,g);
>sol1:=simplify(subs(U=0,eqn1),size);
 c:=solve(sol1,c);
># Collect terms in U^((m-1)/2))
 eqn2:=collect(simplify(lhs(eqn1)-rhs(eqn1),size),U,'recursive');
># Equate coeff of U to zero
 eqn3:=coeff(eqn2,U^((m-1)/2))=0;
>sol2:=solve(eqn3,alpha);
 alpha:=sol2[1];
># Fomulate 1st order ode from [D-f[1]]U=0
 f[1]:=subs(U=U(xi),f[1]);ode2:=diff(U(xi),xi)-f[1]*U(xi);
># Obtain solution to ode2
 sol4:=dsolve(ode2);
># Check solution U(xi) satisfies ode1
 odeCHK:=simplify(eval(subs({U(xi)=rhs(sol4)},rhs(ode1))),symbolic):
 if odeCHK = 0 then
   print('solution PASSES');
 else
   print('solution FAILS!');
 end if;
># Obtain solution to pde1
 sol5:=u=simplify(eval(subs({xi=k*(x-c*t)},rhs(sol4))),size);
># Check solution u satisfies pde1:
 m:=3; # Need to assign a value to m
 pdeCHK:=simplify(pdetest(sol5,pde1),symbolic);
 if pdeCHK = 0 then
   print('solution PASSES');
 else
   print('solution FAILS!');
 end if;
```

LISTING 10.5: Maple code to derive a solution to the the *KPP equation* using the *factorization method.*

Finally, additional Maple codes that obtain different solutions to the KPP equation using *tanh-*, *exp-*, and *Riccati*-based methods are included with the downloads for this book.

References

[1] L.M. Berkovich, Factorization as a method of finding exact invariant solutions of the Kolmogorov–Petrovskii–Piskunov equation and the related Semenov and Zeldovich equations, *Sov. Math. Dokl.* 45 (1992) 162–167.

[2] O. Cornejo-Perez, H.C. Rosu, Nonlinear second order ODE: Factorizations and particular solutions, *Progr. Theo. Phy.* 114(3) (2005) 533–538.

[3] A. Fasano, M.A. Herrero, M.R. Rodrigo, Slow and fast invasion waves in a model of acid-mediated tumour growth, *Math. Biosci.* 220 (2009) 45–56.

[4] A.D. Polyanin, V.F. Zaitsev, *Handbook of Nonlinear Partial Differential Equations*, Chapman & Hall/ CRC, Boca Raton, FL, 2004.

[5] H.C. Rosu, O. Cornejo-Perez, Supersymmetric pairing of kinks for polynomial nonlinearities, http://arxiv.org/PS_cache/math-ph/pdf/0401/0401040v3.pdf, Accessed 23 April, 2010.

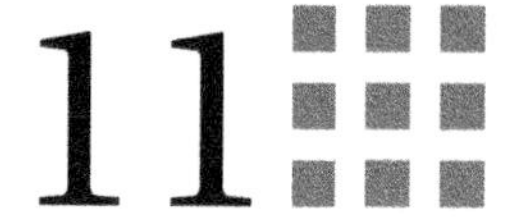

11 Kuramoto–Sivashinsky Equation

The *Kuramoto–Sivashinsky equation* ([6], p593) is

$$\frac{\partial u}{\partial t} = -u\frac{\partial u}{\partial x} - \alpha\frac{\partial^2 u}{\partial x^2} - \beta\frac{\partial^3 u}{\partial x^3} - \gamma\frac{\partial^4 u}{\partial x^4} \tag{11.1}$$

with the analytical solution [6] for the parameter values $\beta = 0, \alpha = \gamma = 1, k = \pm\sqrt{\frac{11}{19}}$.

$$u(x,t) = \frac{15}{19}k\left(11H^3 - 9H + 2\right);\; H = \tanh\left(\frac{1}{2}kx - \frac{15}{19}k^2 t\right). \tag{11.2}$$

Two *Dirichlet* BCs are available by applying eq. (11.2) with $x = -10, 20$. Two more BCs are available by differentiating eq. (11.2)

$$\frac{\partial u}{\partial x} = \frac{15}{19}k\left(33H^2\frac{\partial H}{\partial x} - 9\frac{\partial H}{\partial x}\right) \tag{11.3}$$

with

$$\frac{\partial H}{\partial x} = \frac{1}{2}k\left[1 - \tanh^2\left(\frac{1}{2}kx - \frac{15}{19}k^2 t\right)\right]$$

Equation (11.3) can then be applied as *Neumann* BCs at $x = -10, 20$.

The Matlab routines closely resemble those of Chapters 3–10. Here, we list a few details pertaining to eqns. (11.1)–(11.3). First, the ODE routine `pde_1.m` is

```
  function ut=pde_1(t,u)
%
% Function pde_1 computes the t derivative vector for the Kuramoto-
% Sivashinsky equation
%
  global xl xu x n ncall
%
% Model parameters
  global alpha beta gamma k
%
% BCs at x = 0,1
  u(1)=ua_1(x(1),t);
  u(n)=ua_1(x(n),t);
%
% ux
  ux=dss004(xl,xu,n,u);
%
```

```
% BCs at x = 0,1
  ux(1)=uax_1(x(1),t);
  ux(n)=uax_1(x(n),t);
%
% uxx
  uxx=dss004(xl,xu,n,ux);
%
% uxxx
  uxxx=dss004(xl,xu,n,uxx);
%
% uxxxx
  uxxxx=dss004(xl,xu,n,uxxx);
%
% PDE
  for i=2:n-1
    ut(i)=-u(i)*ux(i)-alpha*uxx(i)-beta*uxxx(i)-gamma*uxxxx(i);
  end
  ut(1)=0;
  ut(n)=0;
  ut=ut';
%
% Increment calls to pde_1
  ncall=ncall+1;
```

LISTING 11.1: Function pde_1.m for eq. (11.1).

We can note the following points about pde_1.m:

- The function and some global parameters are first defined.

```
  function ut=pde_1(t,u)
%
% Function pde_1 computes the t derivative vector for the Kuramoto-
% Sivashinsky equation
%
  global xl xu x n ncall
%
% Model parameters
  global alpha beta gamma k
```

- The first derivative in eq. (11.1), ux, is computed using the function dss004. Equation (11.1) is fourth order in x, and the four required BCs are taken from the analytical solution with $x = -10, 20$ corresponding to the two grid points i=1,n with n=201 set in inital_1.m. Equations (11.2) are used to define two Dirichlet BCs at $x = -10, 20$ and eqs. (11.3) are used to define two Neumann BCs at $x = -10, 20$. Both sets of BCs are programmed in ua_1.m (Dirichlet) and uax_1.m (Neumann) discussed subsequently.

```
%
% BCs at x = 0,1
  u(1)=ua_1(x(1),t);
  u(n)=ua_1(x(n),t);
%
% ux
  ux=dss004(xl,xu,n,u);
```

```
%
% BCs at x = 0,1
  ux(1)=uax_1(x(1),t);
  ux(n)=uax_1(x(n),t);
```

- The second, third, and fourth derivatives in eq. (11.1), uxx, uxxx, and uxxxx, are computed with successive calls to dss004, so-called *stagewise differentiation*.

```
%
% uxx
  uxx=dss004(xl,xu,n,ux);
%
% uxxx
  uxxx=dss004(xl,xu,n,uxx);
%
% uxxxx
  uxxxx=dss004(xl,xu,n,uxxx);
```

- Equation (11.1) is then programmed.

```
%
% PDE
  for i=2:n-1
    ut(i)=-u(i)*ux(i)-alpha*uxx(i)-beta*uxxx(i)-gamma*uxxxx(i);
  end
  ut(1)=0;
  ut(n)=0;
  ut=ut';
%
% Increment calls to pde_1
  ncall=ncall+1;
```

  Since Dirichlet BCs are used, the derivatives in t are set to zero at the boundaries (so that the ODE integrator does not move the boundary values away from their prescribed values). A transpose is included to meet the requirements of the ODE integrator ode15s. Finally, the counter for the number of calls to pde_1.m is incremented.

The IC from eq. (11.2) with $t = 0$ is programmed in inital_1.m (the IC is programmed in ua_1.m with $t = 0$).

```
  function u0=inital_1(t0)
%
% Function inital_1 sets the initial condition for the Kuramoto-
% Sivashinsky equation
%
% Parameters shared with other routines
  global xl xu x n ncall
%
% Spatial domain and initial condition
  xl=-10;
  xu= 20;
  n=201;
  dx=(xu-xl)/(n-1);
%
```

```
% IC from analytical solution
  for i=1:n
    x(i)=xl+(i-1)*dx;
    u0(i)=ua_1(x(i),0.0);
  end
```

LISTING 11.2: Function `inital_1.m` for the IC from eq. (11.2) with $t = 0$.

We can note the following points about `inital_1.m`:

- The function and some global parameters are first defined.

```
  function u0=inital_1(t0)
%
% Function inital_1 sets the initial condition for the Kuramoto-
% Sivashinsky equation
%
% Parameters shared with other routines
  global xl xu x n ncall
```

- The grid in x is then defined over the interval $-10 \leq x \leq 20$ for 201 points.

```
%
% Spatial domain and initial condition
  xl=-10;
  xu= 20;
  n=201;
  dx=(xu-xl)/(n-1);
%
% IC from analytical solution
  for i=1:n
    x(i)=xl+(i-1)*dx;
    u0(i)=ua_1(x(i),0.0);
  end
```

As the grid in x is defined in the `for` loop, function `ua_1` (listed next) is called (for $t = 0$) to define the IC from eq. (11.2) with $t = 0$.

Function `ua_1.m` is a straightforward implementation of the analytical solution, eq. (11.2).

```
  function uanal=ua_1(x,t)
%
% Function uanal computes the exact solution of the Kuramoto-
% Sivashinsky equation for comparison with the numerical solution
%
% Model parameters
  global alpha beta gamma k
%
% Analytical solution
  H=tanh(0.5*k*x-(15/19)*k^2*t);
  uanal=(15/19)*k*(11*H^3-9*H+2);
```

LISTING 11.3: Function `ua_1.m` for analytical solution (11.2).

As noted previously, eq. (11.1) is fourth order in x and therefore requires four BCs. Two Dirichlet BCs are available by applying eq. (11.2) at $x = -10, 20$. Two more (Neumann) BCs are available by differentiating eq. (11.2) with respect to x (to give eq. (11.3)). This derivative is programmed in `uax_1`.

```
  function uax=uax_1(x,t)
%
% Function uax computes the derivative (in x) of the exact solution
% of the Kuramoto-Sivashinsky equation
%
  global alpha beta gamma k
%
% Analytical solution
  arg=0.5*k*x-(15/19)*k^2*t;
  H=tanh(arg);
  Hx=(1-tanh(arg)^2)*0.5*k;
  uax=(15/19)*k*(33*H^2*Hx-9*Hx);
```

LISTING 11.4: Function `uax_1.m` for the analytical derivative (11.3).

Function `uax_1.m` is a straightforward implementation of the analytical derivative, eq. (11.3).

The main program, `pde_1_main`, is essentially the same as `pde_1_main` of Listing 2.1. A selected part of this routine follows.

```
                          .
                          .
                          .
%
% Model parameters
  global alpha beta gamma k
  alpha=1; beta=0; gamma=1; k=(11/19)^0.5;
%
% Independent variable for ODE integration
  t0=0.0;
  tf=10;
  tout=[t0:2:tf]';
  nout=6;
  ncall=0;
%
% Initial condition
  u0=inital_1(t0);
                          .
                          .
                          .
%
% Display selected output
  for it=1:nout
    fprintf('\n     t        x          u(it,i)   u_anal(it,i)      err(it,i)\n');
    for i=1:5:n
      fprintf('%6.2f%8.3f%15.6f%15.6f%15.6f\n',...
            t(it),x(i),u(it,i),u_anal(it,i),err(it,i));
    end
  end
  fprintf('    ncall = %4d\n\n',ncall);
%
% Plot numerical and analytical solutions
  figure(2)
```

```
plot(x,u,'-',x,u_anal,'o')
xlabel('x')
ylabel('u(x,t)')
title('Kuramoto-Sivashinsky equation; solid - numerical; o - analytical')
figure(3)
surf(x,t,u)
shading 'interp', axis 'tight'
xlabel('x'); ylabel('t'); zlabel('u(x,t)');
title('Kuramoto-Sivashinsky equation');
```

LISTING 11.5: Portion of main program `pde_1_main`.

This code consists essentially of two parts.

- The problem parameters are defined numerically and the scale in t is defined as $0 \leq t \leq 10$ with output displayed at $t = 0, 2, 4, \ldots, 10$.
- The solution is plotted in 2D (by a call to `plot`) and 3D (by a call to `surf`).

This main program produces the same three figures and tabulated output as Chapters 3–10, which are now reviewed. The Jacobian matrix routine `jpattern_num_1.m` is the same as `jpattern_num_1.m` in Chapters 3–10 and is therefore not reproduced here.

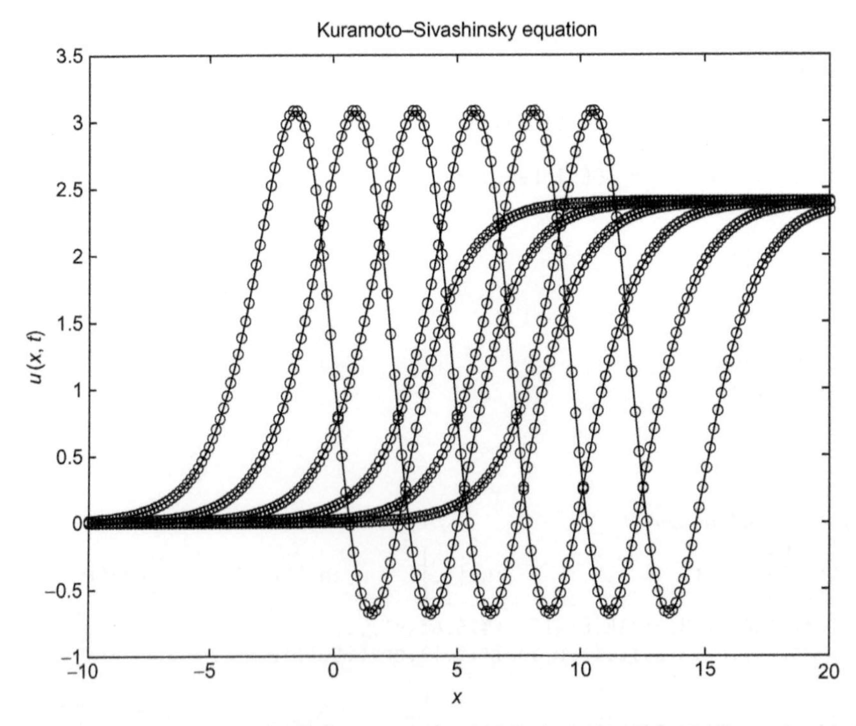

FIGURE 11.1: Numerical solution to eq. (11.1) (lines) with the analytical solution (eq. (11.2)) superimposed (circles) using five-point FD approximations in `dss004` for $t = 0, 2, 4, \ldots 10$ (left to right).

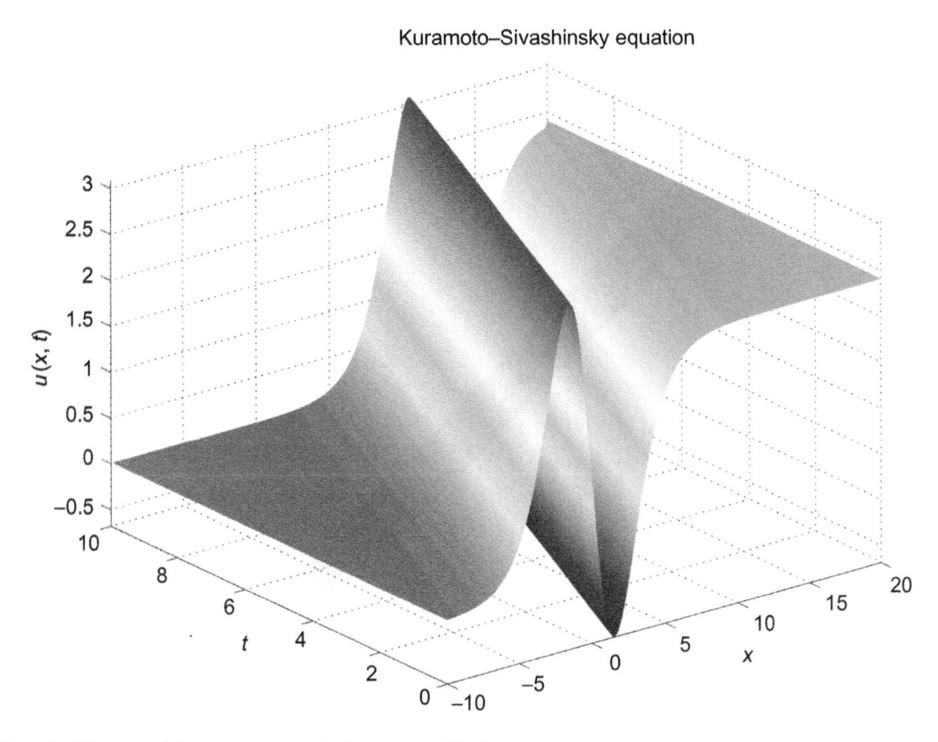

FIGURE 11.2: 3D plot of the numerical solution to eq. (11.1).

Figure 11.1 indicates good agreement between the analytical and numerical solutions. Figure 11.2 is the 3D plot of the numerical solution. The map of the ODE Jacobian matrix, Fig. 11.3, reflects the banded structure of the ODEs produced by `dss004`.

The tabular analytical and numerical solutions indicate good agreement as displayed in Table 11.1. The computational effort reflected in `ncall = 564` is modest.

We note in eq. (11.2) that x and t appear as the linear combination, $\frac{1}{2}kx - \frac{15}{19}k^2t$ so that the analytical solution represents a *traveling wave* as reflected in Fig. 11.1. If we consider the *Lagrangian* variable to be $k(x - vt)$ where k and v are the *wavenumber* and *wave velocity*, respectively, then k and v follow immediately from the linear combination $\frac{1}{2}kx - \frac{15}{19}k^2t$. This point was discussed in earlier chapters in which the wave velocity from the analytical solution was compared with the wave velocity from the numerical solution (see, for example, Chapter 10, for an example of this comparison).

In summary, the solution of eq. (11.1) subject to the IC from eq. (11.2) (with $t = 0$), two Dirichlet BCs from eq. (11.2) (with $x = -10, 20$), and two Neumann BCs from eq. (11.3) (with $x = -10, 20$) is straightforward. Also, eq. (11.1) is nonlinear, yet the programming in `pde_1.m` is straightforward. Consequently, variations in the PDE can easily be made for cases for which an analytical solution might not be available.

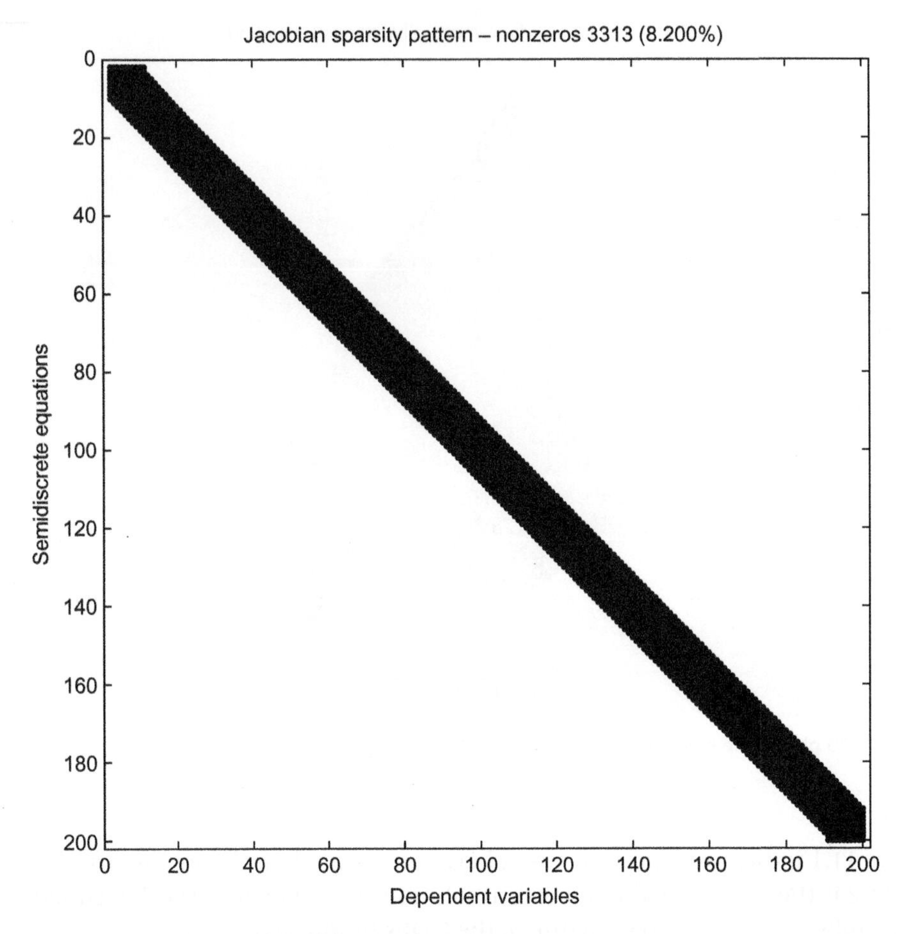

FIGURE 11.3: Jacobian matrix map of the MOL ODEs for $n = 201$.

Appendix

The *Kuramoto–Sivashinsky* equation describes one of the simplest nonlinear systems that exhibit *turbulence*. It has been used to study various reaction-diffusion problems and, in particular, it is used to model the *thermal mechanism of flame propagation* or *combustion waves*. Both Gregory Sivashinsky, whilst studying *laminar flame fronts* [7], and Yoshiki Kuramoto, whilst studying *diffusion-induced chaos* [4], discovered independently the equation now known as the *Kuramoto–Sivashinsky* equation, which is usually presented in a normalized form. This equation models the time evolution of flame-front velocity which is determined by a balance between the quantity of heat released by the combustion reaction and the heat required to preheat the incoming reactants. The eq. (11.1) with constants $\alpha = 1$, $\beta = 0$, $\gamma = 1$ reduces to the form

$$u_t = -\frac{1}{2}\left(u^2\right)_x - u_{xx} - u_{xxxx},$$

Table 11.1: Tabular numerical and analytical solutions

t	x	u(it,i)	u_anal(it,i)	err(it,i)
0.00	−10.000	0.014276	0.014276	0.000000
0.00	−9.250	0.025224	0.025224	0.000000
0.00	−8.500	0.044520	0.044520	0.000000
0.00	−7.750	0.078424	0.078424	0.000000
0.00	−7.000	0.137674	0.137674	0.000000
	.		.	
	.		.	
	.		.	
0.00	17.000	2.402728	2.402728	0.000000
0.00	17.750	2.402758	2.402758	0.000000
0.00	18.500	2.402775	2.402775	0.000000
0.00	19.250	2.402785	2.402785	0.000000
0.00	20.000	2.402791	2.402791	0.000000
	.		.	
	.		.	
	.		.	
	Output for t = 2,4,6,8 removed			
	.		.	
	.		.	
	.		.	

t	x	u(it,i)	u_anal(it,i)	err(it,i)
10.00	−10.000	0.000002	0.000002	0.000000
10.00	−9.250	−0.000003	0.000003	−0.000006
10.00	−8.500	−0.000018	0.000005	−0.000023
10.00	−7.750	−0.000036	0.000008	−0.000045
10.00	−7.000	−0.000048	0.000015	−0.000063
	.		.	
	.		.	
	.		.	
10.00	17.000	1.805714	1.805901	−0.000187
10.00	17.750	2.052825	2.053003	−0.000178
10.00	18.500	2.200865	2.200996	−0.000131
10.00	19.250	2.287359	2.287410	−0.000050
10.00	20.000	2.337154	2.337154	0.000000
ncall=564				

which can be simulated using periodic boundary conditions, $x \in [0, L]$, to give rich examples of chaotic behavior. The results at later times are extremely sensitive to small changes in initial conditions, and the transition from a smooth solution to chaos makes modeling in the time domain very difficult for extended simulated time periods. However, transforming the problem into the Fourier domain greatly reduces the stiffness of the problem and enables good results to be obtained with little computing effort. For example, see Fig. 11.4.

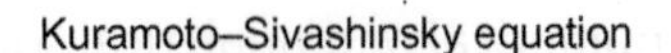

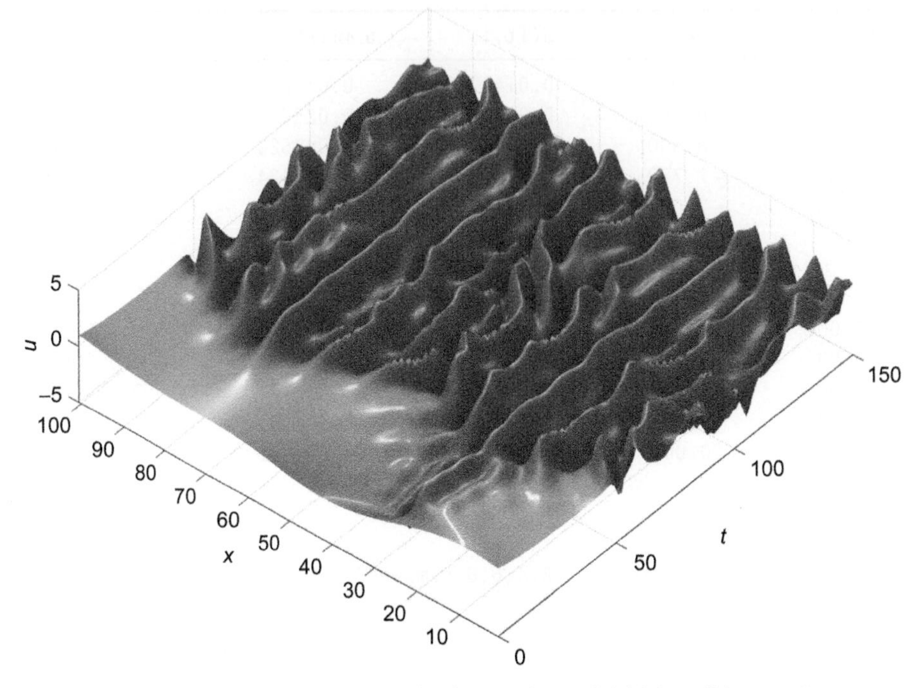

FIGURE 11.4: Time evolution for the *Kuramoto–Sivashinsky* equation with initial condition $u(x,0) = \cos(x/16)(1+\sin(x/16))$ and $x \in [0, 32\pi]$. This image was generated from Matlab code described in the paper by Kassam and Trefethen [3].

The Matlab code, `kursiv.m`, used to generate Fig. 11.4 is based on *spectral methods* and uses the *exponential time differencing* scheme ETDRK4 described in detail by Cox and Matthews [2]. A copy of this code is included with the downloads for this book.

The solution of PDEs by spectral methods is currently an active area of research and can provide outstanding results for some problems. A good general introduction to these methods is given in Trefethen's monograph [9]. This topic will not be considered further here, and for additional information relating to applications and theory of the Kuramoto–Sivashinsky equation, readers are referred to [1, 3, 5, 8].

The Kuramoto–Sivashinsky equation also admits traveling wave solutions, and these can be found using any one of the *tanh, exp,* and *Riccati* methods described in the main Appendix. We choose the tanh method and the Maple code listed in Listing 11.6 finds 11 solutions, one of which corresponds to the original solution given in eq. (11.2).

```
># Kuramoto-Sivashinsky Equation
 # Attempt at Malfliet's tanh solution
 restart; with(PDEtools): with(PolynomialTools):
 with(plots):unprotect(gamma);
>alias(u=u(x,t)):
```

```
>pde1:=diff(u,t)+u*diff(u,x)+alpha*diff(u,x,x)
         +beta*diff(u,x,x,x)+gamma*diff(u,x,x,x,x)=0;
># set parameter values
 beta:=0; alpha:=1;gamma:=1;
>read("tanhMethod.txt");
>intFlg:=1: # No integration of U(xi) needed !
 M:=3; # Set order of approximation
 infoLevOut:=0;
 tanhMethod(M,pde1,intFlg,infoLevOut);
># Animate solution
 zz:=rhs(sol[8]);x0:=0;
 animate(zz,x=-10..35, t=0..20,
   numpoints=100,frames=50, axes=framed,labels=["x","u"],
   thickness=3,title="Kuramoto-Sivashinsky Equation",
   labelfont=[TIMES, ROMAN, 16],axesfont=[TIMES, ROMAN, 16],
   titlefont=[TIMES, ROMAN, 16]);
># Generate a 3D surface plot of solution
 plot3d(zz,x=-10..35,t=0..20,
   axes=framed, grid=[100,100],thickness=0,
   labeldirections=[HORIZONTAL,HORIZONTAL,VERTICAL],
   style=patchnogrid,labels=["x","t","u(x,t)"],
   orientation=[-116,46],title="Kuramoto-Sivashinsky Equation",
   shading=Z, lightmodel=none,
   labelfont=[TIMES, ROMAN, 16],axesfont=[TIMES, ROMAN, 16],
   titlefont=[TIMES, ROMAN, 16]);
```

LISTING 11.6: Maple code to solve eq. (11.1) using the *tanh* method.

It is as equally straightforward to find traveling wave solutions by application of either of the Maple procedures `expMethod()` or `riccatiMethod()` described in the main Appendix. The *exp* method finds eight solutions, and the *Riccati* method finds 11×6 solutions (recall that each solution of the Riccati equation yields six separate traveling wave solutions). They both find solutions that match the original solution of eq. (11.2).

In order to save space, listings of the Maple code implementations of the exp and Riccati methods will not be included here, but they are available in the downloadable software for this book.

References

[1] C. Chandre, F.K. Diakonos, P. Schmelcher, Turbulence, in: *Chaos: Classical and Quantum*, Niels Bohr Institute, Copenhagen, 2009, <http://chaosbook.org/version13/chapters/ChaosBook.pdf>.

[2] S.M. Cox, P.C. Matthews, Exponential time differencing for stiff systems, *J. Comp. Phys.* 176 (2002) 430–455.

[3] A-K. Kassam, L.N. Trefethen, Fourth order time stepping for stiff PDEs, *SIAM J. Sci. Comp.* 26 (2002) 1214–1233.

[4] Y. Kuramoto, Diffusion-induced chaos in reaction systems, *Progr. Theor. Phys. Suppl.* 64 (1978) 346–367.

[5] Y. Lan, *Dynamical systems approach to 1-d spatiotemporal chaos – A cyclist's view*, PhD thesis (version 0.8), School of Physics, Georgia Institute of Technology, May 18, 2004.

[6] A.D. Polyanin, V.F. Zaitsev, *Handbook of Nonlinear Partial Differential Equations*, Chapman & Hall/CRC, Boca Raton, FL, 2004.

[7] G.I. Sivashinsky, Nonlinear analysis of hydrodynamic instability in laminar flames-I. Derivation of basic equations, *Acta Astronautica* 4 (1977) 1177–1206.

[8] G.I. Sivashinsky, Instabilities, pattern formation, and turbulence in flames, *Ann. Rev. Fluid Mech.* 15 (1983) 179–99.

[9] L.N. Trefethen, *Spectral Methods in Matlab*, SIAM, Philadelphia, PA, 2000.

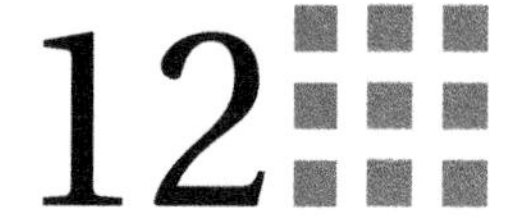

12 Kawahara Equation

The Kawahara equation ([8], p632) is

$$\frac{\partial u}{\partial t}+u\frac{\partial u}{\partial x}+a\frac{\partial^3 u}{\partial x^3}=b\frac{\partial^5 u}{\partial x^5} \tag{12.1}$$

An analytical solution is ([8], p632)

$$u(x,t)=\frac{105a^2}{169b\cosh^4 z}+2C_1;\quad z=\frac{1}{2}kx-(18bk^5+C_1k)t+C_2;\quad k=\sqrt{\frac{a}{13b}};\quad ab>0 \tag{12.2}$$

where C_1, C_2 are arbitrary constants. We study this equation to investigate the use of a fifth-order derivative, which is computed using finite differences (FDs) as explained subsequently. This chapter is also an introduction to Chapter 18 pertaining to a PDE fourth-order in x and second-order in t.

The Matlab routines closely resemble those of previous chapters. Here, we list a few details pertaining to eqns. (12.1) and (12.2). First, the ODE routine `pde_1.m` is

```
  function ut=pde_1(t,u)
%
% Function pde_1 computes the t derivative vector for the fifth-
% order Kawahara equation
%
  global xl xu x n ncall
%
% Model parameters
  global a b k C1 C2
%
% BCs
  u(1)=ua_1(x(1),t);
  u(n)=ua_1(x(n),t);
%
% ux
  ux=dss004(xl,xu,n,u);
%
% uxxx
  uxxx=u3x9p(xl,xu,n,u);
%
% uxxxxx
  uxxxxx=u5x11p(xl,xu,n,u);
%
% PDE
%   For loop
    for i=1:n
      ut(i)=-u(i)*ux(i)-a*uxxx(i)+b*uxxxxx(i);
    end
```

```
      ut=ut';
%
% BCs
  ut(1)=0;
  ut(n)=0;
%
% Increment calls to pde_1
  ncall=ncall+1;
```

LISTING 12.1: Function pde_1.m for eq. (12.1).

We can note the following points about pde_1.m:

- The function and some global parameters are first defined.

```
  function ut=pde_1(t,u)
%
% Function pde_1 computes the t derivative vector for the fifth-
% order Kawahara equation
%
  global xl xu x n ncall
%
% Model parameters
  global a b k C1 C2
```

- Since eq. (12.1) is fifth-order in x, it generally requires five BCs. However, since the solution and its derivatives are essentially zero at the boundaries in x, we will specify only the dependent variable at the boundaries, i.e., Dirichlet BCs. In fact, we will consider a small variation in pde_1.m in which no BCs are programmed, and we will find that the numerical solution is still accurate. This second case will be discussed later. For the first case, we use Dirichlet BCs from eq. (12.1).

```
%
% BCs
  u(1)=ua_1(x(1),t);
  u(n)=ua_1(x(n),t);
```

  where ua_1.m is the code for eq. (12.2). x(1) and x(n) correspond to $x = -30, 30$ and $n = 101$ as subsequently defined in inital_1.m and passed as global variables to pde_1.m

- The first derivative, ux, is then computed by dss004.

```
%
% ux
  ux=dss004(xl,xu,n,u);
```

- The third derivative in eq. (12.1), uxxx, is computed by routine u3x9p. The notation signifies 3x = third derivative, 9p = nine points used in the FD approximation for uxxx. The origin of u3x9p is discussed later.

```
%
% uxxx
  uxxx=u3x9p(xl,xu,n,u);
```

- Similarly, the fifth derivative in eq. (12.1), uxxxxx, is computed by routine u5x11p. The notation signifies 5x = fifth derivative, 11p = 11 points used in the FD approximation for uxxxxx. The origin of u5x11p is discussed later.
- Equation (12.1) is then programmed.

```
%
% PDE
%   For loop
    for i=1:n
      ut(i)=-u(i)*ux(i)-a*uxxx(i)+b*uxxxxx(i);
    end
    ut=ut';
%
% BCs
  ut(1)=0;
  ut(n)=0;
%
% Increment calls to pde_1
  ncall=ncall+1;
```

Since Dirichlet BCs are used, the derivatives in t are set to zero at the boundaries (so that the ODE integrator does not move the boundary values away from their prescribed values). A transpose is included to meet the requirements of the ODE integrator ode15s. Finally, the counter for the number of calls to pde_1.m is incremented.

The IC from eq. (12.2) with $t = 0$ is programmed in inital_1.m.

```
  function u0=inital_1(t0)
%
% Function inital_1 sets the initial condition for the fifth-order
% Kawahara equation
%
% Parameters shared with other routines
  global xl xu x n ncall
%
% Model parameters
  global a b k C1 C2
%
% Spatial domain and initial condition
  xl=-30;
  xu= 30;
  n=101;
  dx=(xu-xl)/(n-1);
%
% Spatial grid
  for i=1:n
    x(i)=xl+(i-1)*dx;
    u0(i)=ua_1(x(i),0.0);
  end
```

LISTING 12.2: Function inital_1.m for IC from eq. (12.2) with $t = 0$.

We can note the following points about inital_1.m:

- The function and some global parameters are first defined.

```
  function u0=inital_1(t0)
%
% Function inital_1 sets the initial condition for the fifth-order
% Kawahara equation
%
% Parameters shared with other routines
  global xl xu x n ncall
%
% Model parameters
  global a b k C1 C2
```

- The grid in x is then defined over the interval $-30 \leq x \leq 30$ for 101 points.

```
%
% Spatial domain and initial condition
  xl=-30;
  xu= 30;
  n=101;
  dx=(xu-xl)/(n-1);
%
% Spatial grid
  for i=1:n
    x(i)=xl+(i-1)*dx;
    u0(i)=ua_1(x(i),0.0);
  end
```

As the grid in x is defined in the for loop, function ua_1 (listed next) is called (for $t = 0$) to define the IC from eq. (12.2) with $t = 0$.

Function ua_1.m is a straightforward implementation of the analytical solution, eq. (12.2).

```
function uanal=ua_1(x,t)
%
% Function uanal computes the exact solution of the fifth-order
% Kawahara equation for comparison with the numerical solution.
%
  global a b k C1 C2
%
% Analytical solution
  z=0.5*k*x-(18*b*k^5+C1*k)*t+C2;
  uanal=105*a^2/(169*b*cosh(z)^4)+2*C1;
```

LISTING 12.3: Function ua_1.m for analytical solution (12.2).

The main program, pde_1_main, is essentially the same as pde_1_main of previous chapters and therefore is not listed here. The parameters of eq. (12.1) are set in pde_1_main and then passed as global variables to the other routines where they are required.

```
%
% Model parameters (C3, C5 are the FD coefficients)
```

```
  global a b k C1 C2 C3 C5
%
  a=1; b=1; C1=1; C2=0; k=(a/(13*b))^0.5;
```

We now consider function `u3x9p.m`, which has a nine-point, centered FD approximation for a third derivative.

```
  function uxxx=u3x9p(xl,xu,n,u)
%
% Function u3x9p computes the derivative uxxx based on nine points
  global ncall C3
%
% For the first call to u3x9p (ncall = 0), compute the FD weighting
% coefficients
%
   if(ncall==0)
%
%   Default points, equally spaced grid
    for i=1:9
      x(i)=i-1;
    end
%
%   Compute FD approximation for up to and including the mth
%   derivative
    m=3;
%
%   Number of grid points
    ng=9;
    nd=ng;
%
%   Compute weighting coefficients for finite differences
%   over ip points
    for ip=1:ng
%
%     Weighting coefficients in array C3
      C3(ip,:,:)=weights(x(ip),x,ng-1,nd-1,m);
    end
%
% Display coefficients for derivatives of orders 0, 1, 2, 3
% for i=1:m+1
%
% Display coefficients for derivative of order 3
  i=m+1;
    fprintf('\n\n Numerical Derivative Order: %d',i-1);
    fprintf('\n============================\n');
%
%   Coefficients in u3x9p.m with m = 3, ng = 9
%   C3(:,:,i)*8
    C3(5,:,i)*8
% end
%
% Calculation of FD weights complete
  end
%
```

```
% uxxx
% Spatial increment
  dx=(xu-xl)/(n-1);
  rdx3=1.0/dx^3;
  for i=1:n
%
%   At the left end, uxxx = 0
    if(i<5)uxxx(i)=0.0;
%
%   At the right end, uxxx = 0
    elseif(i>(n-4))uxxx(i)=0.0;
%
%   Interior points
    else
      uxxx(i)=rdx3*...
      (C3(5,1,4)*u(i-4)...
      +C3(5,2,4)*u(i-3)...
      +C3(5,3,4)*u(i-2)...
      +C3(5,4,4)*u(i-1)...
      +C3(5,5,4)*u(i  )...
      +C3(5,6,4)*u(i+1)...
      +C3(5,7,4)*u(i+2)...
      +C3(5,8,4)*u(i+3)...
      +C3(5,9,4)*u(i+4));
    end
  end
```

LISTING 12.4: Function `u3x9p.m` for the third derivative `uxxx` in eq. (12.1).

We can note the following points about `u3x9p.m`:

- The function and some global parameters are first defined.

```
  function uxxx=u3x9p(xl,xu,n,u)
%
% Function u3x9p computes the derivative uxxx based on nine points
  global ncall C3
```

- For the first call to `u3x9p.m` (with `ncall=0` set in the main program), a uniform grid in x of nine equally spaced points is defined.

```
%
% For the first call to u3x9p (ncall = 0), compute the FD weighting
% coefficients
%
   if(ncall==0)
%
%    Default points, equally spaced grid
     for i=1:9
       x(i)=i-1;
     end
```

The definition of the grid x is for the purpose of calculating the FD weighting coefficients only; it is not the grid used in the method of lines (MOL) solution of eq. (12.1), which can have any grid spacing (and not necessarily integer spacing as used in `x(i)`).

- The weighting coefficients in the FD approximation are then computed. First, the order of the derivative to be approximated is specified, `m=3`. Then, the number of grid points in the FD approximation is specified (`ng=9`). Finally, the weighting coefficients are computed by function `weights` and placed in array `C3`. The details of `weights` will be considered subsequently.

```
%
%   Compute FD approximation for up to and including the mth
%   derivative
    m=3;
%
%   Number of grid points
    ng=9;
    nd=ng;
%
%   Compute weighting coefficients for finite differences
%   over ip points
    for ip=1:ng
%
%     Weighting coefficients in array C3
      C3(ip,:,:)=weights(x(ip),x,ng-1,nd-1,m);
    end
```

The `for` loop steps through the $ng = 9$ grid points for which weighting coefficients are computed (at each of the grid points, `ip=1,2,...,ng`).

- An optional display of the coefficients is included (just for our information).

```
%
% Display coefficients for derivatives of orders 0, 1, 2, 3
% for i=1:m+1
%
% Display coefficients for derivative of order 3
  i=m+1;
    fprintf('\n\n Numerical Derivative Order: %d',i-1);
    fprintf('\n=============================\n');
%
%   Coefficients in u3x9p.m with m = 3, ng = 9
%   C3(:,:,i)*8
    C3(5,:,i)*8
% end
%
% Calculation of FD weights complete
  end
```

We can note the following details about this code:

- `weights` computes the FD weighting cofficients for the derivatives up to and including third order. It also computes the weights for the function to be differentiated, which are, for example, `0,0,0,0,1,0,0,0,0` at grid point 5. In other words, the fifth coefficient is 1 for the function at the fifth point and 0 elsewhere.

Thus, there are four sets of weighting coefficients for the function and its first three derivatives, and the third subscript of `C3` is $i = 1,2,3,4$. Note that the `for` loop is commented, and thus only `i=m+1=4` corresponding to the third derivative is displayed.

- The first and second subscripts of `C3` each range over 1–9 so that, for example, `C3(irow,jcol,4)` corresponds to row *irow* for the FD approximation at grid point `irow`, and column `jcol` specifies the nine weighting coefficients at grid point `irow`. As noted above, the third index 4 specifies the weighting coefficients for the third derivative. In other words, `C3(irow,jcol,4)` includes a `np x np = 9 x 9` array of FD weighting coefficients for the third derivative; `irow=1` corresponds to grid point 1 in the basic grid of nine points, and `irow=9` corresponds to grid point 9 in the basic grid of nine points.
- Thus, using `C3(5,:,i)*8` displays the nine FD weighting coefficients (using :) at grid point 5 for the third derivative (since `i=4`). The multiplication by 8 merely rescales the coefficients to values that are more typical of those reported in the literature. The final `end` completes the `if` for the first call to `u3x9p.m` (`if(ncall==0)`, so that the weighting coefficients are calculated only once, and then used repeatedly as `u3x9p` is called).

- The FD approximation of the third derivative, `uxxx`, is then computed. First, the grid spacing, `dx`, is computed. Then, the `for` loop steps through all of the grid points (`n=101` set in `inital_1.m`).

```
%
% uxxx
% Spatial increment
  dx=(xu-xl)/(n-1);
  rdx3=1.0/dx^3;
  for i=1:n
%
%   At the left end, uxxx = 0
    if(i<5)uxxx(i)=0.0;
%
%   At the right end, uxxx = 0
    elseif(i>(n-4))uxxx(i)=0.0;
%
%   Interior points
    else
      uxxx(i)=rdx3*...
      (C3(5,1,4)*u(i-4)...
      +C3(5,2,4)*u(i-3)...
      +C3(5,3,4)*u(i-2)...
      +C3(5,4,4)*u(i-1)...
      +C3(5,5,4)*u(i  )...
      +C3(5,6,4)*u(i+1)...
      +C3(5,7,4)*u(i+2)...
      +C3(5,8,4)*u(i+3)...
      +C3(5,9,4)*u(i+4));
    end
  end
```

We can note the following details about this code:

- For the first four grid points (`i<5` in the total of `n=101`), the third derivative is set to zero. This does not mean that, in general, the third derivative will be zero or that the weighting coefficients are not set for grid points `i=1,2,3,4`. Rather, experience has indicated that in the MOL solution of PDEs, numerical distortions can occur at the boundaries (this is discussed later). To minimize any such effects, we set `uxxx` to zero at the left boundary, which is acceptable if the solution and its derivatives remain at zero at the left boundary (as they do in the case of eq. (12.2)).
- Similarly, for the last four grid points (`i>(n-4)` in the total of `n=101`), the third derivative is set to zero. Again, experience has indicated that in the MOL solution of PDEs, numerical distortions can occur at the boundaries. To minimize such effects, we set `uxxx` to zero at the right boundary, which is acceptable if the solution and its derivatives remain at zero at the left boundary (as they do in the case of eq. (12.2)).
- For the intermediate points `i = 5` to `i = n-4`, the third derivative is computed by a central FD for the middle grid point (first subscript = 5), using nine weighting coefficients (second subscript = 1–9), for the third deriviative (third subscript = 4).
- The two `end` statements conclude the `if` statement and the `for` loop in `i`.

The function `weights` is an implementation of the algorithm by Fornberg [3] for the calculation of weighting coefficients for FD approximations. This remarkable algorithm computes weighting coefficients for FD approximations:

- Of derivatives of any order, e.g., third order as in `u3x9p.m`.
- Based on any number of grid points, e.g., nine as in `u3x9p.m`. As the number of grid points is increased, the order of the FD approximation also increases. One small disadvantage in using more grid points within the MOL integration of PDE is the increase in the ODE Jacobian bandwidth. This is a small price, however, for the improved accuracy of the FD approximation, as the number of grid points is increased. In fact, for a given accuracy of the approximation, fewer total grid points are often required as the number of grid points in the approximation is increased. For example, $n = 101$ could possibly be reduced to achieve a specified accuracy in the numerical MOL PDE solution, as more points are used in the FD approximation. The additional computational effort in using a higher-order FD approximation is generally more than offset by the reduced number of total grid points to achieve a prescribed accuracy.
- At the boundary and near-boundary points. This is accomplished by using noncentered approximations.
- With arbitrary selection of the grid points, the algorithm computes weighting coefficients for FD approximations of nonuniform grids. This approach is often used in order to concentrate grid points where they are most needed, e.g., where the function to be differentiated varies most rapidly.
- So efficiently that the algorithm can be used to compute new weighting coefficients for FD approximations during the course of the calculation, such as in the adaptive grid solution of PDEs.

The Fornberg algorithm is the centerpiece of much of the reported FD PDE solutions. A listing of function `weights.m` follows without any further explanation. The details of the algorithm can be found in the publications by Fornberg [3].

```
        function [c]=weights(z,x,n,nd,m)
%
%       Input Parameters
%
%            z                location where approximations are to be
%                             accurate
%
%            n                one less than total number of grid points
%
%            x(1:n+1)         grid point locations
%
%            m                highest derivative for which weights are
%                             sought
%       Output Parameter
%
%            c(1:n+1,0:m)     weights at grid locations x(1:n+1) for
%                             derivatives of order 0:m, found in
%                             c(1:n+1,1:m+1)
%
        c1 = 1.0;
        c4 = x(1)-z;
        for k=0:m
          for j=0:n
            c(j+1,k+1) = 0.0;
        end
        end
        c(1,1) = 1.0;
        for i=1:n
           mn = min(i,m);
           c2 = 1.0;
           c5 = c4;
           c4 = x(i+1)-z;
           for j=0:i-1
             c3 = x(i+1)-x(j+1);
             c2 = c2*c3;
             if (j==i-1)
               for k=mn:-1:1
                 c(i+1,k+1) = c1*(k*c(i,k)-c5*c(i,k+1))/c2;
               end
               c(i+1,1) = -c1*c5*c(i,1)/c2;
             end
             for k=mn:-1:1
               c(j+1,k+1) = (c4*c(j+1,k+1)-k*c(j+1,k))/c3;
             end
               c(j+1,1) = c4*c(j+1,1)/c3;
          end
          c1 = c2;
        end
```

LISTING 12.5: Function `weights.m` for the FD approximation of derivatives.

Function `u5x11p.m` called in `pde_1.m` is very similar to `u3x9p.m` in Listing 12.4. The only essential differences are the use of 5 and 11 in place of 3 and 9, respectively. The calculation of the fifth derivative `uxxxxx` (third subscript `6` in `C5`) in `u5x11p.m` at grid point `6` (first subscript `6` in `C5`) is

```
uxxxxx(i)=rdx5*...
(C5(6,1,6)*u(i-5)...
+C5(6,2,6)*u(i-4)...
+C5(6,3,6)*u(i-3)...
+C5(6,4,6)*u(i-2)...
+C5(6,5,6)*u(i-1)...
+C5(6,6,6)*u(i  )...
+C5(6,7,6)*u(i+1)...
+C5(6,8,6)*u(i+2)...
+C5(6,9,6)*u(i+3)...
+C5(6,10,6)*u(i+4)...
+C5(6,11,6)*u(i+5));
```

Again, the fifth derivative at grid points $i = 1,2,3,4,5$ and $i = n-4, n-3, n-2, n-1, n$ is zeroed to minimize unrealistic effects at the boundaries.

The main program produces the same three figures and tabulated output as in previous chapters, which are now reviewed. The Jacobian matrix routine `jpattern_num_1.m` is the same as `jpattern_num_1.m` in previous chapters and is therefore not reproduced here.

Figure 12.1 indicates good agreement between the analytical and numerical solutions. The resolution of the traveling pulses can be increased by decreasing the interval in x from $-30 \le x \le 30$ to, for example, $-10 \le x \le 20$ by inserting a statement like `axis([-10 20 2 2.7]);` after `plot`. This does not affect the numerical solution, but only the plotting of the numerical and analytical solutions. Figure 12.2 is a 3D plot of the numerical solution.

The map of the ODE Jacobian matrix, Fig. 12.3, reflects the banded structure of the ODEs produced by `u3x9p` and `5x11p`.

The tabular analytical and numerical solutions given in Table 12.1 also reflect the good agreement between these two solutions. The computational effort reflected in `ncall = 166` is quite modest.

We can note the following points about this output (Table 12.1):

- The weighting coefficients for the center points, $i = 5$ for a 9-point FD approximation and $i = 6$ for an 11-point FD approximation, are zero. In general, this will be the case for a centered approximation of odd-order derivatives (such as 5th, 11th).
- The weighting coefficients are antisymmetric (of opposite sign) around the center points. This will also be true for centered FD approximations of odd-order derivatives.
- The coefficients sum to zero (the reader might verify this). This is required in order to differentiate a constant function to zero.
- The FD approximations are exact not only for a zeroth-order polynomial (constant function); they are exact for higher-order polynomials as well. Specifically, the nine-point approximation is exact for a polynomial up to and including eighth order (it will differentiate an eighth-order polynomial exactly). Also, the 11-point

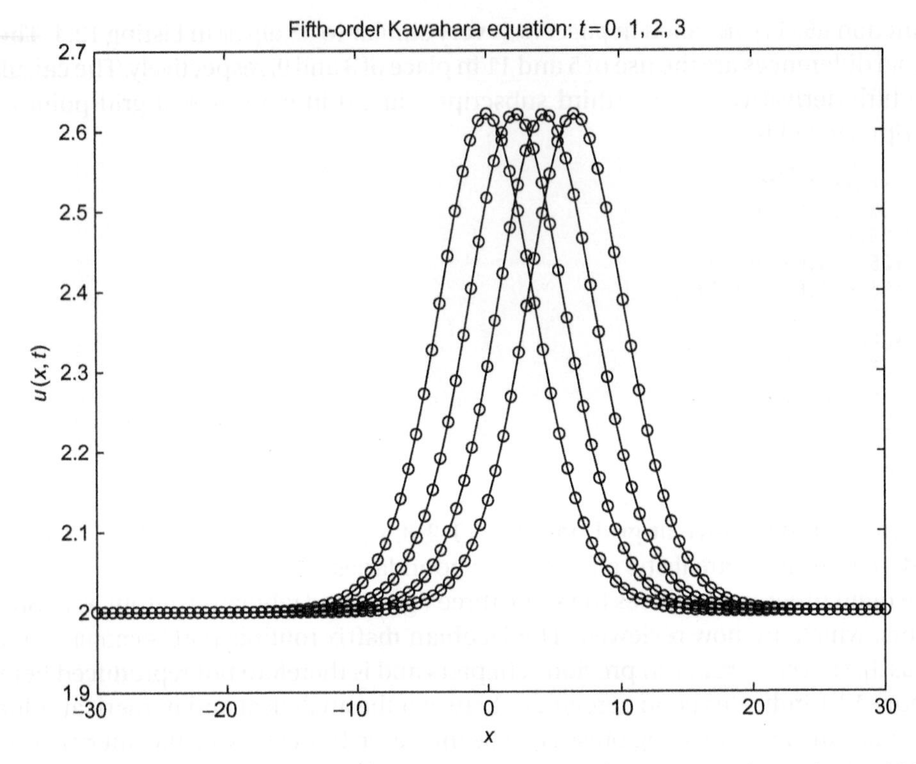

FIGURE 12.1: Numerical solution to eq. (12.1) (lines) with the analytical solution, eq. (12.2), superimposed (circles) using 9- and 11-point approximations in `u3x9p.m` and `u5x11p.m`.

approximation is exact for polynomials up to and including tenth order. These order conditions can be studied numerically using the programs listed in Appendix 1 at the end of this chapter.

- The agreement between the analytical and numerical solution of approximately four figures is acceptable for most applications. In particular, the peak value of the solution of eq. (12.1) (where the solution changes most rapidly in x) is accurately preserved as demonstrated in Fig. 12.1 and the following selected numerical output from Table 12.1.

We can note, however, from Table 12.1 that the maximum error in the numerical solution does not occur at the maximum point in the solution. For example, at $t = 3$, we have from Table 12.1,

```
3.00   6.600        2.621197        2.621265        -0.000069
3.00   7.200        2.613723        2.613835        -0.000113
3.00   7.800        2.589914        2.590046        -0.000132
```

For these three points, the maximum error is `-0.000132`, which is approximately twice the error at the maximum of the solution (at `x=6.600`). However, generally, the error is acceptable (as reflected in Fig. 12.1), and it could be reduced by increasing the number

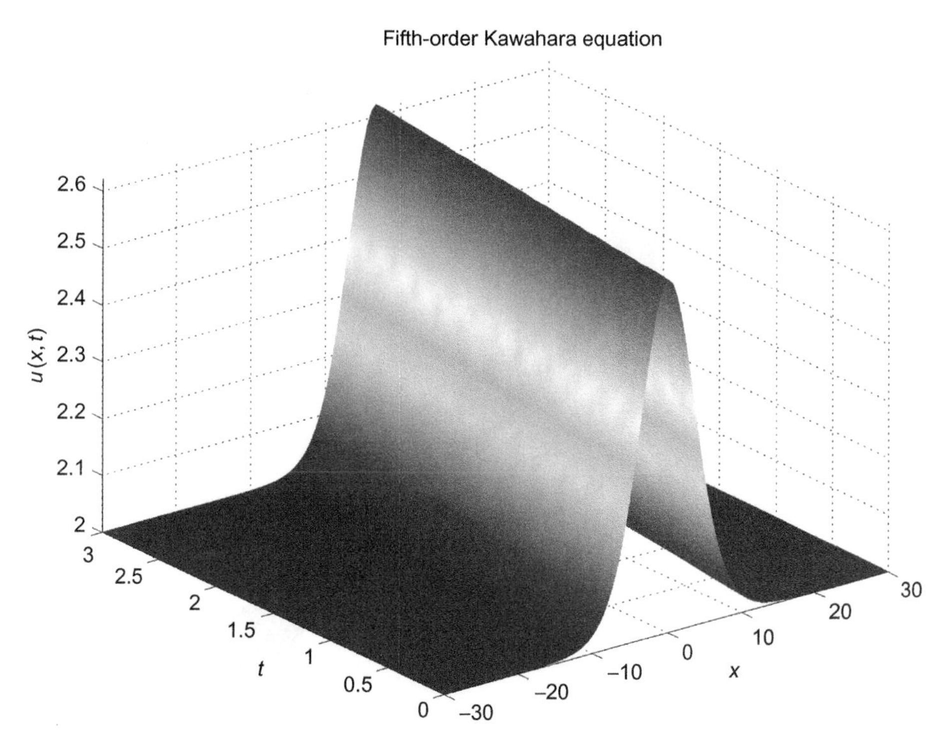

FIGURE 12.2: 3D plot of the numerical solution to eq. (12.1).

of grid points (e.g., `n=301` demonstrates substantially smaller errors in the numerical solution).

We also mentioned previously that BCs are not required for eq. (12.1) since the solution and its derivatives at the boundaries remain essentially at zero. In other words, the IC from eq. (12.2) at ($t = 0$) sets the boundary values of the solution and its derivatives zero, and for subsequent values of t, the boundary values remain zero. To demonstrate this point (no BCs required), the Dirichlet BCs in `pde_1.m` can be commented (deactivated) and the execution of `pde_1_main.m` can be repeated. Abbreviated output analogous to that in Table 12.1 is listed below in Table 12.3.

To compare the output with and without BCs, a portion of the output for $t = 3$ from Tables 12.1 and 12.3 is listed in Table 12.4. We observe that the two outputs are essentially identical indicating that BCs are not required for eq. (12.1) for the particular solution of eq. (12.2).

Of course, this is not a general conclusion for eq. (12.1); BCs would be required if the solution and its derivatives depart from zero near the boundaries. For this case (of significant boundary effects), we can add the following comments:

- The derivatives at the boundary points would have to be calculated using the FD weighting coefficients available from `weights`. For example, in the case of `u3x9p.m`, the

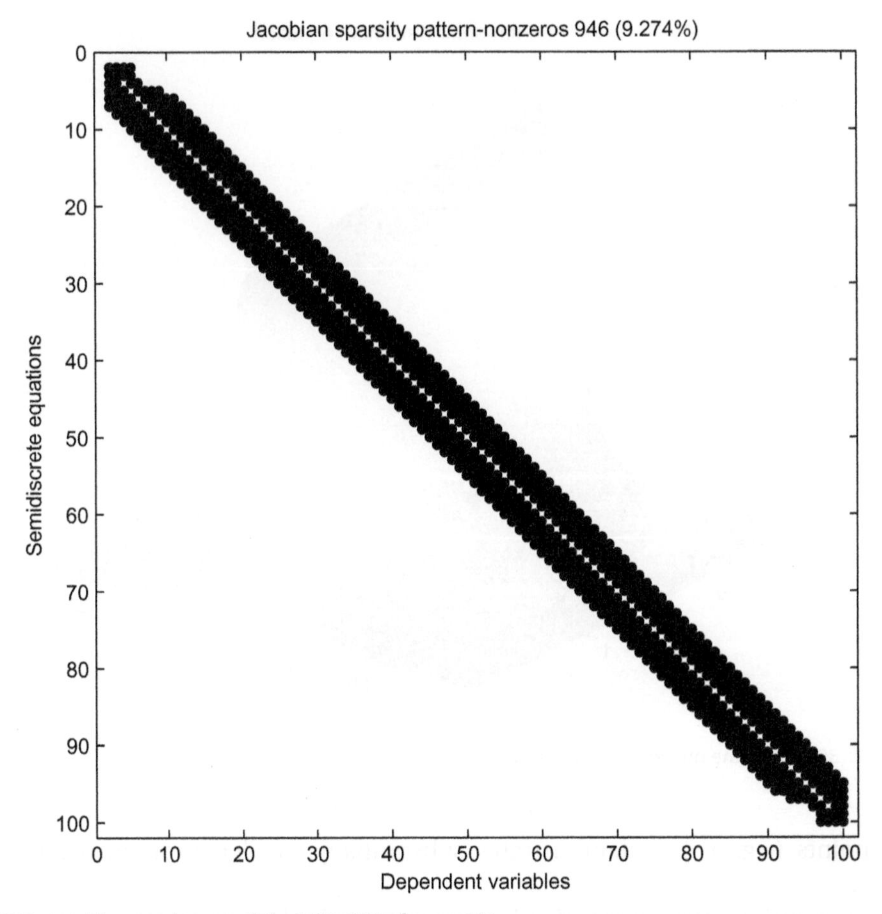

FIGURE 12.3: Jacobian matrix map of the MOL ODEs for $n = 101$.

third derivative `uxxx` would be calculated at points `i=1,2,3,4,n-3,n-2,n-1,n` rather than set to zero as in Listing 12.4.

- Even with the derivative calculated at all the grid points, unexpected boundary effects might be observed, which require additional refinement of the MOL algorithm and programming in the ODE derivative routine (`pde_1.m`). We will not go into the details of what may be required but rather refer to where boundary effects in MOL analysis are discussed in detail [9, 11].

In summary, the solution of eq. (12.1) subject to the IC from eq. (12.2) (with $t = 0$) and without or with two Dirichlet BCs from eq. (12.2) (with $x = -30, 30$) is straightforward. Also, eq. (12.1) is nonlinear, yet the programming in `pde_1.m` is straightforward. Consequently, variations in the PDE can easily be made for cases for which an analytical solution might not be available.

We should also point out that the development of an MOL code for a new PDE problem is not necessarily straightforward. Specifically,

Table 12.1: Abbreviated tabular numerical and analytical solutions for $n = 101$

```
Numerical Derivative Order: 3 (from u3x9p.m)
==============================
ans=
    -0.2333       2.4000     -11.2667     16.2667      -0.0000
   -16.2667      11.2667      -2.4000      0.2333
Numerical Derivative Order: 5 (from u5x11p.m)
==============================
ans=
   -0.3611        4.2222     -21.7500     52.0000     -53.8333     -0.0000
   53.8333      -52.0000      21.7500     -4.2222       0.3611
a=1.000          b=1.000
```

t	x	u(it,i)	u_anal(it,i)	err(it,i)
0.00	−30.000	2.000001	2.000001	0.000000
0.00	−29.400	2.000001	2.000001	0.000000
0.00	−28.800	2.000001	2.000001	0.000000
0.00	−28.200	2.000002	2.000002	0.000000
0.00	−27.600	2.000002	2.000002	0.000000
0.00	−27.000	2.000003	2.000003	0.000000
	.		.	
	.		.	
	.		.	
0.00	−1.200	2.587976	2.587976	0.000000
0.00	−0.600	2.612768	2.612768	0.000000
0.00	0.000	2.621302	2.621302	0.000000
0.00	0.600	2.612768	2.612768	0.000000
0.00	1.200	2.587976	2.587976	0.000000
	.		.	
	.		.	
	.		.	
0.00	27.000	2.000003	2.000003	0.000000
0.00	27.600	2.000002	2.000002	0.000000
0.00	28.200	2.000002	2.000002	0.000000
0.00	28.800	2.000001	2.000001	0.000000
0.00	29.400	2.000001	2.000001	0.000000
0.00	30.000	2.000001	2.000001	0.000000

t	x	u(it,i)	u_anal(it,i)	err(it,i)
1.00	−30.000	2.000001	2.000000	0.000000
1.00	−29.400	2.000001	2.000000	0.000000
1.00	−28.800	2.000000	2.000000	0.000000
1.00	−28.200	2.000001	2.000000	0.000000
1.00	−27.600	2.000001	2.000001	0.000000
1.00	−27.000	2.000001	2.000001	−0.000000
	.		.	
	.		.	
	.		.	

(Continued)

Table 12.1: *(Continued)*

t	x	u(it,i)	u_anal(it,i)	err(it,i)
1.00	1.200	2.597367	2.597334	0.000033
1.00	1.800	2.617241	2.617241	0.000000
1.00	2.400	2.620436	2.620467	−0.000031
1.00	3.000	2.606650	2.606705	−0.000055
1.00	3.600	2.577185	2.577252	−0.000067
	.		.	
	.		.	
	.		.	
1.00	27.000	2.000011	2.000011	0.000000
1.00	27.600	2.000007	2.000008	−0.000000
1.00	28.200	2.000005	2.000005	−0.000000
1.00	28.800	2.000004	2.000004	−0.000000
1.00	29.400	2.000003	2.000003	−0.000000
1.00	30.000	2.000001	2.000002	−0.000001
t	x	u(it,i)	u_anal(it,i)	err(it,i)
2.00	−30.000	2.000001	2.000000	0.000001
2.00	−29.400	2.000001	2.000000	0.000001
2.00	−28.800	2.000001	2.000000	0.000001
2.00	−28.200	2.000001	2.000000	0.000001
2.00	−27.600	2.000000	2.000000	−0.000000
2.00	−27.000	2.000001	2.000000	0.000000
	.		.	
	.		.	
	.		.	
2.00	3.000	2.574920	2.574849	0.000071
2.00	3.600	2.605264	2.605243	0.000021
2.00	4.200	2.620053	2.620082	−0.000029
2.00	4.800	2.617898	2.617970	−0.000072
2.00	5.400	2.599010	2.599108	−0.000099
2.00	6.000	2.565150	2.565257	−0.000108
	.		.	
	.		.	
	.		.	
2.00	27.000	2.000038	2.000036	0.000002
2.00	27.600	2.000025	2.000026	−0.000001
2.00	28.200	2.000017	2.000019	−0.000001
2.00	28.800	2.000012	2.000013	−0.000001
2.00	29.400	2.000008	2.000010	−0.000001
2.00	30.000	2.000001	2.000007	−0.000006
t	x	u(it,i)	u_anal(it,i)	err(it,i)
3.00	−30.000	2.000001	2.000000	0.000001
3.00	−29.400	2.000006	2.000000	0.000006
3.00	−28.800	2.000003	2.000000	0.000003

Table 12.1: *(Continued)*

```
3.00     -28.200     2.000003     2.000000      0.000003
3.00     -27.600     2.000000     2.000000      0.000000
3.00     -27.000     2.000004     2.000000      0.000004
               .                       .
               .                       .
               .                       .
3.00       5.400     2.585907     2.585846      0.000061
3.00       6.000     2.611622     2.611632     -0.000010
3.00       6.600     2.621197     2.621265     -0.000069
3.00       7.200     2.613723     2.613835     -0.000113
3.00       7.800     2.589914     2.590046     -0.000132
               .                       .
               .                       .
               .                       .
3.00      27.000     2.000124     2.000122      0.000002
3.00      27.600     2.000081     2.000088     -0.000007
3.00      28.200     2.000055     2.000063     -0.000008
3.00      28.800     2.000039     2.000045     -0.000006
3.00      29.400     2.000028     2.000032     -0.000005
3.00      30.000     2.000001     2.000023     -0.000023
 ncall=166
```

Table 12.2: Peak values of the solution of eq. (12.1) (from Table 12.1)

t	x	peak value (numerical)	peak value (analytical)	error
0.00	0.000	2.621302	2.621302	0.000000
1.00	2.400	2.620436	2.620467	-0.000031
2.00	4.200	2.620053	2.620082	-0.000029
3.00	6.600	2.621197	2.621265	-0.000069

- It is quite natural in developing a MOL code to include all of the PDE derivatives, and associated ICs and BCs. Typically, there can be a series of such terms of varying complexity. However, initial execution of the code might not proceed as expected. The question then is what caused the failed execution. Unfortunately, in most cases, this is not obvious, even after reviewing the code to correct any programming errors.
- To progress past this unsatisfactory result, we suggest an *incremental approach* to code development. For example, in the case of eq. (12.1), we might start with $a = b = 0$ in which case, we have just the *nonlinear advection equation*. This PDE has been

Table 12.3: Abbreviated tabular numerical and analytical solutions for $n = 101$, no BCs

t	x	u(it,i)	u_anal(it,i)	err(it,i)
0.00	−30.000	2.000001	2.000001	0.000000
0.00	−29.400	2.000001	2.000001	0.000000
0.00	−28.800	2.000001	2.000001	0.000000
0.00	−28.200	2.000002	2.000002	0.000000
0.00	−27.600	2.000002	2.000002	0.000000
0.00	−27.000	2.000003	2.000003	0.000000
	.		.	
	.		.	
	.		.	
0.00	−1.200	2.587976	2.587976	0.000000
0.00	−0.600	2.612768	2.612768	0.000000
0.00	0.000	2.621302	2.621302	0.000000
0.00	0.600	2.612768	2.612768	0.000000
0.00	1.200	2.587976	2.587976	0.000000
	.		.	
	.		.	
	.		.	
0.00	27.000	2.000003	2.000003	0.000000
0.00	27.600	2.000002	2.000002	0.000000
0.00	28.200	2.000002	2.000002	0.000000
0.00	28.800	2.000001	2.000001	0.000000
0.00	29.400	2.000001	2.000001	0.000000
0.00	30.000	2.000001	2.000001	0.000000
t	x	u(it,i)	u_anal(it,i)	err(it,i)
1.00	−30.000	2.000001	2.000000	0.000000
1.00	−29.400	2.000001	2.000000	0.000001
1.00	−28.800	2.000001	2.000000	0.000000
1.00	−28.200	2.000001	2.000000	0.000000
1.00	−27.600	2.000001	2.000001	0.000000
1.00	−27.000	2.000001	2.000001	−0.000000
	.		.	
	.		.	
	.		.	
1.00	1.200	2.597367	2.597334	0.000033
1.00	1.800	2.617241	2.617241	0.000000
1.00	2.400	2.620436	2.620467	−0.000031
1.00	3.000	2.606650	2.606705	−0.000055
1.00	3.600	2.577185	2.577252	−0.000067
	.		.	
	.		.	
	.		.	
1.00	27.000	2.000011	2.000011	0.000001
1.00	27.600	2.000008	2.000008	0.000000
1.00	28.200	2.000005	2.000005	−0.000000

Table 12.3: (*Continued*)

1.00	28.800	2.000003	2.000004	−0.000000
1.00	29.400	2.000003	2.000003	−0.000000
1.00	30.000	2.000001	2.000002	−0.000001
t	x	u(it,i)	u_anal(it,i)	err(it,i)
2.00	−30.000	2.000001	2.000000	0.000001
2.00	−29.400	2.000001	2.000000	0.000001
2.00	−28.800	2.000001	2.000000	0.000001
2.00	−28.200	2.000001	2.000000	0.000001
2.00	−27.600	2.000000	2.000000	0.000000
	.		.	
	.		.	
	.		.	
2.00	3.000	2.574920	2.574849	0.000071
2.00	3.600	2.605264	2.605243	0.000021
2.00	4.200	2.620053	2.620082	−0.000029
2.00	4.800	2.617898	2.617970	−0.000072
2.00	5.400	2.599010	2.599108	−0.000099
2.00	6.000	2.565150	2.565257	−0.000107
	.		.	
	.		.	
	.		.	
2.00	27.000	2.000039	2.000036	0.000003
2.00	27.600	2.000029	2.000026	0.000003
2.00	28.200	2.000021	2.000019	0.000002
2.00	28.800	2.000013	2.000013	−0.000000
2.00	29.400	2.000009	2.000010	−0.000000
2.00	30.000	2.000001	2.000007	−0.000006
t	x	u(it,i)	u_anal(it,i)	err(it,i)
3.00	−30.000	2.000001	2.000000	0.000001
3.00	−29.400	2.000007	2.000000	0.000007
3.00	−28.800	2.000004	2.000000	0.000004
3.00	−28.200	2.000003	2.000000	0.000003
3.00	−27.600	2.000001	2.000000	0.000001
3.00	−27.000	2.000004	2.000000	0.000004
	.		.	
	.		.	
	.		.	
3.00	5.400	2.585907	2.585846	0.000061
3.00	6.000	2.611622	2.611632	−0.000009
3.00	6.600	2.621197	2.621265	−0.000069
3.00	7.200	2.613722	2.613835	−0.000113
3.00	7.800	2.589914	2.590046	−0.000133
	.		.	
	.		.	
	.		.	

(*Continued*)

Table 12.3: *(Continued)*

t	x	u(it,i)	u_anal(it,i)	err(it,i)
3.00	27.000	2.000130	2.000122	0.000008
3.00	27.600	2.000097	2.000088	0.000009
3.00	28.200	2.000070	2.000063	0.000007
3.00	28.800	2.000045	2.000045	0.000000
3.00	29.400	2.000034	2.000032	0.000002
3.00	30.000	2.000001	2.000023	-0.000023
	ncall=162			

Table 12.4: Comparison of the tabular output for $n = 101$, with and without BCs in pde_1.m

With two Dirichlet BCs in pde_1 (from Table 12.2)				
3.00	5.400	2.585907	2.585846	0.000061
3.00	6.000	2.611622	2.611632	-0.000010
3.00	6.600	2.621197	2.621265	-0.000069
3.00	7.200	2.613723	2.613835	-0.000113
3.00	7.800	2.589914	2.590046	-0.000132
Without BCs in pde_1.m (from Table 12.3)				
3.00	5.400	2.585907	2.585846	0.000061
3.00	6.000	2.611622	2.611632	-0.000009
3.00	6.600	2.621197	2.621265	-0.000069
3.00	7.200	2.613722	2.613835	-0.000113
3.00	7.800	2.589914	2.590046	-0.000133

thoroughly studied (as discussed in Chapter 2), and we know that solutions can be computed. Of course, the solutions are not what we require (they are not for eq. (12.1)), but this is a start.

- Then, a solution might be attempted with "small" (nonzero) a and $b = 0$. In other words, this will be a small departure from the case of the nonlinear advection equation. If this is successful, then a might be increased (incrementally) to its final (desired) value.
- If success in computing a numerical solution continues, then b might be changed to a "small" (nonzero) value and increased (assuming continuing success) to its desired value.
- At some point along this incremental development, the code might fail, but at least, we will know the latest change that produced the failure and therefore have some indication of the source of the problem (such as the last change that was made).

- In the case of eq. (12.1), experience has indicated that the inclusion of higher-order spatial derivatives often increases the likelihood of a computational failure and therefore including the higher-order derivative incrementally (by increasing b) can reveal a sensitivity that must then somehow be remedied.

A way to view this incremental approach is to consider it as a form of *continuation* in which the solution to the initial problem such as the nonlinear advection equation is continued (or extended) to the solution of interest (for eq. (12.1)) by small parameter changes (slowly increasing a and b). Experience has demonstrated that this form of continuation can often proceed from a starting problem to the final problem of interest (rather than attempting the final problem at the beginning). This idea can also be applied to systems of PDEs, e.g., start with one PDE with a known solution (and use constant solutions for the other PDEs), then add a second PDE, etc.

The preceding discussion of an incremental approach is general but admittedly also rather vague. The chance of success of this approach is clearly going to depend on the problem, but it is offered as a way to gain some insight into why a prototype MOL code fails and to suggest what alternative approaches might lead to a successful code development. Clearly, this process is not necessarily straightforward with a priori assurance of success.

Appendix 1

In the preceding discussion, we briefly considered the application of `u3x9p.m` and `u5x11p.m` to the differentiation of polynomials. Here, we illustrate this process by differentiation of some polynomials. For this purpose, the following main program computes a third derivative of polynomials of various orders, starting with second-, third-, and fourth-order polynomials.

```
  clear all
  clc
%
% Default points, equally spaced grid
  for i=1:9
    x(i)=i-1;
  end
%
% Compute FD approximation for up to and including the mth
% derivative
  m=3;
%
% Number of grid points
  ng=9;
  nd=ng;
%
% Compute weighting coefficients for finite differences
% over ip points
  for ip=1:ng
%
```

```
%   Weighting coefficients in array C3
    C3(ip,:,:)=weights(x(ip),x,ng-1,nd-1,m);
  end
% size(C3)
%
% Display coefficients for derivative of order 3
% i=m+1;
% fprintf('\n\n Numerical Derivative Order: %d',i-1);
% fprintf('\n============================\n');
%
% Coefficients with m = 3, ng = 9
% C3(:,:,4)
%
% Grid in x
  xu=1; xl=0; dx=(xu-xl)/(ng-1); rdx3=1.0/dx^3;
% xu=2; xl=0; dx=(xu-xl)/(ng-1); rdx3=1.0/dx^3;
%
% Second order polynomial
  a=ones(1,3);
  for i=1:ng
    x(i)=(i-1)*dx;
    term=0;
    for ip=1:3
      term=term+a(ip)*x(i)^(ip-1);
    end
    p2(i)=term;
  end
%
% Differentiate second order polynomial
  fprintf('\n Third derivative of second order polynomial\n');
  fprintf(' irow    x(irow)       p2(irow)      uxxx(irow)   uxxxe(irow)');
  for irow=1:ng
    uxxx=0;
    for icol=1:ng
      uxxx=uxxx+rdx3*C3(irow,icol,4)*p2(icol);
    end
    uxxxe=0;
    fprintf('\n %4d%11.3f%15.5f%16.5f%14.5f',...
            irow,x(irow),p2(irow),uxxx,uxxxe);
  end
  fprintf('\n\n');
%
% Third order polynomial
  a=ones(1,4);
  for i=1:ng
    x(i)=(i-1)*dx;
    term=0;
    for ip=1:4
      term=term+a(ip)*x(i)^(ip-1);
    end
    p3(i)=term;
  end
%
% Differentiate third order polynomial
```

```
  fprintf('\n Third derivative of third order polynomial\n');
  fprintf(' irow    x(irow)       p3(irow)      uxxx(irow)   uxxxe(irow)');
  for irow=1:ng
    uxxx=0;
    for icol=1:ng
      uxxx=uxxx+rdx3*C3(irow,icol,4)*p3(icol);
    end
    uxxxe=3*2*1;
    fprintf('\n %4d%11.3f%15.5f%16.5f%14.5f',...
            irow,x(irow),p3(irow),uxxx,uxxxe);
  end
  fprintf('\n\n');
%
% Fourth order polynomial
  a=ones(1,5);
  for i=1:ng
    x(i)=(i-1)*dx;
    term=0;
    for ip=1:5
      term=term+a(ip)*x(i)^(ip-1);
    end
    p4(i)=term;
  end
%
% Differentiate fourth order polynomial
  fprintf('\n Third derivative of fourth order polynomial\n');
  fprintf(' irow    x(irow)       p4(irow)      uxxx(irow)   uxxxe(irow)');
  for irow=1:ng
    uxxx=0;
    for icol=1:ng
      uxxx=uxxx+rdx3*C3(irow,icol,4)*p4(icol);
    end
    uxxxe=3*2*1+4*3*2*x(irow);
    fprintf('\n %4d%11.3f%15.5f%16.5f%14.5f',...
            irow,x(irow),p4(irow),uxxx,uxxxe);
  end
  fprintf('\n\n');
```

LISTING 12.6: Main program `u3x_poly.m` for the third-order derivative of polynomials of varying order.

We can note the following points about `u3x_poly.m`:

- After clearing files, a uniform grid of nine points is defined.

```
  clear all
  clc
%
% Default points, equally spaced grid
  for i=1:9
    x(i)=i-1;
  end
```

- The input parameters to routine `weights` are defined numerically.

```
%
% Compute FD approximation for up to and including the mth
```

```
% derivative
  m=3;
%
% Number of grid points
  ng=9;
  nd=ng;
```

- The FD weighting coefficients are computed by `weights`. Note that the output statements for `C3` are deactivated (commented) to conserve space in the output. If they are activated (uncommented), they will display the 9×9 array of FD coefficients.

```
%
% Compute weighting coefficients for finite differences
% over ip points
  for ip=1:ng
%
%   Weighting coefficients in array C3
    C3(ip,:,:)=weights(x(ip),x,ng-1,nd-1,m);
  end
% size(C3)
%
% Display coefficients for derivative of order 3
% i=m+1;
% fprintf('\n\n Numerical Derivative Order: %d',i-1);
% fprintf('\n==============================\n');
%
% Coefficients with m = 3, ng = 9
% C3(:,:,4)
```

- A second-order polynomial, $p_2(x)$,

$$p_2(x) = a_0 + a_1 x + a_2 x^2 \tag{12.3}$$

is defined (`p2(i)`). Note that Matlab does not permit a zero subscript for use in the first polynomial coefficient, a_0. Also, the three polynomial coefficients are set to one with the Matlab `ones` utility, but the conclusions to follow apply to any second-order polynomial.

```
%
% Grid in x
  xu=1; xl=0; dx=(xu-xl)/(ng-1); rdx3=1.0/dx^3;
% xu=2; xl=0; dx=(xu-xl)/(ng-1); rdx3=1.0/dx^3;
%
% Second order polynomial
  a=ones(1,3);
  for i=1:ng
    x(i)=(i-1)*dx;
    term=0;
    for ip=1:3
      term=term+a(ip)*x(i)^(ip-1);
    end
    p2(i)=term;
  end
```

- The second-order polynomial is then differentiated by multiplication by the matrix of FD weighting coefficients, `C3` (to produce the third-order derivative `uxxx`).

```
%
% Differentiate second order polynomial
  fprintf('\n Third derivative of second order polynomial\n');
  fprintf(' irow    x(irow)       p2(irow)      uxxx(irow)   uxxxe(irow)');
  for irow=1:ng
    uxxx=0;
    for icol=1:ng
      uxxx=uxxx+rdx3*C3(irow,icol,4)*p2(icol);
    end
    uxxxe=0;
    fprintf('\n %4d%11.3f%15.5f%16.5f%14.5f',...
            irow,x(irow),p2(irow),uxxx,uxxxe);
  end
  fprintf('\n\n');
```

The exact third-order derivative, `uxxxe`, is also set (the third-order derivative of a second-order polynomial is zero). The exact and numerical third derivatives are then displayed for comparison. The output follows.

```
Third derivative of second order polynomial
 irow    x(irow)       p2(irow)      uxxx(irow)   uxxxe(irow)
    1      0.000        1.00000         0.00000       0.00000
    2      0.125        1.14063         0.00000       0.00000
    3      0.250        1.31250         0.00000       0.00000
    4      0.375        1.51563         0.00000       0.00000
    5      0.500        1.75000        -0.00000       0.00000
    6      0.625        2.01563        -0.00000       0.00000
    7      0.750        2.31250         0.00000       0.00000
    8      0.875        2.64063        -0.00000       0.00000
    9      1.000        3.00000        -0.00000       0.00000
```

The exact and numerical third-order derivatives are in agreement.

- The third derivative of a third-order polynomial, `p3(i)`, is then programmed in the same way as for a second-order polynomial.

```
%
% Third order polynomial
  a=ones(1,4);
  for i=1:ng
    x(i)=(i-1)*dx;
    term=0;
    for ip=1:4
      term=term+a(ip)*x(i)^(ip-1);
    end
    p3(i)=term;
  end
%
% Differentiate third order polynomial
  fprintf('\n Third derivative of third order polynomial\n');
  fprintf(' irow    x(irow)       p3(irow)      uxxx(irow)   uxxxe(irow)');
  for irow=1:ng
    uxxx=0;
```

```
    for icol=1:ng
      uxxx=uxxx+rdx3*C3(irow,icol,4)*p3(icol);
    end
    uxxxe=3*2*1;
    fprintf('\n %4d%11.3f%15.5f%16.5f%14.5f',...
            irow,x(irow),p3(irow),uxxx,uxxxe);
  end
  fprintf('\n\n');
```

The output from this code is

```
Third derivative of third order polynomial
 irow     x(irow)        p3(irow)          uxxx(irow)    uxxxe(irow)
    1       0.000         1.00000             6.00000        6.00000
    2       0.125         1.14258             6.00000        6.00000
    3       0.250         1.32813             6.00000        6.00000
    4       0.375         1.56836             6.00000        6.00000
    5       0.500         1.87500             6.00000        6.00000
    6       0.625         2.25977             6.00000        6.00000
    7       0.750         2.73438             6.00000        6.00000
    8       0.875         3.31055             6.00000        6.00000
    9       1.000         4.00000             6.00000        6.00000
```

As expected, the numerical and exact derivatives agree (the third derivative of a third-order polynomial is $(3)(2)(1)a_3$ and in this case, $a_3 = 1$).

- The third derivative of a fourth-order polynomial, `p4(i)`, is then programmed in the same way as for the second- and third-order polynomials.

```
%
% Fourth order polynomial
  a=ones(1,5);
  for i=1:ng
    x(i)=(i-1)*dx;
    term=0;
    for ip=1:5
      term=term+a(ip)*x(i)^(ip-1);
    end
    p4(i)=term;
  end
%
% Differentiate fourth order polynomial
  fprintf('\n Third derivative of fourth order polynomial\n');
  fprintf(' irow     x(irow)         p4(irow)        uxxx(irow)    uxxxe(irow)');
  for irow=1:ng
    uxxx=0;
    for icol=1:ng
      uxxx=uxxx+rdx3*C3(irow,icol,4)*p4(icol);
    end
    uxxxe=3*2*1+4*3*2*x(irow);
    fprintf('\n %4d%11.3f%15.5f%16.5f%14.5f',...
            irow,x(irow),p4(irow),uxxx,uxxxe);
  end
  fprintf('\n\n');
```

The output from this code is

```
 Third derivative of fourth order polynomial
  irow     x(irow)          p4(irow)         uxxx(irow)    uxxxe(irow)
     1       0.000           1.00000            6.00000        6.00000
     2       0.125           1.14282            9.00000        9.00000
     3       0.250           1.33203           12.00000       12.00000
     4       0.375           1.58813           15.00000       15.00000
     5       0.500           1.93750           18.00000       18.00000
     6       0.625           2.41235           21.00000       21.00000
     7       0.750           3.05078           24.00000       24.00000
     8       0.875           3.89673           27.00000       27.00000
     9       1.000           5.00000           30.00000       30.00000
```

The numerical and exact derivatives agree (the third derivative of a fourth-order polynomial is a linear function in x, that is, $(3)(2)(1)a_3 + (4)(3)(2)a_4x$ and in this case, $a_4 = 1, a_4 = 1$).

The preceding results imply that the numerical third derivative is exact for polynomials up to fourth order. In fact, it is exact for polynomials up to eighth order. This is demonstrated by the following extensions to the preceding code. The third derivative of a seventh-order polynomial, `p7(i)`, is computed as

```
%
% Seventh order polynomial
  a=ones(1,8);
  for i=1:ng
    x(i)=(i-1)*dx;
    term=0;
    for ip=1:8
      term=term+a(ip)*x(i)^(ip-1);
    end
    p7(i)=term;
  end
%
% Differentiate seventh order polynomial
  fprintf('\n Third order derivative of seventh order polynomial\n');
  fprintf(' irow     x(irow)          p7(irow)       uxxx(irow)  uxxxe(irow)');
  for irow=1:ng
    uxxx=0;
    for icol=1:ng
      uxxx=uxxx+rdx3*C3(irow,icol,4)*p7(icol);
    end
    uxxxe=3*2*1+4*3*2*x(irow)  +5*4*3*x(irow)^2 ...
               +6*5*4*x(irow)^3+7*6*5*x(irow)^4;
    fprintf('\n %4d%11.3f%15.5f%16.5f%14.5f',...
             irow,x(irow),p7(irow),uxxx,uxxxe);
  end
  fprintf('\n\n');
```

The output from this code is

```
Third order derivative of seventh order polynomial
 irow     x(irow)         p7(irow)        uxxx(irow)    uxxxe(irow)
    1       0.000          1.00000           6.00000        6.00000
    2       0.125          1.14286          10.22314       10.22314
    3       0.250          1.33331          18.44531       18.44531
    4       0.375          1.59937          33.91846       33.91846
    5       0.500          1.99219          61.12500       61.12500
    6       0.625          2.60458         105.77783      105.77783
    7       0.750          3.59955         174.82031      174.82031
    8       0.875          5.25113         276.42627      276.42627
    9       1.000          8.00000         420.00000      420.00000
```

so that the numerical and exact derivatives agree.

The third derivative of an eighth-order polynomial, `p8(i)`, is computed as

```
%
% Eighth order polynomial
  a=ones(1,9);
  for i=1:ng
    x(i)=(i-1)*dx;
    term=0;
    for ip=1:9
      term=term+a(ip)*x(i)^(ip-1);
    end
    p8(i)=term;
  end
%
% Differentiate eighth order polynomial
  fprintf('\n Third order derivative of eighth order polynomial\n');
  fprintf(' irow    x(irow)        p8(irow)       uxxx(irow)   uxxxe(irow)');
  for irow=1:ng
    uxxx=0;
    for icol=1:ng
      uxxx=uxxx+rdx3*C3(irow,icol,4)*p8(icol);
    end
    uxxxe=3*2*1+4*3*2*x(irow)  +5*4*3*x(irow)^2 ...
               +6*5*4*x(irow)^3+7*6*5*x(irow)^4 ...
               +8*7*6*x(irow)^5;
    fprintf('\n %4d%11.3f%15.5f%16.5f%14.5f',...
            irow,x(irow),p8(irow),uxxx,uxxxe);
  end
  fprintf('\n\n');
```

The output from this code is

```
Third order derivative of eighth order polynomial
 irow     x(irow)         p8(irow)        uxxx(irow)    uxxxe(irow)
    1       0.000          1.00000           6.00000        6.00000
    2       0.125          1.14286          10.23340       10.23340
    3       0.250          1.33333          18.77344       18.77344
    4       0.375          1.59977          36.41016       36.41016
    5       0.500          1.99609          71.62500       71.62500
```

```
     6      0.625          2.62786         137.82129      137.82129
     7      0.750          3.69966         254.55469      254.55469
     8      0.875          5.59474         448.76367      448.76367
     9      1.000          9.00000         756.00000      756.00000
```

so that the numerical and exact derivatives agree.

Finally, the third derivative of a ninth-order polynomial, `p9(i)`, is computed as

```
%
% Ninth order polynomial
  a=ones(1,10);
  for i=1:ng
    x(i)=(i-1)*dx;
    term=0;
    for ip=1:10
      term=term+a(ip)*x(i)^(ip-1);
    end
    p9(i)=term;
  end
%
% Differentiate ninth order polynomial
  fprintf('\n Third order derivative of ninth order polynomial\n');
  fprintf(' irow    x(irow)        p9(irow)      uxxx(irow)    uxxxe(irow)');
  for irow=1:ng
    uxxx=0;
    for icol=1:ng
      uxxx=uxxx+rdx3*C3(irow,icol,4)*p9(icol);
    end
    uxxxe=3*2*1+4*3*2*x(irow)  +5*4*3*x(irow)^2 ...
                +6*5*4*x(irow)^3+7*6*5*x(irow)^4 ...
                +8*7*6*x(irow)^5+9*8*7*x(irow)^6;
    fprintf('\n %4d%11.3f%15.5f%16.5f%14.5f',...
                irow,x(irow),p9(irow),uxxx,uxxxe);
  end
  fprintf('\n\n');
```

The output from this code is

```
Third order derivative of ninth order polynomial
 irow    x(irow)        p9(irow)      uxxx(irow)    uxxxe(irow)
    1      0.000         1.00000         3.29636        6.00000
    2      0.125         1.14286        10.23679       10.23532
    3      0.250         1.33333        18.92679       18.89648
    4      0.375         1.59991        37.79013       37.81174
    5      0.500         1.99805        79.51877       79.50000
    6      0.625         2.64241       167.84042      167.86203
    7      0.750         3.77475       344.28616      344.25586
    8      0.875         5.89540       674.95798      674.95651
    9      1.000        10.00000      1257.29636     1260.00000
```

so that the numerical and exact derivatives **do not** agree. The preceding results imply that the third-order FD differentiation is exact for polynomials up to and including eighth order.

Before we go on to consider some additional cases, we first consider the operation of the nested `for` loops that perform the numerical differentiation through the use of `C3`. For example, the third-order differentiation of a third-order polynomial was programmed as

```
%
% Differentiate third order polynomial
  fprintf('\n Third derivative of third order polynomial\n');
  fprintf(' irow    x(irow)        p3(irow)       uxxx(irow)    uxxxe(irow)');
  for irow=1:ng
    uxxx=0;
    for icol=1:ng
      uxxx=uxxx+rdx3*C3(irow,icol,4)*p3(icol);
    end
    uxxxe=3*2*1;
    fprintf('\n %4d%11.3f%15.5f%16.5f%14.5f',...
            irow,x(irow),p3(irow),uxxx,uxxxe);
  end
  fprintf('\n\n');
```

Basically, the inner `for` loop on `icol` produces one row from a matrix multiplication between the vector `p3(icol)` and the matrix `C3(irow,icol,4)`. The outer `for` loop on `irow` then produces the nine rows (`ng=9`) of the matrix–vector product. In other words, the two nested `for` loops produce the following matrix equation (for nine points).

$$\begin{bmatrix} \dfrac{d^3 p_3(x_0)}{dx^3} \\ \\ \dfrac{d^3 p_3(x_1)}{dx^3} \\ \vdots \\ \dfrac{d^3 p_3(x_7)}{dx^3} \\ \\ \dfrac{d^3 p_3(x_8)}{dx^3} \end{bmatrix} = \begin{bmatrix} C_3(1,1,4)p_3(x_0)+C_3(1,2,4)p_3(x_1) & \cdots & +C_3(1,9,4)p_3(x_8) \\ C_3(2,1,4)p_3(x_0)+C_3(2,2,4)p_3(x_1) & \cdots & +C_3(2,9,4)p_3(x_8) \\ & \ddots & \\ C_3(8,1,4)p_3(x_0)+C_3(8,2,4)p_3(x_1) & \cdots & +C_3(8,9,4)p_3(x_8) \\ C_3(9,1,4)p_3(x_0)+C_3(9,2,4)p_3(x_1) & \cdots & +C_3(9,9,4)p_3(x_8) \end{bmatrix} \tag{12.4}$$

Note that the use of the third index as `4` in `C3(irow,jcol,4)` corresponds to the third derivative, since the index 1 corresponds to the function $p_3(x)$, and the indices 2, 3, and 4 correspond to the first-, second-, and third-order derivatives, respectively, as explained previously.

Also, we can easily look at the FD weighting coefficients (by activating the line C3(:,:,4) in the preceding code). The result is the 9×9 matrix, which is listed below.

```
-10.0125   58.1667 -152.9417  239.1000 -242.8333  163.0333  -70.1250   17.5667   -1.9542
 -1.9542    7.5750  -12.1833   11.2083   -7.1250    3.3917   -1.1167    0.2250   -0.0208
 -0.0208   -1.7667    6.8250  -10.4333    8.5833   -4.5000    1.6417   -0.3667    0.0375
  0.0375   -0.3583   -0.4167    3.6750   -5.7083    3.8583   -1.3500    0.2917   -0.0292
 -0.0292    0.3000   -1.4083    2.0333   -0.0000   -2.0333    1.4083   -0.3000    0.0292
```

```
 0.0292   -0.2917    1.3500   -3.8583    5.7083   -3.6750    0.4167    0.3583   -0.0375
-0.0375    0.3667   -1.6417    4.5000   -8.5833   10.4333   -6.8250    1.7667    0.0208
 0.0208   -0.2250    1.1167   -3.3917    7.1250  -11.2083   12.1833   -7.5750    1.9542
 1.9542  -17.5667   70.1250 -163.0333  242.8333 -239.1000  152.9417  -58.1667   10.0125
```

The fifth row is used in function `u3xp9.m` of Listing 12.4 through the code

```
uxxx(i)=rdx3*...
(C3(5,1,4)*u(i-4)...
+C3(5,2,4)*u(i-3)...
+C3(5,3,4)*u(i-2)...
+C3(5,4,4)*u(i-1)...
+C3(5,5,4)*u(i  )...
+C3(5,6,4)*u(i+1)...
+C3(5,7,4)*u(i+2)...
+C3(5,8,4)*u(i+3)...
+C3(5,9,4)*u(i+4));
```

so that, for example, `C3(5,1,4) = -0.0292, C3(5,5,4) = 0, C3(5,9,4) = 0.0292`. The first to fourth rows and sixth to ninth are not used in `u3x9p.m`, as explained previously (the derivative `uxxx` is set to zero).

The above 9×9 matrix used in the matrix–vector multiplication to produce the nine-point FD approximations is an example of a *differentiation matrix*. The elements of a differentiation matrix can be selected to produce FD approximations of varying orders for derivatives of varying orders, including approximations for boundary conditions [4, 10].

The FD approximations over nine points are actually eighth-order correct, so they will compute the numerical eighth-order derivative of an eighth-order polynomial exactly (and we demonstrated previously that they compute a third-order derivative of an eighth-order polynomial exactly). We can demonstrate this conclusion (the FDs are eighth-order correct) by executing the preceding code for the eighth-order derivative of a seventh-, eighth-, and ninth-order polynomial. This requires an easy modification of the preceding code.

```
  clear all
  clc
%
% Default points, equally spaced grid
  for i=1:9
    x(i)=i-1;
  end
%
% Compute FD approximation for up to and including the mth
% derivative
  m=8;
%
% Number of grid points
  ng=9;
  nd=ng;
%
% Compute weighting coefficients for finite differences
% over ip points
```

```
  for ip=1:ng
%
%   Weighting coefficients in array C8
    C8(ip,:,:)=weights(x(ip),x,ng-1,nd-1,m);
  end
% size(C8)
%
% Display coefficients for derivative of order 8
% i=m+1;
% fprintf('\n\n Numerical Derivative Order: %d',i-1);
% fprintf('\n=============================\n');
%
% Coefficients with m = 8, ng = 9
% C8(:,:,9)
%
% Grid in x
  xu=1; xl=0; dx=(xu-xl)/(ng-1); rdx8=1.0/dx^8;
%
% Seventh order polynomial
  a=ones(1,8);
  for i=1:ng
    x(i)=(i-1)*dx;
    term=0;
    for ip=1:8
      term=term+a(ip)*x(i)^(ip-1);
    end
    p7(i)=term;
  end
%
% Differentiate seventh order polynomial
  fprintf('\n Eighth order derivative of seventh order polynomial\n');
  fprintf(' irow    x(irow)       p7(irow)       u8x(irow)    u8xe(irow)');
  for irow=1:ng
    u8x=0;
    for icol=1:ng
      u8x=u8x+rdx8*C8(irow,icol,9)*p7(icol);
    end
    u8xe=0;
    fprintf('\n %4d%11.3f%15.5f%16.5f%14.5f',...
            irow,x(irow),p7(irow),u8x,u8xe);
  end
  fprintf('\n\n');
%
% Eighth order polynomial
  a=ones(1,9);
  for i=1:ng
    x(i)=(i-1)*dx;
    term=0;
    for ip=1:9
      term=term+a(ip)*x(i)^(ip-1);
    end
    p8(i)=term;
  end
%
```

```
% Differentiate eighth order polynomial
  fprintf('\n Eighth order derivative of eighth order polynomial\n');
  fprintf(' irow    x(irow)       p3(irow)       u8x(irow)    u8xe(irow)');
  for irow=1:ng
    u8x=0;
    for icol=1:ng
      u8x=u8x+rdx8*C8(irow,icol,9)*p8(icol);
    end
    u8xe=8*7*6*5*4*3*2*1;
    fprintf('\n %4d%11.3f%15.5f%16.5f%14.5f',...
            irow,x(irow),p8(irow),u8x,u8xe);
  end
  fprintf('\n\n');
%
% Ninth order polynomial
  a=ones(1,10);
  for i=1:ng
    x(i)=(i-1)*dx;
    term=0;
    for ip=1:10
      term=term+a(ip)*x(i)^(ip-1);
    end
    p9(i)=term;
  end
%
% Differentiate ninth order polynomial
  fprintf('\n Eighth order derivative of ninth order polynomial\n');
  fprintf(' irow    x(irow)       p9(irow)       u8x(irow)    u8xe(irow)');
  for irow=1:ng
    u8x=0;
    for icol=1:ng
      u8x=u8x+rdx8*C8(irow,icol,9)*p9(icol);
    end
    u8xe=8*7*6*5*4*3*2*1+9*8*7*6*5*4*3*2*x(irow);
    fprintf('\n %4d%11.3f%15.5f%16.5f%14.5f',...
            irow,x(irow),p9(irow),u8x,u8xe);
  end
  fprintf('\n\n');
```

The numerical output from this code is

```
Eighth order derivative of seventh order polynomial
 irow     x(irow)        p7(irow)        u8x(irow)     u8xe(irow)
    1       0.000         1.00000          0.00000        0.00000
    2       0.125         1.14286          0.00000        0.00000
    3       0.250         1.33331          0.00000        0.00000
    4       0.375         1.59937          0.00000        0.00000
    5       0.500         1.99219          0.00000        0.00000
    6       0.625         2.60458          0.00000        0.00000
    7       0.750         3.59955          0.00000        0.00000
    8       0.875         5.25113          0.00000        0.00000
    9       1.000         8.00000          0.00000        0.00000
```

```
Eighth order derivative of eighth order polynomial
irow     x(irow)         p3(irow)        u8x(irow)      u8xe(irow)
   1       0.000          1.00000      40320.00000     40320.00000
   2       0.125          1.14286      40320.00000     40320.00000
   3       0.250          1.33333      40320.00000     40320.00000
   4       0.375          1.59977      40320.00000     40320.00000
   5       0.500          1.99609      40320.00000     40320.00000
   6       0.625          2.62786      40320.00000     40320.00000
   7       0.750          3.69966      40320.00000     40320.00000
   8       0.875          5.59474      40320.00000     40320.00000
   9       1.000          9.00000      40320.00000     40320.00000

Eighth order derivative of ninth order polynomial
irow     x(irow)         p9(irow)        u8x(irow)      u8xe(irow)
   1       0.000          1.00000     221760.00000     40320.00000
   2       0.125          1.14286     221760.00000     85680.00000
   3       0.250          1.33333     221760.00000    131040.00000
   4       0.375          1.59991     221760.00000    176400.00000
   5       0.500          1.99805     221760.00000    221760.00000
   6       0.625          2.64241     221760.00000    267120.00000
   7       0.750          3.77475     221760.00000    312480.00000
   8       0.875          5.89540     221760.00000    357840.00000
   9       1.000         10.00000     221760.00000    403200.00000
```

This output implies the nine-point, eighth-order FD differentiation is eighth-order correct, i.e., exact for polynomials up to and including eighth order.

We could perform a similar analysis for the fifth-order FD differentiation in `u5x11p.m` and would come to the conclusion that the 11-point, fifth-order differentiation is tenth order correct, i.e., exact for polynomials up to and including tenth order, and also, the 11-point, tenth-order FD differentiation is tenth-order correct.

Finally, the discussion to this point has been entirely about the order (accuracy) of FD differentiation applied to polynomials. However, PDEs in general do not have polynomials as solutions (consider eq. (12.2)). We, therefore, end the discussion with a brief consideration of the accuracy of FD differentiation when applied to nonpolynomial functions. For example, if we consider the function

$$f(x) = e^x \tag{12.5}$$

The nine-point, FD third derivative of this function is programmed as follows.

```
  clear all
  clc
%
% Default points, equally spaced grid
  for i=1:9
    x(i)=i-1;
  end
%
% Compute FD approximation for up to and including the mth
```

```
% derivative
  m=3;
%
% Number of grid points
  ng=9;
  nd=ng;
%
% Compute weighting coefficients for finite differences
% over ip points
  for ip=1:ng
%
%   Weighting coefficients in array C3
    C3(ip,:,:)=weights(x(ip),x,ng-1,nd-1,m);
  end
% size(C3)
%
% Display coefficients for derivative of order 3
% i=m+1;
% fprintf('\n\n Numerical Derivative Order: %d',i-1);
% fprintf('\n==============================\n');
%
% Coefficients with m = 3, ng = 9
% C3(:,:,4)
%
% Grid in x
  xu=1; xl=0; dx=(xu-xl)/(ng-1); rdx3=1.0/dx^3;
%
% Exponential
  for i=1:ng
    x(i)=(i-1)*dx;
    expx(i)=exp(x(i));
  end
%
% Differentiate exponential
  fprintf('\n Exponential\n');
  fprintf(' irow    x(irow)       exp(irow)      uxxx(irow)    uxxxe(irow)');
  for irow=1:ng
    uxxx=0;
    for icol=1:ng
      uxxx=uxxx+rdx3*C3(irow,icol,4)*expx(icol);
    end
    uxxxe=expx(irow);
    fprintf('\n %4d%11.3f%16.5f%16.5f%14.5f',...
            irow,x(irow),expx(irow),uxxx,uxxxe);
  end
  fprintf('\n\n');
```

LISTING 12.7: Main program u3x_exp.m for the third-order derivative of an exponential.

We can note the following details about this code:

- The first part up to the call to weights to compute the FD coefficients is the same as in u3x_poly.m in Listing 12.6.

- The test function of eq. (12.5) is programmed and differentiated as

```
%
% Exponential
  for i=1:ng
    x(i)=(i-1)*dx;
    expx(i)=exp(x(i));
  end
%
% Differentiate exponential
  fprintf('\n Exponential\n');
  fprintf(' irow    x(irow)          exp(irow)        uxxx(irow)    uxxxe(irow)');
  for irow=1:ng
    uxxx=0;
    for icol=1:ng
      uxxx=uxxx+rdx3*C3(irow,icol,4)*expx(icol);
    end
    uxxxe=expx(irow);
    fprintf('\n %4d%11.3f%16.5f%16.5f%14.5f',...
            irow,x(irow),expx(irow),uxxx,uxxxe);
  end
  fprintf('\n\n');
```

The output from this code is

```
Exponential
 irow    x(irow)          exp(irow)        uxxx(irow)    uxxxe(irow)
    1      0.000            1.00000           0.99999        1.00000
    2      0.125            1.13315           1.13315        1.13315
    3      0.250            1.28403           1.28403        1.28403
    4      0.375            1.45499           1.45499        1.45499
    5      0.500            1.64872           1.64872        1.64872
    6      0.625            1.86825           1.86825        1.86825
    7      0.750            2.11700           2.11700        2.11700
    8      0.875            2.39888           2.39888        2.39888
    9      1.000            2.71828           2.71827        2.71828
```

- These results could lead us to the conclusion that the FD differentiation is exact for exponential functions. This is not correct, however. One way to explore this point is to consider the exponential function to be a polynomial of infinite order through its Taylor series

$$e^x = 1 + x/1! + x^2/2! + \cdots \tag{12.6}$$

Since the preceding discussion indicated that FD differentiation is exact only up to a given polynomial of finite order, we could conclude that FD differentiation of the exponential function is not exact. This is a correct conclusion, even though the previous numerical output suggests otherwise. Rather, the numerical output is explained by realizing the exponential function of eq. (12.5) has a small enough variation in x that the FD differentiation appears to be exact, but if more figures were displayed, differences between the numerical (FD) and exact derivatives would be observed (note the difference in the numerical and exact derivatives for x(irow)=1.000).

- To explore this possible explanation, we consider the exponential function

$$f(x) = e^{ax} \tag{12.7}$$

then select a to give the exponential function a greater variation in x. For this purpose, we take a=5 in the following code (at the end of the u3x_exp.m in Listing 12.7)

```
%
% Exponential
  a=5;
  for i=1:ng
    x(i)=(i-1)*dx;
    expx(i)=exp(a*x(i));
  end
%
% Differentiate exponential
  fprintf('\n Exponential\n');
  fprintf(' irow    x(irow)        exp(irow)       uxxx(irow)    uxxxe(irow)');
  for irow=1:ng
    uxxx=0;
    for icol=1:ng
      uxxx=uxxx+rdx3*C3(irow,icol,4)*expx(icol);
    end
    uxxxe=a^3*expx(irow);
    fprintf('\n %4d%11.3f%16.5f%16.5f%14.5f',...
            irow,x(irow),expx(irow),uxxx,uxxxe);
  end
  fprintf('\n\n');
```

Now the exact derivative, uxxxe=a^3*expx(irow), also has a larger variation in x than for a=1.

The output from this code is

```
Exponential
 irow    x(irow)        exp(irow)       uxxx(irow)    uxxxe(irow)
    1      0.000          1.00000        -25.43268      125.00000
    2      0.125          1.86825        233.03952      233.53074
    3      0.250          3.49034        438.36612      436.29287
    4      0.375          6.52082        813.59196      815.10239
    5      0.500         12.18249       1524.17727     1522.81175
    6      0.625         22.75990       2843.35065     2844.98689
    7      0.750         42.52108       5317.46213     5315.13525
    8      0.875         79.43984       9931.33615     9929.97994
    9      1.000        148.41316      18273.62459    18551.64489
```

Clearly, the numerical and exact derivatives **do not** agree. This result provides an example indicating that FD differentiation is generally not exact. The fact that it is exact for polynomials of limited order is to be expected when we realize that the FD approximations are based on polynomials. In the case of the routine weights, a Lagrange interpolation polynomial was used by Fornberg [3]. This suggests that an increase in the number of points on which the FD approximation is based will increase the order of the approximation (since a higher-order polynomial will be the basis for the FD approximation).

In summary, the intent of this discussion in Appendix 1 is to elaborate on the FD third-derivative approximations in `u3xp9.m` and the FD fifth-derivative approximations in `u5x11p.m` used to calculate the third and fifth derivatives, $\frac{\partial^3 u}{\partial x^3}$ and $\frac{\partial^5 u}{\partial x^5}$ in eq. (12.1). Specifically, we have considered the accuracy of the approximations in these two routines in terms of their order, which is relative to polynomials, e.g., the nine-point, third derivative FDs in `u3xp9.m` are eighth-order correct; that is, they differentiate eighth-order polynomials exactly.

However, approximations for derivatives do not have to be based on polynomials, and other approaches to derivative approximations such as spectral methods have been extensively developed and used. Thus, we have presented only one approach based on FDs, and our intent within this context is to demonstrate how approximations of various orders (again, relative to polynomials) for derivatives of various orders can be implemented within the MOL format. Typically, some trial and error is required in the development of a PDE solution, and we have presented one possible approach.

Appendix 2

The Kawahara equation, which we repeat below for convenience,

$$\frac{\partial u}{\partial t} + u\frac{\partial u}{\partial x} + a\frac{\partial^3 u}{\partial x^3} - b\frac{\partial^5 u}{\partial x^5} = 0 \tag{12.1, 12.8}$$

was derived in order to obtain higher-order dispersive equations as corrections to the KdV equation [7]. It describes propagation of signals in electric transmission lines [1], long waves under ice cover in liquids of finite depth [5], and water waves with surface tension [13].

One of the things we notice about eq. (12.8) is that it contains only odd derivatives. Usually, waves propagating in a system described by linear PDEs consisting entirely of odd derivatives, with at least one being a third or higher derivative, exhibit *dispersion*. Systems described by PDEs with both odd and even (with appropriate signs)- order derivatives generally, but not always, exhibit both dispersion and *dissipation*. The degree of dispersion or dissipation is usually determined by the medium in which the wave propagates. Consequently, a medium can be characterized as being dispersive or dissipative. We will illustrate these concepts by example and show how they relate to the Kawahara equation.

Let us first consider a linearized form of the *Korteweg-de Vries* (KdV) equation

$$\frac{\partial u}{\partial t} + \frac{\partial u}{\partial x} + \frac{\partial^3 u}{\partial x^3} = 0 \tag{12.9}$$

and assume that it has a pure harmonic solution $u(x,t) = A\exp(i[kx - \omega t])$, where A represents amplitude, ω represents *frequency*, and k represents *wavenumber* (number of *wave lengths* λ, in 2π, i.e., $k = 2\pi/\lambda$). On substituting this solution into eq. (12.9), we obtain

$$\omega(k) = k - k^3$$

which is known as the *dispersion relation*; this defines the relationship between frequency and wavenumber. A situation where waves of different frequencies move at different speeds without a change in amplitude is called *dispersion*. Generally, this results in higher-frequency components traveling at slower speeds than the lower-frequency components. The effect of dispersion, therefore, is that solutions involving *sharp gradient, discontinuity*, or *shock* effects usually exhibit a changing profile over time.

Thus, our solution becomes $u(x,t) = A\exp\{ik[x - (1-k^2)t]\}$. But, from our discussion on traveling waves, we know that traveling waves are defined by $u(x,t) = u[kx - ct]$, where c represents the speed of propagation. Thus, we have $c_p = \omega/k$, which is known as the *phase velocity* and implies for eq. (12.9) that $c_p = 1 - k^2$. Therefore, we observe that the speed of propagation of a single harmonic wave varies according to the wavenumber k. In other words, waves of different frequencies propagate at different phase velocities in a dispersive medium.

For *wave packets* consisting of a group of harmonic waves, the associated *group velocity* is given by the derivative of the real part of ω with respect to k, i.e., $c_g = d\Re(\omega)/dk$. For this example, $c_g = 1 - 3k^2$, which means that a wave packet or the *wave envelope* will move at a velocity different from the individual harmonic phase velocities from which it is composed. It is assumed that the packet is confined to a finite region of the spatial domain and that it may contain a superposition of many harmonic waves, but with wavenumbers clustered around k [12]. For a nondispersive wave, all frequency components travel at the same speed and therefore we have $c_g = c_p$. Wave *energy* and *information signals* propagate at group velocity.

If we include an even spatial derivative in eq. (12.9), say second, we have

$$\frac{\partial u}{\partial t} + \frac{\partial u}{\partial x} + \frac{\partial^2 u}{\partial x^2} + \frac{\partial^3 u}{\partial x^3} = 0 \qquad (12.10)$$

Then, its dispersion relation becomes

$$\omega(k) = k - k^3 - ik^2$$

and the harmonic solution is therefore

$$\begin{aligned} u(x,t) &= A\exp\left\{i\left[kx - (k-k^3)t\right] - k^2 t\right\} \\ &= A\exp\left\{i\left[kx - (k-k^3)t\right]\right\}\exp(-k^2 t) \end{aligned} \qquad (12.11)$$

The term $\exp\{i[kx - (k-k^3)t]\}$ in eq. (12.11) is a dispersive term and means that the frequency depends upon wavenumber. However, the term $\exp(-k^2 t)$ means that for real

$k \neq 0$, the solution u decays exponentially over time, i.e., $u(x,t) \to 0$ as $t \to \infty$. This latter phenomenon, where waves are *damped,* is called *dissipation.* Generally, higher-frequency components of a wave are damped more than lower-frequency components. The effect of dissipation, therefore, is that *sharp gradients, discontinuities,* or *shocks* in the solution tend to be smoothed over time. Note: odd derivatives in a PDE with real coefficients contribute to the real part of the dispersion relation $\omega(k)$, whereas even derivatives contribute to the imaginary part of $\omega(k)$.

Let us now consider a *perturbed* or *small signal* form of the Kawahara equation, say $u(x,t) = U_0 + \hat{u}(x,t)$, where $\hat{u}$ is a small perturbation on a nominal steady-state solution U_0. Substituting this solution into eq. (12.8) leads to

$$\frac{\partial \hat{u}}{\partial t} + U_0 \frac{\partial \hat{u}}{\partial x} + a \frac{\partial^3 \hat{u}}{\partial x^3} - b \frac{\partial^5 \hat{u}}{\partial x^5} = 0 \qquad (12.12)$$

Without loss of generality, we set $U_0 = 0$ and, following the same approach as used above, we obtain

$$\omega = -ak^3 - bk^5$$

$$\downarrow$$

$$\hat{u} = A\exp\left(ik\left[x - \left(ak^2 + bk^4\right)t\right]\right)$$

$$c_p = -k^2\left(a + bk^2\right)$$

$$c_g = -k^2\left(3a + 5bk^2\right)$$

This means that for small amplitude waves and real k, a, and b, the small signal Kawahara equation is purely dispersive, i.e., there is no dissipation. But from the above discussion, we note that if $c_p = c_g$, then the system will also be nondispersive. This will be true only if

$$k = \pm\sqrt{\frac{-a}{2b}} \qquad (12.13)$$

and would imply $ab < 0$ for real k.

Now, from the traveling wave solution to the nonlinear form, i.e., eq. (12.2) and the associated Fig. (12.2), we see that in addition to exhibiting zero dispersion, it also exhibits zero dissipation. This is one of the remarkable characteristics of nonlinear partial differential equations that admit soliton solutions, such as the Kawahara equation.

Solitons are the result of nonlinear terms in the PDE canceling the nonlinear dispersive effects of the medium in which the traveling waves propagate. See also Chapter 18 for some additional discussion on solitons.

A Maple code that derives ω, c_p, and c_g for the Kawahara equation is given in Listing 12.8. It can be modified easily to derive dispersion relations for other PDEs. For a more in-depth discussion relating to dispersion and dissipation, readers are referred to [2, 6, 12].

```
> # Find dispersion relation for Kawahara equation
  restart;
> # Set up alias for u and v
  alias(u=u(x,t)):alias(v=v(x,t)):
> # Define Kawahara equation
  pde1:=diff(v,t)+v*diff(v,x)+alpha*diff(v,x,x,x)
                        -beta*diff(v,x,x,x,x,x)=0;
> # Derive small signal version of pde1
  pde2:=expand(subs(v=(U[0]+u),pde1));
  # set diff(u,x)*u=0, second order magnitude
  nops(lhs(pde2)); # number of operands
  pde3:=subsop(3=0, lhs(pde2))=0; # set 3rd operand to zero
> # without loss of generality set U[0]=0
  pde4:=subs(U[0]=0,lhs(pde3));
> eqn1:=simplify(eval(subs(u=A*exp(I*(k*x-omega*t)),pde4)),size);
  # dispersion relation
> eqn2:=isolate(eqn1,omega);assign(eqn2);
> # phase velocity
  c_p:=simplify(omega/k);
> # group velocity (wave packet)
  c_g:=diff(omega,k);
> # Solve for k (to be non-dispersive)
  solve({c_p=c_g},k);
```

LISTING 12.8: Maple code to derive the dispersion relation for the Kawahara equation.

Traveling wave solutions are obtained by straightforward application of any of the Maple procedures `tanhMethod`, `expMethod()`, or `riccatiMethod()` described in the main Appendix. The *tanh* method finds 15 solutions, the *exp* method finds 12 solutions, and the *Riccati* method finds 15×6 solutions. They all find solutions that match the original solution of eq. (12.2).

The Maple code implementations of the *tanh*, *exp*, and *Riccati* methods are not detailed here, but they are included with the downloadable software for this book.

References

[1] L. Debnath, *Nonlinear Partial Differential Equations for Scientists and Engineers*, Second ed., Birkhauser, Boston, 2005.

[2] P.G. Drazin, R.S. Johnson, *Solitons: an Introduction*, Cambridge University Press, New York, 1989.

[3] B. Fornberg, Calculation of weights in finite difference formulas, *SIAM Review*, 40 (3) (1998) 685–691.

[4] S. Hamdi, W.E. Schiesser, G.W. Griffiths, Method of Lines, *http://www.scholarpedia.org/article/method_of_lines*, Scholarpedia, 2 (7) (2007) 2859.

[5] T. Ilichev, A.V. Marchenko, Propagation of long nonlinear waves in a ponderable fluid beneath an ice sheet, *Fluid Dyn.* 24–1 (1989) 73–79.

[6] R.S. Johnson, A Modern Introduction to the Mathematical Theory of Water Waves, Cambridge University Press, New York, 1997.

[7] T. Kawahara, Oscillatory solitary waves in dispersive media, *J. Phys. Soc. of Japan*, 33 (1972) 260–264.

[8] A.D. Polyanin, V.F. Zaitsev, *Handbook of Nonlinear Partial Differential Equations*, Chapman & Hall/CRC, Boca Raton, FL, 2004.

[9] P. Saucez, W.E. Schiesser, A. vande Wouwer, Upwinding in the method of lines, *Math. Comput. Simulat.* 56 (2001) 171–185; ibid, 2002, 541.

[10] W.E. Schiesser, G.W. Griffiths, *A Compendium of Partial Differential Equation Models*, Cambridge University Press, Cambridge, UK, 2009.

[11] W.E. Schiesser, C.A. Silebi, *Computational Transport Phenomena*, Cambridge University Press, Cambridge, UK, 1997, 207–288.

[12] G.K. Valis, *Atmospheric and Oceanic Fluid Dynamics: Fundamentals and Large-Scale Circulation*, Cambridge University Press, New York, 2006.

[13] J.A. Zufira, Symmetry breaking in periodic and solitary gravity capillary waves on water of finite depth, *J. Fluid Mech.* 184 (1987) 183–206.

13 Regularized Long-Wave Equation

To illustrate the numerical integration of a PDE with a mixed partial derivative, we consider the following 1D *regularized long-wave equation (RLW)* ([5], p998):

$$\frac{\partial u}{\partial t}+\frac{\partial u}{\partial x}+\varepsilon u\frac{\partial u}{\partial x}-\mu\frac{\partial}{\partial t}\left(\frac{\partial^2}{\partial x^2}\right)=0 \tag{13.1}$$

Equation (13.1) is first order in t and second order in x (through the mixed partial). It therefore requires one *initial condition* (IC) and two *boundary conditions* (BCs). The initial condition is taken as

$$u(x,t=0)=g(x) \tag{13.2}$$

and the two BCs as

$$u(x=x_l,t)=f_1(t) \tag{13.3}$$

$$u(x=x_u,t)=f_2(t) \tag{13.4}$$

An analytical solution to eqns. (13.1)–(13.4) is ([5], p1002)

$$u_a(x,t)=3d\text{sech}^2\left[k(x-x_o-\nu t)\right] \tag{13.5}$$

where $\nu=1+\varepsilon d, k=\frac{1}{2}\sqrt{\frac{\varepsilon d}{\mu\nu}}$. From eq. (13.5), we have the IC and BCs

$$g(x)=3d\text{sech}^2\left[k(x-x_o)\right] \tag{13.6}$$

$$f_1(t)=3d\text{sech}^2\left[k(x_l-x_o-\nu t)\right] \tag{13.7}$$

$$f_2(t)=3d\text{sech}^2\left[k(x_u-x_o-\nu t)\right] \tag{13.8}$$

The method of lines (MOL) solution is computed for $d=0.1, x_o=0, \varepsilon=\mu=1$ ([5], p1002).

The MOL approximation of eq. (13.1) in t (with $\frac{\partial u}{\partial t}=u_t, \frac{\partial u}{\partial x}=u_x$) is based on the following finite difference (FD) approximation for the second derivative in x:

$$\frac{\partial^2 u}{\partial x^2}\approx\frac{u(i+1)-2u(i)+u(i-1)}{\Delta x^2}+O(\Delta x^2) \tag{13.9}$$

where i is a grid index in x, $i = 0, 1, 2, \ldots, n, n+1$ corresponding to $i = 0$ at $x = x_l$ and $i = n+1$ at $x = x_u$, with grid spacing $\Delta x = \dfrac{x_u - x_l}{n+1}$ (so that the interior grid points correspond to $x = i\Delta x, i = 1, 2, \ldots, n$). Application of eq. (13.9) to eq. (13.1) gives

$$
\begin{aligned}
&\left(-\frac{2}{\Delta x^2} - \frac{1}{\mu}\right) u_t(\Delta x, t) \quad \frac{1}{\Delta x^2} u_t(2\Delta x, t) \\
&\frac{1}{\Delta x^2} u_t(\Delta x, t) \quad \left(-\frac{2}{\Delta x^2} - \frac{1}{\mu}\right) u_t(2\Delta x, t) \quad \frac{1}{\Delta x^2} u_t(3\Delta x, t) \\
&\qquad \ddots \qquad \ddots \qquad \ddots \\
&\frac{1}{\Delta x^2} u_t((n-2)\Delta x, t) \quad \left(-\frac{2}{\Delta x^2} - \frac{1}{\mu}\right) u_t((n-1)\Delta x, t) \quad \frac{1}{\Delta x^2} u_t(n\Delta x, t) \\
&\qquad \frac{1}{\Delta x^2} u_t((n-1)\Delta x, t) \quad \left(-\frac{2}{\Delta x^2} - \frac{1}{\mu}\right) u_t(n\Delta x) \\
&= \left(\frac{1}{\mu} u_x + \frac{\varepsilon}{\mu} u u_x\right)_{x=\Delta x} - \frac{1}{\Delta x^2} u_{t,a}(x = x_l, t) \\
&= \left(\frac{1}{\mu} u_x + \frac{\varepsilon}{\mu} u u_x\right)_{x=2\Delta x} \\
&\qquad \vdots \\
&= \left(\frac{1}{\mu} u_x + \frac{\varepsilon}{\mu} u u_x\right)_{x=(n-1)\Delta x} \\
&= \left(\frac{1}{\mu} u_x + \frac{\varepsilon}{\mu} u u_x\right)_{x=n\Delta x} - \frac{1}{\Delta x^2} u_{t,a}(x = x_u, t)
\end{aligned}
\tag{13.10}
$$

or in matrix format

$$
\begin{bmatrix}
\left(-\frac{2}{\Delta x^2} - \frac{1}{\mu}\right) & \frac{1}{\Delta x^2} & & \\
\frac{1}{\Delta x^2} & \left(-\frac{2}{\Delta x^2} - \frac{1}{\mu}\right) & \frac{1}{\Delta x^2} & \\
\ddots & \ddots & \ddots & \\
\frac{1}{\Delta x^2} & \left(-\frac{2}{\Delta x^2} - \frac{1}{\mu}\right) & \frac{1}{\Delta x^2} & \\
& \frac{1}{\Delta x^2} & \left(-\frac{2}{\Delta x^2} - \frac{1}{\mu}\right) &
\end{bmatrix}
\bullet
\begin{bmatrix}
u_t(\Delta x, t) \\
u_t(2\Delta x, t) \\
u_t(3\Delta x, t) \\
\vdots \\
u_t((n-2)\Delta x, t) \\
u_t((n-1)\Delta x, t) \\
u_t(n\Delta x, t)
\end{bmatrix}
$$

$$
= \begin{bmatrix}
\left(\frac{1}{\mu}u_x + \frac{\varepsilon}{\mu}uu_x\right)_{x=\Delta x} - \frac{1}{\Delta x^2}u_{t,a}(x=x_l,t) \\
\left(\frac{1}{\mu}u_x + \frac{\varepsilon}{\mu}uu_x\right)_{x=2\Delta x} \\
\vdots \\
\left(\frac{1}{\mu}u_x + \frac{\varepsilon}{\mu}uu_x\right)_{x=(n-1)\Delta x} \\
\left(\frac{1}{\mu}u_x + \frac{\varepsilon}{\mu}uu_x\right)_{x=n\Delta x} - \frac{1}{\Delta x^2}u_{t,a}(x=x_u,t)
\end{bmatrix} \tag{13.11}
$$

We can then solve eq. (13.11) for the vector of derivatives in t

$$
\begin{bmatrix}
u_t(\Delta x,t) \\
u_t(2\Delta x,t) \\
u_t(3\Delta x,t) \\
\vdots \\
u_t((n-2)\Delta x,t) \\
u_t((n-1)\Delta x,t) \\
u_t(n\Delta x,t)
\end{bmatrix}
=
\begin{bmatrix}
\left(-\frac{2}{\Delta x^2}-\frac{1}{\mu}\right) & \frac{1}{\Delta x^2} & & \\
\frac{1}{\Delta x^2} & \left(-\frac{2}{\Delta x^2}-\frac{1}{\mu}\right) & \frac{1}{\Delta x^2} & \\
\ddots & \ddots & \ddots & \\
\frac{1}{\Delta x^2} & \left(-\frac{2}{\Delta x^2}-\frac{1}{\mu}\right) & \frac{1}{\Delta x^2} & \\
 & \frac{1}{\Delta x^2} & \left(-\frac{2}{\Delta x^2}-\frac{1}{\mu}\right) &
\end{bmatrix}^{-1}
$$

$$
\bullet \begin{bmatrix}
\left(\frac{1}{\mu}u_x + \frac{\varepsilon}{\mu}uu_x\right)_{x=\Delta x} - \frac{1}{\Delta x^2}u_{t,a}(x=x_l,t) \\
\left(\frac{1}{\mu}u_x + \frac{\varepsilon}{\mu}uu_x\right)_{x=2\Delta x} \\
\vdots \\
\left(\frac{1}{\mu}u_x + \frac{\varepsilon}{\mu}uu_x\right)_{x=(n-1)\Delta x} \\
\left(\frac{1}{\mu}u_x + \frac{\varepsilon}{\mu}uu_x\right)_{x=n\Delta x} - \frac{1}{\Delta x^2}u_{t,a}(x=x_u,t)
\end{bmatrix} \tag{13.12}
$$

The RHS terms of eq. (13.12) with u and u_x such as $\left(\frac{1}{\mu}u_x + \frac{\varepsilon}{\mu}uu_x\right)_{x=\Delta x}$ can be accommodated by the usual MOL analysis with application of a library differentiator, e.g., `dss004`, to compute ux. The terms involving the analytical solution $u_a(x,t)$ such as $-\frac{1}{\Delta x^2}u_{t,a}(x = x_l, t)$ require differentiation of the analytical solution with respect to t. However, this may essentially be avoided if the initial condition pulse of eq. (13.6) does not approach the boundaries so that the required derivative $u_{t,a}$ can be taken as zero.

Equation (13.1) has three invariants representing conservation of mass, momentum, and energy, which can be used to check the numerical solution ([5], p1002)

$$c_1 = \int_{-\infty}^{\infty} u(x,t)\,dx \tag{13.13a}$$

$$c_2 = \int_{-\infty}^{\infty} \left[u^2(x,t) + \mu\left(\frac{\partial u(x,t)}{\partial x}\right)^2\right] dx \tag{13.13b}$$

$$c_3 = \int_{-\infty}^{\infty} \left[u^3(x,t) + 3u^2(x,t)\right] dx \tag{13.13c}$$

Equations (13.12) is implemented in the following Matlab routines. The ODE routine `pde_1.m` follows first.

```
  function ut=pde_1(t,u)
%
% Function pde_1 computes the t derivative vector of the RLW PDE
%
  global xl xu x dx n ncall
%
% Model parameters
  global eps mu cm d k xo nu
%
% BCs
  u(1)=ua_1(x(1),t);
  u(n)=ua_1(x(n),t);
%
% ux
  ux=dss004(xl,xu,n,u);
%
% PDE
  for i=1:n
    urhs(i)=(1/mu)*(ux(i)+eps*u(i)*ux(i));
  end
% urhs=(1/mu)*(ux'+eps*u'.*ux');
  ut=cm\urhs';
%
% Increment calls to pde_1
  ncall=ncall+1;
```

LISTING 13.1: Function `pde_1.m` for eq. (13.12).

We can note the following points about pde_1.m:

- The function and global variables are defined.

```
  function ut=pde_1(t,u)
%
% Function pde_1 computes the t derivative vector of the RLW PDE
%
  global xl xu x dx n ncall
%
% Model parameters
  global eps mu cm d k xo nu
```

- Two Dirichlet BCs are defined using the analytical solution in eqns. (13.3), (13.4), (13.7), and (13.8) at $x = x_l, x_u$ (i=1,n).

```
%
% BCs
  u(1)=ua_1(x(1),t);
  u(n)=ua_1(x(n),t);
```

- ux in the RHS of eq. (13.12) is computed by dss004.

```
%
% ux
  ux=dss004(xl,xu,n,u);
```

- The ODEs of eq. (13.12) are programmed in two ways.

```
%
% PDE
  for i=1:n
    urhs(i)=(1/mu)*(ux(i)+eps*u(i)*ux(i));
  end
% urhs=(1/mu)*(ux'+eps*u'.*ux');
  ut=cm\urhs';
%
% Increment calls to pde_1
  ncall=ncall+1;
```

In the first approach, the programming of the RHS terms of eq. (13.12) is explicit over the grid i=1,...n in a for loop. In the second approach, the vector operations of Matlab are used (deactivated as a comment, where ' is a transpose). In both cases, the RHS vector urhs and the coefficient matrix cm (set in the main program discussed subsequently) are used with the Matlab inverse matrix operator \ to compute the derivatives in t, ut, which are also transposed to meet the requirements of the ODE integrator ode45. Finally, the counter for the calls to pde_1.m is incremented before the return from this routine.

The initial condition function inital_1.m implements eqns. (13.2) and (13.6).

```
  function u0=inital_1(t0)
%
% Function inital_1 sets the initial condition for the RLW PDE with
```

```
% a mixed partial derivative
%
  global xl xu x dx n ncall
%
% Model parameters
  global eps mu cm d k xo nu
%
  d=0.1; xo=0; eps=1; mu=1;
  nu=1+eps*d;
  k=0.5*(eps*d/(mu*nu))^0.5;
%
% IC over the spatial grid
  xl=-50;
  xu= 50;
  n=199;
  dx=(xu-xl)/(n+1);
  for i=1:n
%
%   Uniform grid
    x(i)=xl+i*dx;
%
%   Initial condition
    u0(i)=ua_1(0,x(i));
  end
```

LISTING 13.2: Function `inital_1.m` for IC from eqns. (13.2) and (13.6).

We can note the following points about `inital_1.m`:

- The function and some global variables are defined, and the problem parameters are given numerical values.

```
  function u0=inital_1(t0)
%
% Function inital_1 sets the initial condition for the RLW PDE with
% a mixed partial derivative
%
  global xl xu x dx n ncall
%
% Model parameters
  global eps mu cm d k xo nu
%
  d=0.1; xo=0; eps=1; mu=1;
  nu=1+eps*d;
  k=0.5*(eps*d/(mu*nu))^0.5;
```

- The parameters for the spatial grid are defined.

```
%
% IC over the spatial grid
  xl=-50;
  xu= 50;
  n=199;
  dx=(xu-xl)/(n+1);
```

We can note the following points about these parameters:

- 199 grid points are used, and dx is computed by dividing by n+1 (rather than n-1 as in previous cases) since the end points of the grid for the 199-ODE system are at xl+dx and xu-dx (and not at xl, xu as in previous cases). This variation from the usual procedure follows from eq. (13.12), where $u(x = x_l, t)$ and $u(x = x_u, t)$ are set by the analytical solution of eq. (13.5) and therefore ODEs at the end points are not actually required to compute the solution at the end points.
- The values xl=-50, xu=50 were selected so that they are effectively at infinity, and therefore the boundary values of the solution remain at their initial values (from eq. (13.6)) as we shall observe in the numerical and plotted output; however, this condition is not required and BCs from eqns. (13.7) and (13.8) could be used at essentially finite values of x.

• The grid in x and the associated IC (from function ua_1.m at $t = 0$) are set in a for loop.

```
  for i=1:n
%
%   Uniform grid
    x(i)=xl+i*dx;
%
%   Initial condition
    u0(i)=ua_1(0,x(i));
  end
```

Function ua_1.m is a straightforward implementation of the analytical solution, eq. (13.5).

```
  function uanal=ua_1(t,x)
%
% Function ua_1 computes the analytical solution to the RLW PDE with
% a mixed partial derivative
%
% Model parameters
  global eps mu cm d k xo nu
%
% Analytical solution
  uanal=3*d*sech(k*(x-xo-nu*t))^2;
```

LISTING 13.3: Function ua_1.m for analytical solution (13.5).

The routine for calculating the integrals of eqns. (13.13) by Simpson's rule, simp, is

```
  function uint=simp(xl,xu,n,u)
%
% Function simp computes three integral invariants by Simpson's
% rule
%
% Parameter in the integrand of the second integral
  global mu
%
% Step through the three integrals (invariants)
  for int=1:3
    h=(xu-xl)/(n+1);
```

```
%
%   Conservation of mass
    if(int==1)
       uint(1)=u(1)-u(n);
       for i=3:2:n
         uint(1)=uint(1)+4.0*u(i-1)+2.0*u(i);
       end
       uint(1)=h/3.0*uint(1);
    end
%
%   Conservation of momentum
    if(int==2)
       ux=dss004(xl,xu,n,u);
       uint(2)=u(1)^2+mu*ux(1)^2-(u(n)^2+mu*ux(n)^2);
       for i=3:2:n
         uint(2)=uint(2)+4.0*(u(i-1)^2+mu*ux(i-1)^2)...
                        +2.0*(u(i  )^2+mu*ux(i  )^2);
       end
       uint(2)=h/3.0*uint(2);
    end
%
%   Energy conservation
    if(int==3)
       uint(3)=u(1)^3+3*u(1)^2-(u(n)^3+3*u(n)^2);
       for i=3:2:n
         uint(3)=uint(3)+4.0*(u(i-1)^3+3*u(i-1)^2)...
                        +2.0*(u(i  )^3+3*u(i  )^2);
       end
       uint(3)=h/3.0*uint(3);
    end
%
% Next integral
  end
```

LISTING 13.4: Numerical quadrature routine `simp` applied to eqns. (13.13).

`simp` has three parts corresponding to the integrals `c1,c2,c3` of eqns. (13.13).

1. The coding for `c1` is

```
  function uint=simp(xl,xu,n,u)
%
% Function simp computes three integral invariants by Simpson's
% rule
%
% Parameter in the integrand of the second integral
  global mu
%
% Step through the three integrals (invariants)
  for int=1:3
    h=(xu-xl)/(n+1);
%
%   Conservation of mass
    if(int==1)
       uint(1)=u(1)-u(n);
```

```
      for i=3:2:n
        uint(1)=uint(1)+4.0*u(i-1)+2.0*u(i);
      end
      uint(1)=h/3.0*uint(1);
   end
```

After defining the function, the integration interval `h=(xu-xl)/(n+1)` is computed, where `xl=-50, xu=50` are the lower and upper limits of the integral (set in `inital_1`) and `n=199`. The `for` loop for c_1 is an implementation of the weighted sum for Simpson's rule applied to the function $u(x,t)$ (the integrand in eq. (13.13a)).

$$\int_{-\infty}^{\infty} u(x,t)dx \approx \frac{h}{3}\left[u_1 + \sum_{i=2}^{n-2}(4u_i + 2u_{i+1}) + u(n)\right] \tag{13.14}$$

2. Similarly, the coding for c_2 of eq. (13.13b) is

```
%
%  Conservation of momentum
   if(int==2)
      ux=dss004(xl,xu,n,u);
      uint(2)=u(1)^2+mu*ux(1)^2-(u(n)^2+mu*ux(n)^2);
      for i=3:2:n
        uint(2)=uint(2)+4.0*(u(i-1)^2+mu*ux(i-1)^2)...
                        +2.0*(u(i  )^2+mu*ux(i  )^2);
      end
      uint(2)=h/3.0*uint(2);
   end
```

which reflects the difference in the integrands of c_1 and c_2 of eqns. (13.13a) and (13.13b). Also, the derivative in c_2, $\frac{\partial u}{\partial x}$, is computed by `dss004`.

3. Finally, the coding for c_3 of eq. (13.13c) is

```
%
%  Energy conservation
   if(int==3)
      uint(3)=u(1)^3+3*u(1)^2-(u(n)^3+3*u(n)^2);
      for i=3:2:n
        uint(3)=uint(3)+4.0*(u(i-1)^3+3*u(i-1)^2)...
                        +2.0*(u(i  )^3+3*u(i  )^2);
      end
      uint(3)=h/3.0*uint(3);
   end
%
% Next integral
  end
```

The main program, `pde_1_main.m`, is structured in the same way as previous main programs.

```
%
% Clear previous files
  clear all
```

```
  clc
%
  global xl xu x dx n ncall
%
% Model parameters
  global eps mu cm d k xo nu
%
% Initial condition
  t0=0.0;
  u0=inital_1(t0);
%
% Coefficient matrix
%
  for i=1:n
%
% Row 1
  if(i==1)
    for j=1:n
      if(j==1)cm(1,1)=(-2/dx^2-1/mu);
      elseif(j==2)cm(1,2)=1/dx^2;
      else cm(1,j)=0;
      end
    end
%
% Row n
  elseif(i==n)
    for j=1:n
      if(j==n)cm(n,n)=(-2/dx^2-1/mu);
      elseif(j==n-1)cm(n,n-1)=1/dx^2;
      else cm(n,j)=0;
      end
    end
  else
    for j=1:n
      if(i==j)cm(i,j)=(-2/dx^2-1/mu);
      elseif(abs(i-j)==1)cm(i,j)=1/dx^2;
      else cm(i,j)=0;
      end
    end
  end
%
% Next i
  end
%
% Independent variable for ODE integration
  tf=18;
  tout=[t0:6:tf]';
  nout=4;
  ncall=0;
%
% ODE integration
  reltol=1.0e-06; abstol=1.0e-06;
  options=odeset('RelTol',reltol,'AbsTol',abstol);
%
```

```
% Explicit (nonstiff) integration
  [t,u]=ode45(@pde_1,tout,u0,options);
%
% Analytical solution and difference between the numerical and
% analytical solutions at selected points
  for it=1:nout
    fprintf('\n\n     t         x       u(x,t)       u(x,t)          err\n')
    fprintf('                           num          anal                \n')
    for i=1:n
      u_anal(it,i)=ua_1(t(it),x(i));
      err(it,i)=u(it,i)-u_anal(it,i);
    end
    for i=1:5:n
      fprintf('%6.2f%8.3f%12.6f%12.6f%12.6f\n',...
              t(it),x(i),u(it,i),u_anal(it,i),err(it,i));
    end
%
%     Calculate and display three invariants
      ui=u(it,:);
      uint=simp(xl,xu,n,ui);
      fprintf('\n Invariants at t = %5.2f',t(it));
      fprintf('\n    I1 = %10.4f   Mass conservation'    ,uint(1));
      fprintf('\n    I2 = %10.4f   Momentum conservation',uint(2));
      fprintf('\n    I3 = %10.4f   Energy invariant\n\n' ,uint(3));
  end
  fprintf('\n  ncall = %4d\n\n',ncall);
%
% Plot numerical and analytical solutions
  xplot(1)=xl;
  xplot(n+2)=xu;
  for i=1:n
    xplot(i+1)=x(i);
  end
  for it=1:nout
    uplot(it,1)=ua_1(t(it),x(1));
    uplot(it,2:n+1)=u(it,1:n);
    uplot(it,n+2)=ua_1(t(it),x(n));
    uaplot(it,1)=ua_1(t(it),x(1));
    uaplot(it,2:n+1)=u_anal(it,1:n);
    uaplot(it,n+2)=ua_1(t(it),x(n));
  end
  figure(1)
  plot(xplot,uplot,'o',xplot,uaplot,'-')
  xlabel('x')
  ylabel('u(x,t)')
  title('RLW equation; t = 0, 6, 12, 18; o - numerical;
         solid - analytical')
  figure(2)
  surf(uplot)
  shading 'interp', axis 'tight'
  xlabel('x'); ylabel('t'); zlabel('u(x,t)');
  title('RLW equation');
```

LISTING 13.5: Main program pde_1_main.m.

We can note the following points about this main program:

- After previous files are cleared, some global variables are defined.

```
%
% Clear previous files
  clear all
  clc
%
  global xl xu x dx n ncall
%
% Model parameters
  global eps mu cm d k xo nu
```

- The initial condition of eqns. (13.2) and (13.6) are then set by `inital_1.m`.

```
%
% Initial condition
  t0=0.0;
  u0=inital_1(t0);
```

- Before the ODE integration (by `ode45`), the coefficient matrix of eq. (3.12) is defined `cm`. This coding follows directly from eq. (3.12) and illustrates that the first and 199th equations (`i=1,n`) are special cases; all of the intervening equations (`i=2,...,198`) have the same structure and programming. `j` is the column index used in each of the three `for` loops.

```
%
% Coefficient matrix
%
  for i=1:n
%
% Row 1
  if(i==1)
    for j=1:n
      if(j==1)cm(1,1)=(-2/dx^2-1/mu);
      elseif(j==2)cm(1,2)=1/dx^2;
      else cm(1,j)=0;
      end
    end
%
% Row n
  elseif(i==n)
    for j=1:n
      if(j==n)cm(n,n)=(-2/dx^2-1/mu);
      elseif(j==n-1)cm(n,n-1)=1/dx^2;
      else cm(n,j)=0;
      end
    end
  else
    for j=1:n
      if(i==j)cm(i,j)=(-2/dx^2-1/mu);
      elseif(abs(i-j)==1)cm(i,j)=1/dx^2;
      else cm(i,j)=0;
```

```
      end
    end
  end
%
% Next i
  end
```

- The time scale is defined

```
%
% Independent variable for ODE integration
  tf=18;
  tout=[t0:6:tf]';
  nout=4;
  ncall=0;
```

  that is, $t = 0, 6, 12, 18$ with four output points.
- The integration of the 199 ODEs is accomplished by `rkf45`.

```
%
% ODE integration
  reltol=1.0e-06; abstol=1.0e-06;
  options=odeset('RelTol',reltol,'AbsTol',abstol);
%
% Explicit (nonstiff) integration
  [t,u]=ode45(@pde_1,tout,u0,options);
```

  The nonstiff integrator, `rkf45`, performed the ODE integration very efficiently (as demonstrated in the subsequent output). If the execution required a long computer run time (large value of `ncall`), the selection of a stiff integrator such as `ode15s` would be reasonable. However, this example illustrates two ideas:

  1. An MOL solution of a PDE does not necessarily lead to stiff ODEs (a common misconception).
  2. A nonstiff integrator should be considered first since it, if it performs satisfactorily, will require fewer calculations than a stiff integrator. In other words, the enhanced stability of a stiff integrator has a cost, namely, more calculations for each step along the solution than for a nonstiff integrator.

- After the ODE integration by `ode45`, the analytical solution, eq. (13.5), is evaluated by function `ua_1` for comparison with the numerical solution (in array `u(it,i)`), and selected values of the two solutions and their difference (`err(it,i)`) are displayed numerically.

```
%
% Analytical solution and difference between the numerical and
% analytical solutions at selected points
  for it=1:nout
    fprintf('\n\n     t        x        u(x,t)       u(x,t)          err\n')
    fprintf('                          num          anal                \n')
    for i=1:n
      u_anal(it,i)=ua_1(t(it),x(i));
```

```
      err(it,i)=u(it,i)-u_anal(it,i);
    end
    for i=1:5:n
      fprintf('%6.2f%8.3f%12.6f%12.6f%12.6f\n',...
              t(it),x(i),u(it,i),u_anal(it,i),err(it,i));
    end
%
%     Calculate and display three invariants
      ui=u(it,:);
      uint=simp(xl,xu,n,ui);
      fprintf('\n Invariants at t = %5.2f',t(it));
      fprintf('\n    I1 = %10.4f   Mass conservation'    ,uint(1));
      fprintf('\n    I2 = %10.4f   Momentum conservation',uint(2));
      fprintf('\n    I3 = %10.4f   Energy invariant\n\n' ,uint(3));
  end
  fprintf('\n  ncall  = %4d\n\n',ncall);
```

Also, the three invariants of eqns. (13.13) are evaluated by `simp.m` of Listing 13.4 and are displayed at each of the four output points in t. Finally, the counter for calls to `pde_1.m` is displayed as a measure of the overall computational effort.

- Arrays for plotting are set up by including the end point values at $x = x_l$ and $x = x_u$ (the grid for plotting is expanded from 199 to 201 points); the analytical solution is included by using `ua_1.m`.

```
%
% Plot numerical and analytical solutions
  xplot(1)=xl;
  xplot(n+2)=xu;
  for i=1:n
    xplot(i+1)=x(i);
  end
  for it=1:nout
    uplot(it,1)=ua_1(t(it),x(1));
    uplot(it,2:n+1)=u(it,1:n);
    uplot(it,n+2)=ua_1(t(it),x(n));
    uaplot(it,1)=ua_1(t(it),x(1));
    uaplot(it,2:n+1)=u_anal(it,1:n);
    uaplot(it,n+2)=ua_1(t(it),x(n));
  end
```

- 2D and 3D plots are produced by using `plot` and `surf`.

```
  figure(1)
  plot(xplot,uplot,'o',xplot,uaplot,'-')
  xlabel('x')
  ylabel('u(x,t)')
  title('RLW equation; t = 0, 6, 12, 18; o - numerical;
         solid - analytical')
  figure(2)
  surf(xplot,t,uplot)
  xlabel('x'); ylabel('t'); zlabel('u(x,t)');
  title('RLW equation');
```

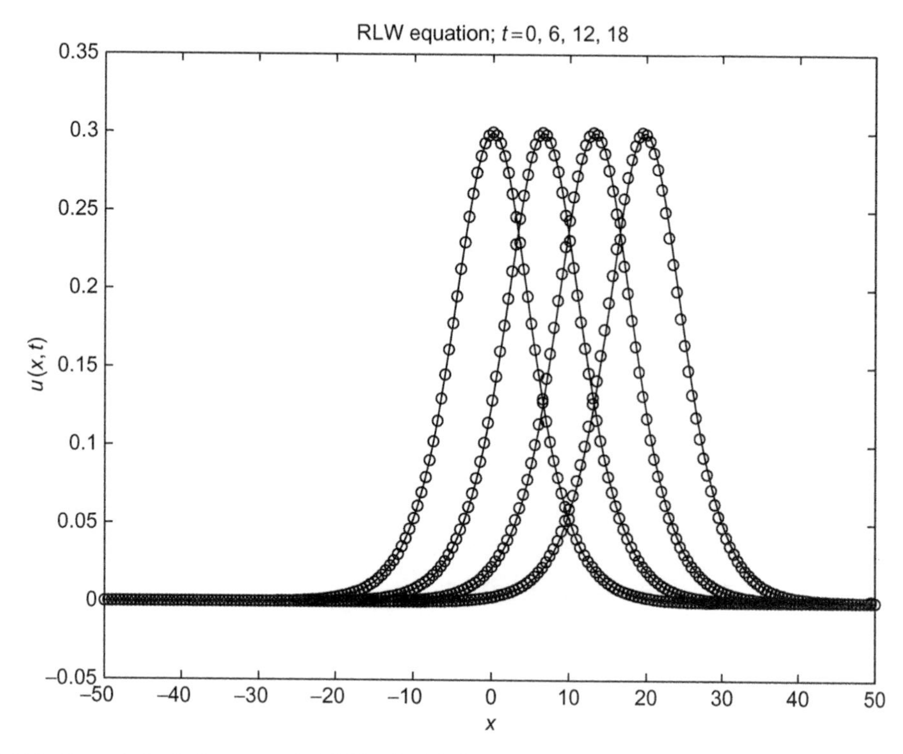

FIGURE 13.1: Numerical solution to eq. (13.1) (lines) with the analytical solution, eq. (13.5), superimposed (circles).

The output from the preceding code appears in Figs. 13.1 and 13.2, and Table 13.1. Figure 13.1 indicates good agreement between the analytical and numerical solutions. Note, in particular, that the *sech* pulse of eq. (13.5) does not reach the boundaries at $x = -50, 50$ so that BCs (13.3), (13.4), (13.7), and (13.8) do not have an effect, but this may not be the case depending on the boundary values of x and the elapsed time t. Figure 13.2 is a 3D plot of the numerical solution.

The tabular analytical and numerical solutions given in Table 13.1 also reflect the good agreement between these two solutions. The three invariants remained constant throughout the solution. Also, the computational effort of `rkf45` reflected in `ncall = 145` is quite modest.

In summary, the numerical solution of a PDE with a mixed partial derivative such as eq. (13.1) is straightforward if the mixed partial derivative includes differentiation with respect to an initial value variable, such as t in the preceding example. PDEs with mixed partial derivatives that have only boundary value independent variables such as $\frac{\partial^2 u}{\partial x \partial y}$ in Cartesian coordinates, in principle, can also be accommodated within the MOL framework. This class of problems is addressed in a forthcoming publication.

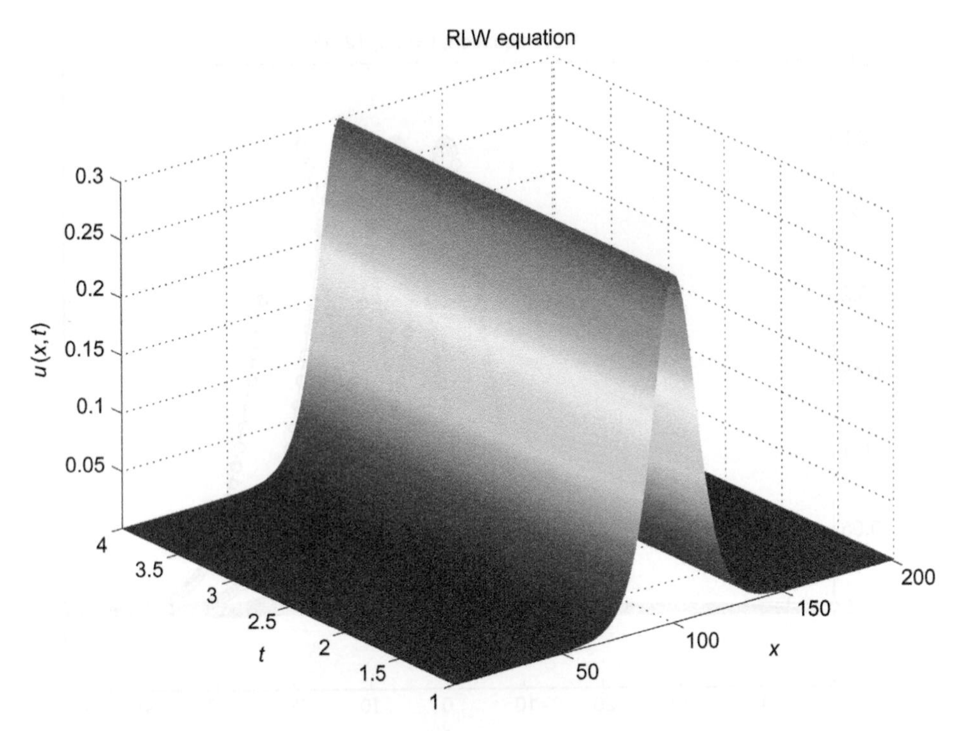

FIGURE 13.2: 3D plot of the numerical solution to eq. (13.1).

Appendix

We conclude this chapter by including a little background information for the *regularized long-wave equation* (RLWE) and outlining some traveling wave solutions that can be obtained using the procedures detailed in the main Appendix. By experimenting with different values for the order of approximations, M (see eqn. (A.8) in the main Appendix), it becomes immediately apparent that there are many nontrivial traveling wave solutions to the RLWE. Higher values of M tend to produce more complex traveling wave solutions, which can reveal interesting behavior when examined by means of an animation.

Brief Background Information

To describe water waves rigorously, in a mathematical and physical sense, requires that the *Navier–Stokes equations* be solved with free boundary condition(s). This poses very difficult problems from both a theoretical and a numerical standpoint. Therefore, it is useful to construct simpler mathematical models that are able to capture the essential details of water waves under a variety of operating conditions. To this end, the regularized long-wave equation (RLW) or *Benjamin–Bona–Mahony equation* (BBM) [1] was derived as a model to describe gravity water waves. It is an alternative model to the well-known *Korteweg–de*

Table 13.1: Abbreviated tabular numerical and analytical solutions for $n = 199$

t	x	u(x,t) num	u(x,t) anal	err
0.00	-49.500	0.000000	0.000000	0.000000
0.00	-47.000	0.000001	0.000001	0.000000
0.00	-44.500	0.000002	0.000002	0.000000
0.00	-42.000	0.000004	0.000004	0.000000
0.00	-39.500	0.000008	0.000008	0.000000
0.00	-37.000	0.000017	0.000017	0.000000
0.00	-34.500	0.000036	0.000036	0.000000
0.00	-32.000	0.000077	0.000077	0.000000
0.00	-29.500	0.000165	0.000165	0.000000
0.00	-27.000	0.000349	0.000349	0.000000
0.00	-24.500	0.000742	0.000742	0.000000
0.00	-22.000	0.001575	0.001575	0.000000
0.00	-19.500	0.003337	0.003337	0.000000
0.00	-17.000	0.007047	0.007047	0.000000
0.00	-14.500	0.014777	0.014777	0.000000
0.00	-12.000	0.030538	0.030538	0.000000
0.00	-9.500	0.061241	0.061241	0.000000
0.00	-7.000	0.115672	0.115672	0.000000
0.00	-4.500	0.195401	0.195401	0.000000
0.00	-2.000	0.274299	0.274299	0.000000
0.00	0.500	0.298302	0.298302	0.000000
0.00	3.000	0.246128	0.246128	0.000000
0.00	5.500	0.161271	0.161271	0.000000
0.00	8.000	0.090587	0.090587	0.000000
0.00	10.500	0.046599	0.046599	0.000000
0.00	13.000	0.022900	0.022900	0.000000
0.00	15.500	0.011002	0.011002	0.000000
0.00	18.000	0.005228	0.005228	0.000000
0.00	20.500	0.002472	0.002472	0.000000
0.00	23.000	0.001166	0.001166	0.000000
0.00	25.500	0.000549	0.000549	0.000000
0.00	28.000	0.000259	0.000259	0.000000
0.00	30.500	0.000122	0.000122	0.000000
0.00	33.000	0.000057	0.000057	0.000000
0.00	35.500	0.000027	0.000027	0.000000
0.00	38.000	0.000013	0.000013	0.000000
0.00	40.500	0.000006	0.000006	0.000000
0.00	43.000	0.000003	0.000003	0.000000
0.00	45.500	0.000001	0.000001	0.000000
0.00	48.000	0.000001	0.000001	0.000000

```
Invariants at t=0.00
    I1=3.9799  Mass conservation
    I2=0.8102  Momentum conservation
    I3=2.5790  Energy invariant
```

(Continued)

Table 13.1: *(Continued)*

t	x	u(x,t) num	u(x,t) anal	err
.				
.				
.				
Solution for t=6, 12 deleted				
.				
.				
.				

t	x	u(x,t) num	u(x,t) anal	err
18.00	−49.500	0.000000	0.000000	0.000000
18.00	−47.000	0.000000	0.000000	0.000000
18.00	−44.500	0.000000	0.000000	0.000000
18.00	−42.000	−0.000000	0.000000	−0.000000
18.00	−39.500	−0.000000	0.000000	−0.000000
18.00	−37.000	−0.000000	0.000000	−0.000000
18.00	−34.500	0.000000	0.000000	−0.000000
18.00	−32.000	0.000000	0.000000	−0.000000
18.00	−29.500	0.000000	0.000000	0.000000
18.00	−27.000	0.000001	0.000001	0.000000
18.00	−24.500	0.000002	0.000002	0.000000
18.00	−22.000	0.000004	0.000004	0.000000
18.00	−19.500	0.000009	0.000009	0.000001
18.00	−17.000	0.000019	0.000018	0.000001
18.00	−14.500	0.000041	0.000039	0.000002
18.00	−12.000	0.000087	0.000082	0.000005
18.00	−9.500	0.000185	0.000175	0.000011
18.00	−7.000	0.000393	0.000371	0.000022
18.00	−4.500	0.000836	0.000788	0.000048
18.00	−2.000	0.001779	0.001673	0.000106
18.00	0.500	0.003746	0.003543	0.000203
18.00	3.000	0.007895	0.007479	0.000416
18.00	5.500	0.016634	0.015671	0.000963
18.00	8.000	0.034332	0.032331	0.002001
18.00	10.500	0.068203	0.064613	0.003590
18.00	13.000	0.126790	0.121227	0.005562
18.00	15.500	0.209005	0.202362	0.006642
18.00	18.000	0.283111	0.278950	0.004161
18.00	20.500	0.294737	0.296684	−0.001947
18.00	23.000	0.233341	0.239741	−0.006400
18.00	25.500	0.148335	0.154717	−0.006381
18.00	28.000	0.081826	0.086113	−0.004287

Table 13.1: *(Continued)*

```
18.00      30.500      0.041696      0.044081      -0.002385
18.00      33.000      0.020397      0.021609      -0.001211
18.00      35.500      0.009780      0.010369      -0.000590
18.00      38.000      0.004644      0.004925      -0.000281
18.00      40.500      0.002195      0.002328      -0.000133
18.00      43.000      0.001035      0.001098      -0.000063
18.00      45.500      0.000488      0.000517      -0.000030
18.00      48.000      0.000251      0.000243       0.000008
Invariants at t=18.00
   I1=3.9799 Mass conservation
   I2=0.8102 Momentum conservation
   I3=2.5790 Energy invariant
 ncall=145
```

Vries equation (KdV) and describes some of the more exotic dynamic behavior of water waves and, in particular, the solitary wave phenomena [4].

In the following dimensionless form, the RLW equation represents a unidirectional propagation model,

$$\frac{\partial u}{\partial t}+\frac{\partial u}{\partial x}+u\frac{\partial u}{\partial x}-\frac{\partial}{\partial t}\left(\frac{\partial^2 u}{\partial x^2}\right)=0 \tag{13.15}$$

This is eq. (13.1) with $\epsilon = 1$ and $\mu = 1$. However, when it is required to analyze bidirectional waves with wave interactions, the following system of coupled equations can be used [2].

$$\frac{\partial \eta}{\partial t}+\frac{\partial u}{\partial x}+\frac{\partial}{\partial x}(u\eta)+\alpha\frac{\partial^3 u}{\partial x^3}-\beta\frac{\partial}{\partial t}\left(\frac{\partial^2 \eta}{\partial x^2}\right)=0 \tag{13.16a}$$

$$\frac{\partial u}{\partial t}+\frac{\partial \eta}{\partial x}+u\frac{\partial u}{\partial x}+\gamma\frac{\partial^3 \eta}{\partial x^3}-\delta\frac{\partial}{\partial t}\left(\frac{\partial^2 u}{\partial x^2}\right)=0 \tag{13.16b}$$

The dimensionless variables $\eta(x,t)$, $u(x,t)$, and x are scaled by the length scale h_0, and the time scale by $\sqrt{h_0/g}$, where h_0 represents the undisturbed water depth and g represents gravitational acceleration. The nondimensionless wave height above the equilibrium water level is represented by η, and the dimensionless horizontal surface wave velocity at a height θh_0 (where $0 \le \theta \le 1$) above the bottom of the channel is represented by $u(x,t)$.

For a physical system, the parameters α, β, γ, and δ represent *dispersion constants* and are subject to the following constraints:

$$\alpha+\beta=\frac{1}{2}\left(\theta^2-\frac{1}{3}\right) \quad \text{and} \quad \gamma+\delta=\frac{1}{2}(1-\theta^2)\ge 0$$

It has been shown[3] that a physically relevant system based on eqns. (13.16a) and (13.16b) is linearly well posed and will generally give rise to smooth solutions if the constants α, β, γ, and δ satisfy the following conditions:

$$\beta \geq 0,\, \delta \geq 0,\, \alpha \leq 0\, \gamma \leq 0 \quad \text{or} \quad \beta \geq 0,\, \delta \geq 0,\, \alpha = \gamma > 0$$

It has also been shown that the above systems are able to model the main characteristics of flow in an *ideal fluid*. Further, when damping due to viscosity effects is comparable with the effects of nonlinearity and/or dispersion, as is likely to occur in laboratory-scale experiments or in the field, the model and its numerical results should correspond well with experimental results. For more details, the reader is referred to [2–4] and the references therein.

Analytical Traveling Wave Solutions for Single RLW Equation

We will now outine some additional traveling wave solutions to eq. (13.1) that can be obtained by using Maple `tanhMethod()`, `expMethod()`, and `riccatiMethod()` procedures detailed in the main Appendix. In order to save space, the application codes will not be included here, but they are included in the downloadable Maple software available for this book.

Solution Using tanh Method

Application of the Maple procedure `tanhMethod()` enables us to derive three trivial solutions and the following two nontrivial traveling wave solutions

$$\begin{aligned} u &= \frac{12\mu k^2}{\epsilon\,(4\mu k^2 - 1)}\left[-1 + \tanh^2\left(\frac{k\,(4x\mu k^2 - x + t)}{4\mu k^2 - 1}\right)\right] \\ &= \frac{-12\mu k^2}{\epsilon\,(4\mu k^2 - 1)}\operatorname{sech}^2\left(\frac{k\,(4x\mu k^2 - x + t)}{4\mu k^2 - 1}\right) \end{aligned} \tag{13.17}$$

and

$$\begin{aligned} u &= \frac{-4\mu k^2}{\epsilon\,(4\mu k^2 + 1)}\left[-1 + 3\tanh^2\left(\frac{k\,(4x\mu k^2 + x - t)}{4\mu k^2 + 1}\right)\right] \\ &= \frac{-4\mu k^2}{\epsilon\,(4\mu k^2 + 1)}\left[2 - 3\operatorname{sech}^2\left(\frac{k\,(4x\mu k^2 + x - t)}{4\mu k^2 + 1}\right)\right] \end{aligned} \tag{13.18}$$

Equation (13.18) is the same solution as eq. (13.5) used in the main body of this chapter for the numerical simulation, where $d = \dfrac{4\mu k^2}{\epsilon(1 - 4\mu k^2)}$.

Solution Using exp Method

Application of the Maple procedure `expMethod()` enables us to derive two trivial solutions and three nontrivial traveling wave solutions, including the following:

$$u(x,t) = -\frac{96k^2\mu b_0}{\epsilon\left(4\mu k^2-1\right)}\left[b_0^2\exp\left(-\frac{2k\left(4x\mu k^2-x+t\right)}{4\mu k^2-1}\right)+4b_0\right.$$
$$\left.+4\exp\left(\frac{2k\left(4x\mu k^2-x+t\right)}{4\mu k^2-1}\right)\right]^{-1} \tag{13.19}$$

where b_0 denotes an arbitrary constant. The other two traveling wave solutions will not be detailed here as they are more complex and can be obtained by simply running the downloadable Maple code.

Solution Using Riccati Method

Application of the Maple procedure `riccatiMethod()` enables us to derive 18 trivial solutions and 12 nontrivial traveling wave solutions, some of which are duplicates. One is the same as the solution, eq. (13.5), used in the main body of this chapter for the numerical simulation. These solutions will not be detailed here as they can be obtained by simply running the downloadable Maple code.

Analytical Traveling Wave Solution for Coupled RLW Equations

One set of values for constants α, β, γ, and δ in eqns. (13.16a) and (13.16b) that satisfies the above constraints for surface waves (where $\theta \approx 0$, i.e., applicable to very small amplitude waves) is: $\alpha = -1/3$, $\beta = 1/6$, $\gamma = -1/3$, and $\delta = 5/6$.

Application of the Maple procedure `tanhMethod2()` enables us to derive two trivial solutions and 10 nontrivial traveling wave solutions, including the following, where for simplicity, we have set the wavenumber equal to unity, i.e., $k = 1$,

$$\eta = \frac{699-35\sqrt{401}}{96} - \frac{9+\sqrt{401}}{8}\tanh^2\left(x+\frac{11-\sqrt{401}}{20}t\right) \tag{13.20a}$$

$$u = -\frac{11+9\sqrt{401}}{120} + \tanh^2\left(x+\frac{11-\sqrt{401}}{20}t\right) \tag{13.20b}$$

A plot of this solution is given in Fig. 13.3 which shows the initial condition (at $t = 0$) for an animation, which then moves left to right when activated.

The Maple code that derives the solution, eqns. (13.20a) and (13.20b), is not included here but is available with the downloads for this book.

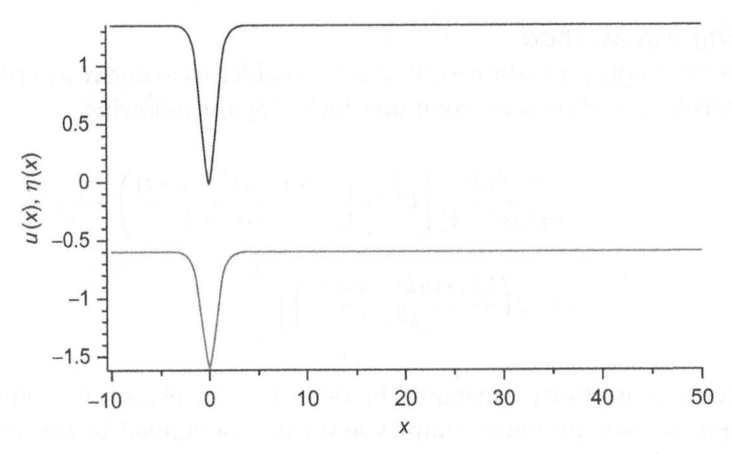

FIGURE 13.3: Plot of traveling wave solution, eqns. (13.20a and b), at $t = 0$ for *coupled regularized long-wave equations*, eqns. (13.16a and b). The top wave η represents wave height, and the bottom wave u represents wave velocity.

References

[1] T.B. Benjamin, J.L. Bona, J.J. Mahony, Model equations for long waves in nonlinear dispersive systems, *Philos. Trans. Roy. Soc. London Ser. A* 272 (1972) 47–78.

[2] J.L. Bona, M. Chen, J.-C. Saut, Boussinesq equations and other systems for small-amplitude long waves in nonlinear dispersive media I: Derivation and the linear theory, *J. Nonlinear Sci.* 12 (2002) 283–318.

[3] J.L. Bona, M. Chen, J.-C. Saut, Boussinesq equations and other systems for small-amplitude long waves in nonlinear dispersive media II: Nonlinear theory, *Nonlinearity*, 17 (2004) 925–952.

[4] M. Chen, O. Goubet, Long-time asymptotic behavior of dissipative Boussinesq system. Available online at `http://arxiv.org/PS_cache/math/pdf/0607/0607708v1.pdf`, 2006, Accessed: 24 March, 2010.

[5] S-u. Islam, S. Haq, A. Ali, A meshfree method for the numerical solution of the RLW equation, *J. Comp. Applied. Math.* 223 (2009) 997–1012.

14 Extended Bernoulli Equation

Chapter 1 includes a discussion of how a PDE can be reduced to an ODE through a change of variable, basically going from an *Eulerian coordinate system* to a *Lagrangian coordinate system* by assuming a *traveling wave solution*. Here, we demonstrate how this procedure can be reversed. To illustrate the method, we start with an ODE, which is a special case of the Bernoulli equation [1, 2]

$$\theta_0 \frac{dU}{d\xi} + bU(1-U) = 0 \tag{14.1}$$

and we impose the auxiliary conditions $U(\xi = -\infty) = 1, U(\xi = \infty) = 0$.

An analytical solution to eq. (14.1) that satisfies the auxiliary conditions is

$$U(\xi) = \frac{1}{(1 + e^{b\xi/\theta_0})} \tag{14.2}$$

We can verify this solution by direct substitution.

$$\begin{array}{lc} \theta_0 \dfrac{dU}{d\xi} & -\theta_0(1+e^{b\xi/\theta_0})^{-2}(e^{b\xi/\theta_0})(b/\theta_0) \\ +bU & b(1+e^{b\xi/\theta_0})^{-1} = b(1+e^{b\xi/\theta_0})^{-2}(1+e^{b\xi/\theta_0}) \\ -bU^2 & -b(1+e^{b\xi/\theta_0})^{-2} \\ 0 & 0 \end{array} \tag{14.3}$$

If $U(\xi)$ in eq. (14.1) is considered a Lagrangian (traveling wave) solution (with $\xi = x - ct$), it can be related to a Eulerian (fixed frame) variable u through

$$\begin{aligned} \frac{dU}{d\xi}\frac{\partial \xi}{\partial t} &= \frac{dU}{d\xi}(-c) = \frac{\partial u}{\partial t} \\ \frac{dU}{d\xi} &= -(1/c)\frac{\partial u}{\partial t} \\ \frac{dU}{d\xi}\frac{\partial \xi}{\partial x} &= \frac{\partial u}{\partial x} \\ \frac{dU}{d\xi} &= \frac{\partial u}{\partial x} \end{aligned} \tag{14.4}$$

If the second and fourth equations are added,

$$2\frac{dU}{d\xi} = \frac{\partial u}{\partial x} - (1/c)\frac{\partial u}{\partial t} \tag{14.5}$$

Then, $\dfrac{dU}{d\xi}$ of eq. (14.5) can be substituted into eq. (14.1) to obtain an equivalent PDE

$$\frac{\theta_0}{2}\left(\frac{\partial u}{\partial x} - (1/c)\frac{\partial u}{\partial t}\right) + bu(1-u) = 0 \tag{14.6}$$

The solution to eq. (14.6) is, from eq. (14.2),

$$u(x,t) = \frac{1}{(1+e^{b(x-ct)/\theta_0})} \tag{14.7}$$

which satisfies the conditions $u(x-ct=-\infty)=1, u(x-ct=\infty)=0$. In particular, as x approaches the limiting values $\pm\infty$ for finite t, $u(x,t)$ approaches the limiting values 1, 0.

To verify eq. (14.7) as the solution to eq. (14.6), we have

$$\begin{array}{ll}
\dfrac{\theta_0}{2}\dfrac{\partial u}{\partial x} & -\dfrac{\theta_0}{2}(1+e^{b(x-ct)/\theta_0})^{-2}(e^{b(x-ct)/\theta_0})(b/\theta_0) \\[2ex]
(-1/c)\dfrac{\theta_0}{2}\dfrac{\partial u}{\partial t} & (1/c)\dfrac{\theta_0}{2}(1+e^{b(x-ct)/\theta_0})^{-2}(e^{b(x-ct)/\theta_0})(-bc/\theta_0) \\[2ex]
+bu & b(1+e^{b(x-ct)/\theta_0})^{-1} = b(1+e^{b(x-ct)/\theta_0})^{-2}(1+e^{b(x-ct)/\theta_0}) \\[2ex]
-bu^2 & -b(1+e^{b(x-ct)/\theta_0})^{-2} \\[2ex]
0 & 0
\end{array} \tag{14.8}$$

In summary, we have illustrated a procedure for going from an ODE (eq. (14.1)) to a PDE (eq. (14.6)) that can be generalized to other examples. Also, we can directly extend the ODE solution (e.g., eq. (14.2)) to a solution to the PDE (eq. (14.7)). This procedure is useful for generating exact PDE solutions that can be used to test numerical procedures such as the *method of lines (MOL)*, particularly, if an analytical solution to the ODE is readily available as a starting point to derive a PDE solution.

Specifically, eq. (14.7) is the starting point for the calculation of a numerical PDE/MOL solution using the following Matlab routines. We consider first the ODE routine pde_1.m.

```
  function ut=pde_1(t,u)
%
% Function pde_1 computes the t derivative vector for the extended
% Bernoulli equation
%
  global xl xu x n ncall
%
```

```
  global th0 c b
%
% ux
  ux=dss004(xl,xu,n,u);
%
% PDE
  for i=1:n
    ut(i)=c*ux(i)+(2*b*c/th0)*u(i)*(1-u(i));
  end
  ut=ut';
%
% Increment calls to pde_1
  ncall=ncall+1;
```

LISTING 14.1: ODE routine `pde_1.m` for eq. (14.1).

We can note the following points about `pde_1.m`:

- The function and some global variables are defined.

```
  function ut=pde_1(t,u)
%
% Function pde_1 computes the t derivative vector for the extended
% Bernoulli equation
%
  global xl xu x n ncall
%
  global th0 c b
```

- The derivative $\dfrac{\partial u}{\partial x}$ in eq. (14.6) is computed by `dss004`.

```
%
% ux
  ux=dss004(xl,xu,n,u);
```

- The PDE, eq. (14.6), is programmed as a set of ODE derivatives in t, `ut`.

```
%
% PDE
  for i=1:n
    ut(i)=c*ux(i)+(2*b*c/th0)*u(i)*(1-u(i));
  end
  ut=ut';
%
% Increment calls to pde_1
  ncall=ncall+1;
```

Here, we have elected not to use BCs since the solution from the exp of IC (14.7) does not depart from the initial values $u(x=-\infty, t=0)=1, u(x=\infty, t=0)=0$; in other words, the boundary values do not have an effect on the solution for $t>0$. If this was not the case, we could program BCs at this point by using eq. (14.7) at the boundary values of x (the reader might try this to extend the numerical solution to finite boundary values for x). Also, only one BC is required since eq. (14.6) is first order in x.

ut is transposed to meet the requirement of the ODE integrator, ode15s, used in the main program. Also, the counter for the calls to pde_1.m is incremented at the end of pde_1.m.

The IC function is listed next.

```
  function u0=inital_1(t0)
%
% Function inital_1 sets the initial condition for the extended
% Bernoulli equation
%
  global xl xu x n
%
% Spatial domain and initial condition
  xl=-10;
  xu= 15;
  n=101;
  dx=(xu-xl)/(n-1);
%
% IC from analytical solution
  for i=1:n
    x(i)=xl+(i-1)*dx;
    u0(i)=ua_1(x(i),0.0);
  end
```

LISTING 14.2: IC routine inital_1.m from eq. (14.7) with $t = 0$.

We can note the following details about inital_1.m:

- After definition of the function and some global variables, the x domain is defined as $-10 \le x \le 15$ with 101 points. This range in x is large enough that the boundaries at $x = -10, 15$ are effectively at $\pm\infty$.
- The IC, eq. (14.7) with $t = 0$, is defined numerically by ua_1 (discussed subsequently).

Function ua_1.m below is a straightforward implementation of eq. (14.7).

```
  function uanal=ua_1(x,t)
%
% Function uanal computes the exact solution of the extended
% Bernoulli equation for comparison with the numerical solution
  global th0 c b
%
% Analytical solution
  uanal=1/(1+exp(b*(x-c*t)/th0));
```

LISTING 14.3: Analytical solution, eq. (14.7), for eq. (14.6).

The main program pde_1_main.m is similar to previous main programs. However, we list it here so the details are clear, particularly, since this is the last example of a PDE application first order in t.

```
%
% Clear previous files
  clear all
  clc
```

```
%
% Parameters shared with other routines
  global xl xu x n ncall
%
  global th0 c b
  th0=1; c=1; b=1;
%
% Initial condition
  t0=0.0;
  u0=inital_1(t0);
%
% Independent variable for ODE integration
  tf=4;
  tout=[t0:2:tf]';
  nout=3;
  ncall=0;
%
% ODE integration
  mf=2;
  reltol=1.0e-06; abstol=1.0e-06;
  options=odeset('RelTol',reltol,'AbsTol',abstol);
%
% Explicit (nonstiff) integration
  if(mf==1)[t,u]=ode45(@pde_1,tout,u0,options); end
%
% Implicit (sparse stiff) integration
  if(mf==2)
    S=jpattern_num_1;
%   pause
    options=odeset(options,'JPattern',S)
    [t,u]=ode15s(@pde_1,tout,u0,options);
  end
%
% Store analytical solution, errors in numerical solution
  for it=1:nout
    for i=1:n
      u_anal(it,i)=ua_1(x(i),t(it));
      err(it,i)=u(it,i)-u_anal(it,i);
    end
  end
%
%   Display selected output
    for it=1:nout
      fprintf('\n     t       x        u(it,i)   u_anal(it,i)
           err(it,i)\n');
      for i=1:10:n
        fprintf('%6.2f%8.3f%15.6f%15.6f%15.6f\n',...
            t(it),x(i),u(it,i),u_anal(it,i),err(it,i));
      end
    end
    fprintf('    ncall = %4d\n\n',ncall);
%
%   Plot numerical and analytical solutions
    figure(2)
```

```
plot(x,u,'-',x,u_anal,'o')
xlabel('x')
ylabel('u(x,t)')
title('Extended Bernoulli equation; t = 0, 2, 4; solid - numerical;
      o - analytical')
figure(3)
surf(x,t,u)
shading 'interp', axis 'tight'
view(21,24);
xlabel('x'); ylabel('t'); zlabel('u(x,t)');
title('Extended Bernoulli equation');
```

LISTING 14.4: Main program `pde_1_main.m` for eq. (14.6).

We can note the following details about `pde_1_main.m`:

- After previous files are cleared, some global variables are defined.

```
%
% Clear previous files
  clear all
  clc
%
% Parameters shared with other routines
  global xl xu x n ncall
%
  global th0 c b
  th0=1; c=1; b=1;
```

- The IC, eq. (14.7) with $t = 0$, is set by `inital_1.m`, and the t scale is defined as $0 \le t \le 4$ with outputs at $t = 0, 2, 4$.

```
%
% Initial condition
  t0=0.0;
  u0=inital_1(t0);
%
% Independent variable for ODE integration
  tf=4;
  tout=[t0:2:tf]';
  nout=3;
  ncall=0;
```

- The 101 ODEs are integrated by the sparse matrix option of `ode15s` (`mf=2`).

```
%
% ODE integration
  mf=2;
  reltol=1.0e-06; abstol=1.0e-06;
  options=odeset('RelTol',reltol,'AbsTol',abstol);
%
% Explicit (nonstiff) integration
  if(mf==1)[t,u]=ode45(@pde_1,tout,u0,options); end
%
```

```
% Implicit (sparse stiff) integration
  if(mf==2)
    S=jpattern_num_1;
%   pause
    options=odeset(options,'JPattern',S)
    [t,u]=ode15s(@pde_1,tout,u0,options);
  end
```

- The analytical solution of eq. (14.7) is computed (`u_anal`), and the difference between the analytical solutions is put into an array `err` for subsequent plotting.

```
%
% Store analytical solution, errors in numerical solution
  for it=1:nout
    for i=1:n
      u_anal(it,i)=ua_1(x(i),t(it));
      err(it,i)=u(it,i)-u_anal(it,i);
    end
  end
%
%   Display selected output
    for it=1:nout
      fprintf('\n      t        x          u(it,i)    u_anal(it,i)
                 err(it,i)\n');
      for i=1:10:n
        fprintf('%6.2f%8.3f%15.6f%15.6f%15.6f\n',...
                t(it),x(i),u(it,i),u_anal(it,i),err(it,i));
      end
    end
    fprintf('     ncall = %4d\n\n',ncall);
```

Selected numerical output is then displayed.

- A 2D plot with the numerical and analytical solutions is produced by `plot`, and a 3D plot of the numerical solution is produced by `surf`.

```
%
%   Plot numerical and analytical solutions
    figure(2)
    plot(x,u,'-',x,u_anal,'o')
    xlabel('x')
    ylabel('u(x,t)')
    title('Extended Bernoulli equation; t = 0, 2, 4; solid - numerical;
           o - analytical')
    figure(3)
    surf(x,t,u)
    xlabel('x'); ylabel('t'); zlabel('u(x,t)');
    title('Extended Bernoulli equation');
```

`jpattern_num_1.m` for the sparse matrix integration of the 101 ODEs called by the ODE integrator, `ode15s`, is similar to earlier versions of this routine and therefore is not listed here. Also, the ODE Jacobian map is not reproduced since it has the expected banded structure for a single PDE first order in t.

Table 14.1: Selected numerical output from the main program `pde_1_main.m` of Listing 14.4

t	x	u(it,i)	u_anal(it,i)	err(it,i)
0.00	-10.000	0.999955	0.999955	0.000000
0.00	-7.500	0.999447	0.999447	0.000000
0.00	-5.000	0.993307	0.993307	0.000000
0.00	-2.500	0.924142	0.924142	0.000000
0.00	0.000	0.500000	0.500000	0.000000
0.00	2.500	0.075858	0.075858	0.000000
0.00	5.000	0.006693	0.006693	0.000000
0.00	7.500	0.000553	0.000553	0.000000
0.00	10.000	0.000045	0.000045	0.000000
0.00	12.500	0.000004	0.000004	0.000000
0.00	15.000	0.000000	0.000000	0.000000
t	x	u(it,i)	u_anal(it,i)	err(it,i)
2.00	-10.000	0.999994	0.999994	0.000000
2.00	-7.500	0.999925	0.999925	0.000000
2.00	-5.000	0.999089	0.999089	0.000000
2.00	-2.500	0.989013	0.989013	-0.000001
2.00	0.000	0.880796	0.880797	-0.000001
2.00	2.500	0.377544	0.377541	0.000003
2.00	5.000	0.047437	0.047426	0.000011
2.00	7.500	0.004071	0.004070	0.000001
2.00	10.000	0.000335	0.000335	0.000000
2.00	12.500	0.000028	0.000028	0.000000
2.00	15.000	0.000014	0.000002	0.000012
t	x	u(it,i)	u_anal(it,i)	err(it,i)
4.00	-10.000	0.999999	0.999999	0.000000
4.00	-7.500	0.999990	0.999990	0.000000
4.00	-5.000	0.999877	0.999877	0.000000
4.00	-2.500	0.998499	0.998499	0.000000
4.00	0.000	0.982014	0.982014	-0.000000
4.00	2.500	0.817600	0.817574	0.000026
4.00	5.000	0.268998	0.268941	0.000057
4.00	7.500	0.029328	0.029312	0.000015
4.00	10.000	0.002474	0.002473	0.000001
4.00	12.500	0.000427	0.000203	0.000224
4.00	15.000	0.010740	0.000017	0.010724
	ncall = 186			

A portion of the numerical output is listed in Table 14.1.
We can note the following points about this output:

- Generally, the agreement between the numerical and analytical solutions is to four figures or better.

- The exception is at $t = 4$ near the right boundary at $x = 15$.

```
4.00  12.500        0.000427        0.000203        0.000224
4.00  15.000        0.010740        0.000017        0.010724
```

  This output suggests that as the solution travels from left to right, it comes close enough to the right boundary that a boundary effect appears to be developing numerically (note the increase in the difference between the numerical and analytical solutions). Thus, if the numerical solution is computed for $t > 4$, the right boundary might be extended beyond $x = 15$, or a boundary condition might be imposed using eq. (14.7) in `pde_1.m`.
- The computational effort is quite modest, with `ncall = 186`.

The plotted output follows. Figure 14.1 demonstrates that the solution is a traveling wave, which follows from eq. (14.7) and the argument $x - ct$. The traveling wave solution of eq. (14.6) is also clear in Fig. 14.2.

Another way to elucidate the traveling wave solution of eq. (14.7) is to consider the argument $x - ct$. With $c = 1$ (set in `pde_1_main.m`), the numerical solution of Fig. 14.1 should be moving left to right with unit velocity. This is confirmed in the following output (given in Table 14.2) produced by displaying the solution at $x = t$ rather than every 10th value as in Table 14.1.

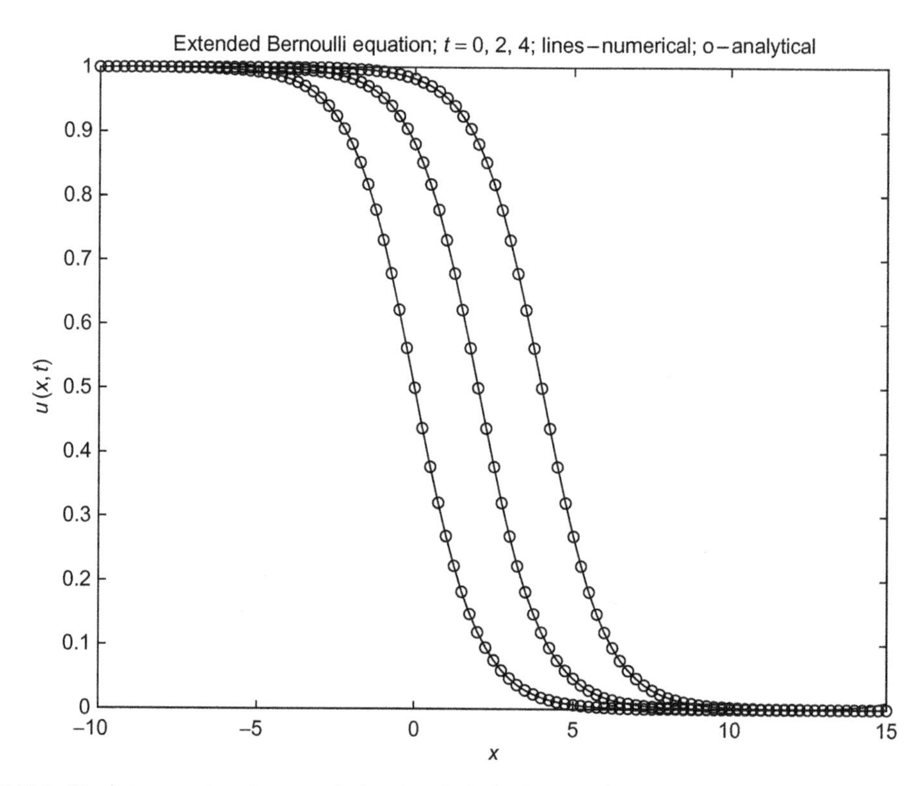

FIGURE 14.1: 2D plot comparing the numerical and analytical solutions of eq. (14.6).

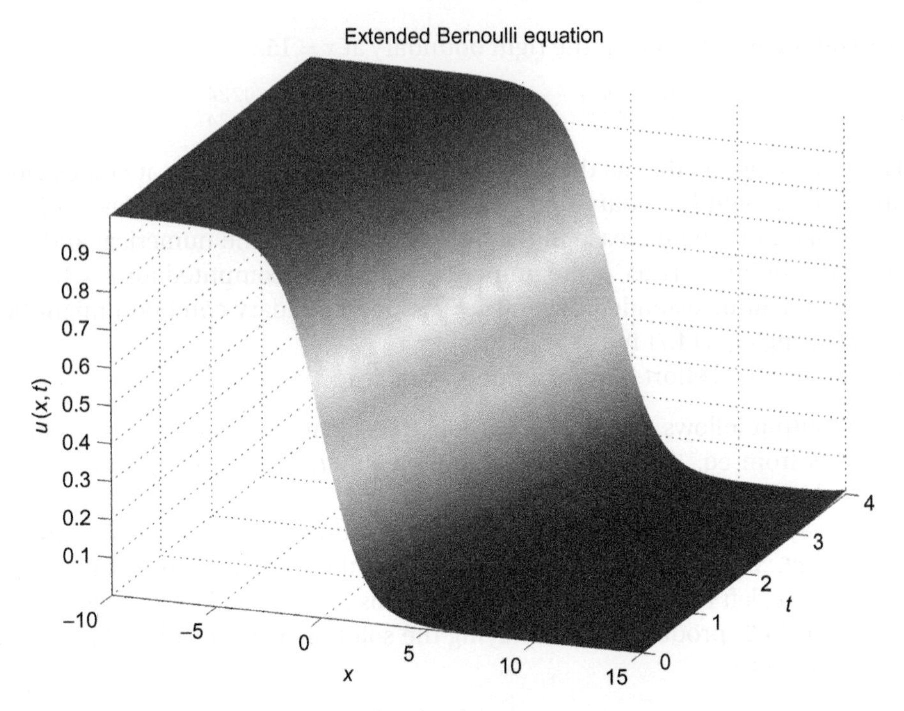

FIGURE 14.2: 3D plot of the numerical solution of eq. (14.6).

Note, in particular, the values of the numerical solution $u(x=0,t=0)=0.500000$, $u(x=2,t=2)=0.500004$, $u(x=4,t=4)=0.500065$. In other words, as t increases by 2, x also increases by 2 for the solution value ≈ 0.5 corresponding to $c=1$. Also, there is an increase in the departure from 0.5 due to accumulating numerical error with increasing t, i.e., 0.000004 and 0.000065 for $t=2$ and 4, respectively. If this accumulating error became excessive, it could be reduced by using more grid points ($n > 101$), generally termed *h refinement*, or a higher-order approximation of the derivative in x in eq. (14.1) (higher-order than in `dss004`) could be used, generally termed *p refinement*.

Also, this accumulating error may originate in the integration with respect to t by the ODE routine `ode15s` in which case the error tolerances for this integrator could be tightened (they are set in `pde_1_main.m` as `abstol,reltol` in Listing 14.4). This discussion illustrates an important conclusion: *the total error in a numerical solution is determined by both the spatial (in x) and the temporal (in t) errors*. In other words, *it is essential to balance the spatial and the temporal errors* to achieve a numerical solution of acceptable accuracy.

In summary, we have observed the following:

- The conversion of an ODE (eq. (14.1)) into an associated PDE (eq. (14.6)).
- The derivation of an exact solution to the PDE (eq. (14.7)) from the analytical solution of the ODE (eq. (14.2)).
- The straightforward MOL solution of the PDE for an essentially infinite domain in x.

Table 14.2: Selected numerical output from the main program `pde_1_main.m` of Listing 14.4

t	x	u(it,i)	u_anal(it,i)	err(it,i)
	.		.	
	.		.	
	.		.	
0.00	-0.250	0.562177	0.562177	0.000000
0.00	0.000	0.500000	0.500000	0.000000
0.00	0.250	0.437823	0.437823	0.000000
	.		.	
	.		.	
	.		.	
2.00	1.750	0.562182	0.562177	0.000005
2.00	2.000	0.500004	0.500000	0.000004
2.00	2.250	0.437827	0.437823	0.000003
	.		.	
	.		.	
	.		.	
4.00	3.750	0.562240	0.562177	0.000063
4.00	4.000	0.500065	0.500000	0.000065
4.00	4.250	0.437887	0.437823	0.000064
	.		.	
	.		.	
	.		.	

Appendix

At the beginning of this chapter, it was demonstrated that the partial differential equation (renumbered here from (14.6) to (14.9) to facilitate the subsequent discussion)

$$\frac{\theta_0}{2}\left(\frac{\partial u}{\partial x}-\frac{1}{c}\frac{\partial u}{\partial t}\right)+bu(1-u)=0, \quad u=u(x,t),\ t>0 \qquad (14.6, 14.9)$$

can be derived from the original ordinary differential equation (14.1). In this appendix, by application of the *tanh method*, we provide simple confirmation that the traveling wave solution of this equation is the same as the original solution given in eq. (14.7).

We begin by using the transformation $u(x,t)=U(\xi)$ where $\xi=k(x-ct)$, which reduces the PDE of eq. (14.9) to the ODE

$$\theta_0\frac{dU}{d\xi}+bU(1-U)=0 \qquad (14.1, 14.10)$$

as expected. Application of the tanh method using the Maple script shown in Listing 14.5 results in the traveling wave solution

$$u = \frac{1}{2} - \frac{1}{2}\tanh\left(\frac{1}{2}\frac{b(x+x_0-ct)}{\theta_0}\right) \tag{14.11}$$

which, using the relationship $\frac{1}{2}\left(1-\tanh\frac{X}{2}\right) = (1+\exp X)^{-1}$ and setting $x_0 = 0$, reduces to the original solution used for the numerical simulation in the main body of this chapter, i.e.,

$$u = \frac{1}{\left(1+e^{b(x-ct)/\theta_0}\right)} \tag{14.7, 14.12}$$

This solution can also be found using the *exp* and/or the *Riccati* methods, but the Maple listings are not included here in order to save space. However, the associated files are included with the downloads provided for this book.

```
>#  Extended Bernoulli Equation
 #  Attempt at Malfliet's tanh solution
 restart; with(PDEtools): with(PolynomialTools):
 with(plots):
>alias(u=u(x,t)):
>#  Define problem PDE
 pde1:=(theta[0]/2)*(diff(u,x)-(1/c)*diff(u,t))+b*u*(1-u)=0;
>read("tanhMethod.txt");
>intFlg:=0: # integration of U(xi) not needed!
 M:=1;  # Set order of approximation
 infoLevOut:=0; # Minimum level of output
 tanhMethod(M,pde1,intFlg,infoLevOut);
>#  Choose solution 4
 zz:=rhs(sol[4]);
>#  NOTE: Following transformation
 x0:=0;
 u2:=simplify(convert(zz, exp));
>#  Plot results
 #  ============
 theta[0]:=1; c:=1; b:=1;
 #  Create animation
 animate(zz,x=-10..30,t=0..20,
   numpoints=100,frames=50, axes=framed,labels=["x","u"],
   thickness=3,title="Extended Bernoulli Equation",
   labelfont=[TIMES, ROMAN, 16],axesfont=[TIMES, ROMAN, 16],
   titlefont=[TIMES, ROMAN, 16]);
>#  Create 3D plot
 plot3d(zz,x=-10..30,t=0..20,axes='framed',
   labels=["x","t","u(x,t)"],
   orientation=[-72,59],grid=[100,100],
   style=patchnogrid,axesfont=[TIMES, ROMAN, 16],
   labeldirections=[HORIZONTAL,HORIZONTAL,VERTICAL],
```

```
labelfont=[TIMES, ROMAN, 16],
shading=Z,title="Extended Bernoulli Equation",
titlefont=[TIMES, ROMAN, 16]);
```

LISTING 14.5: Maple code to obtain the traveling wave solution for eq. (14.9).

References

[1] A. Fasano, M.A. Herrero, M.R. Rodrigo, Slow and fast invasion waves in a model of acid-mediated tumour growth, *Math. Biosci.* 220 (2009) 45–56.

[2] C.R. Wylie, L.C. Barrett, *Advanced Engineering Mathematics*, Sixth ed., McGraw-Hill Inc., New York, 1995, pp. 42–44.

15 Hyperbolic Liouville Equation

To this point, we have considered PDEs first order in the initial value variable generally designated as t. We now consider a series of PDEs second order in the initial value variable. PDEs of this form are quite common in applications and therefore an approach to their numerical solution is of practical importance.

To start, the *hyperbolic Liouville equation* ([3], p213) with an exponential nonlinearity is

$$\frac{\partial^2 u}{\partial t^2} = a^2 \frac{\partial^2 u}{\partial x^2} + be^{\beta u} \tag{15.1}$$

where a, b, and β are arbitrary.

An analytical solution is ([3], p213)

$$u_a(x,t) = \frac{1}{\beta} \ln\left[\frac{2(a^2A^2 - B^2)}{b\beta \cosh^2(Ax + Bt + C)}\right] \tag{15.2}$$

where A, B, and C are arbitrary constants. Note that eq. (15.2) is a traveling wave solution through the combination $Ax + Bt + C$.

One initial condition (IC) for eq. (15.1) is eq. (15.2) with $t = 0$. Since eq. (15.1) is second order in t, it requires two ICs. A second IC is available by differentiating eq. (15.2) with respect to t,

$$\frac{\partial u_a(x,t)}{\partial t} = \frac{1}{\beta}\left[\frac{b\beta \cosh^2(Ax + Bt + C)}{2(a^2A^2 - B^2)}\right]\left[(-2)\frac{2(a^2A^2 - B^2)}{b\beta \cosh^3(Ax + Bt + C)}\right]\sinh(Ax + Bt + C)(B) \tag{15.3}$$

and setting $t = 0$.

BCs for eq. (15.1) are taken as

$$u(x = x_l, t) = u_a(x = x_l, t); \tag{15.4a}$$

$$u(x = x_u, t) = u_a(x = x_u, t) \tag{15.4b}$$

Equations (15.1)–(15.4) constitute the problem programmed in the following Matlab routines.

Equation (15.1) is programmed in function `pde_1.m` listed first.

```
  function ut=pde_1(t,u)
%
% Function pde_1 computes the t derivative vector for the hyperbolic
```

```
% Liouville equation
equation
%
  global xl xu x n ncall
%
% Model parameters
  global a b beta A B C nu
%
% One vector to two vectors
  for i=1:n
    u1(i)=u(i);
    u2(i)=u(i+n);
  end
%
% BCs at x = 0,1
  u1(1)=ua_1(x(1),t);
  u1(n)=ua_1(x(n),t);
%
% u1xx
  nl=1; nu=1;
  u1x(1)=0;
  u1xx=dss044(xl,xu,n,u1,u1x,nl,nu);
%
% PDE
  for i=2:n-1
    u1t(i)=u2(i);
    u2t(i)=a*u1xx(i)+b*exp(beta*u1(i));
  end
  u1t(1)=0; u1t(n)=0;
  u2t(1)=0; u2t(n)=0;
%
% Two vectors to one vector
  for i=1:n
    ut(i)  =u1t(i);
    ut(i+n)=u2t(i);
  end
  ut=ut';
%
% Increment calls to pde_1
  ncall=ncall+1;
```

LISTING 15.1: Function `pde_1.m` for eq. (15.1).

We can note the following details about `pde_1.m`:

- The function and some global variables are first defined.

```
  function ut=pde_1(t,u)
%
% Function pde_1 computes the t derivative vector for the hyperbolic
% Liouville equation
%
  global xl xu x n ncall
%
% Model parameters
  global a b beta A B C nu
```

- The 1D ODE-dependent variable vector of length $2n$, u, is transferred to two 1D arrays, each of length n, u1,u2, to facilitate the numerical solution of eq. (15.1).

```
%
% One vector to two vectors
  for i=1:n
    u1(i)=u(i);
    u2(i)=u(i+n);
  end
```

This use of two 1D arrays is a standard procedure for accommodating a PDE second order in t such as eq. (15.1). The net effect is to write the second-order PDE in t as two first-order PDEs in t. In the case of eq. (15.1), this gives

$$\frac{\partial u_1}{\partial t} = u_2 \tag{15.5a}$$

$$\frac{\partial u_2}{\partial t} = a^2 \frac{\partial^2 u_1}{\partial x^2} + be^{\beta u_1} \tag{15.5b}$$

This procedure of working with two first-order PDEs in t permits the use of library integrators for first-order ODEs such as ode45 and ode15s. Also, the procedure is quite general in the sense that it can usually be applied to any second-order PDE of interest.
- The BCs (15.4) are programmed as

```
%
% BCs at x = 0,1
  u1(1)=ua_1(x(1),t);
  u1(n)=ua_1(x(n),t);
%
% u1xx
  nl=1; nu=1;
  u1x(1)=0;
  u1xx=dss044(xl,xu,n,u1,u1x,nl,nu);
```

Note that BCs (15.4) are designated as *Dirichlet* through nl=1, nu=1. Also, the first derivative in x, u1x(1)=0, is not actually used with the Dirichlet BCs but is included only to satisfy the Matlab requirement that all of the input (RHS) arguments of a function, such as u1x in dss044, must be defined numerically. The second derivative from dss044, u1xx, can then be used in the programming of eq. (15.1) (or eq. (15.5b)).
- Equations (15.5a,5b) are then programmed in a for loop.

```
%
% PDE
  for i=2:n-1
    u1t(i)=u2(i);
    u2t(i)=a*u1xx(i)+b*exp(beta*u1(i));
  end
  u1t(1)=0; u1t(n)=0;
  u2t(1)=0; u2t(n)=0;
```

Note also that the derivatives in t of the ODEs at the boundaries (i=1,n) are zeroed to ensure that the boundary values remain at the values prescribed by eqns. (15.4a,4b).

- The two derivative vectors, u1t,u2t, are then returned to a single derivative vector ut to be returned from pde_1.m with a transpose included to meet the requirement of ode15s. Also, the counter for pde_1.m is incremented.

```
%
% Two vectors to one vector
  for i=1:n
    ut(i)   =u1t(i);
    ut(i+n)=u2t(i);
  end
  ut=ut';
%
% Increment calls to pde_1
  ncall=ncall+1;
```

In summary, pde_1.m receives the dependent variable vector u as an input (along with the independent variable t) and returns the derivative vector ut.

The IC function, inital_1.m, defines initial values for u1 and u2 in eqns. (15.5a,5b).

```
  function u0=inital_1(t0)
%
% Function inital_1 sets the initial condition for the hyperbolic
% Liouville equation
%
% Parameters shared with other routines
  global xl xu x n ncall
%
% Spatial domain and initial condition
  xl=-1;
  xu= 1;
  n=51;
  dx=(xu-xl)/(n-1);
%
% ICs from analytical solution
  for i=1:n
    x(i)=xl+(i-1)*dx;
    u1(i)= ua_1(x(i),t0);
    u2(i)=uat_1(x(i),t0);
    u0(i)   =u1(i);
    u0(i+n)=u2(i);
  end
```

LISTING 15.2: IC function inital_1.m from eqns. (15.2) and (15.3) with $t = 0$.

inital_1.m defines a grid in x of 51 points for $-1 \leq x \leq 1$. In the for loop, the initial values of u1 for eq. (15.5a) are provided by the function ua_1.m (which has the programming for eq. (15.2)); the initial values of u2 for eq. (15.5b) are provided by the function uat_1.m (which has the programming for eq. (15.3)). All $2n$ initial condition values are returned from inital_1.m through the vector u0 to the main program pde_1_main.m discussed subsequently.

Function ua_1.m is a straightforward implementation of eq. (15.2).

```
  function uanal=ua_1(x,t)
%
% Function uanal computes the exact solution of the hyperbolic Liouville
% equation for comparison with the numerical solution
%
% Model parameters
  global a b beta A B C nc
%
% Analytical solution
  xi=A*x+B*t+C;
  uanal=(1/beta)*log(nc/(b*beta*cosh(xi)^2));
```

LISTING 15.3: Function `ua_1.m` for the analytical solution of eq. (15.2).

Function `uat_1.m` is a straightforward implementation of eq. (15.3).

```
  function uanal=uat_1(x,t)
%
% Function uanal computes the time derivative of the exact solution
% of the hyperbolic Liouville equation
%
% Model parameters
  global a b beta A B C nc
%
% Analytical solution derivative
  xi=A*x+B*t+C;
  uanal=(1/beta)*(b*beta*cosh(xi)^2/nc)*(-2*nc/(b*beta*cosh(xi)^3))...
       *sinh(xi)*B;
```

LISTING 15.4: Function `uat_1.m` for the analytical solution of eq. (15.3).

The main program `pde_1_main.m` is listed next.

```
%
% Clear previous files
  clear all
  clc
%
% Parameters shared with other routines
  global xl xu x n ncall
%
% Model parameters
  global a b beta A B C nc
%
% Model parameters
  a=1; b=1; beta=1;
  A=2; B=-1; C=0; nc=2*(a^2*A^2-B^2);
%
% Initial condition
  t0=0.0;
  u0=inital_1(t0);
%
% Independent variable for ODE integration
  tf=0.9;
  tout=[t0:0.3:tf]';
```

```
  nout=4;
  ncall=0;
%
% ODE integration
  mf=2;
  reltol=1.0e-06; abstol=1.0e-06;
  options=odeset('RelTol',reltol,'AbsTol',abstol);
%
% Explicit (nonstiff) integration
  if(mf==1)[t,u]=ode45(@pde_1,tout,u0,options); end
%
% Implicit (sparse stiff) integration
  if(mf==2)
    S=jpattern_num_1;
    options=odeset(options,'JPattern',S);
    [t,u]=ode15s(@pde_1,tout,u0,options);
  end
%
% One vector to two vectors
  for it=1:nout
  for i=1:n
    u1(it,i)=u(it,i);
    u2(it,i)=u(it,i+n);
  end
  end
%
% Store analytical solution, errors in numerical solution
  for it=1:nout
    u1(it,1)=ua_1(x(1),t(it));
    u1(it,n)=ua_1(x(n),t(it));
    for i=1:n
      u1_anal(it,i)=ua_1(x(i),t(it));
      err(it,i)=u1(it,i)-u1_anal(it,i);
    end
  end
%
% Display selected output
  for it=1:nout
    fprintf('\n     t       x        u1(it,i)  u1_anal(it,i)
             err(it,i)\n');
    for i=1:1:n
      fprintf('%6.2f%8.3f%15.6f%15.6f%15.6f\n',...
            t(it),x(i),u1(it,i),u1_anal(it,i),err(it,i));
    end
  end
  fprintf('    ncall = %4d\n\n',ncall);
%
% Plot numerical and analytical solutions
  figure(2)
  plot(x,u1,'-',x,u1_anal,'o')
  xlabel('x')
  ylabel('u1(x,t)')
  title('hyperbolic Liouville equation; t = 0, 0.3, 0.6, 0.9;
         solid - num; o - anal')
```

```
figure(3)
surf(x,t,u1)
xlabel('x'); ylabel('t'); zlabel('u1(x,t)');
title('hyperbolic Liouville equation');
```

LISTING 15.5: Main program for the solution of eq. (15.1) (eqns. (15.5a,5b)).

We can note the following details about `pde_1_main.m`:

- Previous files are cleared, and the problem parameters are defined as global and numerically.

```
%
% Clear previous files
  clear all
  clc
%
% Parameters shared with other routines
  global xl xu x n ncall
%
% Model parameters
  global a b beta A B C nc
%
% Model parameters
  a=1; b=1; beta=1;
  A=2; B=-1; C=0; nc=2*(a^2*A^2-B^2);
```

- The ICs for eqns. (15.5) are set numerically by `inital_1.m`, and the interval in t is defined as $0 \leq t \leq 0.9$ with displayed values at $t = 0, 0.3, 0.6, 0.9$.

```
%
% Initial condition
  t0=0.0;
  u0=inital_1(t0);
%
% Independent variable for ODE integration
  tf=0.9;
  tout=[t0:0.3:tf]';
  nout=4;
  ncall=0;
```

- The ODE integration is by `ode15s` (`mf=2`).

```
%
% ODE integration
  mf=2;
  reltol=1.0e-06; abstol=1.0e-06;
  options=odeset('RelTol',reltol,'AbsTol',abstol);
%
% Explicit (nonstiff) integration
  if(mf==1)[t,u]=ode45(@pde_1,tout,u0,options); end
%
% Implicit (sparse stiff) integration
  if(mf==2)
    S=jpattern_num_1;
```

```
    options=odeset(options,'JPattern',S);
    [t,u]=ode15s(@pde_1,tout,u0,options);
  end
```

- The solution vector from `ode15s`, `u`, is placed in two arrays, `u1`, `u2`, corresponding to eq. (15.5) to facilitate displaying the solution.

```
%
% One vector to two vectors
  for it=1:nout
  for i=1:n
    u1(it,i)=u(it,i);
    u2(it,i)=u(it,i+n);
  end
  end
```

- The numerical solution at the boundaries (`i=1,n`) is set by the analytical solution from `ua_1.m`. Then, the analytical solution is put in array `u1_anal` and the difference between the numerical and analytical solutions is put in array `err`.

```
%
% Store analytical solution, errors in numerical solution
  for it=1:nout
    u1(it,1)=ua_1(x(1),t(it));
    u1(it,n)=ua_1(x(n),t(it));
    for i=1:n
      u1_anal(it,i)=ua_1(x(i),t(it));
      err(it,i)=u1(it,i)-u1_anal(it,i);
    end
  end
%
% Display selected output
  for it=1:nout
    fprintf('\n     t       x        u1(it,i)  u1_anal(it,i)
             err(it,i)\n');
    for i=1:1:n
      fprintf('%6.2f%8.3f%15.6f%15.6f%15.6f\n',...
             t(it),x(i),u1(it,i),u1_anal(it,i),err(it,i));
    end
  end
  fprintf('    ncall = %4d\n\n',ncall);
```

Numerical output is displayed by two nested `for` loops (in t and x), and finally, the counter for the number of calls to `pde_1.m` is displayed as an indication of the computational effort required to produce the numerical solution.

- A 2D plot with the numerical and analytical solutions is produced by `plots`, and a 3D plot of the numerical solution is produced by `surf`.

```
%
% Plot numerical and analytical solutions
  figure(2)
  plot(x,u1,'-',x,u1_anal,'o')
  xlabel('x')
  ylabel('u1(x,t)')
```

```
title('hyperbolic Liouville equation; t = 0, 0.3, 0.6, 0.9;
       solid - num; o - anal')
figure(3)
surf(x,t,u1)
xlabel('x'); ylabel('t'); zlabel('u1(x,t)');
title('hyperbolic Liouville equation');
```

`jpattern_num_1.m` for the sparse matrix integration of the ODEs is not listed here. Rather just the statements for the $2n = 2(51)$ ODEs and the use of `pde_1.m` are listed.

```
%
% Sparsity pattern of the Jacobian matrix based on a
% numerical evaluation
%
% Set independent, dependent variables for the calculation
% of the sparsity pattern
  tbase=0;
  for i=1:n
    ybase(i)=0.5;
    ybase(i+n)=0.5;
  end
  ybase=ybase';
%
% Compute the corresponding derivative vector
  ytbase=pde_1(tbase,ybase);
  fac=[];
  thresh=1e-16;
  vectorized='on';
  [Jac,fac]=numjac(@pde_1,tbase,ybase,ytbase,thresh,fac,vectorized);
```

LISTING 15.6: Statements in `jpattern_num_1.m` for $2n$ ODEs and the use of `pde_1.m`.

The ODE Jacobian map from `jpattern_num_1.m` indicates two bands for PDEs (15.5a, 15.5b) in Figure 15.1.

A portion of the numerical output from `pde_1_main.m` is listed in Table 15.1. We can note the following points about this output:

- Generally, the agreement between the numerical and analytical solutions is to five figures or better.
- The computational effort is quite modest, with `ncall = 183`.

The plotted solutions follow. Figure 15.2 demonstrates that the boundary values at $x = -1, 1$ vary with t according to eqns. (15.2) and (15.4) and as programmed in `pde_1.m`.

Also, the traveling wave solution produced by eq. (15.2) with the argument $Ax + Bt$ is clear, although the estimation of the wave velocity B/A (from $A(x + B/At)$) using the numerical solution (as in Chapter 12) would require greater resolution in x in the numerical solution, for example, to identify the maximum of the solution at $t = 0, 0.3, 0.6, 0.9$. Figure 15.3 is produced by the call to `surf` in the main program of Listing 15.5.

In summary, we have discussed the numerical solution of a PDE second order in t (eq. (15.1)) by restating the PDE as two PDEs first order in t (eq. (15.5)). Generally, the reformulation of an nth-order PDE as n PDEs first order in an initial value independent variable

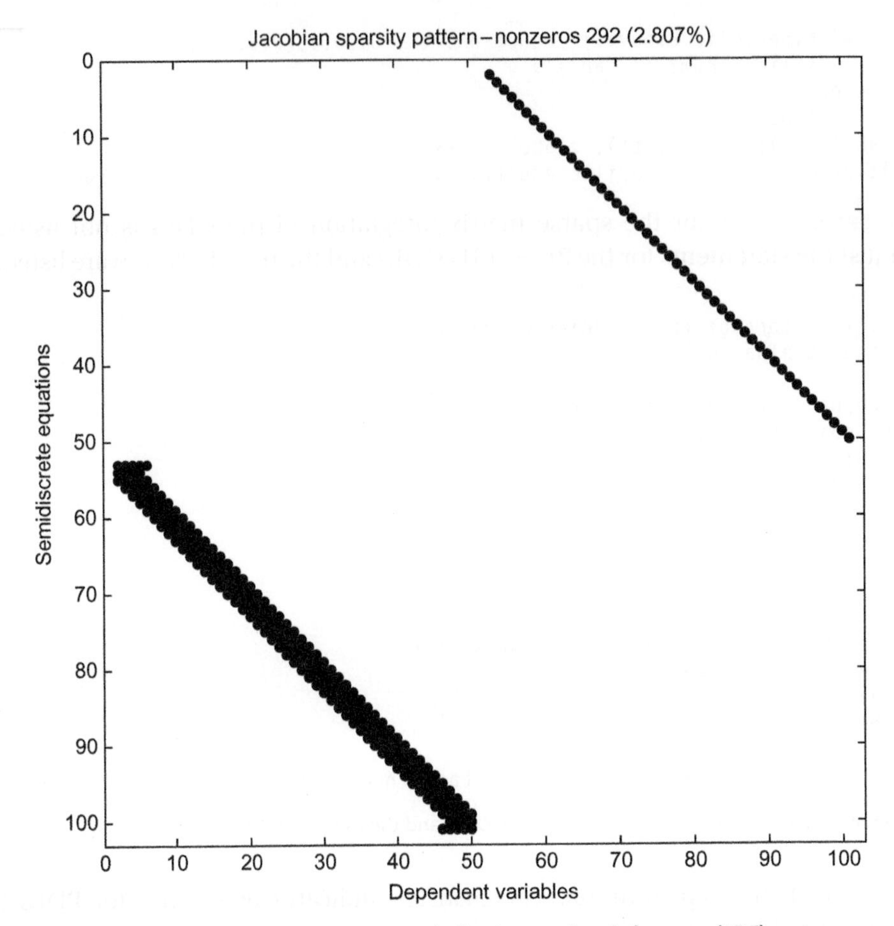

FIGURE 15.1: ODE Jacobian map from `jpattern_num_1.m` indicating two bands for eqns. (15.5).

(t) is straightforward and permits the use of an integrator such as `ode15s` for first-order ODEs. We have also applied this idea of reformulating a higher-order PDE in a boundary value variable as a system of lower-order PDEs in the boundary value variable, but we will not discuss this type of problem here.

Appendix

We conclude this chapter by analyzing the *hyperbolic Liouville* equation [1, 2] using two methods. The first method is very straightforward and leads to a variable separable solution. We start by applying the traveling wave transformation $u(x,t) = U(\xi)/\beta$, where $\xi = k(x - ct)$, k represents wavenumber, and c represents wave velocity (Note the division of $U(\xi)$ by β, which facilitates the calculation). We then have an equation that can be split into separate terms, one a function of U and the other of ξ, that can be integrated

Table 15.1: Selected numerical output from the main program `pde_1_main.m` of Listing 15.5

t	x	u1(it,i)	u1_anal(it,i)	err(it,i)
0.00	−1.000	−0.858246	−0.858246	0.000000
0.00	−0.960	−0.704478	−0.704478	0.000000
0.00	−0.920	−0.551766	−0.551766	0.000000
0.00	−0.880	−0.400286	−0.400286	0.000000
0.00	−0.840	−0.250237	−0.250237	0.000000
0.00	−0.800	−0.101853	−0.101853	0.000000
	.		.	
	.		.	
	.		.	
0.00	−0.200	1.635852	1.635852	0.000000
0.00	−0.160	1.691061	1.691061	0.000000
0.00	−0.120	1.734704	1.734704	0.000000
0.00	−0.080	1.766268	1.766268	0.000000
0.00	−0.040	1.785366	1.785366	0.000000
0.00	0.000	1.791759	1.791759	0.000000
0.00	0.040	1.785366	1.785366	0.000000
0.00	0.080	1.766268	1.766268	0.000000
0.00	0.120	1.734704	1.734704	0.000000
0.00	0.160	1.691061	1.691061	0.000000
0.00	0.200	1.635852	1.635852	0.000000
	.		.	
	.		.	
	.		.	
0.00	0.800	−0.101853	−0.101853	0.000000
0.00	0.840	−0.250237	−0.250237	0.000000
0.00	0.880	−0.400286	−0.400286	0.000000
0.00	0.920	−0.551766	−0.551766	0.000000
0.00	0.960	−0.704478	−0.704478	0.000000
0.00	1.000	−0.858246	−0.858246	0.000000
	.		.	
	.		.	
	.		.	
	output for t = 0.3, 0.6 removed			
	.		.	
	.		.	
	.		.	

t	x	u1(it,i)	u1_anal(it,i)	err(it,i)
0.90	−1.000	−2.627992	−2.627992	0.000000
0.90	−0.960	−2.469039	−2.469039	0.000000
0.90	−0.920	−2.310268	−2.310267	−0.000001
0.90	−0.880	−2.151709	−2.151708	−0.000002
0.90	−0.840	−1.993398	−1.993397	−0.000001
0.90	−0.800	−1.835378	−1.835377	−0.000001

(Continued)

Table 15.1: *(Continued)*

t	x	u1(it,i)	u1_anal(it,i)	err(it,i)
	.		.	
	.		.	
	.		.	
0.90	0.200	1.551529	1.551530	−0.000001
0.90	0.240	1.620313	1.620314	−0.000001
0.90	0.280	1.678320	1.678320	−0.000000
0.90	0.320	1.724907	1.724908	−0.000000
0.90	0.360	1.759533	1.759533	0.000000
0.90	0.400	1.781777	1.781776	0.000001
0.90	0.440	1.791360	1.791359	0.000000
0.90	0.480	1.788161	1.788162	−0.000000
0.90	0.520	1.772224	1.772223	0.000001
0.90	0.560	1.743746	1.743745	0.000001
0.90	0.600	1.703077	1.703078	−0.000001
	.		.	
	.		.	
	.		.	
0.90	0.800	1.337217	1.337219	−0.000002
0.90	0.840	1.236587	1.236588	−0.000001
0.90	0.880	1.128607	1.128608	−0.000001
0.90	0.920	1.014029	1.014030	−0.000001
0.90	0.960	0.893568	0.893568	−0.000000
0.90	1.000	0.767887	0.767887	0.000000
ncall=183				

directly. A solution for u is obtained by the inverse transformation $U(\xi) = \beta u(x,t)$. Thus, the calculation sequence is

$$\frac{\partial^2 u}{\partial t^2} - a^2 \frac{\partial^2 u}{\partial x^2} = b\mathrm{e}^{\beta u} \tag{15.1, 15.6}$$

$\Downarrow$ transformation

$$\frac{d^2 U}{d\xi^2} = \frac{b\beta \mathrm{e}^U}{c^2k^2 - a^2k^2} \tag{15.7}$$

$\Downarrow$ integration

$$\mathrm{e}^{-U} = \frac{b\beta\xi^2}{2(c^2k^2 - a^2k^2)} \tag{15.8}$$

$\Downarrow$ inverse transformation

$$u = \frac{1}{\beta}\ln\left(\frac{2\,(c^2 - a^2)}{b\beta\,(x + x_0 - ct)^2}\right). \tag{15.9}$$

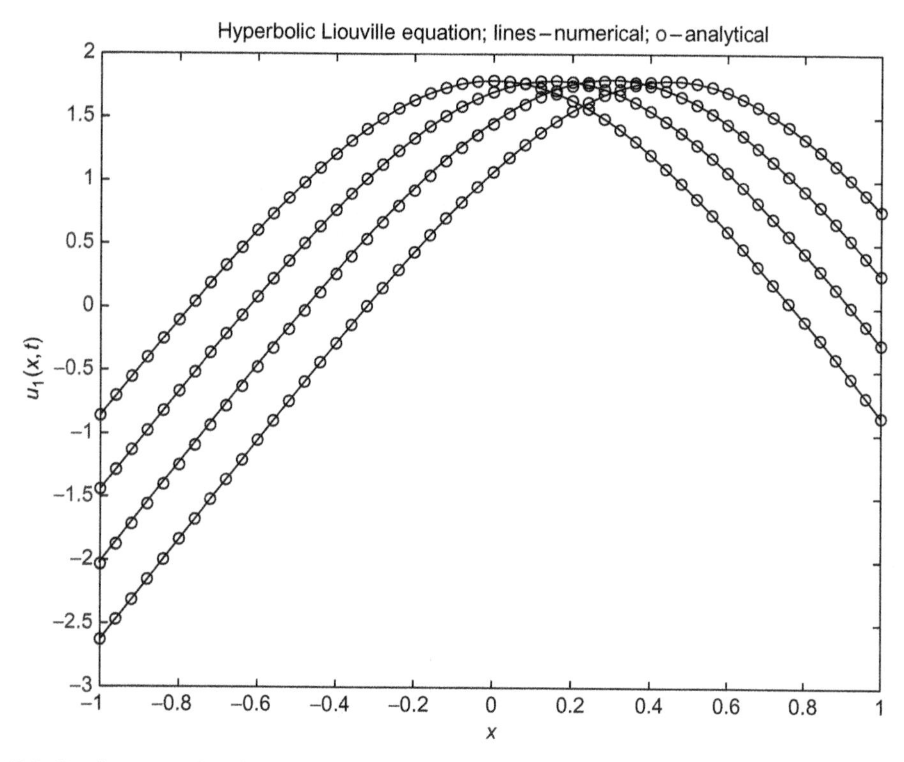

FIGURE 15.2: 2D plot comparing the numerical and analytical solutions of eq. (15.1) (eqns. (15.5a,5b)) moving left to right in x for $t = 0, 0.3, 0.6, 0.9$.

However, although eq. (15.9) is a valid *rational* solution to eq. (5.6), it is not the solution we seek, i.e., the solution used in the main body of this chapter as the basis for the numerical solution, eq. (15.2) (Note this solution becomes singular when $x + x_0 - ct = 0$). The Maple code that generates the solution of eq. (15.9) is given in Listing 15.7.

```
># Hyperbolic Liouville Equation
 restart;
 with(PDEtools): with(PolynomialTools):
>alias(v=v(x,t)):alias(u=u(x,t)):
>pde1:=diff(u,t,t)-a^2*diff(u,x,x)-b*exp(beta*u)=0;
># Assume a travelling wave solution of the form
 # u=U(xi)/beta, xi=k*(x-c*t);
 tr1:={x=(xi/k+c*tau),t=tau,u=U(xi)/beta};
 ode1:=dchange(tr1,pde1,[xi,tau,U(xi)]);
>ode2:=collect(ode1,{diff(U(xi),xi)^2,diff(U(xi),xi,xi)});
># Separate variables and integrate
 eqn1:=int(exp(-U),U,U)-b*beta/(c^2*k^2-a^2*k^2)*int(1,xi,xi)=0;
># Isolate U
 eqn2:=isolate(eqn1,U);
># Inverse transformation xi=k*(x-c*t), U(xi)=u*beta
 pdeSol1:=isolate(subs({U=u*beta,xi=k*(x+x0-c*t)},eqn2),u);
```

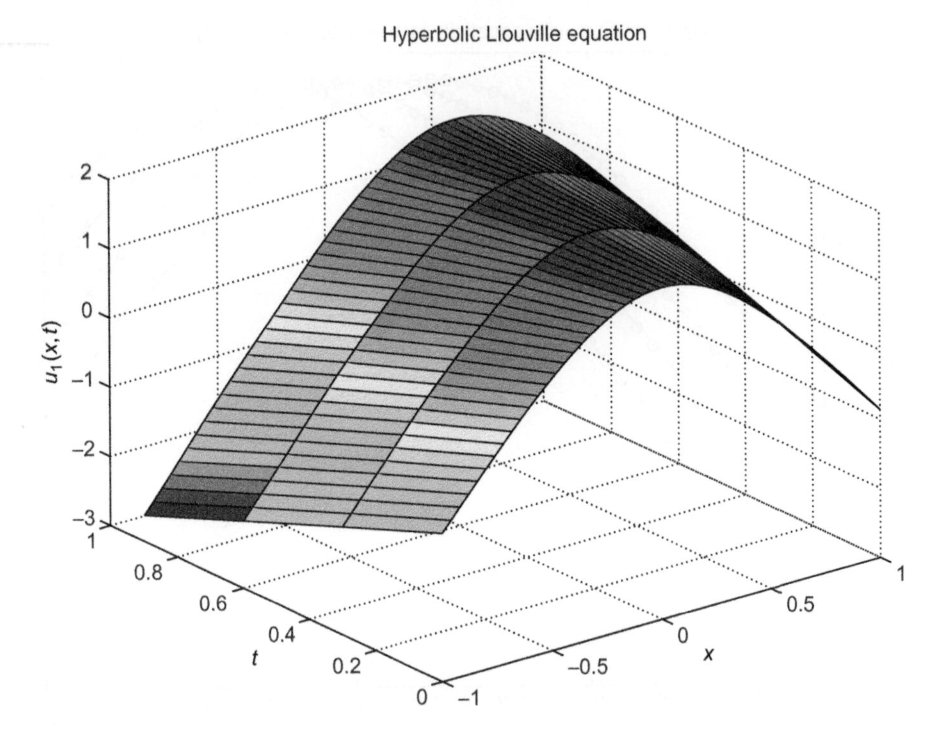

FIGURE 15.3: 3D plot of the numerical solution of eq. (15.1) (eqns. (15.5a,5b)).

```
># Check solution
 pdeCHK:=pdetest(pdeSol1,pde1); # '0' if true
 if pdeCHK <> 0 then
   print("Solution: does not pass pdetest() !");
 else
   print("Solution: passes pdetest() !");
 end if;
```

LISTING 15.7: Maple code used to derive the solution of eq. (15.9).

For the second method, which is slightly more involved, we apply the transformation

$$u = \ln(u)/\beta \tag{15.10}$$

to eq. (15.1), which yields

$$u\left(\frac{\partial^2 u}{\partial t^2} - a^2\frac{\partial^2 u}{\partial x^2}\right) - \left(\frac{\partial u}{\partial t}\right)^2 + a^2\left(\frac{\partial u}{\partial x}\right)^2 - bu^3 = 0 \tag{15.11}$$

While this is a more complex expression than eq. (5.6), the exponential term has been eliminated, which greatly simplifies the solution process. We now apply the *Riccati method* to this modified equation, which generates six trivial solutions and the following six nontrivial solutions:

$$u_1 = \frac{k^2(a^2 - c^2)}{2b\beta}\left(1 - \tanh^2\left[k(x + x_0 - ct)/2\right]\right) \tag{15.12}$$

$$u_2 = \frac{-k^2(a^2 - c^2)}{2b\beta}\left(1 + \tan^2\left[k(x + x_0 - ct)/2\right]\right) \tag{15.13}$$

$$u_3 = \frac{2k^2(a^2 - c^2)}{b\beta}\left(1 - \tanh^2\left[k(x + x_0 - ct)\right]\right) \tag{15.14}$$

$$u_4 = \frac{-2k^2(a^2 - c^2)}{b\beta}\left(1 + \tan^2\left[k(x + x_0 - ct)\right]\right) \tag{15.15}$$

$$u_5 = \frac{8k^2(a^2 - c^2)}{b\beta}\left(1 - \tanh^2\left[2k(x + x_0 - ct)\right]\right) \tag{15.16}$$

$$u_6 = \frac{-8k^2(a^2 - c^2)}{b\beta}\left(1 + \tan^2\left[2k(x + x_0 - ct)\right]\right) \tag{15.17}$$

We choose solution u_3, which we convert from *tanh* form to *cosh* form to give

$$u_3 = \frac{2k^2\left(a^2 - c^2\right)}{b\beta\cosh^2\left[k\left(-x - x_0 + ct\right)\right]} \tag{15.18}$$

The final solution is obtained by applying the inverse transformation, $u = \ln(u_3)/\beta$, i.e.,

$$u = \frac{1}{\beta}\ln\left(\frac{2k^2\left(a^2 - c^2\right)}{b\beta\cosh^2\left[k\left(-x - x_0 + ct\right)\right]}\right) \tag{15.19}$$

On letting $k = -A$, $c = B/k$, and $x_0 = C/k$, we arrive at eq. (15.2), the solution used in the main body of this chapter as the basis for the numerical simulation. If we use the parameter values $a = 1$, $b = 1$, $c = 1/2$, $k = -2$, $x_0 = 0$, and $\beta = 1$, this solution takes the form of a triangular-shaped traveling wave with a curved apex—see Figs. 15.4 and 15.5. Note: while at large scale the traveling wave appears to be a *peakon*, on closer inspection, see Fig. 15.6, we see that this is not the case as the spatial derivative does not have a (finite) discontinuity at the peak—unlike the solution for the *sine-Gordon* equation discussed in Chapter 16.

Setting $t = 0$ and $x = 0$, we see that the peak value is given by $u = \ln(6) = 1.79$. In addition, on plotting the spatial derivative, we observe that the slopes on either side of zero appear to rapidly take on constant values—see Fig. 15.6.

We can check if this is the case by taking the limit of du/dx as $|x| \to \infty$, which gives the traveling wave slopes, i.e.,

$$\left.\frac{du}{dx}\right|_{x=\pm\infty} = \lim_{x \to \pm\infty}\left(\frac{-4\sinh(2x)}{\cosh(2x)}\right) = \mp 4 \tag{15.20}$$

which confirms our visual impression. The Maple code that generates this solution and the associated plots is given in Listing 15.8.

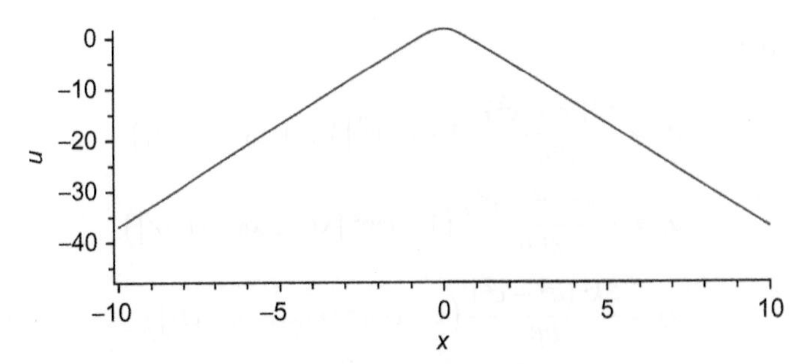

FIGURE 15.4: The initial condition (at $t = 0$) for a 2D animation of the solution to the hyperbolic Liouville equation, which then moves left to right when activated.

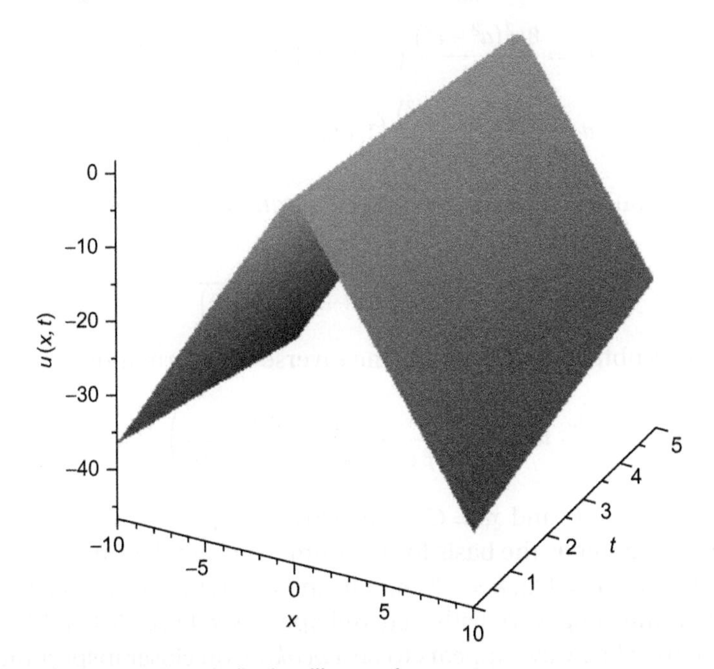

FIGURE 15.5: 3D plot of solution to hyperbolic Liouville equation.

```
># Hyperbolic Liouville Equation
 # Attempt at Riccati solution
 restart; with(PDEtools): with(PolynomialTools):
 with(plots):
>alias(u=u(x,t)):alias(v=v(x,t)):
># Define pde
 pde0:=diff(v,t,t)-alpha^2*diff(v,x,x)-b*exp(beta*v)=0;
># Apply a transformation
 tr0:={v=ln(u)/beta};
 pde1:=simplify(dchange(tr0,pde0,[u]),symbolic);
>pde1 := numer(lhs(pde1))*denom(rhs(pde1)) = numer(rhs(pde1))*denom(lhs(pde1));
>read("riccatiMethod.txt");
>intFlg:=0: # integration not needed!
```

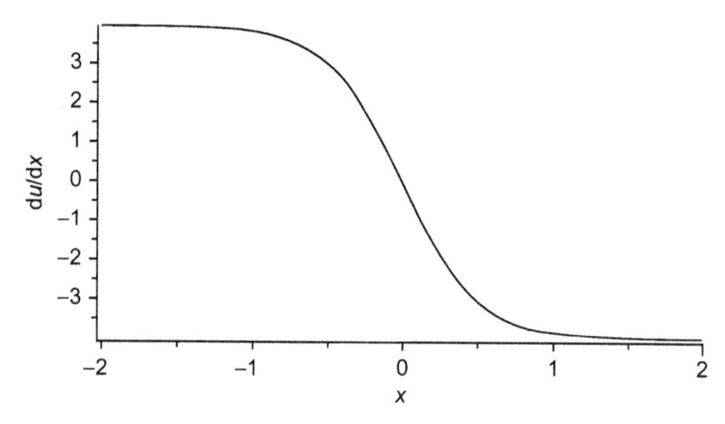

FIGURE 15.6: Plot illustrating how $du/dx \rightarrow \pm 4$ as $x \rightarrow \mp\infty$.

```
 M:=2; # Set order of approximation
 infoLevOut:=0;
 riccatiMethod(M,pde1,intFlg,infoLevOut);
># Apply the inverse transform v=ln(u)
 sol1:=v=ln(rhs(soln[2,3]))/beta;
># Test that the solution solves the original pde
 testSol:=simplify(pdetest(sol1,pde0),symbolic);
 if testSol<> 0 then
   print("Solution: does not pass pdetest() !");
 else
   print("Solution passes pdetest()");
 end if;
># Standard solution constants
 zz:=rhs(sol1);
 a:=1; b:=1; beta:=1; A:=2; B:=-1; C:=0;
 alpha:=a;k:=A;c:=-B/k;x0:=C/k;
># Create animation
 animate(zz,x=-2..2, t=0..1,
   numpoints=100,frames=50, axes=framed,labels=["x","u"],
   thickness=3,title="Hyperbolic Liouville Equation",
   labelfont=[TIMES, ROMAN, 16],axesfont=[TIMES, ROMAN, 16],
   titlefont=[TIMES, ROMAN, 16]);
># Generate a 3D surface plot
 plot3d(zz,x=-2..2, t=0..1,axes='framed',
   labels=["x","t","u(x,t)"],
   orientation=[64,48],grid=[100,100],
   style=patchnogrid,axesfont=[TIMES, ROMAN, 16],
   labeldirections=[HORIZONTAL,HORIZONTAL,VERTICAL],
   labelfont=[TIMES, ROMAN, 16],
   shading=Z,title="Hyperbolic Liouville Equation",
   titlefont=[TIMES, ROMAN, 16]);
```

LISTING 15.8: Maple code used to derive the solution of eq. (15.19) and Figs. 15.4–15.6.

Finally, we mention that eq. (5.6) is also referred to in the literature as the *modified Liouville* equation and that it can be converted, using the transformation $u(x,t) = U(\xi,\eta)$, where $x = a(\xi - \eta)$, $t = (\xi + \eta)$, to the equivalent alternative form

$$\frac{\partial^2 U(\xi,\eta)}{\partial\xi\partial\eta} = be^{\beta U(\xi,\eta)} \tag{15.21}$$

known simply as the *Liouville* equation.

References

[1] C.H. Gu, On the Classification of Initial Data for Nonlinear Wave Equations, in: L. Tatsien (Ed), *Frontiers in Mathematical Analysis and Numerical Methods: In Memory of Jacques-Louis Lions*, World Scientific Publishing Company, Singapore, 2004, pp. 143–148.

[2] J. Liouville, Sur l'équation aux différences partielles $\partial^2 \log\lambda/\partial u\partial v \pm \lambda/(2a^2) = 0$, *J. de Math. Pure et Appliquée*, 18 (1) (1853) 71–72.

[3] A.D. Polyanin, V.F. Zaitsev, *Handbook of Nonlinear Partial Differential Equations*, Chapman & Hall/CRC, Boca Raton, FL, 2004.

16 Sine-Gordon Equation

The sine-Gordon equation ([1], p227) is

$$\frac{\partial^2 u}{\partial t^2} = a^2 \frac{\partial^2 u}{\partial x^2} + b\sin(\lambda u) \tag{16.1}$$

Since the sine-Gordon equation is second order in t, its solution closely parallels that of the hyperbolic Liouville equation of Chapter 15.

An analytical solution is ([1], p227)

$$u_a(x,t) = \frac{4}{\lambda}\tan^{-1}\left\{\exp\left[\frac{b\lambda(kx+\mu t+\theta_0}{\sqrt{b\lambda(\mu^2-ak^2)}}\right]\right\} \tag{16.2}$$

if $b\lambda(\mu^2 - ak^2) > 0$ and where k, μ, and θ_0 are arbitrary. Note that eq. (16.2) defines a traveling wave solution through the combination $kx + \mu t + \theta_0$.

One IC for eq. (16.1) is eq. (16.2) with $t = 0$. Since eq. (16.1) is second order in t, it requires two ICs. A second IC is available by differentiating eq. (16.2) with respect to t. Here, we use

$$\frac{d}{dx}\left(\tan^{-1} x\right) = \frac{1}{1+x^2}$$

$$\frac{\partial u_a(x,t)}{\partial t} = \frac{4}{\lambda}\frac{1}{1+\exp\left[\frac{b\lambda(kx+\mu t+\theta_0)}{\sqrt{b\lambda(\mu^2-ak^2)}}\right]^2}\exp\left[\frac{b\lambda(kx+\mu t+\theta_0)}{\sqrt{b\lambda(\mu^2-ak^2)}}\right](b\lambda\mu)$$

Dividing numerator and denominator by $\exp\left[\frac{b\lambda\left(kx+\mu t+\theta_0\right)}{\sqrt{b\lambda(\mu^2-ak^2)}}\right]$ gives

$$\frac{\partial u_a(x,t)}{\partial t} = \frac{4}{\lambda}\frac{1}{\exp\left[\frac{b\lambda\left(kx+\mu t+\theta_0\right)}{\sqrt{b\lambda(\mu^2-ak^2)}}\right]^{-1} + \exp\left[\frac{b\lambda\left(kx+\mu t+\theta_0\right)}{\sqrt{b\lambda(\mu^2-ak^2)}}\right]}(b\lambda\mu)$$

which simplifies to

$$\frac{\partial u_a(x,t)}{\partial t} = 2b\mu\frac{1}{\cosh\left[\frac{b\lambda\left(kx+\mu t+\theta_0\right)}{\sqrt{b\lambda(\mu^2-ak^2)}}\right]}$$

or

$$\frac{\partial u_a(x,t)}{\partial t} = 2b\mu \operatorname{sech}\left[\frac{b\lambda\,(kx+\mu t+\theta_0)}{\sqrt{b\lambda\,(\mu^2-ak^2)}}\right] \tag{16.3}$$

The required IC then comes from eq. (16.3) with $t = 0$

BCs for eq. (16.1) are taken as

$$u(x=x_l,t) = u_a(x=x_l,t);\; u(x=x_u,t) = u_a(x=x_u,t) \tag{16.4a, 4b}$$

Equations (16.1)–(16.4) constitute the problem programmed in the following Matlab routines.

Equation (16.1) is programmed in function pde_1.m listed first.

```
  function ut=pde_1(t,u)
%
% Function pde_1 computes the t derivative vector for the sine-
% Gordon equation
%
  global xl xu x n ncall
%
% Model parameters
  global a b lambda k mu the0 den
%
% One vector to two vectors
  for i=1:n
    u1(i)=u(i);
    u2(i)=u(i+n);
  end
%
% BCs at x = 0,30
  u1(1)=ua_1(x(1),t);
  u1(n)=ua_1(x(n),t);
%
% u1xx
  nl=1; nu=1;
  u1x(1)=0;
  u1xx=dss044(xl,xu,n,u1,u1x,nl,nu);
%
% PDE
  for i=2:n-1
    u1t(i)=u2(i);
    u2t(i)=a*u1xx(i)+b*sin(lambda*u(i));
  end
  u1t(1)=0; u1t(n)=0;
  u2t(1)=0; u2t(n)=0;
%
% Two vectors to one vector
  for i=1:n
    ut(i)  =u1t(i);
    ut(i+n)=u2t(i);
```

```
  end
  ut=ut';
%
% Increment calls to pde_1
  ncall=ncall+1;
```

LISTING 16.1: Function pde_1.m for eq. (16.1).

We can note the following details about pde_1.m:

- The function and some global variables are first defined.

```
  function ut=pde_1(t,u)
%
% Function pde_1 computes the t derivative vector for the sine-
% Gordon equation
%
  global xl xu x n ncall
%
% Model parameters
  global a b lambda k mu the0 den
```

- The 1D ODE dependent variable vector of length $2n$, u, is transferred to two 1D arrays, each of length n, u1, u2, to facilitate the numerical solution of eq. (16.1).

```
%
% One vector to two vectors
  for i=1:n
    u1(i)=u(i);
    u2(i)=u(i+n);
  end
```

  Again, as for the case of the hyperbolic-Liouville equation of Chapter 15, the use of two 1D arrays is a standard procedure for accommodating a PDE second order in t such as eq. (16.1). In the case of eq. (16.1), this gives

$$\frac{\partial u_1}{\partial t} = u_2 \tag{16.5a}$$

$$\frac{\partial u_2}{\partial t} = a^2 \frac{\partial u_1}{\partial x^2} + b\sin(\lambda u_1) \tag{16.5b}$$

  As indicated in Chapter 15, this procedure of working with two first-order PDEs in t permits the use of library integrators for first-order ODEs such as ode45 and ode15s. Also, the procedure is quite general in the sense that it can usually be applied to any second-order PDE of interest.

- The BCs (16.4) are programmed as

```
%
% BCs at x = 0,30
  u1(1)=ua_1(x(1),t);
  u1(n)=ua_1(x(n),t);
%
% u1xx
```

```
  nl=1; nu=1;
  u1x(1)=0;
  u1xx=dss044(xl,xu,n,u1,u1x,nl,nu);
```

Note that BCs (16.4) are designated as *Dirichlet* through `nl=1`, `nu=1`. Also, the first derivative in x, `u1x(1)=0`, is not actually used with the Dirichlet BCs but is included only to satisfy the Matlab requirement that all of the input (RHS) arguments of a function, such as `u1x` in `dss044`, must be defined numerically. The second derivative from `dss044`, `u1xx`, can then be used in the programming of eq. (16.1) (or eqs. (16.5).

- Equations (16.5) are then programmed in a `for` loop.

```
%
% PDE
  for i=2:n-1
    u1t(i)=u2(i);
    u2t(i)=a*u1xx(i)+b*sin(lambda*u(i));
  end
  u1t(1)=0; u1t(n)=0;
  u2t(1)=0; u2t(n)=0;
```

Note also that the derivatives in t of the ODEs at the boundaries (`i=1,n`) are zeroed to ensure the boundary values remain at the values prescribed by eq. (16.4).

- The two derivative vectors, `u1t,u2t`, are then returned to a single-derivative vector `ut` to be returned from `pde_1.m`, with a transpose included to meet the requirement of `ode15s`. Also, the counter for `pde_1.m` is incremented.

```
%
% Two vectors to one vector
  for i=1:n
    ut(i)  =u1t(i);
    ut(i+n)=u2t(i);
  end
  ut=ut';
%
% Increment calls to pde_1
  ncall=ncall+1;
```

In summary, `pde_1.m` receives the dependent variable vector `u` as an input (along with the independent variable `t`) and returns the derivative vector `ut`.

The IC function `inital_1.m` defines initial values for `u1` and `u2` in eqs. (16.5).

```
  function u0=inital_1(t0)
%
% Function inital_1 sets the initial condition for the sine-
% Gordon equation
%
% Parameters shared with other routines
  global xl xu x n ncall
%
% Spatial domain and initial condition
  xl=0;
  xu=30;
```

```
  n=121;
  dx=(xu-xl)/(n-1);
%
% ICs from analytical solution
  for i=1:n
    x(i)=xl+(i-1)*dx;
    u1(i)= ua_1(x(i),t0);
    u2(i)=uat_1(x(i),t0);
    u0(i)  =u1(i);
    u0(i+n)=u2(i);
  end
```

LISTING 16.2: IC function `inital_1.m` for eqns. (16.2) and (16.3) with $t = 0$.

`inital_1.m` defines a grid in x of 121 points for $0 \le x \le 30$. In the `for` loop, the initial values of `u1` for eq. (16.5a) are provided by the function `ua_1.m` (which has the programming for eq. (16.2)); the initial values of `u2` for eq. (16.5b) are provided by the function `uat_1.m` (which has the programming for eq. (16.3)). All $2n$ initial condition values are returned from `inital_1.m` through the vector `u0` to the main program `pde_1_main.m` discussed subsequently.

Function `ua_1.m` is a straightforward implementation of eq. (16.2).

```
  function uanal=ua_1(x,t)
%
% Function uanal computes the exact solution of the sine-Gordon
% equation for comparison with the numerical solution
%
% Model parameters
  global a b lambda k mu the0 den
%
% Analytical solution
  xi=k*x+mu*t+the0;
  uanal=(4/lambda)*atan(exp(b*lambda*xi/den));
```

LISTING 16.3: Function `ua_1.m` for the analytical solution of eq. (16.2).

Function `ual_1.m` is a straightforward implementation of eq. (16.3).

```
  function uanal=uat_1(x,t)
%
% Function uanal computes the time derivative of the exact solution
% of the sine-Gordon equation
%
% Model parameters
  global a b lambda k mu the0 den
%
% Analytical solution derivative
  xi=k*x+mu*t+the0;
  uanal=2*b*mu/cosh(b*lambda*xi/den);
```

LISTING 16.4: Function `uat_1.m` for the analytical solution of eq. (16.3).

The main program pde_1_main is essentially the same as pde_1_main of Chapter 15 (for PDEs second order in t) and therefore is not listed here. A few details are as follows.

- The parameters in eqns. (16.1) and (16.2) are defined numerically.

```
%
% Model parameters
  a=1; b=0.1; lambda=1; k=4; mu=-5; the0=0;
  den=sqrt(b*lambda*(mu^2-a*k^2));
```

  In particular, $k = 4$, $\mu = -5$, and $\theta_0 = 0$; k was selected to obtain a significant variation in the solution with x.
- The difference between the numerical solution of eq. (16.1) (u_1 of eqs. (16.5)) and the analytical solution of eq. (16.2) is computed as err(it,i) and displayed numerically in the output. The numerical and analytical solutions are then superimposed in a 2D plot from Matlab routine plot, and a 3D plot of the numerical solution is produced by surf.

```
%
% Display selected output
  for it=1:nout
    fprintf('\n      t          x          u1(it,i)  u1_anal(it,i)
                 err(it,i)\n');
    for i=1:5:n
      fprintf('%6.2f%8.3f%15.6f%15.6f%15.6f\n',...
                 t(it),x(i),u1(it,i),u1_anal(it,i),err(it,i));
    end
  end
  fprintf('    ncall = %4d\n\n',ncall);
%
% Plot numerical and analytical solutions
  figure(2)
  plot(x,u1,'-',x,u1_anal,'o')
  xlabel('x')
  ylabel('u1(x,t)')
  title('sine-Gordon equation: t = 0, 3, 6, 9; solid - numerical;
              o - analytical')
  figure(3)
  surf(u1)
  shading 'interp', axis 'tight'
  xlabel('x'); ylabel('t'); zlabel('u1(x,t)');
  title('sine-Gordon equation');
```

jpattern_num_1.m for the sparse matrix integration of the ODEs is not listed here since it is the same as for Chapter 15. The ODE Jacobian map from jpattern_num_1.m indicates two bands for PDEs (16.5), as indicated in Fig. 16.1. A portion of the numerical output from pde_1_main.m is listed in Table 16.1.

We can note the following points about this output:

- The numerical and analytical solutions agree at $t = 0$ as expected, but the difference between the two solutions increases with increasing t. The error in the numerical

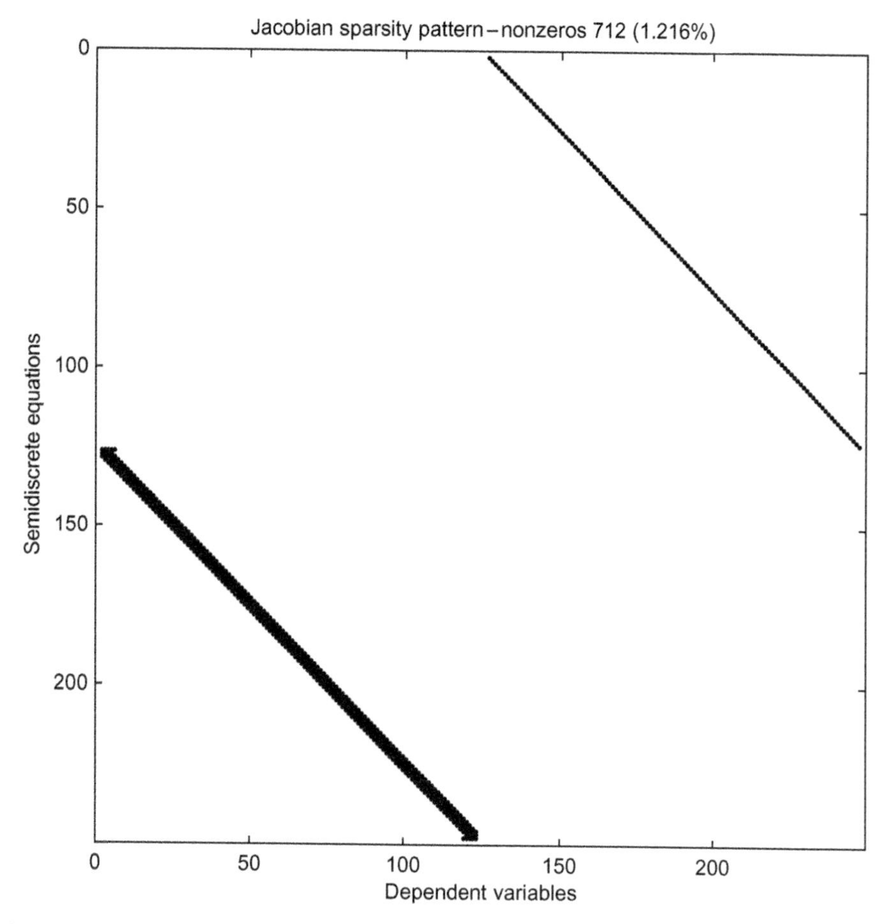

FIGURE 16.1: ODE Jacobian map from `jpattern_num_1.m` indicating two bands for eq. (16.5).

solution could be reduced by using more grid points ($n > 121$), but this results in a greater computational effort (note in `inital_1.m` the values $n = 51,201$ as comments indicating some experimentation with the number of grid points to investigate the accuracy of the numerical solution).

- For the present conditions, the computational effort is still modest, with `ncall = 530`.
- Since the difference between the hyperbolic Liouville equation of Chapter 15 and the sine-Gordon equation of this chapter is rather small, mainly in the nonlinear terms, $be^{\beta u}$ from eq. (15.1) and $b\sin(\lambda u)$ from eq. (16.1) a comparison of the accuracy and computational effort of the solutions in Chapters 15 and 16 (better accuracy and less computational effort in Chapter 15) suggests that the numerical solution of nonlinear PDEs can be very sensitive to the details of the equations. Even for a given PDE, changes in the problem parameters can have unexpectedly large effects in the accuracy and required computational effort. Our experience has indicated that this

Table 16.1: Selected numerical output from the main program `pde_1_main.m`

t	x	u1(it,i)	u1_anal(it,i)	err(it,i)
0.00	0.000	3.141593	3.141593	0.000000
0.00	1.250	4.150022	4.150022	0.000000
0.00	2.500	4.941803	4.941803	0.000000
0.00	3.750	5.471548	5.471548	0.000000
0.00	5.000	5.799721	5.799721	0.000000
0.00	6.250	5.996866	5.996866	0.000000
0.00	7.500	6.113969	6.113969	0.000000
0.00	8.750	6.183251	6.183251	0.000000
0.00	10.000	6.224181	6.224181	0.000000
0.00	11.250	6.248351	6.248351	0.000000
0.00	12.500	6.262621	6.262621	0.000000
0.00	13.750	6.271045	6.271045	0.000000
0.00	15.000	6.276018	6.276018	0.000000
0.00	16.250	6.278954	6.278954	0.000000
0.00	17.500	6.280688	6.280688	0.000000
0.00	18.750	6.281711	6.281711	0.000000
0.00	20.000	6.282315	6.282315	0.000000
0.00	21.250	6.282671	6.282671	0.000000
0.00	22.500	6.282882	6.282882	0.000000
0.00	23.750	6.283006	6.283006	0.000000
0.00	25.000	6.283080	6.283080	0.000000
0.00	26.250	6.283123	6.283123	0.000000
0.00	27.500	6.283148	6.283148	0.000000
0.00	28.750	6.283164	6.283164	0.000000
0.00	30.000	6.283172	6.283172	0.000000

.
.
.

output for t=3, 6 removed

.
.
.

t	x	u1(it,i)	u1_anal(it,i)	err(it,i)
9.00	0.000	0.034834	0.034834	0.000000
9.00	1.250	0.058084	0.059004	-0.000920
9.00	2.500	0.099238	0.099935	-0.000697
9.00	3.750	0.171749	0.169216	0.002533
9.00	5.000	0.291398	0.286319	0.005079
9.00	6.250	0.494642	0.483465	0.011177
9.00	7.500	0.840301	0.811637	0.028664
9.00	8.750	1.400928	1.341382	0.059546
9.00	10.000	2.194166	2.133163	0.061002

Table 16.1: *(Continued)*

9.00	11.250	3.198719	3.141593	0.057126
9.00	12.500	4.196678	4.150022	0.046656
9.00	13.750	4.974121	4.941803	0.032318
9.00	15.000	5.491899	5.471548	0.020351
9.00	16.250	5.812024	5.799721	0.012303
9.00	17.500	6.004191	5.996866	0.007325
9.00	18.750	6.118307	6.113969	0.004337
9.00	20.000	6.185814	6.183251	0.002563
9.00	21.250	6.225695	6.224181	0.001514
9.00	22.500	6.249244	6.248351	0.000893
9.00	23.750	6.263146	6.262621	0.000526
9.00	25.000	6.271353	6.271045	0.000308
9.00	26.250	6.276195	6.276018	0.000177
9.00	27.500	6.279050	6.278954	0.000096
9.00	28.750	6.280730	6.280688	0.000042
9.00	30.000	6.281711	6.281711	0.000000
ncall = 530				

unexpected sensitivity generally precludes any statement a priori about the expected accuracy or required computation of a solution to a PDE system. In other words, each PDE problem should be treated as "new" and experimental.

The plotted solutions follow in figs 16.2 and 16.3. Figure 16.2 demonstrates that the boundary values at $x = 0, 30$ vary with t according to eqns. (16.2) and (16.4) and as programmed in `pde_1.m`. Also, the traveling wave solution produced by eq. (16.2) with the argument $kx + \mu t + \theta_0$ is clear.

In summary, we have again discussed the numerical solution of a PDE second order in t (eq. (16.1)) by restating the PDE as two PDEs first order in t (eq. (16.5)). As noted in Chapter 15, the reformulation of an *nth*-order PDE as n PDEs first order in an initial value independent variable (t) is straightforward and permits the use of an integrator such as `ode15s` for first-order ODEs.

Appendix

The *sine-Gordon* equation is so-called as a result of a wordplay on the similar *Klein–Gordon* equation. It occurs in the study of geometrical surfaces of constant Gaussian curvature, quantum mechanics field theory, and other areas of physics and engineering. It has been found to be particularly useful in applications that give rise to solitons, for example, in the study of fiber optics.

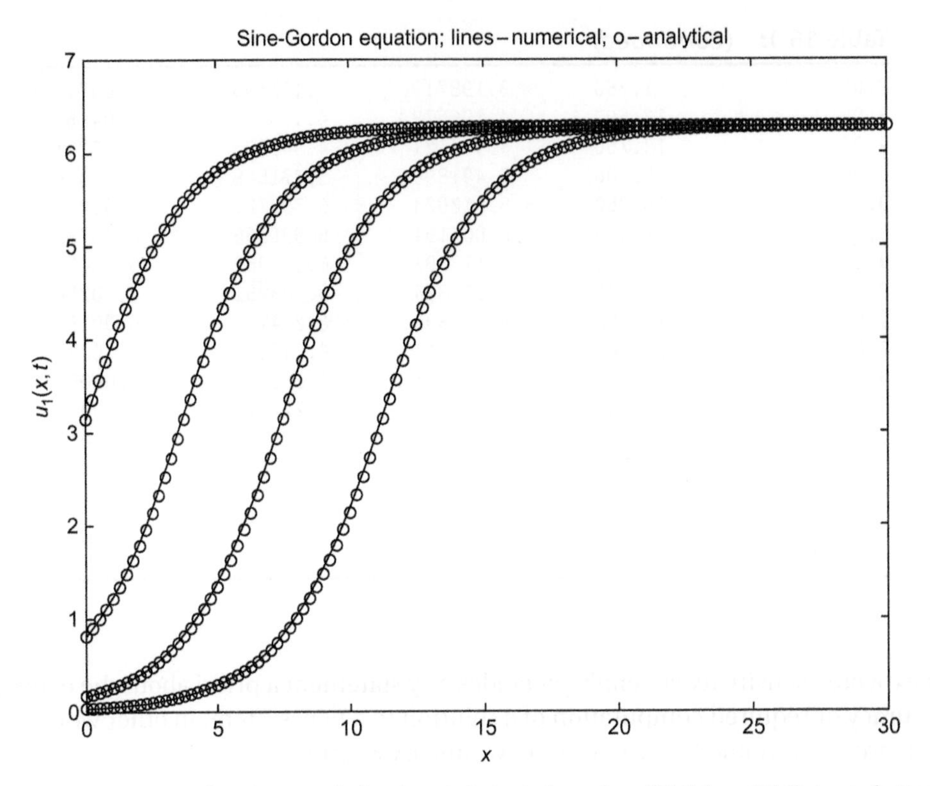

FIGURE 16.2: 2D plot comparing the numerical and analytical solutions of eq. (16.1) (eqs. (16.5); top to bottom for $t = 0,3,6,9$).

There are two common forms used in studies of relativity. The first form is defined in *laboratory coordinates*[1] by

$$\frac{\partial^2 u}{\partial t^2} - \alpha^2 \frac{\partial^2 u}{\partial x^2} = \beta \sin(u) \tag{16.6}$$

This is a simplified form of eq. (16.1), where we have set $\lambda = 1$.

Following Drazon and Johnson [2], eq. (16.6) can be transformed to an alternative form by changing the variables,

$$x = \alpha(\eta - \xi), \quad t = \eta + \xi \tag{16.7}$$

After some algebraic manipulation, we arrive at the second form in *light-cone coordinates*[2]

$$\frac{\partial^2 u}{\partial \xi \partial \eta} = \beta \sin(u) \tag{16.8}$$

[1] *Laboratory coordinates* apply where the observer is located within an inertial laboratory.
[2] A *light cone* is formed by all past and future events that can be connected by light rays.

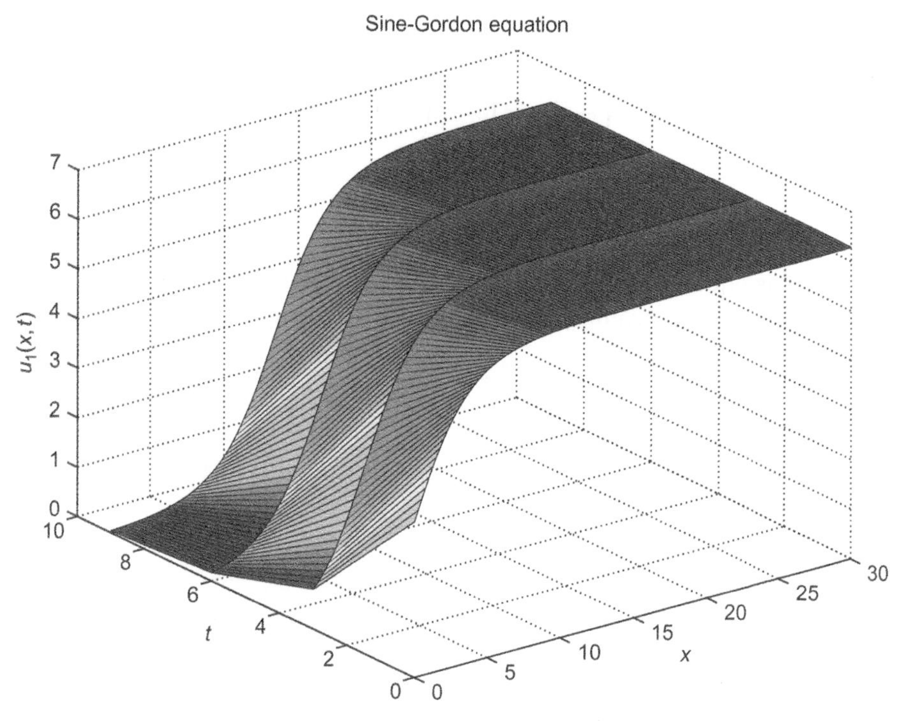

FIGURE 16.3: 3D plot of the numerical solution of eq. (16.1) (eqs. (16.5)).

We now seek an analytical solution (adapted from [3]) and start by introducing the following new variables

$$w = \frac{\partial u}{\partial \xi} \tag{16.9}$$

$$z = \beta \cos(u) - 1 \tag{16.10}$$

from which we note that

$$\frac{\partial^2 w}{\partial \xi \partial \eta} - w - wz = \frac{\partial}{\partial \xi}[\beta \sin(u)] - \frac{\partial u}{\partial \xi} - \frac{\partial u}{\partial \xi}[\beta \cos(u) - 1] = 0 \tag{16.11}$$

$$2z + z^2 + \left(\frac{\partial w}{\partial \eta}\right)^2 - \beta^2 + 1 = 2[\beta \cos(u) - 1] + [\beta \cos(u) - 1]^2$$
$$+ \beta^2 \sin^2(u) - \beta^2 + 1 = 0 \tag{16.12}$$

On rearranging eq. (16.11), we obtain $z = \frac{1}{w}\frac{\partial^2 w}{\partial \xi \partial \eta} - 1$, which we use to eliminate z from eq. (16.12) and obtain the following new equation in w,

$$\left(\frac{\partial^2 w}{\partial \xi \partial \eta}\right)^2 + \left(\frac{\partial w}{\partial \eta}\right)^2 w^2 - \beta^2 w^2 = 0 \tag{16.13}$$

We now make a further transformation by letting $v = w^2$, which gives

$$\left(\frac{\partial v}{\partial \eta}\right)^2 \left(\frac{\partial v}{\partial \xi}\right)^2 - 4\left(\frac{\partial v}{\partial \eta}\right)\left(\frac{\partial v}{\partial \xi}\right)\left(\frac{\partial^2 v}{\partial \eta \partial \xi}\right) v$$
$$+4\left(\frac{\partial^2 v}{\partial \eta \partial \xi}\right)^2 v^2 + 4\left(\frac{\partial v}{\partial \eta}\right)^2 v^3 - 16\beta^2 v^4 = 0. \tag{16.14}$$

Now, while this equation may seem more complex than the above equations, it does readily yield the following traveling wave solution by application of the tanh method,

$$v = 4k^2\left[1 - \tanh^2\left(\frac{\xi k^2 + \beta\eta}{k}\right)\right] = 4k^2 \mathrm{sech}\left(\frac{\xi k^2 + \beta\eta}{k}\right), \tag{16.15}$$

where k is an arbitrary constant. We continue with the tanh form as it leads to a more compact final solution.

By applying the inverse transformation $w = \sqrt{v}$, we obtain

$$w = 2k\sqrt{1 - \tanh^2\left(\frac{\xi k^2 + \beta\eta}{k}\right)} \tag{16.16}$$

and, again using $z = \dfrac{1}{w}\dfrac{\partial^2 w}{\partial \xi \partial \eta} - 1$ from eq. (16.11), we obtain

$$z = 2\beta \tanh^2\left(\frac{\xi k^2 + \beta\eta}{k}\right) - \beta - 1 \tag{16.17}$$

Applying the inverse transformation

$$u = \arccos\left[(z+1)/\beta\right], \tag{16.18}$$

we obtain

$$u = \arccos\left[2\tanh^2\left(\frac{\xi k^2 + \beta\eta}{k}\right) - 1\right] \tag{16.19}$$

This is a traveling wave solution to the sine-Gordon equation in *light cone coordinates*, i.e., to eq. (16.8).

Finally, we use the inverse transformation $\xi = (-x/\alpha + t)/2$, $\eta = (x/\alpha + t)/2$ to obtain the desired result, i.e.,

$$u = \arccos\left[2\tanh^2\left(\frac{(-x/\alpha + t)k^2 + \beta(x/\alpha + t)}{2k}\right) - 1\right]. \tag{16.20}$$

Equation (16.20) is a soliton traveling wave solution to the sine-Gordon eq. (16.6) in *laboratory coordinates*—see Fig. 16.4 where we have used the following parameter values: $\alpha = 1$,

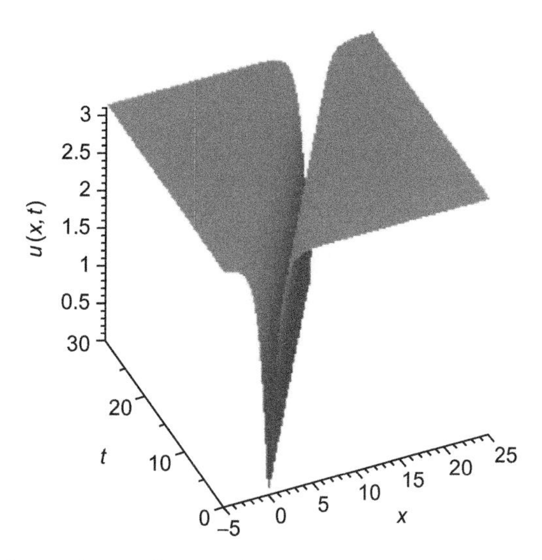

FIGURE 16.4: 3D plot of solution to sine-Gordon equation. Note, the wave peak value is $u_{\max} = \pi$.

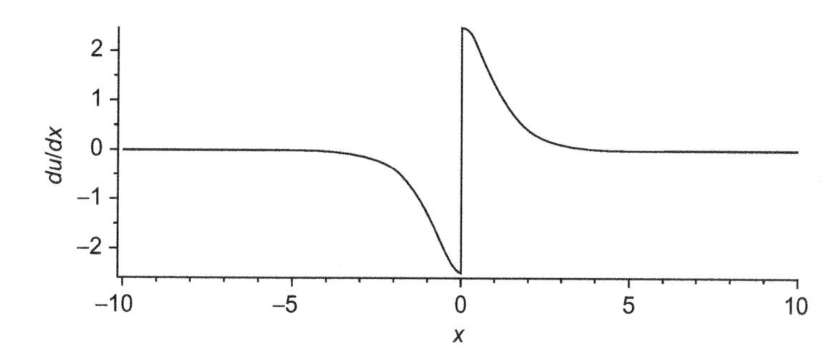

FIGURE 16.5: Plot illustrating the discontinuous first derivative at the peak of the *peakon* solution to the sine-Gordon equation.

$\beta = 1$, and $k = 2$. However, it is not the familiar *kink*, or *hump*-type soliton, it is a *peakon*. A peakon (peaked soliton) is a special type of soliton which has a discontinuity in the first derivative at its peak—see Fig. 16.5. Peakons with a negative peak are sometimes referred to as *antipeakons*. It is, of course, not the only solution as there are an infinity of traveling wave solutions to this famous equation.

A Maple code that performs the above calculations and generates animated and 3D plots is given in Listing 16.5.

```
># Sine-Gordon Equation
 # Attempt at Malfliet's tanh solution
 # See: Solitons: An Introduction, Drazin and Johnson, p110-112
 restart; with(PDEtools): with(PolynomialTools):
 with(plots):
>alias(u=u(x,t)): alias(v=v(x,t)):
 alias(w=w(x,t)): alias(z=z(x,t)):
># Sine-Gordon equation in laboratory coordinates
```

```
pde:=diff(v,t,t)-alpha^2*diff(v,x,x)-beta*sin(v)=0;
># Apply a transformation to convert to 'light cone' co-ordinates
tr0:={v=W(xi,eta),x=alpha*(-xi+eta),t=(xi+eta)};
 pde0:=simplify(dchange(tr0,pde,[W(xi,eta),xi,eta]),symbolic);
># Revert back to original variables
 tr1:={W(xi,eta)=v,xi=x,eta=t};
 pde0:=simplify(dchange(tr1,pde0,[v,x,t]),symbolic);
># Now apply a change of variables:
z:=beta*cos(v)-1; w:=diff(v,x); unassign('w','z');
># From w, z and pde0 we obtain
 eqn1:=simplify(diff(w,x,t)-w-w*z=0);
 eqn2:=2*z+z^2+diff(w,t)^2-beta^2+1=0;
># solve for z using eqn1
 solZ:=solve(eqn1,z);
># Apply solZ and cross multiply eqn2
 # to eliminate denominators
 z:=solZ;
 pde1:=numer(lhs(eqn2))*denom(rhs(eqn2))=0;
># Apply a transformation to facilitate use of tanh method
 tr3:={w=u^(1/2)};
 pde1:=simplify(dchange(tr3,pde1,[u]),symbolic);
>read("tanhMethod.txt");
># Calculate solution for transformed problem
 intFlag:=0: # No integration of U(xi) needed!
 M:=2; # Set order of approximation
 infoLevOut:=0;
 tanhMethod(M,pde1,intFlag,infoLevOut);
># Apply inverse transformation: w=sqrt(u)
 w:=simplify(rhs(sol[3])^(1/2),symbolic);
># Calculate z
 z:=simplify(diff(w,x,t)/w-1,symbolic);
># Apply the inverse transform to obtain
 # solution in 'light cone' co-ordinates
 sol1:=v=arccos((z+1)/beta);
># Test solution in 'light cone' co-ordinates
 simplify(pdetest(sol1,pde0),symbolic);
># Animate the final solution
 # NOTE: u(max)=3.142 (pi)
 sol1a:=subs({beta=1,k=2,x0=0},rhs(sol1));
 animate(sol1a,x=-5..25, t=0..80,
   numpoints=300,frames=50, axes=framed,
   labels=["x","u"],thickness=3,
   title="Sine-Gordon Equation\n Light cone Co-ordinates",
   labelfont=[TIMES, ROMAN, 16],axesfont=[TIMES, ROMAN, 16],
   titlefont=[TIMES, ROMAN, 16]);
># Transform solution back to original co-ordinates
 sol2:=v=subs({x=(t-x/alpha)/2,t=(t+x/alpha)/2},rhs(sol1));
># Test final solution in original pde
 simplify(subs(sol2,pde),symbolic);
># Animate the final solution
 # NOTE: u(max)=3.142 (pi)
 sol2a:=subs({alpha=1,beta=1,k=2,x0=0},rhs(sol2));
 animate(sol2a,x=-5..25, t=0..30,
   numpoints=300,frames=50, axes=framed,
```

```
   labels=["x","u"],thickness=3,
   title="Sine-Gordon Equation\n Laboratory Co-ordinates",
   labelfont=[TIMES, ROMAN, 16],axesfont=[TIMES, ROMAN, 16],
   titlefont=[TIMES, ROMAN, 16]);
># Generate a 3D surface plot
 plot3d(sol2a,x=-5..25, t=0..30,axes='framed',
   labels=["x","t","u(x,t)"],
   orientation=[-114,52],grid=[100,100],
   style=patchnogrid,shading=Z,
   labeldirections=[HORIZONTAL,HORIZONTAL,VERTICAL],
   title="Sine-Gordon Equation\n Laboratory Co-ordinates",
   labelfont=[TIMES, ROMAN, 16],axesfont=[TIMES, ROMAN, 16],
   titlefont=[TIMES, ROMAN, 16]);
># Now show that the solution is actually a 'Peakon',
 # a peaked soliton, which is a solitary wave with
 # discontinuous first derivative
 diffSol2:=diff(rhs(sol2),x);
 diffSol2:=subs({alpha=1,beta=1,k=2,x0=0,t=0},diffSol2);
 plot(diffSol2,x=-10..10, axes=framed,
   numpoints=200,labels=["x","du/dx"],thickness=3,
   title="Sine-Gordon Equation - First Derivative
   Laboratory Co-ordinates",
   labeldirections=[HORIZONTAL,VERTICAL],
   labelfont=[TIMES, ROMAN, 16],axesfont=[TIMES, ROMAN, 16],
   titlefont=[TIMES, ROMAN, 16]);
```

LISTING 16.5: Maple listing to derive traveling wave solutions to the sine-Gordon equation using the `tanhMethod()` procedure.

Traveling *wave* solutions can also be found to the sine-Gordon equation using the *exp* and/or the *Riccati* methods, but the Maple listings are not included here in order to save space. However, the associated files are included with the downloads provided for this book.

References

[1] A.D. Polyanin, V.F. Zaitsev, *Handbook of Nonlinear Partial Differential Equations*, Chapman & Hall/CRC, Boca Raton, FL, 2004.

[2] P.G. Drazin, R.S. Johnson, *Solitons: An Introduction*, Cambridge University Press, Cambridge, UK, 1992.

[3] W. Hereman, *Exact Solutions of Nonlinear Partial Differential Equations: The Tanh/Sech Method*, notes from invited lecture, Wolfram Research, Inc., Champaign, Illinois, November, 2000.

17 Mth-Order Klein–Gordon Equation

We start with a PDE which we term the *mth-order Klein–Gordon equation.*

$$\frac{\partial^2 u}{\partial t^2} + \alpha\frac{\partial^2 u}{\partial x^2} + \beta u + \gamma u^m = f(x,t) \tag{17.1}$$

Equation (17.1) is a generalization of the *quadratic Klein–Gordon* equation with $m = 2$ and the *cubic Klein–Gordon equation* with $m = 3$ ([3]). We will consider the method of lines (MOL) numerical solution of eq. (17.1) for three cases:

Case 1: Linear PDE ($\alpha = -c^2, \beta = \gamma = f(x,t) = 0$)

For Case 1, $\alpha = -c^2, \beta = \gamma = f(x,t) = 0$, and eq. (17.1) reduces to the *linear wave equation*

$$\frac{\partial^2 u}{\partial t^2} = c^2\frac{\partial^2 u}{\partial x^2} \tag{17.2}$$

In the subsequent programming, we consider the analytical solution

$$u_a(x,t) = \sin(x)\cos(ct) = \frac{1}{2}[\sin(x+ct) + \sin(x-ct)],\ t \geq 0 \tag{17.3}$$

Note that eq. (17.3) is a superposition of two traveling waves moving left and right with velocity c. Equation (17.3) is therefore a special case of the *d'Alembert solution* to the wave equation [2]. The verification of this solution follows directly from substitution in eq. (17.2)

Term in eq. (17.2)	Term from eq. (17.3)
$\frac{\partial^2 u}{\partial t^2}$	$-c^2\sin(x)\cos(ct)$
$-c^2\frac{\partial^2 u}{\partial x^2}$	$c^2\sin(x)\cos(ct)$
Sum of terms	Sum of terms
0	0

For the ICs of eq. (17.2), we use eq. (17.3) at $t = 0$

$$u(x,t=0) = \sin(x),\quad \frac{\partial u(x,t=0)}{\partial t} = 0 \tag{17.4a,4b}$$

and the Dirichlet BCs,

$$u(x = x_l, t) = f_{b1}(t), u(x = x_u, t) = f_{b2}(t) \tag{17.5a,5b}$$

where x_l, x_u and f_{b1}, f_{b2} (two BC functions) are specified for three cases.

In Case 1, we consider two subcases: (1) Case 1.1 - basic scale in x defined as $0 \leq x \leq 2\pi$ and (2) Case 1.2 - expanded scale in x defined as $-3\pi \leq x \leq 4\pi$. Case 1.1 illustrates the oscillatory properties of eq. (17.3) in x and t (the first form of this solution, $\sin(x)\cos(ct)$). Case 1.2 illustrates the traveling wave properties of eq. (17.3) in x and t (the second form of this solution, $\frac{1}{2}[\sin(x+ct)+\sin(x-ct)]$). In the Matlab routines that follow, Case 1.1 is selected by using `ncase=11`, and Case 1.2 is selected by using `ncase=12`. The details of these two subcases will be apparent from the Matlab code and the associated numerical and plotted output.

The ODE routine for eqns. (17.1)–(17.5) is as follows.

```
  function ut=pde_1(t,u)
%
% Function pde_1 computes the t derivative vector for the Klein-
% Gordon equation
%
  global xl xu x n ncall ncase
%
% Model parameters
  global c c2 m r a b g
%
% One vector to two vectors
  for i=1:n
    u1(i)=u(i);
    u2(i)=u(i+n);
  end
%
% BCs
  if(ncase==11)
    u1(1)=ua_11(x(1),t);
    u1(n)=ua_11(x(n),t);
  end
  if(ncase==12)
    u1(1)=ua_12(x(1),t);
    u1(n)=ua_12(x(n),t);
  end
  if(ncase==2)
    u1(1)=ua_2(x(1),t);
    u1(n)=ua_2(x(n),t);
  end
  if(ncase==3)
    u1(1)=ua_3(x(1),t);
    u1(n)=ua_3(x(n),t);
  end
%
% u1xx
  nl=1; nu=1;
  u1x(1)=0;
```

```
  u1xx=dss044(xl,xu,n,u1,u1x,nl,nu);
%
% PDE
  if(ncase==11)|(ncase==12)
    for i=2:n-1
      u1t(i)=u2(i);
      u2t(i)=c2*u1xx(i);
    end
  end
  if(ncase==2)|(ncase==3)
    for i=2:n-1
      u1t(i)=u2(i);
        if(ncase==2)f=f_2(x(i),t);end
        if(ncase==3)f=0;end
      u2t(i)=-a*u1xx(i)-b*u1(i)-g*u1(i)^m+f;
    end
  end
  u1t(1)=0; u1t(n)=0;
  u2t(1)=0; u2t(n)=0;
%
% Two vectors to one vector
  for i=1:n
    ut(i)  =u1t(i);
    ut(i+n)=u2t(i);
  end
  ut=ut';
%
% Increment calls to pde_1
  ncall=ncall+1;
```

LISTING 17.1: Function pde_1.m for eq. (17.1) (ncase=11,12,2,3).

We can note the following details about pde_1.m:

- The function and some global variables are first defined.

```
  function ut=pde_1(t,u)
%
% Function pde_1 computes the t derivative vector for the Klein-
% Gordon equation
%
  global xl xu x n ncall ncase
%
% Model parameters
  global c c2 m r a b g
```

- Since eq. (17.1) and the special case, eq. (17.2), are second order in t, the 1D ODE dependent variable vector of length $2n$, u, is transferred to two 1D arrays each of length n, u1, u2 to facilitate the numerical solution of eqns. (17.1) and (17.2).

```
%
% One vector to two vectors
  for i=1:n
    u1(i)=u(i);
    u2(i)=u(i+n);
  end
```

Again, as for the case of the hyperbolic Liouville equation of Chapter 15 and the sine-Gordon equation of Chapter 16, the use of two 1D arrays is a standard procedure for accommodating a PDE second order in t such as eqns. (17.1) and (17.2). In the case of eq. (17.1), this gives

$$\frac{\partial u_1}{\partial t} = u_2 \tag{17.6a}$$

$$\frac{\partial u_2}{\partial t} = -\alpha\frac{\partial^2 u_1}{\partial x^2} - \beta u_1 - \gamma u_1^m + f(x,t) \tag{17.6b}$$

As indicated in Chapters 15 and 16, this procedure of working with two first-order PDEs in t permits the use of library integrators for first-order ODEs such as `ode45` and `ode15s`. Also, the procedure is quite general that it can usually be applied to any second-order PDE of interest.

- Four sets of BCs are programmed for `ncase=11,12,2,3` (`ncase` is set in the main program discussed subsequently).

```
%
% BCs
  if(ncase==11)
    u1(1)=ua_11(x(1),t);
    u1(n)=ua_11(x(n),t);
  end
  if(ncase==12)
    u1(1)=ua_12(x(1),t);
    u1(n)=ua_12(x(n),t);
  end
  if(ncase==2)
    u1(1)=ua_2(x(1),t);
    u1(n)=ua_2(x(n),t);
  end
  if(ncase==3)
    u1(1)=ua_3(x(1),t);
    u1(n)=ua_3(x(n),t);
  end
```

Each of these cases will be considered, one at a time. To start, we consider `ncase=11`. Also, the analytical solutions in functions `ua_11.m,ua_12.m,ua_2.m,ua_3.m` (discussed subsequently) are used to set the BCs.

- In all of the cases, *Dirichlet boundary conditions* (eqns. (17.5a,5b)) are specified since the corresponding analytical solutions are used to define the BCs.

```
%
% u1xx
  nl=1; nu=1;
  u1x(1)=0;
  u1xx=dss044(xl,xu,n,u1,u1x,nl,nu);
```

Note that the Dirichlet BCs are designated through `nl=1, nu=1`. Also, the first derivative in x, `u1x(1)=0`, is not actually used with the Dirichlet BCs but is included only to satisfy the Matlab requirement that all of the input (RHS) arguments of a function, such as `u1x` in `dss044`, must be defined numerically. The second derivative from `dss044`, `u1xx`, can then be used in the programming of eqns. (17.1) (or eqns. (17.5a,5b)) and (17.2).

- Equations (17.1) and (17.2) are then programmed in a `for` loop. The code is selected according to `ncase=11,12` (for eq. (17.2)) or `ncase=2,3` (for eq. (17.1)).

```
%
% PDE
  if(ncase==11)|(ncase==12)
    for i=2:n-1
      u1t(i)=u2(i);
      u2t(i)=c2*u1xx(i);
    end
  end
  if(ncase==2)|(ncase==3)
    for i=2:n-1
      u1t(i)=u2(i);
        if(ncase==2)f=f_2(x(i),t);end
        if(ncase==3)f=0;end
      u2t(i)=-a*u1xx(i)-b*u1(i)-g*u1(i)^m+f;
    end
  end
  u1t(1)=0; u1t(n)=0;
  u2t(1)=0; u2t(n)=0;
```

Note also that the derivatives in t of the ODEs at the boundaries (`i=1,n`) are zeroed to ensure the boundary values remain at the values prescribed by the analytical solutions. Specifically, the coding of eqns. (17.6a,6b) is

```
      u1t(i)=u2(i);
      u2t(i)=c2*u1xx(i);
```

where c^2= `c2` (c is set in the main program and passed to `pde_1.m` as a global variable).

- The two derivative vectors, `u1t,u2t`, are then returned to a single derivative vector `ut` to be returned from `pde_1.m` with a transpose included to meet the requirement of `ode15s`. Also, the counter for `pde_1.m` is incremented.

```
%
% Two vectors to one vector
  for i=1:n
    ut(i)  =u1t(i);
    ut(i+n)=u2t(i);
  end
  ut=ut';
%
% Increment calls to pde_1
  ncall=ncall+1;
```

In summary, `pde_1.m` receives the dependent variable vector `u` as an input (along with the independent variable `t`) and returns the derivative vector `ut`.

We now consider the other routines for `ncase=11`. After looking at the output for this case, we continue the discussion for `ncase=12,2,3`. The IC function, `inital_1.m`, defines initial values for `u1` and `u2` in eqns. (17.6a,6b).

```
  function u0=inital_1(t0)
%
% Function inital_1 sets the initial condition for the Klein-
% Gordon equation
%
% Parameters shared with other routines
  global xl xu x n ncall ncase
%
% Spatial domain and initial condition
  if(ncase==11)
    xl=0;
    xu=2*pi;
    n=101;
  end
  if(ncase==12)
    xl=-3*pi;
    xu= 4*pi;
    n=151;
  end
  if(ncase==2)
    xl=0;
    xu=1;
    n=51;
  end
  if(ncase==3)
    xl=0;
    xu=1;
    n=51;
  end
  dx=(xu-xl)/(n-1);
%
% ICs from analytical solution
  for i=1:n
    x(i)=xl+(i-1)*dx;
    if(ncase==11)
      u1(i)= ua_11(x(i),t0);
      u2(i)=uat_11(x(i),t0);
    end
    if(ncase==12)
      u1(i)= ua_12(x(i),t0);
      u2(i)=uat_12(x(i),t0);
    end
    if(ncase==2)
      u1(i)= ua_2(x(i),t0);
      u2(i)=uat_2(x(i),t0);
    end
    if(ncase==3)
      u1(i)= ua_3(x(i),t0);
      u2(i)=uat_3(x(i),t0);
    end
```

```
    u0(i)   =u1(i);
    u0(i+n)=u2(i);
  end
```

LISTING 17.2: IC function `inital_1.m` for eqns. (17.1) and (17.2) with $t = 0$ (for `ncase=11,12,2,3`).

We can note the following points about `inital_1.m`:

- The function and some global variables are defined.

```
  function u0=inital_1(t0)
%
% Function inital_1 sets the initial condition for the Klein-
% Gordon equation
%
% Parameters shared with other routines
  global xl xu x n ncall ncase
```

- Considering first `ncase=11`, `inital_1.m` defines a grid in x of 101 points for $0 \le x \le 2\pi$.

```
%
% Spatial domain and initial condition
  if(ncase==11)
    xl=0;
    xu=2*pi;
    n=101;
  end
```

- In the `for` loop, the initial values of `u1` for eq. (17.6a) are provided by the function `ua_11.m` (which has the programming for the first form of the solution in eq. (17.3), $\sin(x)\cos(ct)$); the initial values of `u2` for eq. (17.6b) are provided by the function `uat_11.m` (which has the programming for the derivative of eq. (17.3)).

```
%
% ICs from analytical solution
  for i=1:n
    x(i)=xl+(i-1)*dx;
    if(ncase==11)
      u1(i)= ua_11(x(i),t0);
      u2(i)=uat_11(x(i),t0);
    end
```

- All $2n$ initial condition values are returned from `inital_1.m` through the vector `u0` to the main program `pde_1_main.m` discussed subsequently.

```
    u0(i)   =u1(i);
    u0(i+n)=u2(i);
  end
```

Function `ua_11.m` is a straightforward implementation of the first form of the solution of eq. (17.3), $\sin(x)\cos(ct)$.

```
  function uanal=ua_11(x,t)
%
% Function uanal computes the exact solution of the Klein-Gordon
```

```
% equation for comparison with the numerical solution
%
% Model parameters
  global c
%
% Analytical solution
  uanal=sin(x)*cos(c*t);
```

LISTING 17.3: Function `ua_11.m` for the analytical solution of eq. (17.3).

Function `uat_11.m` is a straightforward implementation of the derivative (in t) of the first form of the solution of eq. (17.3), $\sin(x)\cos(ct)$.

```
  function uanal=uat_11(x,t)
%
% Function uanal computes the time derivative of the exact solution
% of the Klein-Gordon equation
%
% Model parameters
  global c
%
% Analytical solution derivative
  uanal=-c*sin(x)*sin(c*t);
```

LISTING 17.4: Function `uat_11.m` for the derivative of the analytical solution of eq. (17.3).

Main program `pde_1_main.m` accommodates the cases `ncase=11,12,3,4` for eqns. (17.1) and (17.2). This main program closely parallels `pde_1_main.m` of Chapter 15 and therefore only a few features are discussed here.

- The parameters are set for `ncase=11,12,3,4`.

```
%
% Model parameters
  global c c2 m r a b g B K
%
% Select case
  ncase=11;
% ncase=12;
% ncase=2;
% ncase=3;
%
% Model parameters
  if(ncase==11)|(ncase==12)c=1;c2=c^2;end
  if(ncase==2)m=2;r=3;a=-1;b=0;g=1;end
  if(ncase==3)m=3;a=-2.5;b=1;g=1.5;c=0.5;
      B=(b/g)^0.5;K=(-b/(2*(a+c^2)))^0.5;end
```

- The range in t varies for `ncase=11,12,3,4`.

```
%
% Independent variable for ODE integration
  if(ncase==11)
    tf=pi; tout=[t0:pi/3:tf]'; nout=4; ncall=0;
  end
  if(ncase==12)
```

```
    tf=2*pi; tout=[t0:pi:tf]'; nout=3; ncall=0;
  end
  if(ncase==2)
    tf=1; tout=[t0:0.25:tf]'; nout=5; ncall=0;
  end
  if(ncase==3)
    tf=4; tout=[t0:1:tf]'; nout=5; ncall=0;
  end
```

- The difference in the numerical and analytical solutions is computed as `err(it,i)` and displayed. These solutions are then plotted in 2D by `plot` and the numerical solution is plotted in 3D by `surf`.

```
%
% Display selected output
  for it=1:nout
    fprintf('\n      t        x        u1(it,i)  u1_anal(it,i)
             err(it,i)\n');
    for i=1:5:n
      fprintf('%6.2f%8.3f%15.6f%15.6f%15.6f\n',...
            t(it),x(i),u1(it,i),u1_anal(it,i),err(it,i));
    end
  end
  fprintf('    ncall = %4d\n\n',ncall);
%
% Plot numerical and analytical solutions
  figure(2)
  plot(x,u1,'-',x,u1_anal,'o')
  xlabel('x')
  ylabel('u1(x,t)')
  if(ncase==11)
    title('Klein-Gordon equation; ncase = 11; solid - numerical;
           o - analytical')
  elseif(ncase==12)
    title('Klein-Gordon equation; ncase = 12; solid - numerical;
           o - analytical')
  elseif(ncase==2)
    title('Klein-Gordon equation; ncase = 2; solid - numerical;
           o - analytical')
  elseif(ncase==3)
    title('Klein-Gordon equation; ncase = 3; solid - numerical;
           o - analytical')
  end
  figure(3)
  surf(x,t,u1)
  xlabel('x'); ylabel('t'); zlabel('u1(x,t)');
  title('Klein-Gordon equation');
```

`jpattern_num_1.m` for the sparse matrix integration of the ODEs is not listed here since it is the same as for Chapters 15 and 16. The ODE Jacobian map from `jpattern_num_1.m` indicates two bands for PDEs (17.6) as expected, and the map is therefore not included in the discussion of the output that follows.

A portion of the numerical output from `pde_1_main.m` is listed in Table 17.1.

Table 17.1: Selected numerical output from the main program `pde_1_main.m` for `ncase=11`

t	x	u1(it,i)	u1_anal(it,i)	err(it,i)
0.00	0.000	0.000000	0.000000	0.000000
0.00	0.314	0.309017	0.309017	0.000000
0.00	0.628	0.587785	0.587785	0.000000
0.00	0.942	0.809017	0.809017	0.000000
0.00	1.257	0.951057	0.951057	0.000000
0.00	1.571	1.000000	1.000000	0.000000
0.00	1.885	0.951057	0.951057	0.000000
0.00	2.199	0.809017	0.809017	0.000000
0.00	2.513	0.587785	0.587785	0.000000
0.00	2.827	0.309017	0.309017	0.000000
0.00	3.142	−0.000000	−0.000000	0.000000
0.00	3.456	−0.309017	−0.309017	0.000000
0.00	3.770	−0.587785	−0.587785	0.000000
0.00	4.084	−0.809017	−0.809017	0.000000
0.00	4.398	−0.951057	−0.951057	0.000000
0.00	4.712	−1.000000	−1.000000	0.000000
0.00	5.027	−0.951057	−0.951057	0.000000
0.00	5.341	−0.809017	−0.809017	0.000000
0.00	5.655	−0.587785	−0.587785	0.000000
0.00	5.969	−0.309017	−0.309017	0.000000
0.00	6.283	0.000000	0.000000	0.000000

.
.
.
output for t = 1.05, 2.09 removed
.
.
.

t	x	u1(it,i)	u1_anal(it,i)	err(it,i)
3.14	0.000	0.000000	0.000000	0.000000
3.14	0.314	−0.309016	−0.309017	0.000001
3.14	0.628	−0.587783	−0.587785	0.000002
3.14	0.942	−0.809014	−0.809017	0.000003
3.14	1.257	−0.951055	−0.951057	0.000002
3.14	1.571	−0.999997	−1.000000	0.000003
3.14	1.885	−0.951054	−0.951057	0.000002
3.14	2.199	−0.809015	−0.809017	0.000002
3.14	2.513	−0.587784	−0.587785	0.000001
3.14	2.827	−0.309016	−0.309017	0.000001
3.14	3.142	−0.000000	0.000000	−0.000000
3.14	3.456	0.309016	0.309017	−0.000001
3.14	3.770	0.587784	0.587785	−0.000001

Table 17.1: *(Continued)*

3.14	4.084	0.809015	0.809017	−0.000002
3.14	4.398	0.951054	0.951057	−0.000002
3.14	4.712	0.999997	1.000000	−0.000003
3.14	5.027	0.951055	0.951057	−0.000002
3.14	5.341	0.809014	0.809017	−0.000003
3.14	5.655	0.587783	0.587785	−0.000002
3.14	5.969	0.309016	0.309017	−0.000001
3.14	6.283	−0.000000	−0.000000	0.000000
ncall=341				

We can note the following points about this output:

- The numerical and analytical solutions agree at $t = 0$ as expected (since both solutions are produced by `ua_11.m` with $t = 0$), and the difference between the two solutions at $t = 3.14$ is better than five figures.
- The computational effort is modest, with `ncall = 341`. The plotted solutions follow. Figures 17.1 and 17.2 demonstrate that the solution is simply a standing sin wave in x from eq. (17.3) (starting from $t = 0$) according to $\sin(x)\cos(ct)$.

The preceding discussion was for `ncase=11`, that is, the linear wave equation (17.2) over the domain $0 \le x \le 2\pi$. We now consider `ncase=12`, which is again for eq. (17.2), but with the x domain extended to $-3\pi \le x \le 4\pi$ in order to demonstrate the traveling wave solution of eq. (17.3), $\frac{1}{2}[\sin(x+ct)+\sin(x-ct)]$. This solution indicates the IC ($t = 0$) is $\sin(x)$, which then separates into two sine functions, $\frac{1}{2}\sin(x+ct)$ traveling right to left with velocity c and $\frac{1}{2}\sin(x-ct)$ traveling left to right with velocity c ($c > 0$).

Also, initially, only the positive portion of $\sin(x)$ is used ($0 \le x \le \pi$) in order to simplify the plotted output; the negative portion of the $\sin(x)$ could easily be included through a minor modification of `ua_12.m` and `uat_12.m` to $0 \le x \le 2\pi$.

Here we reproduce only the code for `ncase=12` indicated in Listings 17.1 and 17.2. The BCs for `ncase=12` in `pde_1.m` are provided by `ua_12.m` (refer to Listing 17.1 for the complete listing of `pde_1.m`).

```
if(ncase==12)
  u1(1)=ua_12(x(1),t);
  u1(n)=ua_12(x(n),t);
end
```

The programming of eq. (17.2) in `pde_1.m` is the same for `ncase=11,12` (since the PDE is the same).

```
%
% PDE
  if(ncase==11)|(ncase==12)
    for i=2:n-1
```

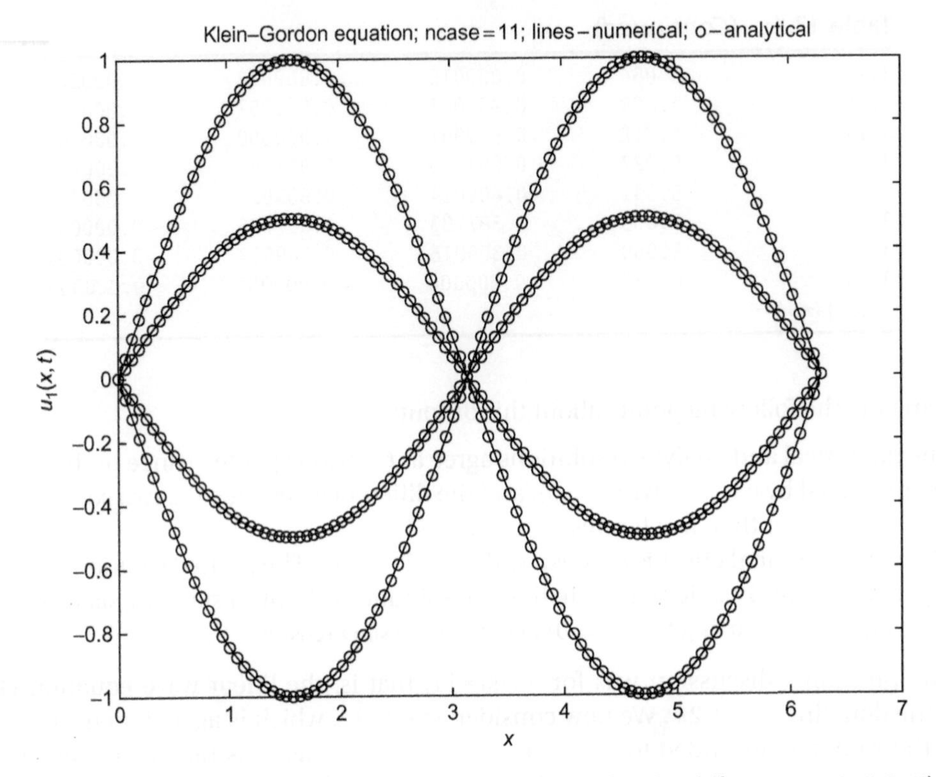

FIGURE 17.1: 2D plot comparing the numerical and analytical solutions of eq. (17.2) (for $t = 0, \pi/3, 2\pi/3, 3\pi/3$).

```
      u1t(i)=u2(i);
      u2t(i)=c2*u1xx(i);
    end
  end
```

In `inital_1.m` for `ncase=12`, we use $-3\pi \leq x \leq 4\pi$ on 151 points (refer to Listing 17.2 for the complete listing of `inital_1.m`).

```
  if(ncase==12)
    xl=-3*pi;
    xu= 4*pi;
    n=151;
  end
```

The increase in the number of grid points ($n = 101$ for `ncase=11` to $n = 151$ for `ncase=12`) was determined by trial and error to gain an improved spatial resolution of the numerical solution.

`ua_12.m` and `uat_12.m` are then used to set the ICs (with `t0=0` from `pde_1_main.m`).

```
    if(ncase==12)
      u1(i)= ua_12(x(i),t0);
```

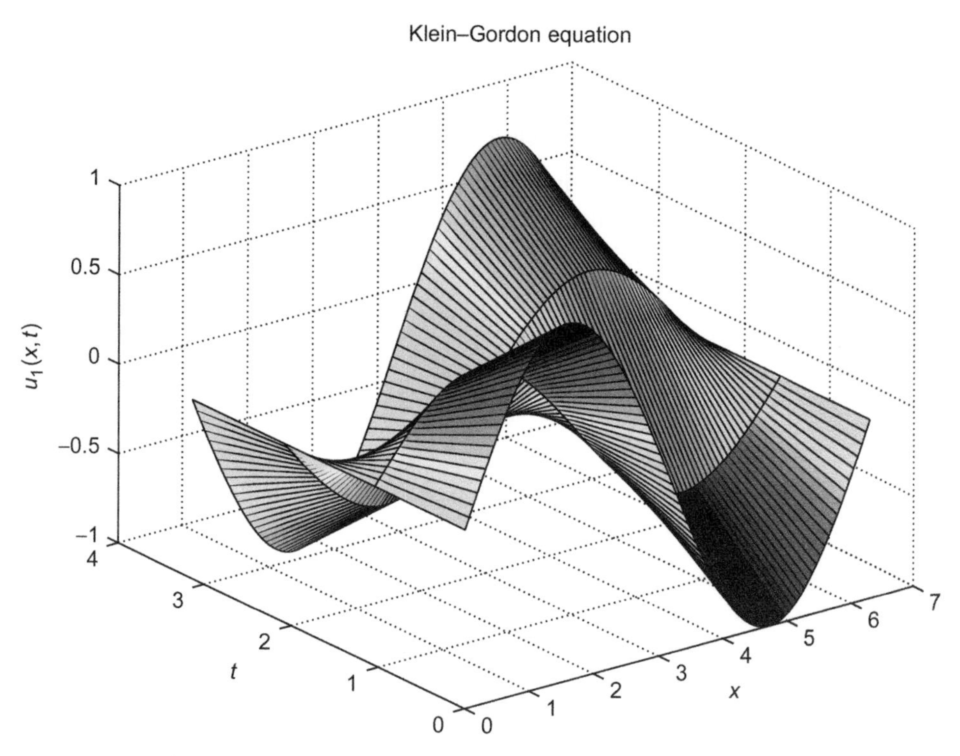

FIGURE 17.2: 3D plot of the numerical solution of eq. (17.2).

```
    u2(i)=uat_12(x(i),t0);
  end
```

Function `ua_12.m` is a straightforward implementation of the second form of the solution of eq. (17.3), $\frac{1}{2}[\sin(x+ct)+\sin(x-ct)]$.

```
  function uanal=ua_12(x,t)
%
% Function uanal computes the exact solution of the Klein-Gordon
% equation for comparison with the numerical solution
%
% Model parameters
  global c
%
% Analytical solution
  if((x-c*t>0)&(x-c*t)<=pi)
    sin1=sin(x-c*t);
  else
    sin1=0;
  end
  if((x+c*t>0)&(x+c*t)<=pi)
    sin2=sin(x+c*t);
  else
```

```
    sin2=0;
  end
  uanal=0.5*(sin1+sin2);
```

LISTING 17.5: Function `ua_12.m` for the analytical solution of eq. (17.3).

Note that here we use only the positive portion of the $\sin(x)$ function, e.g., $0 \leq x \leq \pi$ for $t = 0$; this restriction is imposed only to simplify the plotted solution.

Function `ua_12.m` is a straightforward implementation of the derivative (in t) of the second form of the solution of eq. (17.3), $\frac{1}{2}[\sin(x+ct)+\sin(x-ct)]$.

```
  function uanal=uat_12(x,t)
%
% Function uanal computes the derivative of the exact solution of the
% Klein-Gordon equation
%
% Model parameters
  global c
%
% Analytical solution
  if((x-c*t>0)&(x-c*t)<=pi)
    cos1=-c*cos(x-c*t);
  else
    cos1=0;
  end
  if((x+c*t>0)&(x+c*t)<=pi)
    cos2=c*cos(x+c*t);
  else
    cos2=0;
  end
  uanal=0.5*(cos1+cos2);
```

LISTING 17.6: Function `uat_12.m` for the derivative of the analytical solution of eq. (17.3).

The additional coding in the main program `pde_1_main.m` follows. The velocity c in eq. (17.2) is the same for `ncase=11,12`.

```
%
% Model parameters
  if(ncase==11)|(ncase==12)c=1;c2=c^2;end
```

t is defined over the interval $0 \leq t \leq 2\pi$ for three outputs in the numerical and plotted output at $t = 0, \pi, 2\pi$.

```
  if(ncase==12)
    tf=2*pi; tout=[t0:pi:tf]'; nout=3; ncall=0;
  end
```

The analytical solution is included in the output by using `ua_12.m` (for a comparison of the numerical and analytical solutions).

```
  if(ncase==12)
    u1(it,1)=ua_12(x(1),t(it));
```

```
    u1(it,n)=ua_12(x(n),t(it));
    for i=1:n
      u1_anal(it,i)=ua_12(x(i),t(it));
      err(it,i)=u1(it,i)-u1_anal(it,i);
    end
  end
```

The plot of the numerical and analytical solutions includes a label for the `ncase=12`.

```
elseif(ncase==12)
  title('Klein-Gordon equation; ncase = 12; solid - numerical; o - analytical')
```

A portion of the numerical output from `pde_1_main.m` is listed in Table 17.2. We can note the following points about this output:

- The numerical and analytical solutions agree at $t = 0$ as expected, and the agreement between the two solutions at $t = 6.28$ is about three figures.
- The computational effort is modest, with `ncall = 840`.

The plotted solutions follow in Figs. 17.3 and 17.4. Figure 17.3 demonstrates that the solution for the IC ($t = 0$) is $\sin(x)$, which then separates into two sine functions, $\frac{1}{2}\sin(x + ct)$ traveling right to left with velocity c and $\frac{1}{2}\sin(x - ct)$ traveling left to right with velocity c.

To summarize, we assumed eq. (17.3) as a solution to eq. (17.2), then demonstrated that it satisfies eq. (17.2) by direct substitution. We also developed two ICs from eq. (17.3) by setting $t = 0$ and two BCs from eq. (17.3) by setting $x_l = 0, x_u = 2\pi$ (`ncase=11`) or $x_l = -3\pi, x_u = 4\pi$ with $u_a(x, t = 0) = 0, x < 0, x > \pi$ (`ncase=12`).

To extend this procedure, we could assume a solution to the PDE that will in general not be correct (it was correct in the case of eq. (17.3), which satisfies eq. (17.2), but this was possible mainly because eq. (17.2) is quite straightforward, i.e, it is linear with constant coefficients and $f(x, t) = 0$). If we substitute the assumed solution into the PDE, a *residual function* will result that can then be included in the PDE to make the assumed solution exact. Also, the required ICs and BCs can then be obtained from the assumed solution. The advantage of this approach as we shall demonstrate with the next case (`ncase=2`) is that it can in principle be *applied to any PDE (linear and nonlinear)* to obtain an analytical solution to a related PDE; this *method of residual functions* can also be *applied to systems of PDEs*.

Case 2: Nonlinear PDE (`ncase=2`, $m = 2, r = 3, \alpha = -1, \beta = 0, \gamma = 1, f(x, t) \neq 0$)

We now consider `ncase=2` for eq. (17.1) with $m = 2, r = 3, \alpha = -1, \beta = 0, \gamma = 1$ (which is nonlinear for $m \neq 1$). To apply the method of assumed solutions or residual functions, we start with an assumed solution to eq. (17.1)

$$u_a(x, t) = x^n t^n,\ 0 \leq x \leq 1;\ t \geq 0 \tag{17.7}$$

Table 17.2: Selected numerical output from the main program `pde_1_main.m` for `ncase=12`

t	x	u1(it,i)	u1_anal(it,i)	err(it,i)
0.00	-9.425	0.000000	0.000000	0.000000
0.00	-8.692	0.000000	0.000000	0.000000
0.00	-7.959	0.000000	0.000000	0.000000
0.00	-7.226	0.000000	0.000000	0.000000
0.00	-6.493	0.000000	0.000000	0.000000
0.00	-5.760	0.000000	0.000000	0.000000
0.00	-5.027	0.000000	0.000000	0.000000
0.00	-4.294	0.000000	0.000000	0.000000
0.00	-3.560	0.000000	0.000000	0.000000
0.00	-2.827	0.000000	0.000000	0.000000
0.00	-2.094	0.000000	0.000000	0.000000
0.00	-1.361	0.000000	0.000000	0.000000
0.00	-0.628	0.000000	0.000000	0.000000
0.00	0.105	0.104528	0.104528	0.000000
0.00	0.838	0.743145	0.743145	0.000000
0.00	1.571	1.000000	1.000000	0.000000
0.00	2.304	0.743145	0.743145	0.000000
0.00	3.037	0.104528	0.104528	0.000000
0.00	3.770	0.000000	0.000000	0.000000
0.00	4.503	0.000000	0.000000	0.000000
0.00	5.236	0.000000	0.000000	0.000000
0.00	5.969	0.000000	0.000000	0.000000
0.00	6.702	0.000000	0.000000	0.000000
0.00	7.435	0.000000	0.000000	0.000000
0.00	8.168	0.000000	0.000000	0.000000
0.00	8.901	0.000000	0.000000	0.000000
0.00	9.634	0.000000	0.000000	0.000000
0.00	10.367	0.000000	0.000000	0.000000
0.00	11.100	0.000000	0.000000	0.000000
0.00	11.833	0.000000	0.000000	0.000000
0.00	12.566	0.000000	0.000000	0.000000

.
.
.
output for t=3.14 removed
.
.
.

t	x	u1(it,i)	u1_anal(it,i)	err(it,i)
6.28	-9.425	0.000000	0.000000	0.000000
6.28	-8.692	0.000000	0.000000	0.000000
6.28	-7.959	0.000000	0.000000	0.000000
6.28	-7.226	0.000017	0.000000	0.000017
6.28	-6.493	-0.000793	0.000000	-0.000793

Table 17.2: *(Continued)*

t	x	ul(it,i)	ul_anal(it,i)	err(it,i)
6.28	−5.760	0.258354	0.250000	0.008354
6.28	−5.027	0.474751	0.475528	−0.000777
6.28	−4.294	0.458250	0.456773	0.001477
6.28	−3.560	0.200618	0.203368	−0.002750
6.28	−2.827	−0.009532	0.000000	−0.009532
6.28	−2.094	0.001694	0.000000	0.001694
6.28	−1.361	−0.000967	0.000000	−0.000967
6.28	−0.628	0.002879	0.000000	0.002879
6.28	0.105	0.000210	0.000000	0.000210
6.28	0.838	0.000868	0.000000	0.000868
6.28	1.571	0.002498	0.000000	0.002498
6.28	2.304	0.000868	0.000000	0.000868
6.28	3.037	0.000210	0.000000	0.000210
6.28	3.770	0.002879	0.000000	0.002879
6.28	4.503	−0.000967	0.000000	−0.000967
6.28	5.236	0.001694	0.000000	0.001694
6.28	5.969	−0.009532	0.000000	−0.009532
6.28	6.702	0.200618	0.203368	−0.002750
6.28	7.435	0.458250	0.456773	0.001477
6.28	8.168	0.474751	0.475528	−0.000777
6.28	8.901	0.258354	0.250000	0.008354
6.28	9.634	−0.000793	0.000000	−0.000793
6.28	10.367	0.000017	0.000000	0.000017
6.28	11.100	0.000000	0.000000	0.000000
6.28	11.833	0.000000	0.000000	0.000000
6.28	12.566	0.000000	0.000000	0.000000
	ncall=840			

The choice of eq. (17.7) is motivated primarily by the ease of computing analytically the partial derivatives in eq. (17.1). Note that this choice is *not* motivated by a physical application although this may be a possibility. In other words, here we assume a solution for ease of use in deriving an analytical solution. Also, we take $n > 2$ so that the second-order partial derivatives in eq. (17.1) evaluated according to eq. (17.7) are not constant (or zero) and therefore provides a relatively rigorous test of the numerical calculation of these second-order derivatives.

Substitution of eq. (17.7) in eq. (17.1) gives

Term in eq. (17.1)	Term from eq. (17.7)
$\frac{\partial^2 u}{\partial t^2}$	$n(n-1)x^n t^{n-2}$
$+\alpha\frac{\partial^2 u}{\partial x^2}$	$+\alpha n(n-1)x^{n-2}t^n$
$+\beta u$	$+\beta x^n t^n$

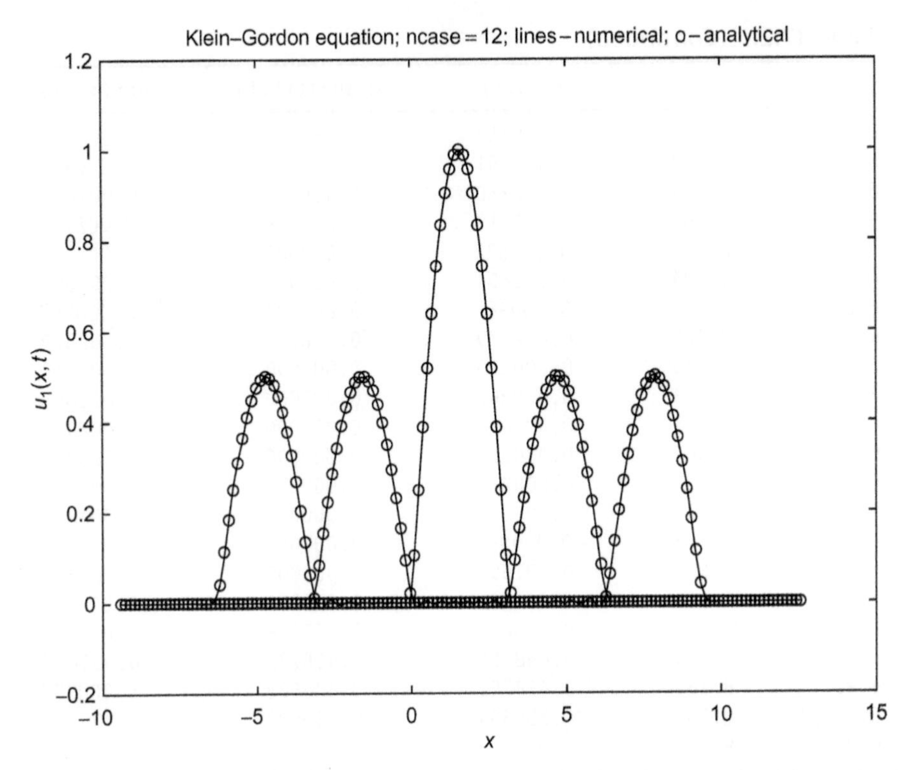

FIGURE 17.3: 2D plot comparing the numerical and analytical solutions of eq. (17.2) (eq. (17.3) for $t = 0, \pi, 2\pi$).

$$
\begin{aligned}
&+\gamma u^m && +\gamma x^{nm} t^{nm} \\
&= f(x,t) && = n(n-1)x^n t^{n-2} \\
& && +\alpha n(n-1)x^{n-2} t^n \\
& && +\beta x^n t^n + \gamma x^{nm} t^{nm}
\end{aligned}
$$

Thus, the residual function (also termed an *inhomogeneous function* or *nonhomogeneous function*), defined for eq. (17.1) so that eq. (17.7) is an exact solution, becomes

$$f(x,t) = n(n-1)x^n t^{n-2} + \alpha n(n-1)x^{n-2} t^n + \beta x^n t^n + \gamma x^{nm} t^{nm} \tag{17.8}$$

We now consider an MOL solution to eqns. (17.1) and (17.8), with ICs and BCs provided by eq. (17.7). Again, we reproduce only the code for `ncase=2` indicated in Listings 17.1 and 17.2. The BCs for `ncase=2` in `pde_1.m` are provided by `ua_2.m` (refer to Listing 17.1 for the complete listing of `pde_1.m`).

```
if(ncase==2)
  u1(1)=ua_2(x(1),t);
  u1(n)=ua_2(x(n),t);
end
```

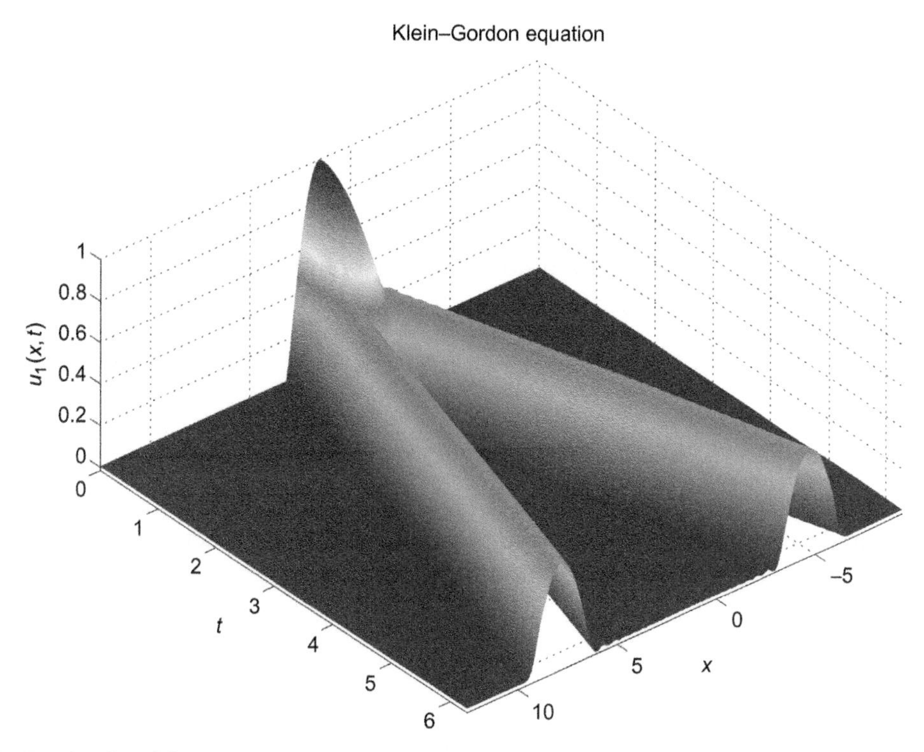

FIGURE 17.4: 3D plot of the numerical solution of eq. (17.2).

The programming of eq. (17.1) in `pde_1.m` is

```
if(ncase==2)|(ncase==3)
  for i=2:n-1
    u1t(i)=u2(i);
      if(ncase==2)f=f_2(x(i),t);end
      if(ncase==3)f=0;end
    u2t(i)=-a*u1xx(i)-b*u1(i)-g*u1(i)^m+f;
  end
end
```

Since eq. (17.1) is also integrated for `ncase=3`, the coding for `ncase=2` and `ncase=3` is the same (and follows directly from eq. (17.1)).

Function `f_2.m` for the residual function of eq. (17.8) called in `pde_1.m` (for `ncase=2`) is straightforward (note that n in eq. (17.7) is programmed as `r` since `n` is the number of grid points).

```
  function rhs=f_2(x,t)
%
% Function rhs computes the inhomogenenous RHS function of
% the mth order Klein-Gordon equation
%
% Model parameters
```

```
  global c c2 m r a b g
%
% Residual function
  rhs=r*(r-1)*x^r*t^(r-2)...
     +a*r*(r-1)*x^(r-2)*t^r...
     +b*x^r*t^r+g*x^(r*m)*t^(r*m);
```

LISTING 17.7: Function `f_2.m` for the residual function of eq. (17.8).

In `inital_1.m` for `ncase=2`, we use $0 \le x \le 1$ on 51 points (refer to Listing 17.2 for the complete listing of `inital_1.m`).

```
  if(ncase==2)
    xl=0;
    xu=1;
    n=51;
  end
```

Because the solution to eq. (17.1) for `ncase=2` is relatively smooth, only $n = 51$ grid points were required to achieve a numerical solution of acceptable accuracy. `ua_2.m` and `uat_2.m` are then used to set the ICs (with `t0=0` from `pde_1_main.m`).

```
    if(ncase==2)
      u1(i)= ua_2(x(i),t0);
      u2(i)=uat_2(x(i),t0);
    end
```

`ua_2.m` is a straightforward implementation of eq. (17.7).

```
  function uanal=ua_2(x,t)
%
% Function uanal computes the exact solution of the mth order
% Klein-Gordon equation for comparison with the numerical solution
%
% Model parameters
  global c c2 m r
%
% Analytical solution
  uanal=x^r*t^r;
```

LISTING 17.8: Function `ua_2.m` for the analytical solution of eq. (17.7).

Function `uat_2.m` is a straightforward implementation of the derivative of the solution (in t) of eq. (17.7).

```
  function uanal=uat_2(x,t)
%
% Function uanal computes the derivative of the exact solution of the
% mth order Klein-Gordon equation
%
% Model parameters
  global c c2 m r
%
% Derivative of the analytical solution
  uanal=x^r*r*t^(r-1);
```

LISTING 17.9: Function `uat_2.m` for the derivative of the analytical solution of eq. (17.7).

The additional coding in the main program `pde_1_main.m` follows.

```
%
% Model parameters
  if(ncase==2)m=2;r=3;a=-1;b=0;g=1;end
```

t is defined over the interval $0 \leq t \leq 1$ for five outputs in the numerical and plotted output at $t = 0, 0.25, 0.5, 0.75, 1$.

```
  if(ncase==2)
    tf=1; tout=[t0:0.25:tf]'; nout=5; ncall=0;
  end
```

The analytical solution is included in the output by using `ua_2.m` (for a comparison of the numerical and analytical solutions).

```
    if(ncase==2)
      u1(it,1)=ua_2(x(1),t(it));
      u1(it,n)=ua_2(x(n),t(it));
      for i=1:n
        u1_anal(it,i)=ua_2(x(i),t(it));
        err(it,i)=u1(it,i)-u1_anal(it,i);
      end
    end
```

The plot of the numerical and analytical solutions includes a label for `ncase=2`.

```
  elseif(ncase==2)
    title('Klein-Gordon equation; ncase = 2; solid - numerical; o - analytical')
```

A portion of the numerical output from `pde_1_main.m` is listed in Table 17.3. We can note the following points about this output:

- The numerical and analytical solutions agree at $t = 0$ as expected, and the agreement between the two solutions at $t = 1$ is about five figures.
- The computational effort is modest, with `ncall = 285`.

The plotted solutions follow in Figs. 17.5 and 17.6. Figure (17.5) indicates the close agreement between the numerical and analytical solutions of Table 17.3, which is due in part to the smooth assumed solution of eq. (17.7). The use of an assumed solution (such as eq. (17.7)) demonstrates that if a residual function, such as $f(x,t)$ in eq. (17.1), is included in the PDE, then the derivation of an exact solution based on the assumed solution is straightforward, even for nonlinear PDEs (such as eq. (17.1) with $m \neq 1$).

Case 3: Nonlinear PDE (`ncase=3`, $m = 3, \alpha = -2.5, \beta = 1, \gamma = 1.5, c = 0.5, f(x,t) = 0$)

Finally, we proceed to `ncase=3` with $m = 3, \alpha = -2.5, \beta = 1, \gamma = 1.5, c = 0.5$, using an analytical solution to eq. (17.1) with $f(x,t) = 0$ ([1]). This is an example of the *cubic Klein–Gordon*

Table 17.3: Selected numerical output from the main program `pde_1_main.m` for `ncase=2`

t	x	u1(it,i)	u1_anal(it,i)	err(it,i)
0.00	0.000	0.000000	0.000000	0.000000
0.00	0.100	0.000000	0.000000	0.000000
0.00	0.200	0.000000	0.000000	0.000000
0.00	0.300	0.000000	0.000000	0.000000
0.00	0.400	0.000000	0.000000	0.000000
0.00	0.500	0.000000	0.000000	0.000000
0.00	0.600	0.000000	0.000000	0.000000
0.00	0.700	0.000000	0.000000	0.000000
0.00	0.800	0.000000	0.000000	0.000000
0.00	0.900	0.000000	0.000000	0.000000
0.00	1.000	0.000000	0.000000	0.000000

.
.
.

output for t=0.25, 0.5, 0.75 removed

.
.
.

t	x	u1(it,i)	u1_anal(it,i)	err(it,i)
1.00	0.000	0.000000	0.000000	0.000000
1.00	0.100	0.000999	0.001000	−0.000001
1.00	0.200	0.008000	0.008000	−0.000000
1.00	0.300	0.027000	0.027000	0.000000
1.00	0.400	0.063998	0.064000	−0.000002
1.00	0.500	0.125000	0.125000	−0.000000
1.00	0.600	0.216000	0.216000	0.000000
1.00	0.700	0.342999	0.343000	−0.000001
1.00	0.800	0.512000	0.512000	−0.000000
1.00	0.900	0.729000	0.729000	0.000000
1.00	1.000	1.000000	1.000000	0.000000

ncall=285

equation (with $m=3$ in eq. (17.1)).

$$\frac{\partial^2 u}{\partial t^2}+\alpha\frac{\partial^2 u}{\partial x^2}+\beta u+\gamma u^3=f(x,t) \tag{17.9}$$

An analytical solution ([1]) is

$$u_a(x,t)=B\tan[K(x+ct)],\quad 0\le x\le 1 \tag{17.10}$$

for the particular values $\alpha=-2.5,\beta=1,\gamma=1.5,c=0.5$ with $B=\sqrt{\frac{\beta}{\gamma}},K=\sqrt{\frac{-\beta}{2(\alpha+c^2)}}$

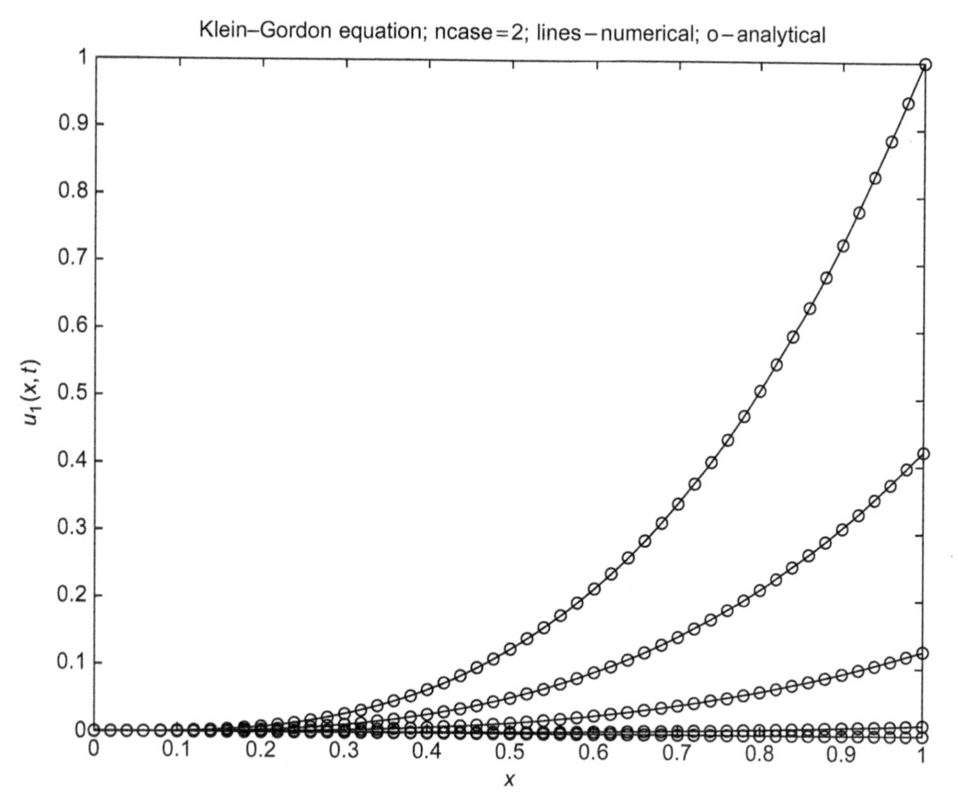

FIGURE 17.5: 2D plot comparing the numerical and analytical solutions of eq. (17.1) for $t = 0, 0.25, 0.5, 0.75, 1$ (bottom to top).

We now consider an MOL solution to eq. (17.9) with ICs and BCs provided by eq. (17.10). Again, we reproduce only the code for `ncase=3` indicated in Listings 17.1 and 17.2. The BCs for `ncase=3` in `pde_1.m` are provided by `ua_3.m` (refer to Listing 17.1 for the complete listing of `pde_1.m`).

```
  if(ncase==3)
    u1(1)=ua_3(x(1),t);
    u1(n)=ua_3(x(n),t);
end
```

The programming of eq. (17.9) in `pde_1.m` is

```
if(ncase==2)|(ncase==3)
  for i=2:n-1
    u1t(i)=u2(i);
      if(ncase==2)f=f_2(x(i),t);end
      if(ncase==3)f=0;end
    u2t(i)=-a*u1xx(i)-b*u1(i)-g*u1(i)^m+f;
  end
end
```

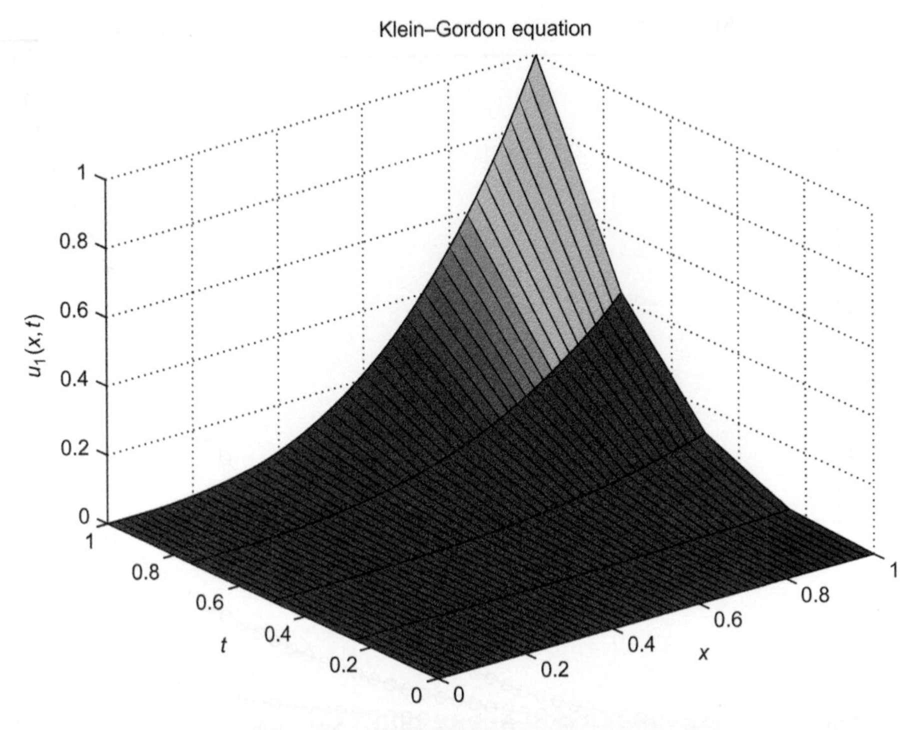

FIGURE 17.6: 3D plot of the numerical solution of eq. (17.1).

Since eq. (17.1) is also integrated for `ncase=3`, the coding for `ncase=2` ($f(x,t)$ from eq. (17.8)) and `ncase=3` ($f(x,t)=0$) is the same (and follows directly from eqns. (17.1) and (17.9)).

In `inital_1.m` for `ncase=3`, we use $0 \leq x \leq 1$ on 51 points (refer to Listing 17.2 for the complete listing of `inital_1.m`).

```
if(ncase==3)
  xl=0;
  xu=1;
  n=51;
end
```

Because the solution, eq. (17.10), for `ncase=3` is relatively smooth, only $n = 51$ grid points were required to achieve a numerical solution of acceptable accuracy. `ua_3.m` and `uat_3.m` are then used to set the ICs (with `t0=0` from `pde_1_main.m`).

```
if(ncase==3)
  u1(i)= ua_3(x(i),t0);
  u2(i)=uat_3(x(i),t0);
end
```

`ua_3.m` is a straightforward implementation of eq. (17.10).

```
  function uanal=ua_3(x,t)
%
% Function uanal computes the exact solution of the cubic
% Klein-Gordon equation for comparison with the numerical solution
%
% Model parameters
  global a b g B K c
%
% Analytical solution
  uanal=B*tan(K*(x+c*t));
```

LISTING 17.10: Function ua_2.m for the analytical solution of eq. (17.10).

Function uat_3.m is a straightforward implementation of the derivative of the solution (in t) of eq. (17.10).

```
  function uanal=uat_3(x,t)
%
% Function uanal computes the time derivative of the exact solution
% of the cubic Klein-Gordon equation
%
% Model parameters
  global c c2 m r a b g B K
%
% Analytical solution derivative
  uanal=B*sec(K*(x+c*t))^2*(K*c);
```

LISTING 17.11: Function uat_3.m for the derivative of the analytical solution of eq. (17.10).

The additional coding in the main program pde_1_main.m follows.

```
%
% Model parameters
  if(ncase==3)m=3;a=-2.5;b=1;g=1.5;c=0.5;
      B=(b/g)^0.5;K=(-b/(2*(a+c^2)))^0.5;end
```

t is defined over the interval $0 \le t \le 4$ for five outputs in the numerical and plotted output at $t = 0, 1, 2, 3, 4$.

```
  if(ncase==3)
    tf=4; tout=[t0:1:tf]'; nout=5; ncall=0;
  end
```

The analytical solution, eq. (17.10), is included in the output by using ua_3.m (for a comparison of the numerical and analytical solutions).

```
    if(ncase==3)
      u1(it,1)=ua_3(x(1),t(it));
      u1(it,n)=ua_3(x(n),t(it));
      for i=1:n
        u1_anal(it,i)=ua_3(x(i),t(it));
        err(it,i)=u1(it,i)-u1_anal(it,i);
      end
    end
```

The plot of the numerical and analytical solutions includes a label for ncase=3.

```
elseif(ncase==3)
  title('Klein-Gordon equation; ncase = 3; solid - numerical; o - analytical')
end
```

A portion of the numerical output from `pde_1_main.m` is listed in Table 17.4. We can note the following points about this output:

- The numerical and analytical solutions agree at $t = 0$ as expected, and the agreement between the two solutions at $t = 4$ is about five figures.

Table 17.4: Selected numerical output from the main program `pde_1_main.m` for `ncase=3`

t	x	u1(it,i)	u1_anal(it,i)	err(it,i)
0.00	0.000	0.000000	0.000000	0.000000
0.00	0.100	0.038519	0.038519	0.000000
0.00	0.200	0.077209	0.077209	0.000000
0.00	0.300	0.116246	0.116246	0.000000
0.00	0.400	0.155811	0.155811	0.000000
0.00	0.500	0.196095	0.196095	0.000000
0.00	0.600	0.237302	0.237302	0.000000
0.00	0.700	0.279655	0.279655	0.000000
0.00	0.800	0.323399	0.323399	0.000000
0.00	0.900	0.368809	0.368809	0.000000
0.00	1.000	0.416196	0.416196	0.000000
	.		.	
	.		.	
	.		.	
	output for t=1, 2, 3 removed			
	.		.	
	.		.	
	.		.	

t	x	u1(it,i)	u1_anal(it,i)	err(it,i)
4.00	0.000	1.124594	1.124594	0.000000
4.00	0.100	1.243940	1.243940	−0.000000
4.00	0.200	1.381770	1.381769	0.000000
4.00	0.300	1.543516	1.543515	0.000001
4.00	0.400	1.736934	1.736935	−0.000002
4.00	0.500	1.973506	1.973508	−0.000002
4.00	0.600	2.270974	2.270974	−0.000000
4.00	0.700	2.658292	2.658293	−0.000001
4.00	0.800	3.186174	3.186178	−0.000004
4.00	0.900	3.952267	3.952272	−0.000005
4.00	1.000	5.171787	5.171787	0.000000
ncall=1581				

- The computational effort is above previous values (but still modest), with `ncall = 1581`.
- The variation in `ncall` from `285 (ncase = 2)` to `1581 (ncase = 3)` indicates that the computational effort can vary substantially, even within apparently minor variations in the same problem; thus, for `ncase = 2,3`, the PDE is the same (eq. (17.1)), and only m changed (from $m = 2$ to $m = 3$), and the ICs and BCs were different. Generally, the expected computational effort for a particular PDE is essentially unpredictable.

The plotted solutions follow in Figs. 17.7 and 17.8. Figure (17.7) indicates the close agreement between the numerical and analytical solutions of Table 17.3, which is due in part to the smooth solution of eq. (17.10).

In summary, we have again discussed the numerical solution of a PDE second order in t (eq. (17.1)) by restating the PDE as two PDEs first order in t (eq. (17.6)). As noted in Chapters 15 and 16, the reformulation of an nth-order PDE as n PDEs first order in an initial value independent variable (t) is straightforward and permits the use of an integrator such as `ode15s` for first-order ODEs.

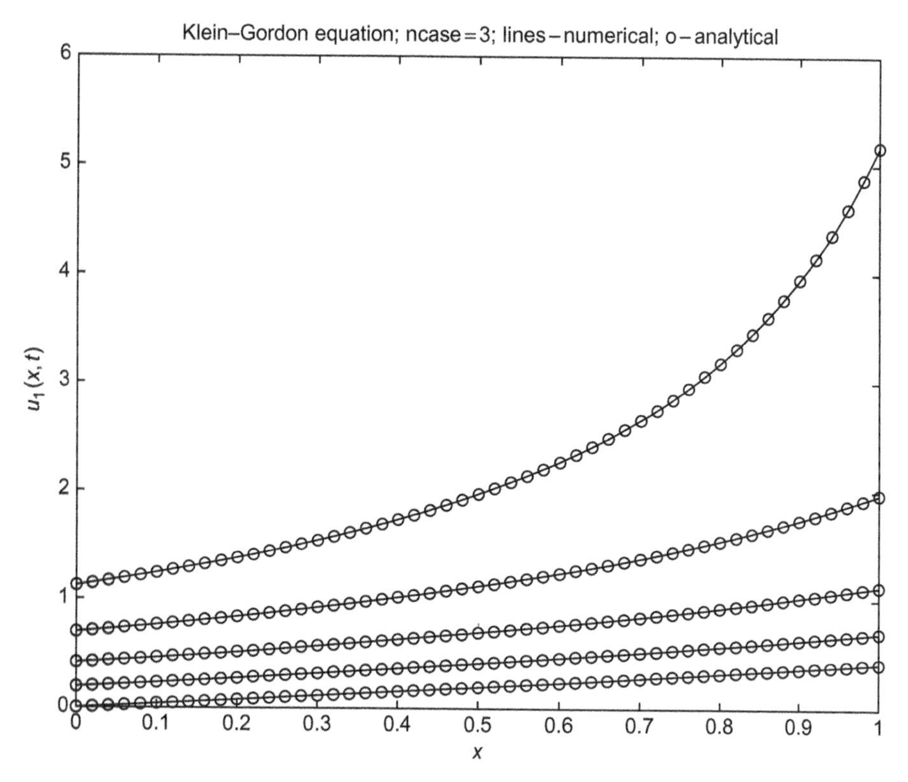

FIGURE 17.7: 2D plot comparing the numerical and analytical solutions of eq. (17.9) for $t = 0,1,2,3,4$ (bottom to top).

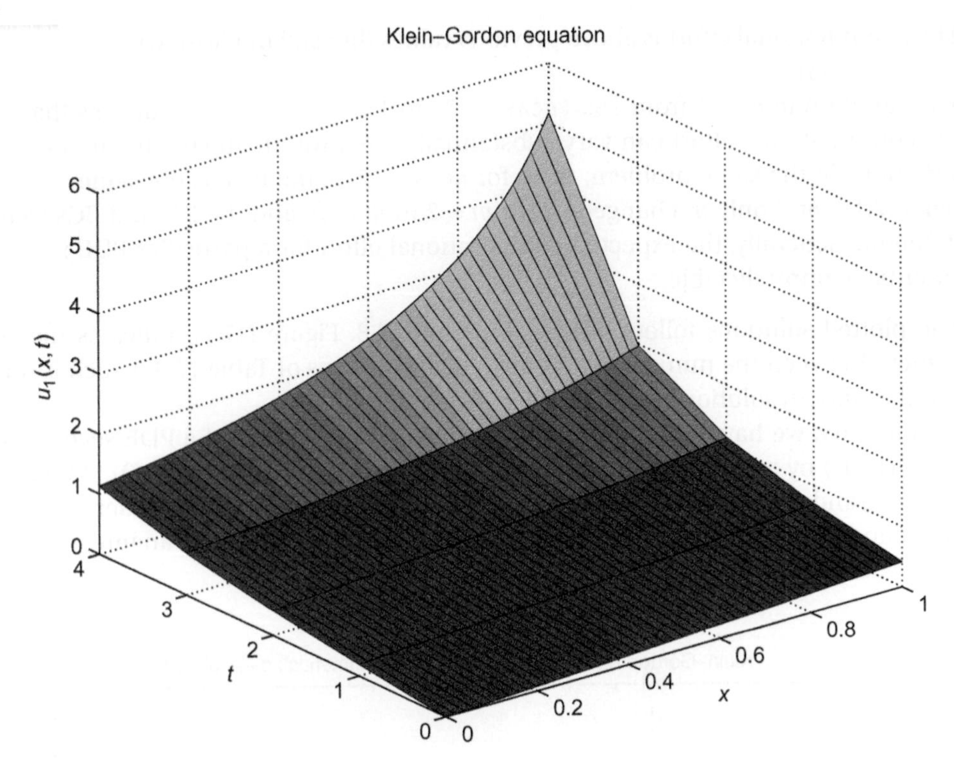

FIGURE 17.8: 3D plot of the numerical solution of eq. (17.9).

Appendix

We conclude this chapter with a traveling wave analysis of eq. (17.1). The conditions for *Case 1* reduce the problem to that of a *linear wave*, the solution of which is well known, and therefore will not be detailed here. The conditions for *Case 2* are such that (intentionally) we do not know how to obtain an analytical solution. Thus, we are left with *Case 3* conditions, which reduce the problem to

$$\frac{\partial^2 u}{\partial t^2} + \alpha \frac{\partial^2 u}{\partial x^2} + \beta u + \gamma u^3 = 0 \tag{17.11}$$

where we have set $f(x) = 0$ and $m = 3$, but have not set values for α, β, or γ.

By inspection, we see that application of the transformation $u(x,t) = U(\xi)$, where $\xi = k(x - ct)$, reduces the PDE of eq. (17.11) to the ODE

$$\left(c^2k^2 + \alpha k^2\right) \frac{d^2}{d\xi^2} U + \beta U + \gamma U^3 = 0 \tag{17.12}$$

which can, in principle, be solved by the separation of variables method. However, if we separate the variables and integrate twice, i.e.,

$$\iint \frac{-dUdU}{\beta U+\gamma U^3}=\frac{1}{c^2k^2+\alpha k^2}\iint d\xi\, d\xi, \tag{17.13}$$

we obtain

$$\frac{U\ln(\beta+\gamma U^2)}{2\beta}+\arctan\left(\frac{\gamma U}{\sqrt{\beta\gamma}}\right)\frac{1}{\sqrt{\beta\gamma}}-\frac{U\ln(U)}{\beta}=\frac{\xi^2}{2(c^2k^2-\alpha k^2)}+K\xi, \tag{17.14}$$

where K is an arbitrary constant. While eq. (17.14) can be simplified slightly, it cannot be readily solved for U, so we abandon this analytical approach. However, use of the built-in Maple function `dsolve` finds the solution

$$U(\xi)=\_C2\sqrt{\frac{-2\beta}{-\gamma-2\beta+\gamma\_C2^2}}\,\mathrm{JacobiSN}\left[\left(\frac{\sqrt{(c^2+\alpha)(\gamma+2\beta)}\xi}{\sqrt{2}(c^2+\alpha)k}+\_C1\right)\right.$$
$$\left.\sqrt{\frac{-2\beta}{-\gamma-2\beta+\gamma\_C2^2}},\frac{\_C2\gamma}{\sqrt{-(\gamma+2\beta)\gamma}}\right],$$

where $\_C1$ and $\_C2$ are arbitrary constants, and JacobiSN represents the elliptic function $\mathrm{sn}(z,m)$, with $0<m<1$ being the elliptic modulus. We will not discuss this solution here, but more details can be found from the Maple help system and, also, some additional discussion on this form of solution can be found in [4]. Instead, we will obtain a traveling wave solution using our Maple procedure `tanhMethod`.

After setting up the problem in Maple, we call `tanhMethod` and set the information level variable `infoLevOut` equal to `2` in order to provide some additional information at certain steps in the calculation. The *tanh method* is successful and finds seven solutions, three of which are trivial. We choose the following solution

$$u=i\sqrt{\beta}\tanh\left(\frac{\sqrt{2}\sqrt{\beta}(x+x_0-ct)}{2\sqrt{\alpha+c^2}}\right)\frac{1}{\sqrt{\gamma}}, \tag{17.15}$$

which on conversion to the *tan* function and rearranging becomes

$$u=\sqrt{\frac{\beta}{\gamma}}\tan\left(i\sqrt{\frac{\beta}{2(\alpha+c^2)}}(x+x_0-ct)\right). \tag{17.16}$$

If we bring i under the square-root sign, set $x_0=0$ and define $c=-c$, eq. (17.16) matches the Case 3 solution, eq. (17.10).

The Maple code that derives solution eq. (17.16) and generates an animation and 3D plot is included under Listing 17.12.

```
> # Klein-Gordon Equation
  # Attempt at Malfliet's tanh solution
  restart; with(PDEtools): with(PolynomialTools): with(plots):
> alias(u=u(x,t)):unprotect(gamma);
> pde1:=diff(u,t,t)+alpha*diff(u,x,x)+beta*u+gamma*u^m=f(x,t);
> pde1:=subs({m=3,f(x,t)=0},pde1);
> read("tanhMethod.txt");
  intFlag:=0: # No integration of U(xi) needed !
  M:=2; # Set order of approximation
  infoLevOut:=2;
  tanhMethod(M,pde1,intFlag, infoLevOut);
> zz:=convert(rhs(sol[7]),tan);
> x0:=0;alpha:=-2.5;beta:=1;gamma:=1.5;c:=-0.5;
> animate(zz,x=0..1, t=0..4,axes=framed,
    thickness=3,frames=50,numpoints=100,
    axesfont=[TIMES, ROMAN, 16],titlefont=[TIMES, ROMAN, 16],
    labelfont=[TIMES, ROMAN, 16],
    title="Klein-Gordon Equation");
> plot3d(zz,x=0..1,t=0..4,axes='framed',
    labels=["x","t","u(x,t)"],
    labeldirections=[HORIZONTAL,HORIZONTAL,VERTICAL],
    orientation=[-132,77],grid=[100,100], style=patchnogrid,
    axesfont=[TIMES, ROMAN, 16],titlefont=[TIMES, ROMAN, 16],
    labelfont=[TIMES, ROMAN, 16],
    shading=Z,title="Klein-Gordon Equation")
```

LISTING 17.12: Maple code used to generate analytical solutions to the the Klein–Gordon equation (Case 3), an animation and 3D plot.

Application of the Maple `riccatiMethod` procedure finds 7×6 traveling wave solutions (recall that the Riccati method finds 6 different forms for each solution), including eqns. (17.10 and 17.16). However, the code is not included here, but it is included with the downloads for this book.

References

[1] M. Dehghan, A. Shokri, Numerical solution of the nonlinear Klein-Gordon equation using radial basis functions, *J. Comp. App. Math.* 230 (2009) 400–410.

[2] S.J. Farlow, *Partial Differential Equations for Scientists and Engineers*, Dover Publications, New York, 1993 (Chapter 17).

[3] A.D. Polyanin, V.F. Zaitsev, *Handbook of Nonlinear Partial Differential Equations*, Chapman & Hall/CRC, Boca Raton, FL, 2004.

[4] Z. Zhang, New exact traveling wave solutions for the nonlinear Klein-Gordon equation, *Turk. J. Phys.* 32 (2008) 235–240.

18

Boussinesq Equation

Several variants of the *Boussinesq equation* have been reported and discussed in the literature. Here we consider the particular equation [3, 6]

$$\frac{\partial^2 u}{\partial t^2} = \frac{\partial^2 u}{\partial x^2} - \frac{\partial^4 u}{\partial x^4} - \frac{\partial^2}{\partial x^2}\left(u^2\right) \tag{18.1}$$

with the reported analytical solution [8]

$$\begin{aligned} u_a(x,t) &= (3/2)(1-c^2)\text{sech}^2\left[\frac{\sqrt{1-c^2}}{2}(x-ct)\right] \\ &= (3/2)(1-c^2)\cosh^{-2}\left[\frac{\sqrt{1-c^2}}{2}(x-ct)\right] \end{aligned} \tag{18.2}$$

where c is an arbitrary constant. We note that eq. (18.2) is a traveling wave solution from the argument $x - ct$.

An initial condition (IC) follows directly from eq. (18.2) with $t = 0$. Since eq. (18.1) is second order in t, we require a second IC, which follows from the derivative (in t) of eq. (18.2).

$$\frac{\partial u_a(x,t)}{\partial t} = (3/2)c(1-c^2)^{3/2}\sinh\left[\frac{\sqrt{1-c^2}}{2}(x-ct)\right]\cosh^{-3}\left[\frac{\sqrt{1-c^2}}{2}(x-ct)\right] \tag{18.3}$$

Equation (18.1) is fourth order in x, and the four required boundary conditions (BCs) are taken as

$$u(x = x_l, t) = u(x = x_u, t) = 0 \tag{18.4a,4b}$$

$$\frac{\partial^2 u(x = x_l, t)}{\partial x^2} = \frac{\partial^2 u(x = x_u, t)}{\partial x^2} = 0 \tag{18.4c,4d}$$

Equations (18.1) to (18.4) constitute the problem of interest and a method of lines (MOL) numerical solution is produced with the following Matlab routines.

The ODE routine for eqs. (18.1)–(18.4) is as follows.

```
  function ut=pde_1(t,u)
%
% Function pde_1 computes the t derivative vector for the Boussinesq
% equation
%
```

```
  global xl xu x n ncall
%
% Model parameters
  global c ncase
%
  if(ncase==1)ut=pde_1a(t,u);end
  if(ncase==2)ut=pde_1b(t,u);end
  if(ncase==3)ut=pde_1c(t,u);end
  if(ncase==4)ut=pde_1d(t,u);end
```

LISTING 18.1: Function `pde_1.m` for eq. (18.1) (`ncase=1,2,3,4`).

We can note the following details about `pde_1.m`:

- The function and some global variables are first defined.

```
  function ut=pde_1(t,u)
%
% Function pde_1 computes the t derivative vector for the Boussinesq
% equation
%
  global xl xu x n ncall
%
% Model parameters
  global c ncase
```

- Since four cases will be considered in the subsequent discussion, these cases are programmed as a series of self-contained routines. This arrangement is offered as a way to investigate various mathematical and MOL formulations in a separated and clearly defined format (rather than, for example, including all of the cases in `pde_1.m` as in Listing 17.1).

```
%
  if(ncase==1)ut=pde_1a(t,u);end
  if(ncase==2)ut=pde_1b(t,u);end
  if(ncase==3)ut=pde_1c(t,u);end
  if(ncase==4)ut=pde_1d(t,u);end
```

  Then we can discuss the individual cases by examining the four routines `pde_1a.m` (for `ncase=1`) to `pde_1d.m` (for `ncase=4`). To select a particular case, `ncase` is set to one of the values `1,2,3,4` in the main program, `pde_1_main.m`, discussed subsequently and passed as a global variable.

Each of the four cases will be considered, one at a time. To start, we consider `pde_1a.m` for `ncase=1`.

Case 1: Direct Calculation of $\dfrac{\partial^4 u}{\partial x^4}$ via `u4x11p`

```
  function ut=pde_1a(t,u)
%
% Function pde_1a computes the t derivative vector for the Boussinesq
% equation
%
```

```
  global xl xu x n ncall
%
% One vector to two vectors
  for i=1:n
    u1(i)=u(i);
    u2(i)=u(i+n);
  end
%
% BCs at x = -15,25
  u1(1)=0;
  u1(n)=0;
%
% u1xx
  nl=1; nu=1;
  u1x(1)=0;
  u1xx=dss044(xl,xu,n,u1,u1x,nl,nu);
%
% BCs at x = -15,25 (not required by u4x11p)
% u1xx(1)=0;
% u1xx(n)=0;
%
% u1xxxx
  u1xxxx=u4x11p(xl,xu,n,u1);
%
% (u1^2)xx
  u1s=u1.^2;
  nl=1; nu=1;
  u1sx(1)=0;
  u1sxx=dss044(xl,xu,n,u1s,u1sx,nl,nu);
%
% PDE
  u1t=u2;
  u2t=u1xx-u1xxxx-u1sxx;
%
% Two vectors to one vector
  for i=1:n
    ut(i)  =u1t(i);
    ut(i+n)=u2t(i);
  end
  ut=ut';
%
% Increment calls to pde_1a
  ncall=ncall+1;
```

LISTING 18.1a: Function `pde_1a.m` for eq. (18.1) (`ncase=1`).

We can note the following details about `pde_1a.m`:

- The function and some global variables are defined.

```
  function ut=pde_1a(t,u)
%
% Function pde_1a computes the t derivative vector for the Boussinesq
% equation
%
  global xl xu x n ncall
```

- As in the preceding chapters (e.g., 16 and 17) for PDEs second order in t, the solution vector u is placed in two arrays u1,u2.

```
%
% One vector to two vectors
  for i=1:n
    u1(i)=u(i);
    u2(i)=u(i+n);
  end
```

u1,u2 are dependent variables first order in t, so a library ODE integrator such as ode45 or ode15s can be used to move the equations through t.

- BCs (18.4a) and (18.4b) are programmed and the second derivative $\frac{\partial^2 u}{\partial x^2}$ is computed with dss044 (with *Dirichlet* BCs designated by nl=1,nu=1). u1x(1)=0 is not actually used in the calculations in dss044 but rather is programmed to meet the Matlab requirement that input arguments to functions must be given a value.

```
%
% BCs at x = -15,25
  u1(1)=0;
  u1(n)=0;
%
% u1xx
  nl=1; nu=1;
  u1x(1)=0;
  u1xx=dss044(xl,xu,n,u1,u1x,nl,nu);
```

- $\frac{\partial^4 u}{\partial x^4}$ is computed by u4x11p.m (the origin of this routine is discussed in Chapter 12 and in further detail next).

```
%
% BCs at x = -15,25 (not required by u4x11p)
% u1xx(1)=0;
% u1xx(n)=0;
%
% u1xxxx
  u1xxxx=u4x11p(xl,xu,n,u1);
```

BCs (18.4c) and (18.4d) are not used (note the comments) since they are not required by u4x11p.m.

- The nonlinear term $\frac{\partial^2}{\partial x^2}(u^2)$ in eq. (18.1) is computed by first squaring the solution ($u^2 =$ u1s), then taking its second derivative with dss044 (note that u1 is the RHS dependent variable u of eq. (18.1)).

```
%
% (u1^2)xx
  u1s=u1.^2;
  nl=1; nu=1;
```

```
u1sx(1)=0;
u1sxx=dss044(xl,xu,n,u1s,u1sx,nl,nu);
```

The nonlinear term of eq. (18.1), $\frac{\partial^2}{\partial x^2}(u^2)$, is placed in array `u1sxx`.

- Equation (18.1) is programmed as two first-order PDEs in t.

```
%
% PDE
  u1t=u2;
  u2t=u1xx-u1xxxx-u1sxx;
```

- The two derivative vectors, `u1t`,`u2t` are placed in a single derivative vector `ut` to be returned from `pde_1a.m` and integrated by `ode45` or `ode15s` in the main program `pde_1_main.m`; a transpose is required by these integrators (so that `ut` is a column vector).

```
%
% Two vectors to one vector
  for i=1:n
    ut(i)  =u1t(i);
    ut(i+n)=u2t(i);
  end
  ut=ut';
%
% Increment calls to pde_1a
  ncall=ncall+1;
```

The number of calls to `pde_1a.m` is incremented for display at the end of the solution in the main program `pde_1_main.m` (note that `ncall` is a global variable).

In summary, `pde_1a.m` receives the dependent variable vector `u` as an input (along with the independent variable `t`) and returns the derivative vector `ut`.

`u4x11p.m` for the calculation of $\frac{\partial^4 u}{\partial x^4}$ based on an 11-point FD approximation follows.

```
  function uxxxx=u4x11p(xl,xu,n,u)
%
% Function u4x11p computes the derivative uxxxx based on 11 points
  global ncall CC
%
% For the first call to u4x11p (ncall = 0), compute the FD weighting
% coefficients
%
   if(ncall==0)
%
%   Default points, equally spaced grid
    for i=1:11
      x(i)=i-1;
    end
%
```

```
%   Compute FD approximation for up to and including the mth
%   derivative
    m=4;
%
%   Number of grid points
    ng=11;
    nd=ng;
%
%   Compute weighting coefficients for finite differences
%   over ip points
    for ip=1:ng
%
%   Weighting coefficients in array CC
      CC(ip,:,:)=weights(x(ip),x,ng-1,nd-1,m);
    end
%
% Display coefficients for derivatives of orders 0, 1, 2, 3, 4
% for i=1:m+1
%
% Display coefficients for derivative of order 4
  i=m+1;
    fprintf('\n\n Numerical Derivative Order: %d',i-1);
    fprintf('\n==============================\n');
%
%   Coefficients in u5x11p.m with m = 4, ng = 11
%   CC(:,:,i)*8
    CC(6,:,i)*8
% end
%
% Calculation of FD weights complete
  end
%
% uxxx
% Spatial increment
  dx=(xu-xl)/(n-1);
  rdx4=1.0/dx^4;
  for i=1:n
%
%   At the left end, uxxxx = 0
    if(i<6)uxxxx(i)=0.0;
%
%   At the right end, uxxxx = 0
    elseif(i>(n-5))uxxxx(i)=0.0;
%
%   Interior points
    else
      uxxxx(i)=rdx4*...
      (CC(6,1,5)*u(i-5)...
      +CC(6,2,5)*u(i-4)...
      +CC(6,3,5)*u(i-3)...
      +CC(6,4,5)*u(i-2)...
      +CC(6,5,5)*u(i-1)...
      +CC(6,6,5)*u(i  )...
      +CC(6,7,5)*u(i+1)...
```

```
        +CC(6,8,5)*u(i+2)...
        +CC(6,9,5)*u(i+3)...
        +CC(6,10,5)*u(i+4)...
        +CC(6,11,5)*u(i+5));
    end
  end
```

LISTING 18.2: Function `u4x11p.m` for $\frac{\partial^4 u}{\partial x^4}$ in eq. (18.1).

The origin of routines for the calculation of higher order derivatives in x based on function `weights.m` is discussed in some detail in Chapter 12. Here we note just a few details specific to the calculation of $\frac{\partial^4 u}{\partial x^4}$ on 11-points.

- After the function is defined, the 11-point FD computational grid is defined during the first call to `u4x11p`.

```
    function uxxxx=u4x11p(xl,xu,n,u)
                              .
                              .
                              .
%
% For the first call to u4x11p (ncall = 0), compute the FD weighting
% coefficients
%
     if(ncall==0)
%
%     Default points, equally spaced grid
      for i=1:11
        x(i)=i-1;
      end
%
%     Compute FD approximation for up to and including the mth
%     derivative
      m=4;
```

A fourth-order derivative is then specified.

- The finite difference (FD) weighting coefficients are computed by `weights` over the 11-point grid (`ng=11`).

```
      for ip=1:ng
%
%     Weighting coefficients in array CC
        CC(ip,:,:)=weights(x(ip),x,ng-1,nd-1,m);
      end
```

- The fourth derivative, `uxxxx`, is computed over n grid points (n is an input parameter to `u4x11p`).

```
%
% uxxx
% Spatial increment
  dx=(xu-xl)/(n-1);
  rdx4=1.0/dx^4;
```

```
  for i=1:n
%
%   At the left end, uxxxx = 0
    if(i<6)uxxxx(i)=0.0;
%
%   At the right end, uxxxx = 0
    elseif(i>(n-5))uxxxx(i)=0.0;
%
%   Interior points
    else
      uxxxx(i)=rdx4*...
      (CC(6,1,5)*u(i-5)...
      +CC(6,2,5)*u(i-4)...
      +CC(6,3,5)*u(i-3)...
      +CC(6,4,5)*u(i-2)...
      +CC(6,5,5)*u(i-1)...
      +CC(6,6,5)*u(i  )...
      +CC(6,7,5)*u(i+1)...
      +CC(6,8,5)*u(i+2)...
      +CC(6,9,5)*u(i+3)...
      +CC(6,10,5)*u(i+4)...
      +CC(6,11,5)*u(i+5));
    end
```

Note that for the grid points `i=1,2,3,4,5` and `i=n-4,n-3,n-2,n-1,n`, `uxxxx` is set to zero. This was done to minimize any spurious end effects at `x=xl,xu`, but this also means `u4x11p` can be used only when the solution satisfies this condition. Generally, this requires that the solution at the end points does not move away from its IC values for $t > 0$, and the derivatives in x of the solution are effectively zero. These conditions are satisfied by the solution of eq. (18.2) as will be observed in the computed output.

We now consider the other routines for the case `ncase=1`. After looking at the output for this case, we continue the discussion for `ncase=2,3,4`. The IC function `inital_1.m` uses one of four initial condition routines, `inital_1a.m`, `inital_1b.m`, `inital_1c.m`, and `inital_1d.m` for `ncase = 1,2,3,` and 4, respectively.

```
  function u0=inital_1(t0)
%
% Function inital_1 sets the initial condition for the Boussinesq
% equation
%
% Parameters shared with other routines
  global ncase
%
  if(ncase==1)u0=inital_1a(t0);end
  if(ncase==2)u0=inital_1b(t0);end
  if(ncase==3)u0=inital_1c(t0);end
  if(ncase==4)u0=inital_1d(t0);end
```

LISTING 18.3a: IC function `inital_1.m` from eq. (18.2) with $t = 0$ (for `ncase=1,2,3,4`).

For `ncase=1`, function `inital_1a.m` is used.

```
  function u0=inital_1a(t0)
%
% Function inital_1a sets the initial condition for the Boussinesq
% equation
%
% Parameters shared with other routines
  global xl xu x n ncall
%
% Spatial domain and initial condition
  xl=-15;
  xu= 25;
  n=101;
  dx=(xu-xl)/(n-1);
%
% ICs from analytical solution
  for i=1:n
    x(i)=xl+(i-1)*dx;
    u1(i)= ua_1a(x(i),t0);
    u2(i)=uat_1a(x(i),t0);
    u0(i)   =u1(i);
    u0(i+n)=u2(i);
  end
```

LISTING 18.3b: IC function `inital_1a.m` from eq. (18.2) with $t = 0$ (for `ncase=1`).

We can note the following details about `inital_1a.m`:

- The function and some global variables are defined.

```
  function u0=inital_1a(t0)
%
% Function inital_1a sets the initial condition for the Boussinesq
% equation
%
% Parameters shared with other routines
  global xl xu x n ncall
```

- A grid in x of 101 points is defined for $-15 \leq x \leq 25$.

```
%
% Spatial domain and initial condition
  xl=-15;
  xu= 25;
  n=101;
  dx=(xu-xl)/(n-1);
```

- In the `for` loop, the initial values of `u1` for eq. (18.1) are provided by the function `ua_1a.m` (which has the programming of eq. (18.2) with $t = 0$); the initial values of `u2` are provided by the function `uat_1a.m` (which has the programming for the derivative of eq. (18.2) with respect to t, eq. (18.3)).

```
%
% ICs from analytical solution
  for i=1:n
    x(i)=xl+(i-1)*dx;
```

```
    u1(i)= ua_1a(x(i),t0);
    u2(i)=uat_1a(x(i),t0);
    u0(i)  =u1(i);
    u0(i+n)=u2(i);
  end
```

All $2n$ initial condition values are returned from inital_1a.m through the vector u0 to the main program pde_1_main.m discussed subsequently.

Three other IC functions inital_1b.m, inital_1c.m, and inital_1d.m called by inital_1.m in Listing 18.3a are the same as inital_1a.m in Listing 18.3b and are therefore not listed in the subsequent discussion. The repetition of these IC routines is done in order to make the coding for each case (ncase=1,2,3,4) self-contained as explained previously. Of course, if the ICs change from case to case, this can easily be included in the four IC routines.

Function ua_1a.m is a straightforward implementation of eq. (18.2).

```
  function uanal=ua_1a(x,t)
%
% Function uanal computes the exact solution of the Boussinesq equation
% for comparison with the numerical solution
%
% Model parameters
  global c
%
% Analytical solution
  xi=(1-c^2)^(0.5)/2*(x-c*t);
  uanal=(3/2)*(1-c^2)*(cosh(xi))^(-2);
```

LISTING 18.4a: Function ua_1a.m for the analytical solution of eq. (18.2).

Function uat_1a.m is a straightforward implementation of the derivative (in t) of eq. (18.3).

```
  function uanal=uat_1a(x,t)
%
% Function uanal computes the time derivative of the exact solution
% of the Boussinesq equation
%
% Model parameters
  global c
%
% Analytical solution derivative
  xi=(1-c^2)^(0.5)/2*(x-c*t);
  uanal=(3/2)*c*(1-c^2)^(3/2)*sinh(xi)*(cosh(xi))^(-3);
```

LISTING 18.4b: Function uat_1a.m for the derivative of the analytical solution, eq. (18.3).

Again, three other IC functions ua_1b.m, ua_1c.m, and ua_1d.m called by inital_1b.m, inital_1c.m, and inital_1d.m, respectively, are identical to ua_1.m in Listing 18.4a, and three functions uat_1b.m, uat_1c.m, and uat_1d.m called by inital_1b.m, inital_1c.m, and inital_1d.m, respectively, are identical to uat_1a.m in Listing 18.4b. The repetition of these IC routines is also done in order to make the coding for each case (ncase=1,2,3,4) self-contained as explained previously. If the analytical solution changes from case to case, this can easily be included in the four analytical solution routines and the four derivative routines.

Main program pde_1_main.m accommodates the cases ncase=1,2,3,4 for eqs. (18.1)–(18.4). This main program closely parallels pde_1_main.m of Chapter 15 and therefore only a few features are discussed here.

- The parameters are set for ncase=1,2,3,4.

```
%
% Model parameters
  global c ncase
%
% Select case
% ncase=1;
% ncase=2;
% ncase=3;
  ncase=4;
%
% Model parameters
  if(ncase==1)c=0.9;end
  if(ncase==2)c=0.9;end
  if(ncase==3)c=0.9;end
  if(ncase==4)c=0.9;end
```

- The range in t is the same for ncase=1,2,3,4, but the range could easily be changed from case to case.

```
%
% Independent variable for ODE integration
  if(ncase==1)
    tf=9;tout=[t0:3:tf]';nout=4;ncall=0;
  end
  if(ncase==2)
    tf=9;tout=[t0:3:tf]';nout=4;ncall=0;
  end
  if(ncase==3)
    tf=9;tout=[t0:3:tf]';nout=4;ncall=0;
  end
  if(ncase==4)
    tf=9;tout=[t0:3:tf]';nout=4;ncall=0;
  end
```

- The difference in the numerical and analytical solutions is computed as err(it,i) and displayed. These solutions are then plotted in 2D by plot, and the numerical solution is plotted in 3D by surf.

```
%
%   Display selected output
    for it=1:nout
      fprintf('\n     t        x        u1(it,i)  u1_anal(it,i)
                 err(it,i)\n');
```

```
      for i=1:5:n
        fprintf('%6.2f%8.3f%15.6f%15.6f%15.6f\n',...
                t(it),x(i),u1(it,i),u1_anal(it,i),err(it,i));
      end
    end
    fprintf('    ncall = %4d\n\n',ncall);
%
% Plot numerical and analytical solutions
    figure(2)
    plot(x,u1,'-',x,u1_anal,'o')
    xlabel('x')
    ylabel('u1(x,t)')
    if(ncase==1)
      title('Boussinesq equation; ncase = 1, t = 0, 3, 6, 9;
             solid - num; o - anal')
    end
    if(ncase==2)
      title('Boussinesq equation; ncase = 2, t = 0, 3, 6, 9;
             solid - num; o - anal')
    end
    if(ncase==3)
      title('Boussinesq equation; ncase = 3, t = 0, 3, 6, 9;
             solid - num; o - anal')
    end
    if(ncase==4)
      title('Boussinesq equation; ncase = 4, t = 0, 3, 6, 9;
             solid - num; o - anal')
    end
    figure(3)
    surf(x,t,u1)
    xlabel('x'); ylabel('t'); zlabel('u1(x,t)');
    title('Boussinesq equation');
```

jpattern_num_1.m for the sparse matrix integration of the ODEs is not listed here since it is the same as for Chapters 15–17. The ODE Jacobian map from jpattern_num_1.m indicates two bands for PDE (18.1) as expected, and the map is therefore not included in the discussion of the output that follows.

A portion of the numerical output from pde_1_main.m is listed in Table 18.1a for ncase=1.

We can note the following points about this output:

- The FD weighting coefficients produced by weights.m are displayed first (these can easily be suppressed by a small change in u4x11p.m). Also, these are the coefficients only for the *center point* in the spatial grid, i.e., grid point 6 in the total of 11 points (recall that uxxxx is set to zero in u4x11p.m for $i = 1,2,3,4,5$ and $i = 7,8,9,10,11$ in the basic grid; also, the center coefficients are used throughout the grid with more than 11 points, such as $n = 101$ in the preceding example for ncase=1, except at these 10 end values that are zeroed).
- Another way to visualize the FD weighting coefficients is to note that the coefficients are symmetric about the $n = 6$ value 101.9333, which is to be expected for a FD approximation of a derivative of even order (in this case, fourth order).

Table 18.1a: Selected numerical output from the main program `pde_1_main.m` for `ncase=1`

```
Numerical Derivative Order: 4
=============================
ans=
  Columns 1 through 9
   -0.0434     0.6672    -5.1524    27.7397    -74.1778    101.9333    -74.1778
   27.7397    -5.1524
  Columns 10 through 11
   0.6672    -0.0434
```

t	x	u1(it,i)	u1_anal(it,i)	err(it,i)
0.00	-15.000	0.001645	0.001645	0.000000
0.00	-13.000	0.003917	0.003917	0.000000
0.00	-11.000	0.009277	0.009277	0.000000
0.00	-9.000	0.021685	0.021685	0.000000
0.00	-7.000	0.049162	0.049162	0.000000
0.00	-5.000	0.104066	0.104066	0.000000
0.00	-3.000	0.191019	0.191019	0.000000
0.00	-1.000	0.271880	0.271880	0.000000
0.00	1.000	0.271880	0.271880	0.000000
0.00	3.000	0.191019	0.191019	0.000000
0.00	5.000	0.104066	0.104066	0.000000
0.00	7.000	0.049162	0.049162	0.000000
0.00	9.000	0.021685	0.021685	0.000000
0.00	11.000	0.009277	0.009277	0.000000
0.00	13.000	0.003917	0.003917	0.000000
0.00	15.000	0.001645	0.001645	0.000000
0.00	17.000	0.000689	0.000689	0.000000
0.00	19.000	0.000288	0.000288	0.000000
0.00	21.000	0.000121	0.000121	0.000000
0.00	23.000	0.000050	0.000050	0.000000
0.00	25.000	0.000021	0.000021	0.000000

.
.
.
output for t=3, 6 removed
.
.
.

t	x	u1(it,i)	u1_anal(it,i)	err(it,i)
9.00	-15.000	0.000048	0.000048	0.000000
9.00	-13.000	-0.000249	0.000115	-0.000365
9.00	-11.000	0.000565	0.000276	0.000289
9.00	-9.000	-0.000019	0.000660	-0.000678
9.00	-7.000	0.001035	0.001575	-0.000540

(Continued)

Table 18.1a: *(Continued)*

t	x	ul(it,i)	ul_anal(it,i)	err(it,i)
9.00	−5.000	0.003283	0.003751	−0.000468
9.00	−3.000	0.008324	0.008887	−0.000563
9.00	−1.000	0.022112	0.020794	0.001318
9.00	1.000	0.046565	0.047247	−0.000682
9.00	3.000	0.100279	0.100497	−0.000218
9.00	5.000	0.187166	0.186238	0.000928
9.00	7.000	0.267848	0.269227	−0.001379
9.00	9.000	0.275694	0.274310	0.001384
9.00	11.000	0.194903	0.195798	−0.000895
9.00	13.000	0.108027	0.107725	0.000302
9.00	15.000	0.051283	0.051146	0.000136
9.00	17.000	0.022371	0.022612	−0.000240
9.00	19.000	0.009739	0.009683	0.000056
9.00	21.000	0.004000	0.004090	−0.000090
9.00	23.000	0.001454	0.001718	−0.000264
9.00	25.000	0.000720	0.000720	0.000000
ncall = 963				

- The numerical and analytical solutions agree at $t = 0$ as expected, and the agreement between the two solutions at $t = 9$ is about three figures.
- The computational effort is modest, with `ncall = 963`.

The plotted solutions follow in Figs. 18.1a and 18.2a. Note plots are organized as follows:

Case	Letter designation	2D plot	3D plot
1	a	Figure 18.1a	Figure 18.2a
2	b	Figure 18.1b	Figure 18.2b
3	c	none	none
4	d	Figure 18.1d	Figure 18.2d

Figure 18.1a demonstrates that the solution is a traveling wave in x from eq. (18.2) (starting from $t = 0$) from the argument of $x - ct$ in eq. (18.2).

We can note the following details about Fig. 18.1a:

- The agreement of the numerical and analytical solutions as reflected in Table 18.1a is evident.
- The numerical and analytical solutions agree exactly at the boundaries $x = x_l = -15, x = x_u = 25$, which results from setting the two solutions equal at the boundaries in the main program, `pde_1_main.m`. This is done because although the *Dirichlet* BCs in `pde_1.m` set the boundary values equal to the analytical solution

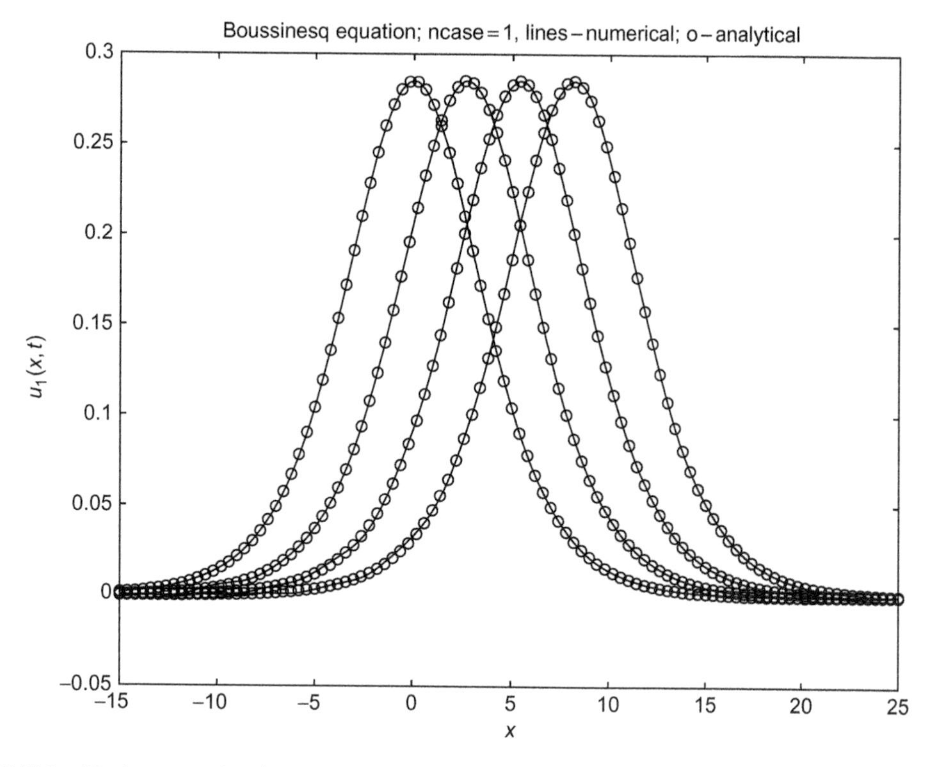

FIGURE 18.1a: 2D plot comparing the numerical and analytical solutions of eq. (18.1) for $t = 0,3,6,9$ (left to right).

(by using `ua_1a.m`), these boundary values are not returned to the main program (from `ode15s`), which is a characteristic of the Matlab ODE integrators. In other words, if a dependent variable is set in the ODE routine (e.g., a boundary value), this value is not returned by `ode15s`; rather an ODE dependent variable can only be computed from its associated ODE programmed in the ODE routine.

- The slope of the solution at the boundaries $x = x_l, x = x_u$ is small (effectively zero) because the nonzero portion of the solution traveling left to right at velocity $c = 0.9$ does not reach the right boundary at $t = 9$ ($ct = (0.9)(9) = 8.1$ units in x while the right boundary is at $x_u = 25$). This is taken as justification for setting `uxxxx=0` in `u4x11p.m` of Listing 18.2 as discussed previously; in other words, the derivatives in x of the solution at $x_u = 25$, including the fourth derivative, remain essentially at zero for $t \leq 9$.

The preceding discussion was for `ncase=1`, that is, eq. (18.1) with $c = 0.9$ in eq. (18.2) and for BCs (18.4), with the MOL calculations in `pde_1a.m` of Listing 18.1a. We now consider the other three cases `ncase=2,3,4`, which basically reflect variations in the MOL coding in the ODE routine to illustrate alternate approaches, not all of which are successful (we report the negative results as well as the positive to provide some experience with approaches that seem reasonable, but do not actually work).

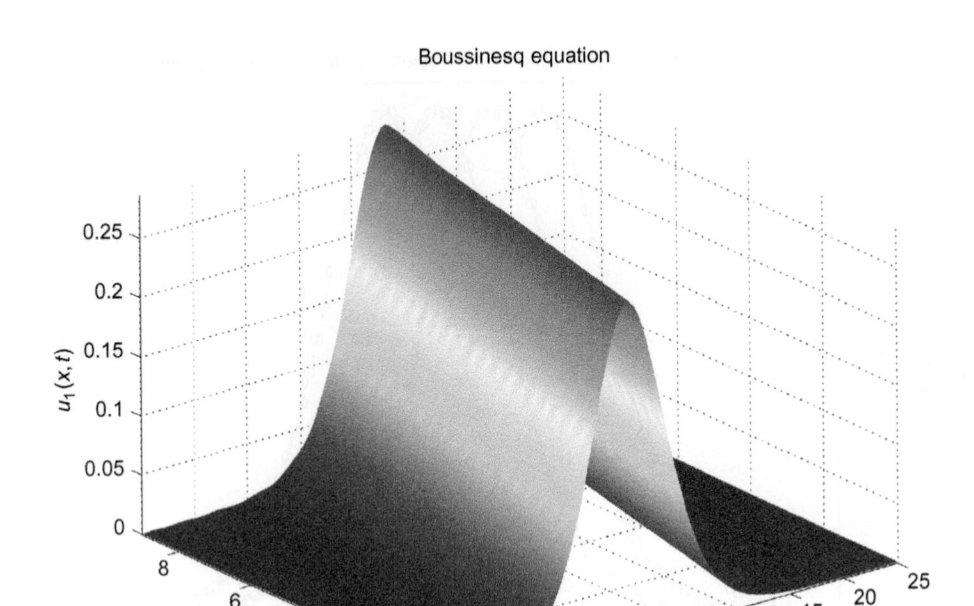

FIGURE 18.2a: 3D plot of the numerical solution of eq. (18.1).

Case 2: Calculation of $\dfrac{\partial^4 u}{\partial x^4}$ via two-stage Differentiation of $\dfrac{\partial^2 u}{\partial x^2}$

We start with the case `ncase=2` in `pde_1b.m`.

```
  function ut=pde_1b(t,u)
%
% Function pde_1b computes the t derivative vector for the Boussinesq
% equation
%
  global xl xu x n ncall
%
% Model parameters
  global c ncase
%
% One vector to two vectors
  for i=1:n
    u1(i)=u(i);
    u2(i)=u(i+n);
  end
%
% BCs at x = -15,25
  u1(1)=0;
  u1(n)=0;
%
% u1xx
```

```
  nl=1; nu=1;
  u1x(1)=0;
  u1xx=dss044(xl,xu,n,u1,u1x,nl,nu);
%
% BCs at x = -15,25 (used by dss044)
  u1xx(1)=0;
  u1xx(n)=0;
%
% u1xxxx
  u1xxx(1)=0;
  u1xxxx=dss044(xl,xu,n,u1xx,u1xxx,nl,nu);
%
% (u1^2)xx
  u1s=u1.^2;
  nl=1; nu=1;
  u1sx(1)=0;
  u1sxx=dss044(xl,xu,n,u1s,u1sx,nl,nu);
%
% PDE
  u1t=u2;
  u2t=u1xx-u1xxxx-u1sxx;
%
% Two vectors to one vector
  for i=1:n
    ut(i)  =u1t(i);
    ut(i+n)=u2t(i);
  end
  ut=ut';
%
% Increment calls to pde_1b
  ncall=ncall+1;
```

LISTING 18.1b: Function pde_1b.m for eq. (18.1) (ncase=2).

We can note the following details about pde_1b.m:

- The initial coding is the same as in pde_1a.m of Listing 18.1a to define the function and some global variables and to place the single dependent variable vector u of length $2n$ in two vectors u1, u2 of length n.

```
  function ut=pde_1b(t,u)
%
% Function pde_1b computes the t derivative vector for the Boussinesq
% equation
%
  global xl xu x n ncall
%
% Model parameters
  global c ncase
%
% One vector to two vectors
  for i=1:n
    u1(i)=u(i);
    u2(i)=u(i+n);
  end
```

- Dirichlet BCs (18.4a) and (18.4b) are set, and the second derivative $\frac{\partial^2 u}{\partial x^2}$ is computed by dss044.

```
%
% BCs at x = -15,25
  u1(1)=0;
  u1(n)=0;
%
% u1xx
  nl=1; nu=1;
  u1x(1)=0;
  u1xx=dss044(xl,xu,n,u1,u1x,nl,nu);
```

- BCs (18.4c) and (18.4d) are set, and the fourth derivative $\frac{\partial^4 u}{\partial x^4}$ is computed by a second application of dss044 to the second derivative, so-called *stagewise differentiaion.*

```
%
% BCs at x = -15,25 (used by dss044)
  u1xx(1)=0;
  u1xx(n)=0;
%
% u1xxxx
  u1xxx(1)=0;
  u1xxxx=dss044(xl,xu,n,u1xx,u1xxx,nl,nu);
```

  Note that BCs (18.4c) and (18.4d) are Dirichlet for the fourth derivative (nl=nu=1 in using dss044 the second time).
- The remainder of pde_1b is the same as pde_1a.

```
%
% PDE
  u1t=u2;
  u2t=u1xx-u1xxxx-u1sxx;
%
% Two vectors to one vector
  for i=1:n
    ut(i)  =u1t(i);
    ut(i+n)=u2t(i);
  end
  ut=ut';
%
% Increment calls to pde_1b
```

The other routines for ncase=2 are the same as for ncase=1, that is, inital_1b.m, ua_1b.m, and uat_1b.m, are the same as inital_1a.m, ua_1a.m, and uat_1a.m, respectively; this arrangement of a complete set of subordinate routines for each case provides for maximum flexibility and clarity. Finally, the coding in the main program pde_1_main.m for ncase=2 directly parallels that for ncase=1 and therefore it will not be discussed.

A portion of the numerical output from pde_1_main.m is listed in Table 18.1b for ncase=2.

Table 18.1b: Selected numerical output from the main program `pde_1_main.m` for `ncase=2`

t	x	u1(it,i)	u1_anal(it,i)	err(it,i)
0.00	−15.000	0.001645	0.001645	0.000000
0.00	−13.000	0.003917	0.003917	0.000000
0.00	−11.000	0.009277	0.009277	0.000000
0.00	−9.000	0.021685	0.021685	0.000000
0.00	−7.000	0.049162	0.049162	0.000000
0.00	−5.000	0.104066	0.104066	0.000000
0.00	−3.000	0.191019	0.191019	0.000000
0.00	−1.000	0.271880	0.271880	0.000000
0.00	1.000	0.271880	0.271880	0.000000
0.00	3.000	0.191019	0.191019	0.000000
0.00	5.000	0.104066	0.104066	0.000000
0.00	7.000	0.049162	0.049162	0.000000
0.00	9.000	0.021685	0.021685	0.000000
0.00	11.000	0.009277	0.009277	0.000000
0.00	13.000	0.003917	0.003917	0.000000
0.00	15.000	0.001645	0.001645	0.000000
0.00	17.000	0.000689	0.000689	0.000000
0.00	19.000	0.000288	0.000288	0.000000
0.00	21.000	0.000121	0.000121	0.000000
0.00	23.000	0.000050	0.000050	0.000000
0.00	25.000	0.000021	0.000021	0.000000

.
.
.
output for t=3, 6 removed
.
.
.

t	x	u1(it,i)	u1_anal(it,i)	err(it,i)
9.00	−15.000	0.000048	0.000048	0.000000
9.00	−13.000	0.000015	0.000115	−0.000101
9.00	−11.000	0.000079	0.000276	−0.000197
9.00	−9.000	0.000298	0.000660	−0.000361
9.00	−7.000	0.001043	0.001575	−0.000532
9.00	−5.000	0.003215	0.003751	−0.000536
9.00	−3.000	0.008681	0.008887	−0.000207
9.00	−1.000	0.021082	0.020794	0.000288
9.00	1.000	0.047473	0.047247	0.000225
9.00	3.000	0.100097	0.100497	−0.000400
9.00	5.000	0.186493	0.186238	0.000255
9.00	7.000	0.269387	0.269227	0.000160
9.00	9.000	0.274227	0.274310	−0.000083
9.00	11.000	0.195772	0.195798	−0.000026

(Continued)

Table 18.1b: *(Continued)*

t	x	u1(it,i)	u1_anal(it,i)	err(it,i)
9.00	13.000	0.107275	0.107725	−0.000450
9.00	15.000	0.051484	0.051146	0.000338
9.00	17.000	0.022524	0.022612	−0.000087
9.00	19.000	0.009621	0.009683	−0.000062
9.00	21.000	0.004207	0.004090	0.000117
9.00	23.000	0.001272	0.001718	−0.000445
9.00	25.000	0.000720	0.000720	0.000000
ncall = 1004				

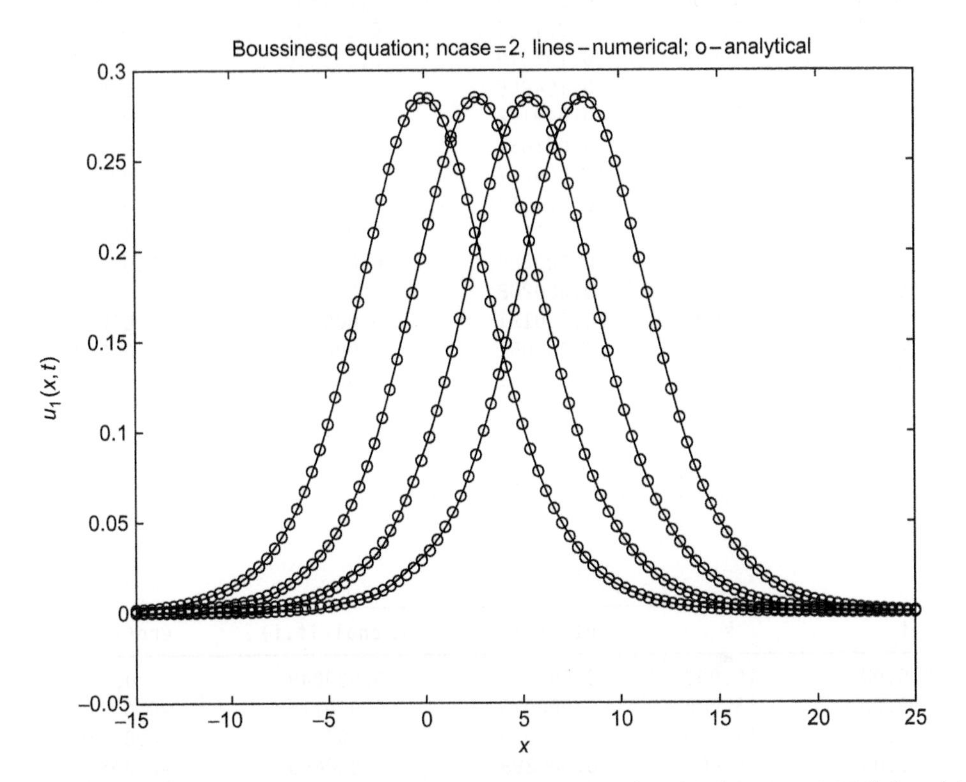

FIGURE 18.1b: 2D plot comparing the numerical and analytical solutions of eq. (18.1) for $t = 0, 3, 6, 9$ (left to right).

We can note the following points about this output:

- The numerical and analytical solutions agree at $t = 0$ as expected, and the agreement between the two solutions at $t = 9$ is about three figures.
- The computational effort is modest, with `ncall = 1004`.

The plotted solutions are included in Figs. 18.1b and 18.2b. Figure 18.1b again demonstrates that the solution is a traveling wave in x from eq. (18.2) (starting from $t = 0$) from the argument of $x - ct$ in eq. (18.2); this plot is very similar to Fig. 18.1a.

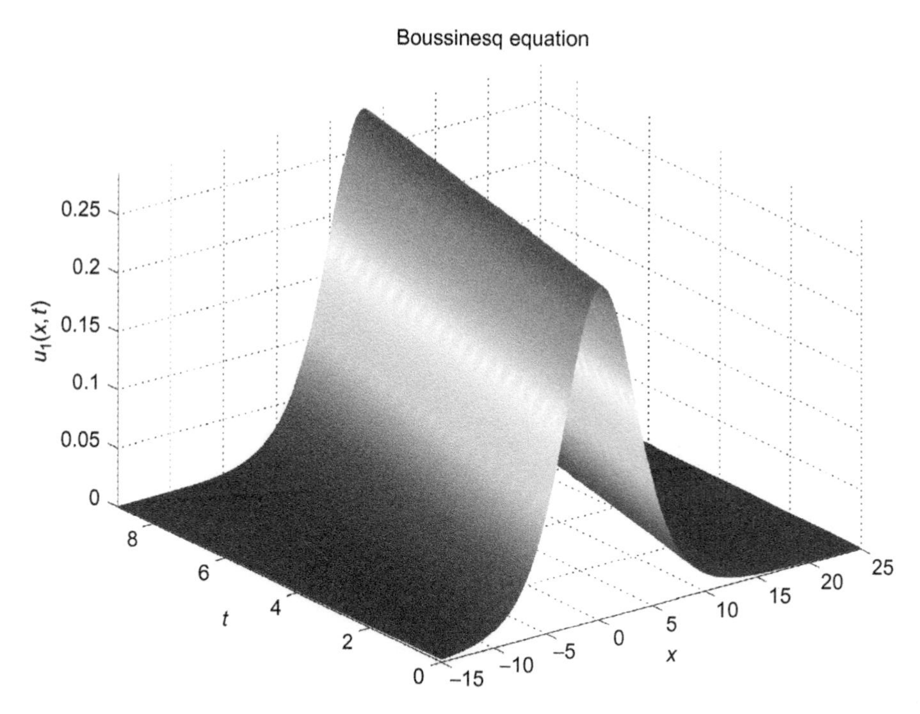

FIGURE 18.2b: 3D plot of the numerical solution of eq. (18.1).

We can note the following details about Fig. 18.1b:

- The agreement of the numerical and analytical solutions as reflected in Table 18.1b is evident.
- The discussion of Fig. 18.1a applies also to Fig. 18.1b. In particular, the slope of the solution at the end points $x = x_l = -15, x = x_u = 25$ is essentially zero.

  This plot is very similar to Fig. 18.2a.

Case 3: Calculation of $\dfrac{\partial^4 u}{\partial x^4}$ via Four-Stage Differentiation of $\dfrac{\partial u}{\partial x}$

As a third approach to a numerical solution (`ncase=3`), we now use four successive (stage-wise) derivative calculations to arrive at the fourth derivative in eq. (18.1) using the first derivative routine `dss004`. All of the preceding routines remain unchanged, except the ODE derivative routine, which is `pde_1c.m` listed below.

```
  function ut=pde_1c(t,u)
%
% Function pde_1c computes the t derivative vector for the Boussinesq
% equation
%
  global xl xu x n ncall
```

```
%
% Model parameters
  global c ncase
%
% One vector to two vectors
  for i=1:n
    u1(i)=u(i);
    u2(i)=u(i+n);
  end
%
% BCs at x = -15,25
  u1(1)=0;
  u1(n)=0;
%
% u1x
  u1x=dss004(xl,xu,n,u1);
%
% u1xx
  u1xx=dss004(xl,xu,n,u1x);
%
% BCs at x = -15,25
  u1xx(1)=0;
  u1xx(n)=0;
%
% u1xxx
  u1xxx=dss004(xl,xu,n,u1xx);
%
% u1xxxx
  u1xxxx=dss004(xl,xu,n,u1xxx);
%
% (u1^2)xx
  u1s=u1.^2;
  nl=1; nu=1;
  u1sx(1)=0;
  u1sxx=dss044(xl,xu,n,u1s,u1sx,nl,nu);
%
% PDE
  u1t=u2;
  u2t=u1xx-u1xxxx-u1sxx;
%
% Two vectors to one vector
  for i=1:n
    ut(i)  =u1t(i);
    ut(i+n)=u2t(i);
  end
  ut=ut';
%
% Increment calls to pde_1c
  ncall=ncall+1;
```

LISTING 18.1c: Function `pde_1c.m` for eq. (18.1) (`ncase=3`).

We can note the following details about pde_1c.m:

- The initial coding is the same as in pde_1a.m of Listing 18.1a to define the function and some global variables and to place the single dependent variable vector u of length $2n$ in two vectors u1, u2 of length n.

```
  function ut=pde_1c(t,u)
%
% Function pde_1c computes the t derivative vector for the Boussinesq
% equation
%
  global xl xu x n ncall
%
% Model parameters
  global c ncase
%
% One vector to two vectors
  for i=1:n
    u1(i)=u(i);
    u2(i)=u(i+n);
  end
```

- Dirichlet BCs (18.4a), (18.4b) are set and the second derivative $\frac{\partial^2 u}{\partial x^2}$ is computed by using dss004 twice.

```
%
% BCs at x = -15,25
  u1(1)=0;
  u1(n)=0;
%
% u1x
  u1x=dss004(xl,xu,n,u1);
%
% u1xx
  u1xx=dss004(xl,xu,n,u1x);
```

- BCs (18.4c) and (18.4d) are set and the fourth derivative $\frac{\partial^4 u}{\partial x^4}$ is computed by two additional applications of dss004 differentiation.

```
%
% BCs at x = -15,25
  u1xx(1)=0;
  u1xx(n)=0;
%
% u1xxx
  u1xxx=dss004(xl,xu,n,u1xx);
%
% u1xxxx
  u1xxxx=dss004(xl,xu,n,u1xxx);
```

Note that BCs (18.4c) and (18.4d) reset the boundary values of the second derivative, `uxx(1)` and `uxx(n)`, before `dss004` is applied a third time to compute the third derivative, `uxxx`.

- The remainder of `pde_1c` is the same as `pde_1a` and `pde_1b.m`.

```
%
% PDE
  u1t=u2;
  u2t=u1xx-u1xxxx-u1sxx;
%
% Two vectors to one vector
  for i=1:n
    ut(i)  =u1t(i);
    ut(i+n)=u2t(i);
  end
  ut=ut';
%
% Increment calls to pde_1c
  ncall=ncall+1;
```

The other routines for `ncase=3` are the same as for `ncase=1,2`; for example, `inital_1c.m`, `ua_1c.m`, and `uat_1c.m`, are the same as `inital_1a.m`, `ua_1a.m`, and `uat_1a.m`, respectively; as mentioned previously, this arrangement of a complete set of subordinate routines for each case provides for maximum flexibility and clarity. Finally, the coding in the main program `pde_1_main.m` for `ncase=3` directly parallels that for `ncase=1,2` and therefore will not be discussed.

When the routines for `ncase=3` (with the designation "c") were executed, the numerical integration in t using `ode15s` failed when t was between 6 and 9 (an error message was reported by `ode15s` that the error tolerances specified in the main program, `pde_1_main.m` could not be satisfied when the integration step t was reduced to the minimum allowable value). When the routines were executed with a final time 6, the numerical solution developed an oscillation (that apparently grew in magnitude to cause an integration failure for $t > 6$).

Various attempts to overcome the integration error, such as reducing the error tolerances for `ode15s`, failed in the same way. Thus we concluded that although the use of stagewise differentiation four times to arrive at the fourth derivative of eq. (18.1) seems logical (at least mathematically), in fact, it failed. This example illustrates some generalizations we have experienced in the analysis of PDE systems:

- Numerical methods (algorithms) that seem logical may in fact not work for reasons that are not apparent. In the present case, we cannot offer an explanation for why the four-stage differentiation failed.
- Some experimentation with various numerical approaches may be required before a successful method is developed. In other words, at least within the MOL context, a successful outcome cannot be assured in advance; rather each new problem must be approached numerically as an experiment that may fail, in which case an alternative approach must be developed. In other words, the MOL is not a mechanical procedure that is guaranteed in advance to produce a numerical solution to a new problem.

On the other hand, our experience has indicated that MOL analysis can be applied successfully to a broad spectrum of PDE problems.

- The use of established quality routines, such as `ode15s`, increases the chances of a successful outcome. This conclusion follows from the careful coding and testing of library routines by experts who anticipate and try to circumvent algorithm limitations and computational difficulties that might not occur to a less experienced analyst.
- PDEs with higher derivatives, such as eq. (18.1) with the fourth-order derivative in x, are increasingly challenging as the order is increased, at least within the MOL framework based on finite differences.

Case 4: Preceding Problem with Second Derivative BCs Replaced by (First Derivative) Neumann BCs

To conclude this discussion, we consider one more case (`ncase=4`) for eq. (18.1) with the homogeneous second derivative BCs of eqs. (18.4c) and (18.4d) replaced with the homogeneous *Neumann* BCs

$$\frac{\partial u(x=x_l,t)}{\partial x} = \frac{\partial u(x=x_u,t)}{\partial x} = 0 \tag{18.4e,4f}$$

The intention in using this modified problem is essentially to demonstrate another application of MOL analysis and, specifically, to demonstrate the ease in programming this modification as an indication of the flexibility of the MOL approach. All of the preceding routines remain unchanged except the ODE derivative routine, which is `pde_1d.m` listed next (only the ODE routine for `ncase=4` is different than the preceding routines for `ncase=1,2,3`).

```
  function ut=pde_1d(t,u)
%
% Function pde_1d computes the t derivative vector for the Boussinesq
% equation
%
  global xl xu x n ncall
%
% Model parameters
  global c ncase
%
% One vector to two vectors
  for i=1:n
    u1(i)=u(i);
    u2(i)=u(i+n);
  end
%
% BCs at x = -15,25
  u1(1)=0;
  u1(n)=0;
%
% u1x
  u1x=dss004(xl,xu,n,u1);
```

```
%
% BCs at x = -15,25
  u1x(1)=0;
  u1x(n)=0;
%
% u1xx
  nl=2;
  nu=2;
  u1xx=dss044(xl,xu,n,u1,u1x,nl,nu);
%
% u1xxxx
  u1xxxx=uxxx7c(xl,xu,n,u1x);
%
% (u1^2)xx
  u1s=u1.^2;
  nl=1; nu=1;
  u1sx(1)=0;
  u1sxx=dss044(xl,xu,n,u1s,u1sx,nl,nu);
%
% PDE
  u1t=u2;
  u2t=u1xx-u1xxxx-u1sxx;
%
% Two vectors to one vector
  for i=1:n
    ut(i)   =u1t(i);
    ut(i+n)=u2t(i);
  end
  ut=ut';
%
% Increment calls to pde_1d
  ncall=ncall+1;
```

LISTING 18.1d: Function `pde_1d.m` for eq. (18.1) (`ncase=4`).

We can note the following details about `pde_1d.m`:

- The initial coding is the same as in `pde_1a.m` of Listing 18.1a to define the function and some global variables and to place the single dependent variable vector `u` of length $2n$ in two vectors `u1,u2` of length n.

```
  function ut=pde_1d(t,u)
%
% Function pde_1d computes the t derivative vector for the Boussinesq
% equation
%
  global xl xu x n ncall
%
% Model parameters
  global c ncase
%
% One vector to two vectors
  for i=1:n
    u1(i)=u(i);
    u2(i)=u(i+n);
  end
```

- Dirichlet BCs (18.4a) and (18.4b) are set, and the first derivative $\frac{\partial u}{\partial x}$ is computed by dss004. *Neumann* BCs (18.4e) and (18.4f) are then set, and the second derivative $\frac{\partial^2 u}{\partial x^2}$ is computed by dss044. For the latter, note the specification of *Neumann* BCs (nl=nu=2 for BCs (18.4e) and (18.4f).

```
%
% BCs at x = -15,25
  u1(1)=0;
  u1(n)=0;
%
% u1x
  u1x=dss004(xl,xu,n,u1);
%
% BCs at x = -15,25
  u1x(1)=0;
  u1x(n)=0;
%
% u1xx
  nl=2;
  nu=2;
  u1xx=dss044(xl,xu,n,u1,u1x,nl,nu);
```

- The fourth derivative in eq. (18.1), $\frac{\partial^4 u}{\partial x^4}$, is then computed by taking the third derivative of $\frac{\partial u}{\partial x}$ with uxxx7c; this routine for a third derivative based on seven-point centered finite differences was used previously in the MOL analysis of the third-order Korteweg-de Vries (KdV) equation [7], and it is listed subsequently.

```
%
% u1xxxx
  u1xxxx=uxxx7c(xl,xu,n,u1x);
```

- The remainder of pde_1d is the same as pde_1a, pde_1b.m, and pde_1c.m.

```
%
% (u1^2)xx
  u1s=u1.^2;
  nl=1; nu=1;
  u1sx(1)=0;
  u1sxx=dss044(xl,xu,n,u1s,u1sx,nl,nu);
%
% PDE
  u1t=u2;
  u2t=u1xx-u1xxxx-u1sxx;
%
% Two vectors to one vector
  for i=1:n
    ut(i)  =u1t(i);
    ut(i+n)=u2t(i);
  end
  ut=ut';
%
```

```
  % Increment calls to pde_1d
    ncall=ncall+1;
```

Function uxxx7c.m is listed below.

```
  function uxxx=uxxx7c(xl,xu,n,u)
%
% Function uxxx7c computes the derivative uxxx
%
% Spatial increment
  dx=(xu-xl)/(n-1);
  r8dx3=1.0/(8.0*(dx^3));
%
% uxxx
  for i=1:n
%
%   At the left end, uxxx = 0
    if(i<4)uxxx(i)=0.0;
%
%   At the right end, uxxx = 0
    elseif(i>(n-3))uxxx(i)=0.0;
%
%   Interior points
    else
    uxxx(i)=r8dx3*...
        (   1.0*u(i-3)...
           -8.0*u(i-2)...
          +13.0*u(i-1)...
           +0.0*u(i  )...
          -13.0*u(i+1)...
           +8.0*u(i+2)...
           -1.0*u(i+3));
    end
  end
```

LISTING 18.5: uxxx7c for the calculation of the third derivative uxxx.

uxxx7c is structured in the same way as u4x11p.m of Listing 18.2. Note in particular that the weighting coefficients are *antisymmetric* (opposite in sign) around the center term +0.0*u(i) since the third derivative is odd order (for even-order derivatives, the weighting coefficient are symmetric).

The other routines for ncase=4 are the same as for ncase=1,2,3; for example, inital_1d.m, ua_1d.m, and uat_1d.m are the same as inital_1a.m, ua_1a.m, and uat_1a.m, respectively; as mentioned previously, this arrangement of a complete set of subordinate routines for each case provides for maximum flexibility and clarity. Finally, the coding in the main program pde_1_main.m for ncase=4 directly parallels that for ncase=1,2,3 and therefore will not be discussed here.

Execution of the routines for ncase=4 (with the designation "d") produced the following numerical and plotted output in Table 18.1d, and Figs. 18.1d and 18.2d.

Table 18.1d: Selected numerical output from the main program `pde_1_main.m` for `ncase=4`

t	x	u1(it,i)	u1_anal(it,i)	err(it,i)
0.00	−15.000	0.001645	0.001645	0.000000
0.00	−13.000	0.003917	0.003917	0.000000
0.00	−11.000	0.009277	0.009277	0.000000
0.00	−9.000	0.021685	0.021685	0.000000
0.00	−7.000	0.049162	0.049162	0.000000
0.00	−5.000	0.104066	0.104066	0.000000
0.00	−3.000	0.191019	0.191019	0.000000
0.00	−1.000	0.271880	0.271880	0.000000
0.00	1.000	0.271880	0.271880	0.000000
0.00	3.000	0.191019	0.191019	0.000000
0.00	5.000	0.104066	0.104066	0.000000
0.00	7.000	0.049162	0.049162	0.000000
0.00	9.000	0.021685	0.021685	0.000000
0.00	11.000	0.009277	0.009277	0.000000
0.00	13.000	0.003917	0.003917	0.000000
0.00	15.000	0.001645	0.001645	0.000000
0.00	17.000	0.000689	0.000689	0.000000
0.00	19.000	0.000288	0.000288	0.000000
0.00	21.000	0.000121	0.000121	0.000000
0.00	23.000	0.000050	0.000050	0.000000
0.00	25.000	0.000021	0.000021	0.000000

.
.
.
output for t=3, 6 removed
.
.
.

t	x	u1(it,i)	u1_anal(it,i)	err(it,i)
9.00	−15.000	0.000048	0.000048	0.000000
9.00	−13.000	0.000161	0.000115	0.000046
9.00	−11.000	0.000069	0.000276	−0.000207
9.00	−9.000	0.000718	0.000660	0.000058
9.00	−7.000	0.000915	0.001575	−0.000660
9.00	−5.000	0.003697	0.003751	−0.000053
9.00	−3.000	0.008551	0.008887	−0.000336
9.00	−1.000	0.021796	0.020794	0.001002
9.00	1.000	0.046933	0.047247	−0.000314
9.00	3.000	0.100035	0.100497	−0.000462
9.00	5.000	0.186944	0.186238	0.000706
9.00	7.000	0.268486	0.269227	−0.000742
9.00	9.000	0.275063	0.274310	0.000753
9.00	11.000	0.195041	0.195798	−0.000756

(Continued)

Table 18.1d: *(Continued)*

t	x	u1(it,i)	u1_anal(it,i)	err(it,i)
9.00	13.000	0.108345	0.107725	0.000620
9.00	15.000	0.050767	0.051146	-0.000379
9.00	17.000	0.022820	0.022612	0.000209
9.00	19.000	0.009687	0.009683	0.000004
9.00	21.000	0.004091	0.004090	0.000001
9.00	23.000	0.001762	0.001718	0.000044
9.00	25.000	0.000720	0.000720	0.000000
ncall=554				

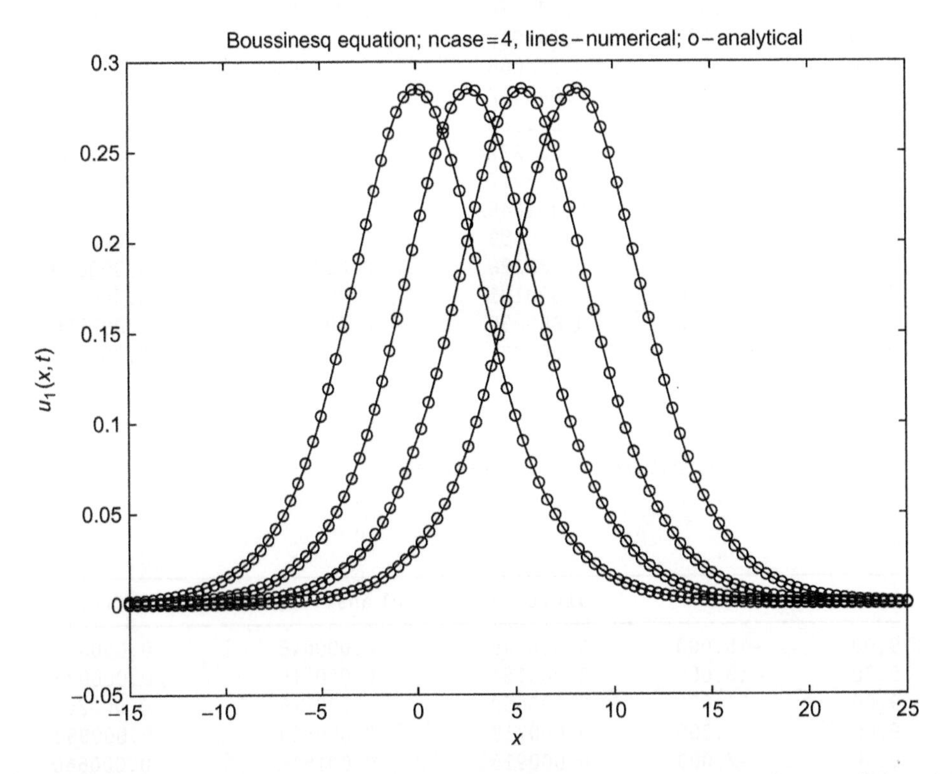

FIGURE 18.1d: 2D plot comparing the numerical and analytical solutions of eq. (18.1) for $t = 0,3,6,9$ (left to right).

We can note the following points about this output:

- The numerical and analytical solutions agree at $t = 0$ as expected, and the agreement between the two solutions at $t = 9$ is about four figures.
- The computational effort is modest, with `ncall = 554`.

The plotted solutions are included in Figs. 18.1d and 18.2d. Figure 18.1d again demonstrates that the solution is a traveling wave in x from eq. (18.2) (starting from $t = 0$) from the argument of $x - ct$ in eq. (18.2); this plot is very similar to Figs. 18.1a and 18.1b.

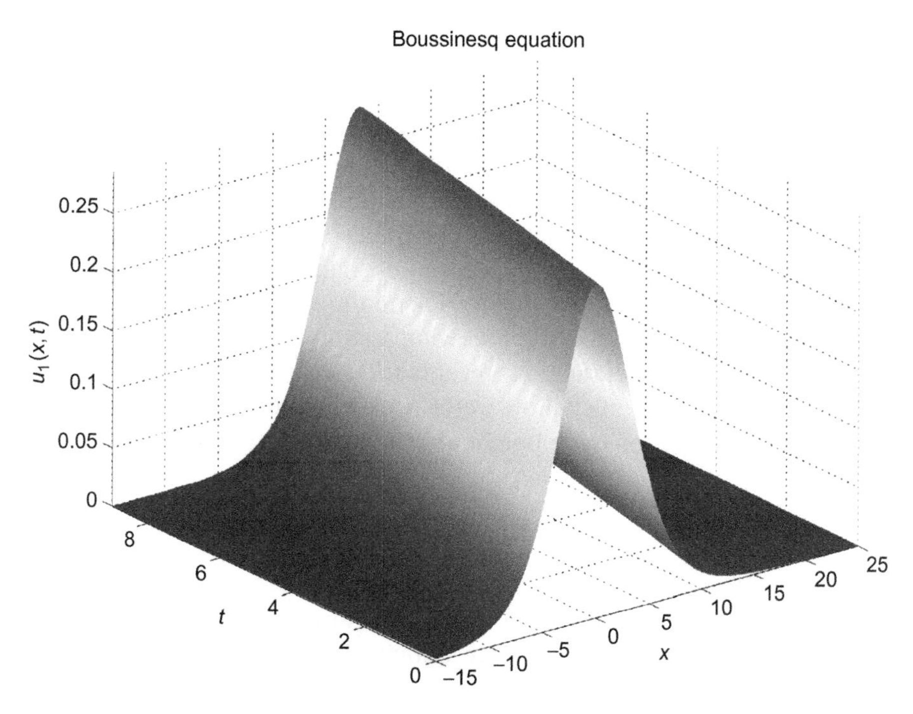

FIGURE 18.2d: 3D plot of the numerical solution of eq. (18.1).

We can note the following details about Fig. 18.1d:

- The agreement of the numerical and analytical solutions as reflected in Table 18.1d is evident.
- The discussion of Figs. 18.1a and 18.1b applies also to Fig. 18.1d. In particular, the slope of the solution at the end points $x = x_l = -15, x = x_u = 25$ is essentially zero. This plot is very similar to Figs. 18.2a and 18.2b.

This last case `ncase=4` illustrates an alternative MOL solution of eq. (18.1) based on BCs (18.4e) and (18.4f) rather than (18.4c) and (18.4d), and more generally it demonstrates the flexibility of the numerical approach when a set of library routines is used (e.g., `ods15s`, `dss004`, `dss044`, `u4x11p.m uxxx7c.m`) for both linear and nonlinear PDEs.

In summary, we have again discussed the numerical solution of a PDE second order in t (eq. (18.1)) by restating the PDE as two PDEs first order in t (for dependent variables `u1` and `u2`). As noted in Chapters 15–17, the reformulation of an *nth* order PDE as n PDEs first order in an initial value independent variable (t) is straightforward and permits the use of an integrator such as `ode15s` for first-order ODEs.

We have also considered four alternative approaches to a higher–order derivative, in this case, the fourth-order derivative of eq. (18.1). We observed in `ncase=3` that an MOL formulation that seems logical (at least mathematically) may, in fact, not execute and therefore some experimentation with alternate formulations may be required to arrive at a workable MOL code.

Appendix

We conclude this chapter by first providing some background to the *Boussinesq equation* based on de Jager's essay *On the Origin of the Korteweg–de Vries Equation* [5] and then deriving traveling wave solutions using *direct integration* and the *Riccati-based* method.

One of the many equations used to describe the propagation of surface long waves on water is the Boussinesq equation, which was first published by J. Boussinesq [1] in a series of papers [2–4] during the years 1871–1872. It assumes shallow water conditions and is generally considered to be the first model to describe nonlinear, dispersive wave propagation. Shallow water models usually make the following assumptions $h/H \ll 1$ and $h\lambda^2/H^3 \ll 1$ where h represents *wave amplitude*, H represents *still water depth*, and λ represents *wavelength*. The standard 1D form of shallow water wave equation is derived by truncating a Taylor series approximation of wave height after the first derivative to give the *classical wave equation*

$$\frac{\partial^2 h}{\partial t^2} = gH\frac{\partial^2 h}{\partial x^2}$$

where x, t, and g represent distance, time, and gravitational acceleration, respectively. The term $gH = c^2$ represents the square of wave velocity. In order to obtain a more accurate model for c, Boussinesq retained the second-order term in the Taylor expansion to obtain his well-known equation

$$\frac{\partial^2 h}{\partial t^2} = gH\frac{\partial^2 h}{\partial x^2} + gH\frac{\partial^2}{\partial x^2}\left(\frac{3h^2}{2H} + \frac{H^2}{3}\frac{\partial^2 h}{\partial x^2}\right) \tag{18.5}$$

One of the advantages over the related Korteweg–de Vries equation for representing waves is that the Boussinesq equation admits solutions with waves traveling in opposite directions.

After assuming a traveling wave solution with $h(x,t) = h(\xi)$ and $\xi = x - ct$, followed by some analysis, we arrive at the Boussinesq improved expression for wave velocity

$$c = \sqrt{gH} + \sqrt{gH}\left(\frac{3}{4}h + \frac{1}{6}\frac{H^3}{h}\frac{\partial^2 h}{\partial \xi^2}\right)$$

Now, Boussinesq was also interested in solitary wave solutions and observed that for a wave shape to remain constant over time, the term $\left(\frac{3}{4}h + \frac{1}{6}\frac{H^3}{h}\frac{\partial^2 h}{\partial \xi^2}\right)$ in the above

[1] *Joseph Valentin Boussinesq* was born in 1842 in Southern France and died in 1929 in Paris. In 1873 he was appointed as Professor of Differential and Integral Calculus at the Faculty of Science in Lille. He taught there until 1886 when he was appointed as Professor of Physical and Experimental Mechanics at the Sorbonne. He remained in this post for 10 years and then held the prestigious chair in Mathematical Physics and Theory of Probabilities at the same institution until he retired in 1918.

equation must be constant. Therefore, we have

$$c=\sqrt{gH}+\frac{1}{2}\sqrt{gH}h_1$$

where the constant factor of a half has been taken outside the brackets, and $h_1=\left(\frac{3}{2}h+\frac{1}{3}\frac{H^3}{h}\frac{\partial^2 h}{\partial \xi^2}\right)$ is an unknown constant to be determined. We also have

$$\frac{\partial^2 h}{\partial \xi^2}=\frac{3h}{2H^3}(2h_1-3h)$$

which on multiplying by $\frac{\partial h}{\partial \xi}$ and integrating leads to the solution

$$h=h_1\text{sech}\left(\sqrt{\frac{3h_1}{4H^3}}\xi\right)^2$$

The variable h_1 represent the wave amplitude, and further analysis leads to the relationship $h_1=\left(\frac{c^2-gH}{g}\right)$. Also, from the above discussion, we have $\xi=x-ct$; therefore, the final solution becomes

$$h=\left(\frac{c^2-gH}{g}\right)\text{sech}\left[\sqrt{\frac{3}{4H^3}\left(\frac{c^2-gH}{g}\right)}(x-ct)\right]^2 \tag{18.6}$$

The Maple code given in Listing 18.6 demonstrates that eq. (18.6) is actually a solution to the original Boussinesq eq. (18.5).

```
># Test of solution to original Boussinesq equation
 restart; with(PDEtools):
>alias(h=h(x,t)):
># Define the Boussinesq PDE
 pde1:=diff(h,t,t)-g*H*diff(h,x,x)-g*H*diff((3*h^2/(2*H)+(H^2/3)
              *diff(h,x,x)),x,x)=0;
># Define sech solution
 Sol:=h=(c^2/g-H)*sech(sqrt((c^2/g-H)*3/(4*H^3))*(x-c*t))^2;
># Use PDE test function to confirm solution is correct
 testSol:=pdetest(Sol,pde1); # = 0 if correct!
 if testSol<> 0 then
   print("Solution: does not pass pdetest() !");
 else
   print("Solution passes pdetest()");
 end if;
```

LISTING 18.6: Maple code used to verify analytical solutions to the original Boussinesq equation.

Analytical Solution Using the Direct Integration Method

The Boussinesq equation in its canonical form is written as

$$\frac{\partial^2}{\partial t^2}u(x,t)-\frac{\partial^2}{\partial x^2}u(x,t)+\frac{\partial^2}{\partial x^2}u(x,t)^2+\frac{\partial^4}{\partial x^4}u(x,t)=0 \tag{18.7}$$

We seek a closed-form *single soliton* solution to eq. (18.7) using *direct integration* as follows. Assume a traveling wave solution of the form $u(x,t)=f(\xi)$, where $\xi=k(x-ct)$, c represents *wave speed*, and k represents *wavenumber*. Then on substituting $f(\xi)$ into eq. (18.7), the PDE is transformed into the following ODE.

$$k^2\left(c^2-1\right)\frac{d^2f(\xi)}{d\xi^2}+2k^2\left(\frac{df(\xi)}{d\xi}\right)^2+2k^2f(\xi)\frac{d^2f(\xi)}{d\xi^2}+k^4\frac{d^4f(\xi)}{d\xi^4}=0$$

Now we integrate again with respect to ξ to obtain

$$3k^4\left(\frac{df(\xi)}{d\xi}\right)^2+3k^2\left(c^2-1\right)f^2(\xi)+2k^2f(\xi)^3=0$$

This leads to the following integral

$$\int d\xi=\int\frac{3k}{f(\xi)\sqrt{9-9c^2-6f(\xi)}}df(\xi)$$

which we integrate to obtain

$$\xi=\frac{-2k}{\sqrt{c^2-1}}\arctan\left(\sqrt{\frac{\left(\frac{2}{3}\right)f(\xi)}{(1-c^2)}-1}\right)$$

$$\downarrow$$

$$f(\xi)=\frac{3}{2}\left(1-c^2\right)\left[1+\tan^2\left(\frac{1}{2k}\sqrt{c^2-1}\xi\right)\right]$$

Recall that $\tan^2(X)=-\tanh^2(iX)$ and $\sec^2(X)=\text{sech}^2(iX)$. Therefore, on back substituting $f(\xi)=u(x,t)$ and $\xi=k(x-ct)$, we obtain the required traveling wave solutions

$$u(x,t)=\frac{3}{2}\left(1-c^2\right)\text{sech}^2\left(\frac{1}{2}\sqrt{1-c^2}(x-ct)+x_0\right),\quad c<1 \tag{18.8}$$

$$u(x,t)=\frac{3}{2}\left(1-c^2\right)\sec^2\left(\frac{1}{2}\sqrt{c^2-1}(x-ct)+x_0\right),\quad c>1 \tag{18.9}$$

where x_0 is an arbitrary constant.

The derivation of eqs. (18.8) and (18.9) can easily be performed in Maple. However, the code is not listed here but is included with the downloads for the book.

Analytical Solution Using the Riccati-Based Method

We now apply the Maple routine `riccatiMethod()` (see Listing 18.7) and obtain a number of traveling wave solutions, one of which is

$$u(x,t) = \frac{1}{2}\left[1-\left(k^2+c^2\right)\right]+\frac{3}{2}k^2\left[1-\tanh^2\left(\frac{k}{2}(x-ct)\right)\right]$$

On letting the wavenumber $k=\sqrt{1-c^2}$, we obtain the same solutions as given by eqs. (18.8) and (18.9) derived using direct integration.

A Maple code that performs the calculations for Figs. 18.3 and 18.4 is given in Listing 18.7.

```
># Boussinesq Equation
 # Attempt at Riccati equation based solution method
 restart; with(PDEtools): with(PolynomialTools):
 with(plots):
>alias(u=u(x,t)):
># Define pde
 pde1:=diff(u,t,t)-diff(u,x,x)+beta*diff(u^2,x,x)+alpha*diff(u,x,x,x,x)=0;
>read("riccatiMethod.txt");
>intFlg:=0: # integration of U(xi) needed!
 M:=2; # Set order of approximation
 infoLevOut:=0;
 riccatiMethod(M,pde1,intFlg,infoLevOut);
># Set constants
 zz:=rhs(soln[3,3]);
 alpha:=1; beta:= 1;
 k:=sqrt(1-c^2);c:=0.9;x0:=0;
># Animate solution
 animate(zz,x=-10..30, t=0..20,
   numpoints=300,frames=50, axes=framed,
   labels=["x","u"],thickness=3,
   title="Boussinesq Equation",
   labelfont=[TIMES, ROMAN, 16],axesfont=[TIMES, ROMAN, 16],
   titlefont=[TIMES, ROMAN, 16]);
># Generate a 3D surface plot
 plot3d(zz,x=-10..30, t=0..20,axes='framed',
   labels=["x","t","u(x,t)"],
   orientation=[-107,40],grid=[100,100],
   style=patchnogrid,shading=Z,
   labeldirections=[HORIZONTAL,HORIZONTAL,VERTICAL],
   title="Boussinesq Equation",
   labelfont=[TIMES, ROMAN, 16],axesfont=[TIMES, ROMAN, 16],
   titlefont=[TIMES, ROMAN, 16]);
```

LISTING 18.7: Maple code used to derive an analytical solution to the Boussinesq equation using the `riccatiMethod()` procedure.

2D and 3D plots of this solution for $c=0.9$ are given in Figs. 18.3 and 18.4. Figure 18.3 is the initial condition (at $t=0$), which then moves left to right when the animation (see Listing 18.7) is activated. Figure 18.4 is a 3D plot that demonstrates the movement of the solution through x and t.

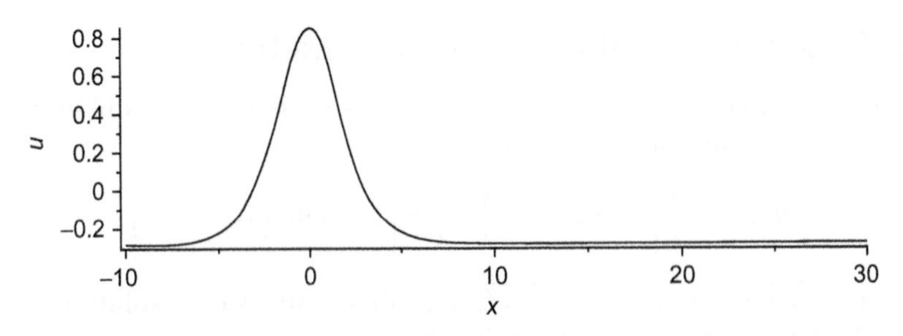

FIGURE 18.3: 2D plot of the solution to Boussinesq equation at $t = 0$.

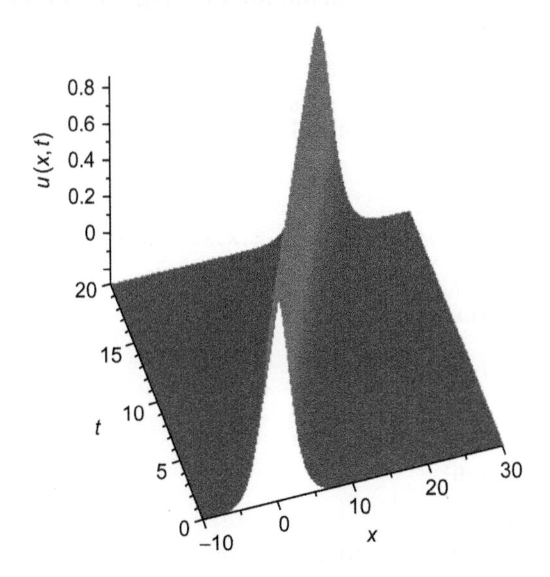

FIGURE 18.4: 3D plot of the solution to Boussinesq equation.

Traveling wave solutions can also be found using the `tanhMethod` and `expMethod` Maple procedures. The codes are not listed here but are included with the downloads for the book.

References

[1] A. Biswas, D. Milovic, A. Ranasinghe, Solitary waves of Boussinesq equation in a power law media, *Commun. Nonlinear Sci. Numer. Simulat.* 14 (2009) 3738–3742.

[2] J. Boussinesq, Théorie de l'intumescence liquide appelée "onde solitaire" ou "de translation", se propageant dans un canal rectangulaire, *C. R. Acad. Sci. Paris*, 72 (1871) 755–759.

[3] J. Boussinesq, Théorie générale des mouvements, qui sont propagés dans un canal rectangulaire horizontal, *C. R. Acad. Sci. Paris*, 73 (1871) 256–260.

[4] J. Boussinesq, Théorie des ondes et des remous qui se propagent le long d'un canal rectangulaire horizontal, en communiquant au liquide continu dans ce canal des vitesses sensiblement pareilles de la surface au fond, *J. Math. Pures Appl.* 17, (1872) 55–108.

[5] E.M. de Jager, On the origin of the Korteweg-de Vries equation, available on-line at `arXiv:math/0602661v1 [math.HO] 28 Feb 2006`.

[6] A.D. Polyanin, V.F. Zaitsev, *Handbook of Nonlinear Partial Differential Equations*, Chapman & Hall/CRC, Boca Raton, FL, 2004.

[7] W.E. Schiesser, Method of Lines Solution of the Korteweg-de Vries Equation, *Comput. Math. Applic.* 28 (10–12) (1994) 147–154.

[8] A. vande Wouwer, Polytechnique Mons, Belgium, private communication, 2002.

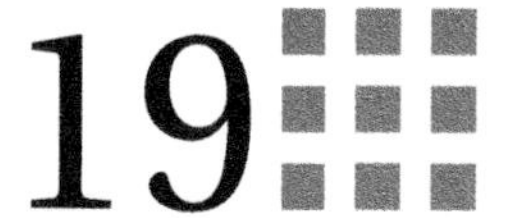

19 Modified Wave Equation

In Chapter 14, we considered the conversion of an ODE to a PDE using the concept of a traveling wave equation. The result was a PDE first order in t. We now consider the same procedure that leads to a PDE second order in t, which we term the *modified wave equation*. We start with the following ODE that can be considered a *Lagrangian* (traveling wave) form of a PDE in an *Eulerian* (fixed) coordinate system.

$$\frac{d^2U}{d\xi^2} + 2U - 2U^3 = 0 \tag{19.1}$$

An analytical solution of eq. (19.1) is [1]

$$U(\xi) = \tanh(\xi) \tag{19.2}$$

which satisfies the BCs

$$U(\xi = -\infty) = -1; \quad U(\xi = \infty) = 1 \tag{19.3a,3b}$$

The significance of BCs (19.3) is discussed subsequently.

To extend this solution to a PDE, we first consider the second derivative $\frac{d^2U}{d\xi^2}$ with $\xi = x - ct$. Therefore, the second derivative $\frac{d^2U}{d\xi^2}$ is (from the addition of the final two equations in Table 19.1)

$$\frac{d^2U}{d\xi^2} = \frac{1}{2}\left(\frac{\partial^2 u}{\partial x^2} + \frac{1}{c^2}\frac{\partial^2 u}{\partial t^2}\right) \tag{19.4}$$

The PDE corresponding to eq. (19.1) is, therefore, from eq. (19.4)

$$\frac{1}{2}\left(\frac{\partial^2 u}{\partial x^2} + \frac{1}{c^2}\frac{\partial^2 u}{\partial t^2}\right) + 2u - 2u^3 = 0 \tag{19.5}$$

We term eq. (19.5) the *modified wave equation* (since it bears a resemblance to the classical, second-order, linear wave equation).

The solution to eq. (19.5) is from eq. (19.2)

$$u(x,t) = \tanh(x - ct) \tag{19.6}$$

Table 19.1: Derivatives for the Lagrangian variable ξ and Eulerian variables x, t

$$\frac{\partial^2 u}{\partial x^2} = \frac{\partial(\partial u/\partial x)}{\partial x} = \frac{\partial[(dU/d\xi)(\partial\xi/\partial x)]}{\partial x} = \frac{d[(dU/d\xi)(\partial\xi/\partial x)]}{d\xi}(\partial\xi/\partial x)$$

$$= \frac{d^2U}{d\xi^2}(\partial\xi/\partial x)^2 = \frac{d^2U}{d\xi^2}(1)^2$$

$$\frac{\partial^2 u}{\partial t^2} = \frac{\partial(\partial u/\partial t)}{\partial t} = \frac{\partial[(dU/d\xi)(\partial\xi/\partial t)]}{\partial t} = \frac{d[(dU/d\xi)(\partial\xi/\partial t)]}{d\xi}(\partial\xi/\partial t)$$

$$= \frac{d^2U}{d\xi^2}(\partial\xi/\partial t)^2 = \frac{d^2U}{d\xi^2}(-c)^2$$

Table 19.2: Verification of eq. (19.6) as the solution to eq. (19.5)

PDE, eq. (19.5)	Solution, eq. (19.6)
(first) $\frac{\partial u}{\partial x}$	$\text{sech}^2(x-ct)(1)$
(second) $\frac{1}{2}\frac{\partial^2 u}{\partial x^2}$	$\text{sech}(x-ct)[-\text{sech}(x-ct)\tanh(x-ct)](1)$
(first) $\frac{\partial u}{\partial t}$	$\text{sech}^2(x-ct)(-c)$
(second) $\frac{1}{2c^2}\frac{\partial^2 u}{\partial t^2}$	$\frac{1}{c^2}\text{sech}(x-ct)[-\text{sech}(x-ct)\tanh(x-ct)](-c)^2$
$+2u$	$+2\tanh(x-ct)$
$-2u^3$	$-2\tanh^3(x-ct)$
	$= -2[1-\text{sech}^2(x-ct)]\tanh(x-ct)$
Sum of terms	Sum of terms
0	0

To verify this solution, we substitute eq. (19.6) into eq. (19.5) (see Table 19.2)

We now consider the MOL solution of eq. (19.5) with the analytical solution (19.6) used to evaluate the numerical solution. The ODE routine, `pde_1.m`, is listed first.

```
  function ut=pde_1(t,u)
%
% Function pde_1 computes the t derivative vector for the modified
% wave equation
%
  global xl xu x n ncall
%
% Model parameters
  global c cs
%
% One vector to two vectors
```

```
  for i=1:n
    u1(i)=u(i);
    u2(i)=u(i+n);
  end
%
% BCs at x = xl,xu
  u1(1)=ua_1(x(1),t);
  u1(n)=ua_1(x(n),t);
%
% u1xx
  nl=1; nu=1;
  u1x(1)=0;
  u1xx=dss044(xl,xu,n,u1,u1x,nl,nu);
%
% PDE
  for i=1:n
    u1t(i)=u2(i);
    u2t(i)=-cs*u1xx(i)-4*cs*(u1(i)-u1(i)^3);
  end
%
% Two vectors to one vector
  for i=1:n
    ut(i)   =u1t(i);
    ut(i+n)=u2t(i);
  end
  ut=ut';
%
% Increment calls to pde_1
  ncall=ncall+1;
```

LISTING 19.1: Function pde_1.m for eq. (19.5).

We can note the following details about pde_1.m:

- The function and some global variables are first defined.

```
  function ut=pde_1(t,u)
%
% Function pde_1 computes the t derivative vector for the modified
% wave equation
%
  global xl xu x n ncall
%
% Model parameters
  global c cs
```

- As in the preceding chapters (i.e., 15–18) for PDEs second order in t, the solution vector u is placed in two arrays u1,u2.

```
%
% One vector to two vectors
  for i=1:n
    u1(i)=u(i);
    u2(i)=u(i+n);
  end
```

u1,u2 are dependent variables first order in t, so a library ODE integrator such as ode45 or ode15s can be used to move the equations through t.

- BCs (19.3a) and (19.3b) are programmed and the second derivative $\frac{\partial^2 u}{\partial x^2}$ is computed with dss044 (with *Dirichlet* BCs designated by nl=1,nu=1). u1x(1)=0 is not actually used in the calculations in dss044 but rather is programmed to meet the Matlab requirement that input arguments to functions must be given a value.

```
%
% BCs at x = xl,xu
  u1(1)=ua_1(x(1),t);
  u1(n)=ua_1(x(n),t);
%
% u1xx
  nl=1; nu=1;
  u1x(1)=0;
  u1xx=dss044(xl,xu,n,u1,u1x,nl,nu);
```

The values of xl=x(1),xu=x(n) are set in IC routine inital_1.m (discussed subsequently) to −8,8 which are effectively $x = \pm\infty$ for the solution of eq. (19.5) (this will be demonstrated in the numerical and plotted solutions); note that array x is a global variable and is therefore available from inital_1.m. The function ua_1.m has the analytical solution, eq. (19.6). Since the boundary values at $x = x_l, x = x_u$ are defined, *Dirichlet* BCs are specified with nl=nu=1.

- Equation (19.5) is programmed as two first-order PDEs in t.

```
%
% PDE
  for i=1:n
    u1t(i)=u2(i);
    u2t(i)=-cs*u1xx(i)-4*cs*(u1(i)-u1(i)^3);
  end
```

- The two derivative vectors u1t,u2t are placed in a single derivative vector ut to be returned from pde_1.m and integrated by ode45 or ode15s in the main program pde_1_main.m; a transpose is required by these integrators (so that ut is a column vector).

```
%
% Two vectors to one vector
  for i=1:n
    ut(i)  =u1t(i);
    ut(i+n)=u2t(i);
  end
  ut=ut';
%
% Increment calls to pde_1
  ncall=ncall+1;
```

The number of calls to pde_1.m is incremented for display at the end of the solution in the main program pde_1_main.m (note that ncall is a global variable).

In summary, pde_1.m receives the dependent variable vector u as an input (along with the independent variable t) and returns the derivative vector ut.

The IC routine inital_1.m is listed next.

```
  function u0=inital_1(t0)
%
% Function inital_1 sets the initial condition for the modified
% wave equation
%
% Parameters shared with other routines
  global xl xu x n ncall
%
% Spatial domain and initial condition
  xl=-8; xu= 8; n=101;
  dx=(xu-xl)/(n-1);
%
% ICs from analytical solution
  for i=1:n
    x(i)=xl+(i-1)*dx;
    u1(i)= ua_1(x(i),t0);
    u2(i)=uat_1(x(i),t0);
    u0(i)  =u1(i);
    u0(i+n)=u2(i);
  end
```

LISTING 19.2: IC function inital_1.m from eq. (19.6) with $t = 0$.

We can note the following details about inital_1a.m:

- The function and some global variables are defined.

```
  function u0=inital_1(t0)
%
% Function inital_1 sets the initial condition for the modified
% wave equation
%
% Parameters shared with other routines
  global xl xu x n ncall
```

- A grid in x of 101 points is defined for $-8 \leq x \leq 8$.

```
%
% Spatial domain and initial condition
  xl=-8;
  xu= 8;
  n=101;
  dx=(xu-xl)/(n-1);
```

- In the for loop, the initial values of u1 for eq. (19.5) are provided by the function ua_1.m (which has the programming of eq. (19.6) with $t = 0$); the initial values of u2 are provided by the function uat_1.m (which has the programming for the derivative of eq. (19.6) with respect to t).

```
%
% ICs from analytical solution
```

```
for i=1:n
  x(i)=xl+(i-1)*dx;
  u1(i)= ua_1(x(i),t0);
  u2(i)=uat_1(x(i),t0);
  u0(i)   =u1(i);
  u0(i+n)=u2(i);
end
```

All $2n$ initial condition values are returned from `inital_1.m` through the vector `u0` to the main program `pde_1_main.m` discussed subsequently.

Function `ua_1.m` is a straightforward implementation of eq. (19.6).

```
  function uanal=ua_1(x,t)
%
% Function uanal computes the exact solution of the modified
% wave equation for comparison with the numerical solution
%
% Model parameters
  global c cs
%
% Analytical solution
  uanal=tanh(x-c*t);
```

LISTING 19.3a: Function `ua_1.m` for the analytical solution of eq. (19.6).

Function `uat_1.m` is a straightforward implementation of the derivative (in t) of eq. (19.6).

```
  function uanal=uat_1(x,t)
%
% Function uanal computes the time derivative of the exact solution
% of the modified wave equation
%
% Model parameters
  global c cs
%
% Analytical solution derivative
  uanal=(-c)*sech(x-c*t)^2;
```

LISTING 19.3b: Function `uat_1.m` for the derivative of the analytical solution of eq. (19.6).

Main program `pde_1_main.m` closely parallels `pde_1_main.m` of Chapter 15 and therefore only a few features are discussed here.

- The problem parameters are set.

```
%
% Model parameters
  global c cs
%
% Model parameters
  c=1; cs=c^2;
```

- The difference in the numerical and analytical solutions is computed as `err(it,i)` and displayed. These solutions are then plotted in 2D by `plot`, and the numerical solution is plotted in 3D by `surf`.

```
%
% Display selected output
  for it=1:nout
    fprintf('\n     t       x       u1(it,i)  u1_anal(it,i)
             err(it,i)\n');
    for i=1:5:n
      fprintf('%6.2f%8.3f%15.6f%15.6f%15.6f\n',...
              t(it),x(i),u1(it,i),u1_anal(it,i),err(it,i));
    end
  end
  fprintf('    ncall = %4d\n\n',ncall);
%
% Plot numerical and analytical solutions
  figure(2)
  plot(x,u1,'-',x,u1_anal,'o')
  axis([-4 4 -1 1]);
  xlabel('x')
  ylabel('u1(x,t)')
  title('Modified wave equation; t = 0, 0.5 1; solid - numerical;
         o - analytical')
  figure(3)
  surf(x,t,u1)
  shading 'interp', axis 'tight'
  xlabel('x'); ylabel('t'); zlabel('u1(x,t)');
  title('Modified wave equation');
```

`jpattern_num_1.m` for the sparse matrix integration of the ODEs is not listed here since it is the same as for Chapters 15–18. The ODE Jacobian map from `jpattern_num_1.m` indicates two bands for PDEs (19.5) as expected, and the map is therefore not included in the discussion of the output that follows.

A portion of the numerical output from `pde_1_main.m` is listed in Table 19.3. We can note the following points about this output:

- The numerical and analytical solutions agree at $t = 0$ as expected, and the agreement between the two solutions at $t = 1$ is about four figures.
- The numerical and analytical solutions agree exactly at the boundaries $x = x_l = -8, x = x_u = 8$, which results from setting the two solutions equal at the boundaries in the main program, `pde_1_main.m`. This is done because although the *Dirichlet* BCs in `pde_1.m` set the boundary values equal to the analytical solution (by using `ua_1.m`), these boundary values are not returned to the main program (from `ode15s`), which is a characteristic of the Matlab ODE integrators. In other words, if a dependent variable is set in the ODE routine (e.g., a boundary value), this value is not returned by `ode15s`; rather, an ODE dependent variable can only be computed from its associated ODE programmed in the ODE routine.
- The computational effort is modest, with `ncall = 267`.

Table 19.3: Selected numerical output from the main program `pde_1_main.m`

t	x	u1(it,i)	u1_anal(it,i)	err(it,i)
0.00	-8.000	-1.000000	-1.000000	0.000000
0.00	-7.200	-0.999999	-0.999999	0.000000
0.00	-6.400	-0.999994	-0.999994	0.000000
0.00	-5.600	-0.999973	-0.999973	0.000000
0.00	-4.800	-0.999865	-0.999865	0.000000
0.00	-4.000	-0.999329	-0.999329	0.000000
0.00	-3.200	-0.996682	-0.996682	0.000000
0.00	-2.400	-0.983675	-0.983675	0.000000
0.00	-1.600	-0.921669	-0.921669	0.000000
0.00	-0.800	-0.664037	-0.664037	0.000000
0.00	0.000	0.000000	0.000000	0.000000
0.00	0.800	0.664037	0.664037	0.000000
0.00	1.600	0.921669	0.921669	0.000000
0.00	2.400	0.983675	0.983675	0.000000
0.00	3.200	0.996682	0.996682	0.000000
0.00	4.000	0.999329	0.999329	0.000000
0.00	4.800	0.999865	0.999865	0.000000
0.00	5.600	0.999973	0.999973	0.000000
0.00	6.400	0.999994	0.999994	0.000000
0.00	7.200	0.999999	0.999999	0.000000
0.00	8.000	1.000000	1.000000	0.000000

```
        .         .
        .         .
        .         .
output for t=0.5 removed
        .         .
        .         .
        .         .
```

t	x	u1(it,i)	u1_anal(it,i)	err(it,i)
1.00	-8.000	-1.000000	-1.000000	0.000000
1.00	-7.200	-1.000000	-1.000000	-0.000000
1.00	-6.400	-0.999999	-0.999999	0.000000
1.00	-5.600	-0.999996	-0.999996	0.000000
1.00	-4.800	-0.999982	-0.999982	0.000000
1.00	-4.000	-0.999909	-0.999909	0.000000
1.00	-3.200	-0.999550	-0.999550	0.000001
1.00	-2.400	-0.997772	-0.997775	0.000003
1.00	-1.600	-0.989019	-0.989027	0.000008
1.00	-0.800	-0.946983	-0.946806	-0.000177
1.00	0.000	-0.760778	-0.761594	0.000816
1.00	0.800	-0.197325	-0.197375	0.000050
1.00	1.600	0.537071	0.537050	0.000022
1.00	2.400	0.885348	0.885352	-0.000004

Table 19.3: (*Continued*)

1.00	3.200	0.975741	0.975743	−0.000002
1.00	4.000	0.995054	0.995055	−0.000000
1.00	4.800	0.998999	0.999000	−0.000000
1.00	5.600	0.999798	0.999798	−0.000000
1.00	6.400	0.999959	0.999959	−0.000000
1.00	7.200	0.999992	0.999992	0.000000
1.00	8.000	0.999998	0.999998	0.000000

ncall = 267

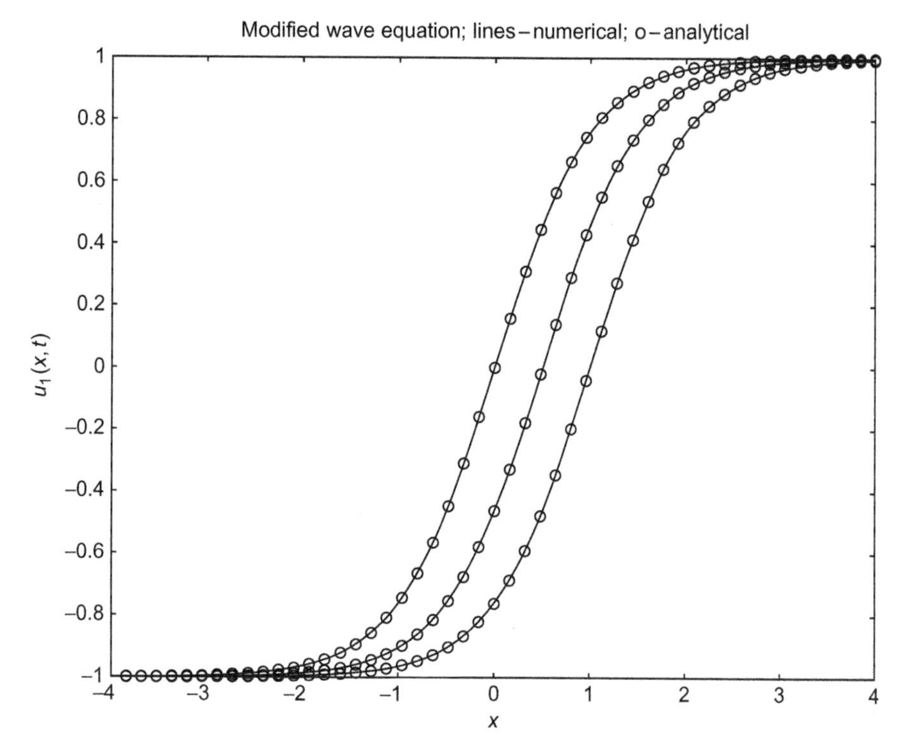

FIGURE 19.1: 2D plot comparing the numerical and analytical solutions of eq. (19.5) for $t = 0, 0.5, 1$ (left to right).

The plotted solutions follow in Figs. 19.1 and 19.2. Figure 19.1 demonstrates that the solution is a traveling wave in x from eq. (19.6) (starting from $t = 0$) from the argument of $x - ct$ in eq. (19.6).

We can note the following details about Fig. 19.1:

- The agreement of the numerical and analytical solutions as reflected in Table 19.1 is evident.
- The slope of the solution at the boundaries $x = x_l, x = x_u$ is small (effectively zero) because the nonzero portion of the solution traveling left to right at velocity $c = 1$ does

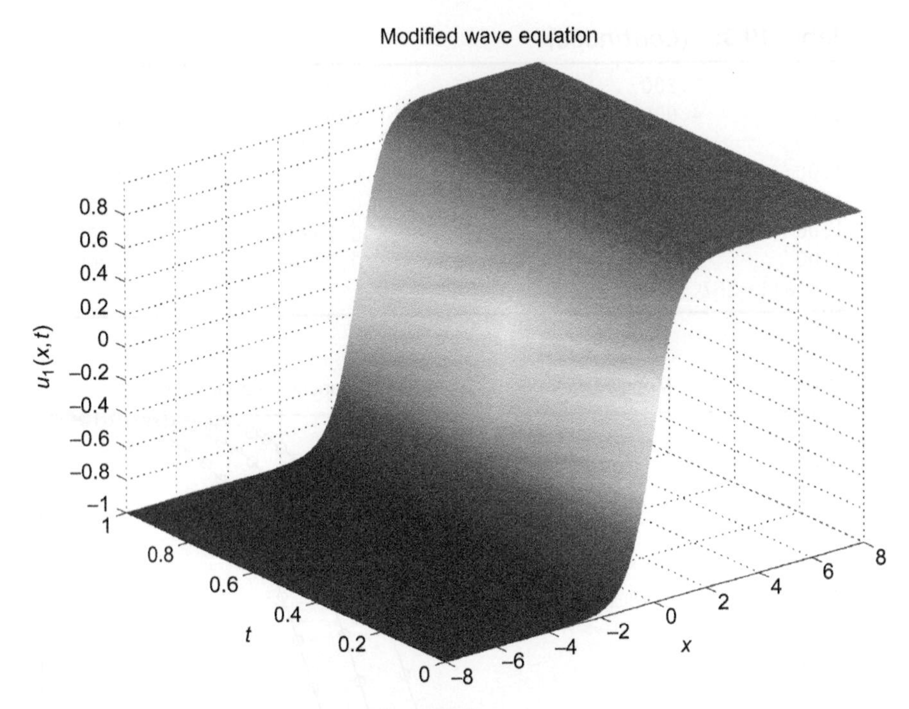

FIGURE 19.2: 3D plot of the numerical solution of eq. (19.5).

not reach the right boundary at $t = 1$ ($ct = (1)(1) = 1$ units in x, whereas the right boundary is at $x_u = 8$). This is taken as justification for assuming that $-8 \leq x \leq 8$ is essentially $-\infty \leq x \leq \infty$ if $t \leq 1$.

We can summarize the procedure for using traveling wave analysis (using $\xi = x - ct$) to produce a PDE from an ODE (with independent variable ξ):

- We can start with a solution to the ODE ($U(\xi)$), which should have the following properties:
  - It has finite limiting values as BCs at $\xi = \pm\infty$ such as the tanh function of eq. (19.2).
  - It can be differentiated so that the solution can be (a) verified as a solution to the ODE as illustrated in Table 19.2 or (b) used to construct an associated ODE (such as eq. (19.1)), possibly including a residual function as illustrated in Chapter 17.
- Once the ODE and its solution have been formulated, a change of variable through $\xi = x - ct$ and substitution for the ODE derivatives gives an associated PDE as was done in the preceding example through the equations of Table 19.1.

The requirement for an assumed solution with finite values for BCs at $\xi = \pm\infty$ provides for the possible use of a variety of functions, such as, for example, $\tanh^m(\xi)$ and $\text{sech}^m(\xi)$. However, functions such as $\sin(\xi)$ and $\cos(\xi)$ are precluded since they are undefined at $\xi = \pm\infty$. By requiring these boundary values, the resulting PDE test problem will be valid for $0 \leq t \leq \infty$ and $-\infty \leq x \leq \infty$ corresponding to $\xi = \pm\infty$; in other words, t and x are unbounded in the use of the PDE and its exact solution. Also, the numerical solution of the PDE can be viewed in the Lagrangian sense (as a function of ξ) or the Eulerian sense (as a function of x and t); the latter case can be used to evaluate numerical methods for PDEs (such as the MOL).

In summary, the traveling wave analysis provides a way for developing PDEs and associated exact solutions that can be used as test problems for PDE numerical methods. However, the PDEs that result from this approach will not necessarily have a physical interpretation (although this possibility is not ruled out, as new PDE applications are developed).

Also, the method can be applied to nonlinear PDEs, such as eq. (19.5), as well as linear PDEs. Once a numerical method has been tested with an exact solution, the PDE can easily be modified to investigate the (numerical) solution of other PDEs (although the exact solution will not apply) by relatively straightforward changes in the MOL ODE routine (such as `pde_1.m` of Listing 19.1). For example, the coding for eq. (19.5) could be extended to include nonlinear terms such as u^m and $f(u)$ by including these terms in the MOL routine. Thus, we have the opportunity for almost unlimited numerical investigations of linear and nonlinear PDEs, and associated numerical methods, through this approach of developing a PDE test problem starting with an ODE.

Appendix

We conclude this chapter by illustrating that a traveling wave solution to the *modified wave* equation, which corresponds to the original solution given in eq. (19.6), can be found using the *tanh*, *exp*, and *Riccati* methods. The Maple code to use the Riccati method to obtain a solution is shown in Listing 19.4 and the major computational steps are as follows:

1. Specify the PDE equation, i.e., eq. (19.5), which was derived from the original ODE eq. (19.1).
2. Set the order of approximation to $M = 1$.
3. Read file `riccatiMethod.txt` which contains Maple procedure `riccatiMethod()`, that will be used to obtain traveling wave solutions.
4. Set integration flag to zero (no integration needed) and call `riccatiMethod()`, which generates $7 \times 6 = 42$ traveling wave solutions: 18 trivial and 24 nontrivial. Of the 24 nontrivial solutions, most are duplicates. All solutions are verified as satisfying eq. (19.5) by application of the Maple function `pdetest()`, which is called within the procedure `riccatiMethod()`.
5. Check each of the derived solutions against the standard traveling wave solution of eq. (19.6). This step utilizes the very useful Maple function `testeq()` to perform the comparison. An exact match is found!
6. Display animation and 3D plot of the solution.

```
># Modified Wave Equation
 # Attempt at Riccati equation based solution method
```

```
restart; with(PDEtools): with(PolynomialTools):
with(plots):
>alias(u=u(x,t)):
>pde1:=(diff(u,t,t)/c^2+diff(u,x,x))/2+2*u-2*u^3=0;
>read("riccatiMethod.txt");
>intFlg:=0: # integration of U(xi) not needed!
 M:=1:# Set order of approximation
 infoLevOut:=0;
 riccatiMethod(M,pde1,intFlg,infoLevOut);
># Check for standard solution match
 u1:=tanh(x+x0-c*t): testFlag:=0:
 for i from 1 to N do
   for j from 1 to 6 do
     u2:=rhs(soln[i,j]);
     del:= testeq(u1=u2):
     if del=true then
       ii:=i; jj:=j;
       testFlg:=1;
       print("soln:",ii,jj," ... MATCH FOUND!");
       break;
     end if;
   end do;
     if testFlg=1 then
       break;
     end if;
 end do;
 if testFlg=0 then
   print("NO MATCH FOUND!");
 end if;
>#Set solution for display
 if testFlg=1 then
   zz:=rhs(soln[ii,jj]);
 else
   zz:=u1;
 end if;
 c:=1; x0:=0;
># Animate solution
 animate(zz,x=-5..25, t=0..20,
   numpoints=300,frames=50, axes=framed,
   labels=["x","u"],thickness=3,
   title="Modified Wave Equation",
   labelfont=[TIMES, ROMAN, 16],axesfont=[TIMES, ROMAN, 16],
   titlefont=[TIMES, ROMAN, 16]);
># Generate a 3D surface plot
 plot3d(zz,x=-5..25, t=0..20,axes='framed',
   labels=["x","t","u(x,t)"],
   orientation=[122,42],grid=[100,100],
   style=patchnogrid,shading=Z,
   labeldirections=[HORIZONTAL,HORIZONTAL,VERTICAL],
   title="Modified Wave Equation",
   labelfont=[TIMES, ROMAN, 16],axesfont=[TIMES, ROMAN, 16],
   titlefont=[TIMES, ROMAN, 16]);
```

LISTING 19.4: Maple code to derive traveling wave solutions to the *modified wave equation* by implementation of the *Riccati method*.

It is equally straightforward to find traveling wave solutions by application of either the Maple procedure `tanhMethod()` or `expMethod()` described in the main Appendix. These two methods each find solutions that match the original solution eq. (19.6). The tanh method finds the exact solution

$$u = \tanh(x - ct) \tag{19.7}$$

and the exp method finds the solution

$$u = -\frac{e^{-x+ct} - e^{x-ct}}{e^{-x+ct} + e^{x-ct}} \tag{19.8}$$

which, of course, transforms exactly to the original solution, eq. (19.6).

In order to save space, listings of the Maple code implementations of the tanh and exp methods will not be included here, but they are available in the downloadable software.

Reference

[1] C.R. Wylie, L.C. Barrett, *Advanced Engineering Mathematics*, Sixth edition, McGraw-Hill, New York, 1995, p. 46.

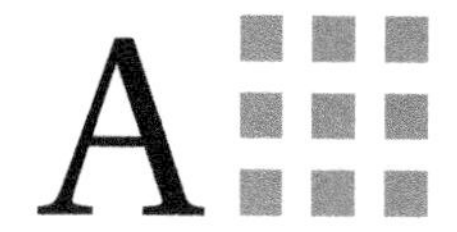

Analytical Solution Methods for Traveling Wave Problems

A.1 Introduction

In this appendix, we discuss some analytical methods for deriving traveling wave solutions to PDEs. These include three fairly new developments, namely *tanh-*, *exp-*, and *Riccati-*based methods. In addition, we outline the *factorization* method and illustrate how solutions may be obtained by direct integration. All the methods are intuitively easy to understand and straightforward to apply. They all benefit greatly from being implemented in a *computer algebra system* (CAS) such as Maple.

The methods are first described in detail and then illustrated by an example. In addition, the tanh-, exp-, and Riccati-based sections also include a computer implementation based on procedures programmed for Maple. Although these procedures solve many problems and demonstrate the way the methods work, they should be viewed as educational/research tools. They will not be able to solve all problems.

We mention that the tanh-, exp-, and Riccati-based procedures are not intended to rival the Maple built-in procedure *TWSolutions*—see Section A.9.

A.2 Tanh Method

We will first discuss the method applied to a problem defined in terms of a single equation having one spatial dimension x, plus the time dimension t. Subsequently, it will be shown that the arguments extend naturally to coupled equations and also to problems defined in terms of two or more spatial dimensions, plus time.

Consider the following evolutionary equation for which we wish to find traveling wave solutions

$$\psi\left(u, u_t, u_x, u_{tt}, u_{xx}, u_{tx}, \ldots\right) = 0 \tag{A.1}$$

The tanh method is simple to implement. We start by introducing the transformation

$$u(x,t) \rightarrow U(\xi), \quad \xi = k(x - ct) \tag{A.2}$$

where U has yet to be defined and, typically, $k > 0$ represents *wavenumber* and c represents *wave velocity*. Partial differentiation operations with respect to t and x are, therefore, transformed to the following equivalent ordinary differentiation operations

$$\frac{\partial}{\partial t} \to -kc\frac{d}{d\xi} \tag{A.3a}$$

$$\frac{\partial}{\partial x} \to k\frac{d}{d\xi} \tag{A.3b}$$

Thus, partial differential equation (A.1) is transformed to the following ordinary differential equation

$$\Psi\left(U, -kc\frac{dU}{d\xi}, k\frac{dU}{d\xi}, k^2c^2\frac{d^2U}{d\xi^2}, k^2\frac{d^2U}{d\xi^2}, -k^2c^3\frac{d^3U}{d\xi^3}, +k^3\frac{d^3U}{d\xi^3}\cdots\right) = 0$$

or in *canonical form*

$$\Psi\left(U, \frac{dU}{d\xi}, \frac{d^2U}{d\xi^2}, \frac{d^3U}{d\xi^3}\cdots\right) = 0 \tag{A.4}$$

where c and k are subsumed into Ψ.

We now apply *Malfliet's tanh method* [13] and introduce the new function

$$Y = \tanh(\xi) \tag{A.5}$$

from which we note that

$$\frac{dY}{d\xi} = \text{sech}^2(\xi) = 1 - \tanh^2(\xi) = 1 - Y^2 \tag{A.6}$$

It therefore follows from the chain rule for differentiation that, for a given function $F(Y)$, we have

$$\frac{dF(Y)}{d\xi} = \frac{dF(Y)}{dY}\frac{dY}{d\xi} = \frac{dF(Y)}{dY}\left(1 - Y^2\right)$$

Thus, by repeating this process, we are able to define a set of *differential operators* L, L^2, $L^3, \ldots$, etc. with respect to ξ, as follows

$$\begin{aligned}
\frac{d}{d\xi} &= L = \left(1 - Y^2\right)\frac{d}{dY} \\
\frac{d^2}{d\xi^2} &= \frac{d}{d\xi}\left(\frac{d}{d\xi}\right) = L^2 \\
&= \left(1 - Y^2\right)\frac{d}{dY}\left[\left(1 - Y^2\right)\frac{d}{dY}\right] \\
&= -2\left(1 - Y^2\right)Y\frac{d}{dY} + \left(1 - Y^2\right)^2\frac{d^2}{dY^2}
\end{aligned} \tag{A.7}$$

$$\frac{d^3}{d\xi^3} = \frac{d}{d\xi}\left(\frac{d^2}{d\xi^2}\right) = L^3$$

$$= \left(1 - Y^2\right)\frac{d}{dY}\left[\left(1 - Y^2\right)\frac{d}{dY}\left(\left(1 - Y^2\right)\frac{d}{dY}\right)\right]$$

$$= \left(1 - Y^2\right)\left(6Y^2 - 2\right)\frac{d}{dY} - 6\left(1 - Y^2\right)^2 Y\frac{d^2}{dY^2} + \left(1 - Y^2\right)^3\frac{d^3}{dY^3}$$

etc.

It is readily seen that there is a simple discernible pattern for higher-order differentials. Thus we have found a set of algebraic functions to represent various orders of derivatives. Consequently, we are now able to convert the ODE of equation (A.4) to an equivalent algebraic equation, as follows.

We start by setting $U = F(Y)$ and introduce the following ansatz (informed guess)

$$F(Y) = \sum_{i=0}^{M} a_i Y^i \tag{A.8}$$

where coefficients a_i are real constants to be determined, and M is a positive integer found by balancing (equating) the largest exponent of Y in the highest order linear term with the largest exponent of Y in the highest order nonlinear term. The idea of balancing is explained in more detail in the following example. Malfleit originally specified the tanh method summation from 0 to M, as above. However the *extended tanh method* [28, 29] defines the summation from $-M$ to M.

Substituting eqns. (A.7) and (A.8) into the ordinary differential eq. (A.4) gives a polynomial in Y for which we have to solve for the unknown parameters. The associated coefficient for each power of Y in the polynomial is represented by an expression consisting of parameters a_i, k, and c. Now, for the problem to be consistent for all values of Y, the coefficient expressions must each equate to zero. We therefore use the method of *Lagrange (undetermined) multipliers* to solve the resulting set of equations for the required unknowns a_i, k, and c. The final solution then becomes the polynomial, whose coefficients are now known, where $\tanh\left[k(x - ct)\right]$ has been substituted for Y.

For anything other than trivial or simple problems, the process of solving for the unknown coefficients is a tedious process and usually requires the use of a computer algebra system if the solution is to be found in a reasonable time. In fact, for many problems, it is totally impractical to attempt a manual solution. For additional information, the reader is referred to [13–15].

A.2.1 Example - KdV Equation

We will apply the above solution method to the canonical form of the KdV equation, i.e.,

$$\frac{\partial u}{\partial t} + 6u\frac{\partial u}{\partial x} + \frac{\partial^3 u}{\partial x^3} = 0 \tag{A.9}$$

for which we have the well-known traveling wave single soliton solution,

$$u = 2k^2 \operatorname{sech}\left[k\left(x - ct\right)\right], \quad c = 4k^2 \tag{A.10}$$

This is a problem with historical significance because it admits *solitary wave*[1] solutions. Solitary waves first came to the attention of the scientific world when, in 1844, John Scott-Russell reported observing their physical occurrence on the Union Canal near Edinburgh [24]. Then in 1895, Korteweg and de Vries published a paper that explained these rare events and detailed the first mathematical equation to exhibit solitary wave behavior [10]. This equation now bears their name. Research into this phenomenon waned until 1965 when it was proven that the KdV equation also admitted *soliton* solutions. A soliton is a solitary wave with the additional property that other solitons can pass through it without changing its shape. The only difference after they emerge from a collision is that each exhibits a small phase shift (displacement or translation in time). This breakthrough was reported in the seminal paper by Kruskal and Zabusky [31], which detailed the results of their numerical solutions and heralded in an era of intense research into this interesting phenomenon. It was also in this paper that the term *soliton* was first coined. A very readable overview relating to the subject of solitons is available at *Scholarpedia,* the online encyclopedia [19].

The following five steps summarize the preceding method when applied to solve the KdV equation:

1. *Apply the traveling wave transformation.*
Apply the transformation of eq. (A.2) to eq. (A.9) to obtain

$$-kc\frac{dU\left(\xi\right)}{d\xi} + 6kU\left(\xi\right)\frac{dU\left(\xi\right)}{d\xi} + k^3\frac{dU^3\left(\xi\right)}{d\xi^3} = 0 \tag{A.11}$$

which can be integrated once directly with respect to $U\left(\xi\right)$ to give

$$-kcU\left(\xi\right) + 3kU\left(\xi\right)^2 + k^3\frac{dU^2\left(\xi\right)}{d\xi^2} = 0 \tag{A.12}$$

Note: For this method, we set the constant of integration to zero.

2. *Apply the tanh approximation.*
Substitute $U = F\left(Y\right)$ into eq. (A.12), and using eq. (A.7), we obtain

$$\begin{aligned} &-kcF\left(Y\right) + 3kF\left(Y\right)^2 + k^3\left(1 - Y^2\right) \\ &\times\left[-2Y\frac{dF\left(Y\right)}{dY} + \left(1 - Y^2\right)\frac{d^2F\left(Y\right)}{dY^2}\right] = 0 \end{aligned} \tag{A.13}$$

[1]The term *solitary wave* was first coined by Scottish engineer John Scott-Russell (1808-82).

We can now substitute into eq. (A.13) the summation of eq. (A.8) to give, after division by k,

$$
\begin{aligned}
&-c\left(\sum_{i=0}^{M} a_i Y^i\right) + 3\left(\sum_{i=0}^{M} a_i Y^i\right)^2 + k^2\left(1 - Y^2\right) \times \\
&\quad\left[-2Y \times \left(a_1 + 2a_2 Y + 3a_3 Y^2 + \cdots + M a_M Y^{M-1}\right) + \right. \\
&\quad\left.\left(1 - Y^2\right) \times \left(2a_2 + 6a_3 Y + \cdots + M(M-1) a_M Y^{M-2}\right)\right] = 0
\end{aligned}
\tag{A.14}
$$

3. *Determine the polynomial approximation order, M.*
For eq. (A.14), we have to balance the largest exponent of Y in the highest order nonlinear term with the largest exponent of Y in the highest order linear term. The largest exponent of Y in the highest order nonlinear term occurs in the second term and is equal to $2M$. The largest exponent of Y in the highest order linear term occurs in the highest derivative term, the third term, and is equal to $4 + (M - 2)$. Therefore on balancing these exponents, we have

$$
2M = 4 + (M - 2) \rightarrow M = 2 \tag{A.15}
$$

4. *Equate coefficients to zero and solve for unknown parameters.*
On rearranging, eq. (A.14) results in the following polynomial in Y:

$$
\begin{aligned}
&\left(3ka_2^2 + 6k^3 a_2\right) Y^4 + \left(2k^3 a_1 + 6k a_1 a_2\right) Y^3 \\
&\quad + \left(-kca_2 + 3ka_1^2 + 6ka_0 a_2 - 8k^3 a_2\right) Y^2 \\
&\quad + \left(-kca_1 + 6ka_0 a_1 - 2k^3 a_1\right) Y - kca_0 + 3ka_0^2 + 2k^3 a_2 = 0
\end{aligned}
\tag{A.16}
$$

We now have a fourth degree polynomial in Y, which is identically equal to zero, and for this to be true, it follows that the coefficients for each power of Y must also be identically equal to zero. Equating the coefficients to zero results in five simultaneous equations, i.e.,

$$
\begin{aligned}
&Y^0: \; -kca_0 + 3ka_0^2 + 2k^3 a_2 &&= 0 \\
&Y^1: \; -kca_1 + 6ka_0 a_1 - 2k^3 a_1 &&= 0 \\
&Y^2: \; -kca_2 + 3ka_1^2 + 6ka_0 a_2 - 8k^3 a_2 &&= 0 \\
&Y^3: \; 2k^3 a_1 + 6ka_1 a_2 &&= 0 \\
&Y^4: \; 3ka_2^2 + 6k^3 a_2 &&= 0.
\end{aligned}
\tag{A.17}
$$

which we solve using Maple. The following five parameter solution sets are produced by Maple:

$$
\begin{aligned}
&a_0 = a_0,\ a_1 = a_1,\ a_2 = a_2,\ k = 0,\ c = c && : \text{trivial solution}\\
&a_0 = 0,\ a_1 = 0,\ a_2 = 0,\ c = c,\ k = k && : \text{trivial solution}\\
&a_0 = a_0,\ a_1 = 0,\ a_2 = 0,\ c = 3a_0,\ k = k && : \text{trivial solution}\\
&a_0 = 2k^2,\ a_1 = 0,\ a_2 = -2k^2,\ c = 4k^2,\ k = k && : \text{real solution}\\
&a_0 = (2/3)k^2,\ a_1 = 0,\ a_2 = -2k^2,\ c = -4k^2,\ k = k && : \text{real solution}
\end{aligned}
\tag{A.18}
$$

5. *Make back substitutions to obtain final solutions.*
The final results are therefore found by substituting the parameter solution sets (A.18) into eq. (A.17) and then setting $Y = \tanh(k[x - ct])$ in eq. (A.16), which yields the following five solutions to eq. (A.9):

$$
\begin{aligned}
u_1 &= a_0\\
u_2 &= 0\\
u_3 &= a_0\\
u_4 &= 2k^2\left\{1 - \tanh^2\left[k\left(x - 4k^2t\right)\right]\right\} = 2k^2\operatorname{sech}^2\left[k\left(x - 4k^2t\right)\right]\\
u_5 &= \frac{2}{3}k^2\left\{1 - 3\tanh^2\left[k\left(x + 4k^2t\right)\right]\right\}
\end{aligned}
\tag{A.19}
$$

Where constants a_0 or k appear in the above solutions, they may take arbitrary values. Note that u_4 is the same solution as given in eq. (A.10).

Aside:

For most problems, the above procedure is best performed using a computer algebra system such as Maple. However, as it happens for this particular problem, we note that $F(Y)$ should be proportional to $(1 - Y^2)$ because each term of eq. (A.11) includes a derivative - refer to eqns. (A.7). Therefore, we can take a short cut and set

$$F(Y) = \alpha\left(1 - Y^2\right)$$

On dividing equation (A.11) through by $\alpha(1 - Y^2)$, we obtain

$$-c + 3\alpha\left(1 - Y^2\right) + k^2\left(-2Y\frac{d(1 - Y^2)}{dY} + \left(1 - Y^2\right)\frac{d^2(1 - Y^2)}{dY^2}\right) = 0$$

which reduces to

$$-c+3\alpha\left(1-Y^2\right)+k^2\left[4Y^2-2\left(1-Y^2\right)\right]=0$$

Thus on equating coefficients of Y^0 and Y^2 to zero, we obtain two equations with three unknowns, i.e.,

$$3\alpha-2k^2-c=0$$

$$6k^2-3\alpha=0$$

and choose k to be a free parameter. The remaining parameters are found to be

$$\alpha=2k^2 \quad \text{and} \quad c=4k^2$$

Recalling from eqns. (A.5) and (A.6) that $Y=\tanh(\xi)=\tanh\left[k(x-ct)\right]$ and that $u(x,t)=U(\xi)=F(Y)$, we can now write the final solution as

$$u(x,t)=2k^2\left\{1+\tanh^2\left[k\left(x-4k^2t\right)\right]\right\}=2k^2\mathrm{sech}^2\left[k\left(x-4k^2t\right)\right] \tag{A.20}$$

which is eq. (A.10). However, using Maple, we also found the additional nontrivial solution $u=\dfrac{2}{3}k^2\left\{1-3\tanh^2\left[k\left(x+4k^2t\right)\right]\right\}$.

It must be stressed that although the tanh method can be used successfully on many problems to find traveling wave solutions, it is not guaranteed to find all traveling wave solutions and may fail altogether to find solutions to some problems.

A.2.2 A Maple tanh Method Procedure

We now illustrate the use of a computer algebra system to solve problems that require the use of the tanh method. This greatly reduces the effort required and, in many cases, is the only practical way of arriving at a solution using this type of approach.

The Maple procedure below generates a solution automatically, and we will demonstrate its use by an application example. However we must stress that this procedure will not solve all problems and is only provided to show what can be done with Maple using a fairly short section of code.

```
tanhMethod:=proc(M,pde,intFlag,infoLev)
   local F, FF, F1, F2, F3, F4,
         ode1, str, tr1, tr2,
         i, j, vars, testFlag;
   global _a, k, c, sol, testSol, N, F5;
   unprotect(_a);
   # Assume a travelling wave solution of the form
   # U(xi), xi=k*(x-c*t);
   tr1:={x=(xi/k+c*tau),t=tau,u=U(xi)};
   ode1:=dchange(tr1,pde,[xi,tau,U(xi)]);
   print('ode1=', ode1);
```

```
   if intFlag > 0 then
     for i from 1 to intFlag do
       ode1:=int(lhs(ode1),xi)=0;
     end do;
     print('After integration, ode1= ', ode1);
   end if;
   tr2:={xi=arctanh(Y),U(xi)=F(Y)};
   F1:=dchange(tr2,ode1,[Y,F(Y)]);
#print('F1=',F1);
   F(Y):=add( _a[i]*Y^i, i=0..M );
   print('F(Y) = ', F(Y));
   F2:=eval(F1);
#print('F2=',F2);
   F3:=collect(expand(F2),{Y}):
#print('F3=',F3);
   # This line included incase there are quotient terms
   F3 := numer(lhs(F3))*denom(rhs(F3)) =
               numer(rhs(F3))*denom(lhs(F3)):
#print('F3=',F3);
   # Get coeff's of Y^m terms
   F4:=CoefficientList(lhs(F3),Y):
   F4:=convert(F4,set):
#print('F4=',F4);
   # Define variables to be solved for
   vars:={seq(_a[r],r=0..M),'k','c'};
   if infoLev>0 then print('Unknowns =', vars) end if;
   if infoLev>1 then print('Coefficient List =',F4) end if;
   # Note: Polynomial coefficients for each power of Y must = 0
   F5:=[solve(F4,vars)]:
   # This line needed to eliminate 'RootOf' terms
   # from the solve() answer
   F5:=map(allvalues,F5);
#print('F5=',F5);
   N:=nops(F5): # No of solution sets found by solve()
   print(' Number of solution sets found = ', N);
   if infoLev>1 then print(F5) end if;
   for i from 1 to N do
     assign(F5[i]): FF[i]:=F(Y):
     # print(FF[i]);
     sol[i]:=u=eval(simplify(subs(Y=tanh(k*(x+x0-c*t)),FF[i]),symbolic));
     str:=sprintf("Solution %2d:\n===========", i): printf(str);
     print(sol[i]);
     # Unassign F(Y) coefficients
     for j from 0 to M do
#print(j);
       unassign('_a[j]');
     end do;
     unassign('k','c');
   end do;
   # Test Solutions!!
   testFlag:=0;
   if infoLev > 0 then
     print("About to run pdetest!");
   end if;
```

```
  for i from 1 to N do
    if infoLev > 0 then
      print("About to run pdetest() on solution: ", i);
    end if;
    testSol[i]:=timelimit(20,pdetest(sol[i],pde) ); # '0' if true
    if testSol[i] <> 0 then
      str:=sprintf(" Solution: %2d does not pass pdetest() !", i):print(str);
      testFlag:=1;
    end if;
  end do;
  # Final check to see if all solutions found solve 'pde'
  if testFlag = 0 then
    print("All solutions pass pdetest() !");
  end if;
end proc:
```

LISTING A.1: Maple procedure `tanhMethod()`.

This procedure is saved in the separate file `tanhMethod.txt` so that it can be reused with any application. Note, because this procedure is general for use with any PDE, we now use `_a[i]` rather than simply `a[i]` so that there is minimum likelihood of a name conflict with constants in the PDE. Also, `tanhMethod` introduces the arbitrary constant x_0 into the solution - refer to Section A.7 for an explanation.

We will now use procedure `tanhMethod()` to solve the Korteweg-de Vries (KdV) equation (A.9), in which we present code fragments followed by the associated Maple output.

- *Initialize Maple and load required packages: PDEtools, PolynomialTools and plots.*

```
> # KdV Equation
  # Attempt at Malfliet's tanh solution
  restart; with(PDEtools): with(PolynomialTools):
  with(plots):
```

- *Define an alias for $u(x,t)$ to simplify writing code and then define the KdV problem PDE.*

```
> alias(u=u(x,t)):
  pde1:=diff(u,t)+6*u*diff(u,x)+diff(u,x,x,x)=0;
```

$$pde1 := \frac{\partial}{\partial t}u + 6u\left(\frac{\partial}{\partial x}u\right) + \frac{\partial^3}{\partial x^3}u = 0$$

- *Read in the procedure* `tanhMethod` *so that it can be called as required.*

```
> read("tanhMethod.txt");
```

- *Call the procedure* `tanhMethod` *with arguments* `M`, `pde1`, `intFlg`, *and* `infoLevOut`. Note the use of $\_a_i$ rather than a_i, as mentioned above. The first three solutions found are trivial, but the fourth and fifth are valid nontrivial solutions. All the solutions found satisfy the original PDE, eq. (A.9).

```
> intFlg:=1: # one integration of U(xi) needed!
  infoLevOut:=2: # set output information to level 2
```

```
M := 2: # Set order of approximation
tanhMethod(M,pde1,intFlg);
```

$$F(Y) := \_a_0 + \_a_1 Y + \_a_2 Y^2$$

$$\text{ode1} =, -ck\frac{d}{d\xi}U(\xi) + 6U(\xi)k\left(\frac{d}{d\xi}U(\xi)\right) + k^3\left(\frac{d^3}{d\xi^3}U(\xi)\right) = 0$$

$$\textit{After integration, } \text{ode1} =, -ckU(\xi) + 3kU(\xi)^2 + k^3\left(\frac{d^2}{d\xi^2}U(\xi)\right) = 0$$

$$\text{Unknowns} =, \{c, k, \_a_0, \_a_1, \_a_2\}$$

$$\text{Coefficient list} =,$$

$$\left[k(3\_a_2^2 + 6k^2\_a_2),\ k(2k^2\_a_1 + 6\_a_1\_a_2),\ k(-c\_a_0 + 3\_a_0^2 + 2k^2\_a_2),\right.$$

$$\left.k(-c\_a_1 + 6\_a_0\_a_1 - 2k^2\_a_1),\ k(-c\_a_2 + 3\_a_1^2 + 6\_a_0\_a_2 - 8k^2\_a_2)\right]$$

$$\textit{"Number of solution sets found} = 5\text{"}$$

$$\left[(\_a_0 = \_a_0,\ \_a_1 = \_a_1,\ \_a_2 = \_a_2,\ k = 0,\ c = c)\right.$$

$$(\_a_0 = 0,\ \_a_1 = 0,\ \_a_2 = 0,\ k = k,\ c = c)$$

$$(\_a_0 = \_a_0,\ \_a_1 = 0,\ \_a_2 = 0,\ k = k,\ c = 3a_0)$$

$$(\_a_0 = 2k^2,\ \_a_1 = 0,\ \_a_2 = -2k^2,\ c = 4k^2,\ k = k)$$

$$\left.\left(\_a_0 = \frac{2}{3}k^2,\ \_a_1 = 0,\ \_a_2 = -2k^2,\ c = -4k^2,\ k = k\right)\right]$$

Solution 1:

=======

$$u = \_a_0$$

Solution 2:

=======

$$u = 0$$

Solution 3:

=======

$$u = \_a_0$$

Solution 4:

=======

$$u = -\frac{2}{3}\frac{k^2\{2\cosh[k(x + x0 + 4k^2t)]^2 - 3\}}{\cosh[k(x + x0 + 4k^2t)]^2}$$

Solution 5:

=======

$$u = \frac{2k^2}{\cosh\left[k(-x-x0+4k^2t)\right]^2}$$

"About to run pdetest!"
"About to run pdetest() on solution: ", 1
"About to run pdetest() on solution: ", 2
"About to run pdetest() on solution: ", 3
"About to run pdetest() on solution: ", 4
"About to run pdetest() on solution: ", 5
"All solutions pass pdetest() !"!

- *Check that the known standard solution passes the Maple procedure* `pdetest()` *and print an error message if it does not.*

```
> # Check derived solutions for match with 'standard' solution
  x0:=0: # Set arbitrary constant to zero!
  sol1:=u=2*k^2*sech(k*(x-4*k^2*t))^2: # standard solution
  s1:=rhs(sol1): testFlg:=1:
  for i from 1 to N do
    s2:=rhs(sol[i]);
    del:=simplify(s1-s2, symbolic);
    if del=0 then
      str:=sprintf(' sol[\%d] matches 'standard' solution ',i):
      print('str');
      print(sol[i]); testFlg:=0;
      break;
    end if:
  end do:
  if testFlg = 1 then print('No match to standard solution!') end if;
```

"sol[5] matches 'standard' solution"

$$u = \frac{2k^2}{\cosh\left[k(-x+4k^2t)\right]^2}$$

- *Set the wavenumber to 0.5 (gives a wave speed of 1), and plot the solution as an animation and as a 3D surface.*

```
># Plot results
 # ============
 k:=0.5: # Set value for wave number
 # Animate solution
 animate(rhs(sol[i]),x=-10..50,t=0..40,
   numpoints=300,frames=50, axes=framed,
   labels=["x","u"],thickness=3,
   title="KdV Equation",
   labelfont=[TIMES, ROMAN, 16],axesfont=[TIMES, ROMAN, 16],
   titlefont=[TIMES, ROMAN, 16]);
># Generate a 3D surface plot
```

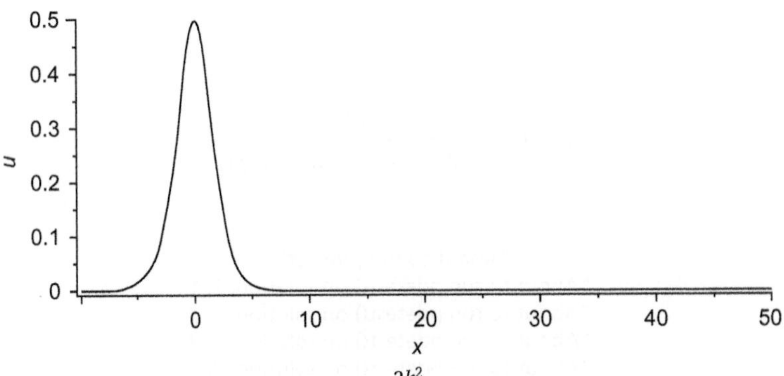

FIGURE A.1: Plot of the derived solution $u(x,t) = \dfrac{2k^2}{\cosh\left[k\left(-x+4k^2t\right)\right]^2}, k = 0.5, t = 0.$

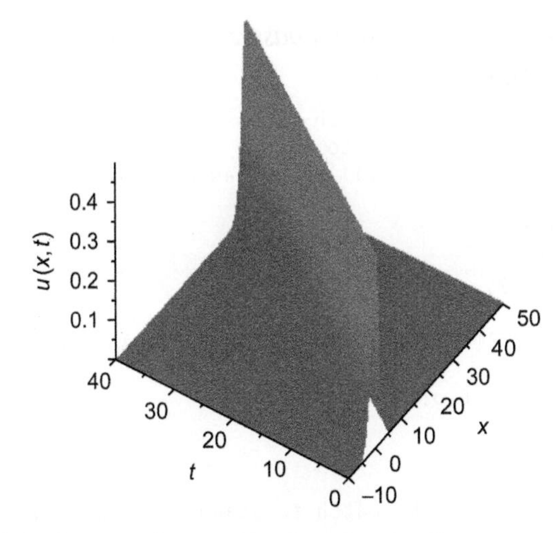

FIGURE A.2: 3D surface of derived solution for $x = -10 \ldots 50$ and $t = 0 \ldots 40$.

```
plot3d(rhs(sol[i]),x=-10..50,t=0..40,axes='framed',
  labels=["x","t","u(x,t)"],
  orientation=[-145,43],grid=[100,100],
  style=patchnogrid,shading=Z,
  labeldirections=[HORIZONTAL,HORIZONTAL,VERTICAL],
  title="KdV Equation",
  labelfont=[TIMES, ROMAN, 16],axesfont=[TIMES, ROMAN, 16],
  titlefont=[TIMES, ROMAN, 16]);
```

Plotted output from the preceding code is shown above as Figs. A.1 and A.2.

Finally, when using the `tanhMethod` procedure, care must be taken when selecting variable names in the main body of the application code. It is advisable to avoid using any of the variable names specified as *local* or *global* in the procedure, except as shown in this example. Otherwise, unexpected behavior may result. If in doubt, try to follow the example given here or the additional download examples.

A.2.3 Extension to Coupled Equations

Application of the tanh method to two or more coupled equations is a natural extension to the one-dimension case. However, we will restrict our discussion here to a system with two coupled PDEs in one spatial dimension (plus time). Problems with more than two coupled equations follow in a straightforward manner.

Consider the following two coupled evolutionary equations for which we wish to find traveling wave solutions

$$\psi\left(u, v, u_t, v_t, u_x, v_x, u_{tt}, v_{tt}, u_{xx}, v_{xx}, u_{tx}, v_{tx}, \ldots\right) = 0$$

$$\phi\left(u, v, u_t, v_t, u_x, v_x, u_{tt}, v_{tt}, u_{xx}, v_{xx}, u_{tx}, v_{tx}, \ldots\right) = 0$$

The tanh method is simple to implement, and we start by introducing the transformations

$$u(x,t) \to U(\xi), \quad v(x,t) \to V(\xi), \quad \xi = k(x - ct)$$

where U and V have yet to be defined and, typically, $k > 0$ represents *wavenumber* and c represents *wave velocity*. Partial differentiation operations with respect to t and x are, therefore, transformed to the following equivalent ordinary differentiation operations

$$\frac{\partial}{\partial t} \to -kc\frac{d}{d\xi}, \quad \frac{\partial}{\partial x} \to k\frac{d}{d\xi}$$

Thus the above partial differential equations are transformed to the following ordinary differential equations

$$\Psi\left(U, V, \frac{dU}{d\xi}, \frac{dV}{d\xi}, \frac{d^2U}{d\xi^2}, \frac{d^2V}{d\xi^2}, \ldots\right) = 0$$

$$\Phi\left(U, V, \frac{dU}{d\xi}, \frac{dV}{d\xi}, \frac{d^2U}{d\xi^2}, \frac{d^2V}{d\xi^2}, \ldots\right) = 0$$

We are now able to convert these ODE equations to equivalent algebraic equations by setting $U = F(Y)$ and $V = G(Y)$ (recall that $Y = \tanh(\xi)$) then introducing the following ansatz

$$F(Y) = \sum_{i=0}^{M1} \_a_i Y^i, \quad G(Y) = \sum_{i=0}^{M2} \_b_i Y^i$$

where coefficients $\_a_i$ and $\_b_i$ are real constants to be determined. $M1$ is a positive integer (associated with PDE [1]) to be found by balancing (equating) the largest exponent of Y in the highest order linear term with the largest exponent of Y in the highest order nonlinear term. Similarly, $M2$ is a positive integer (associated with PDE_2) to be found by balancing (equating) the largest exponent of Y in the highest order linear term with the largest exponent of Y in the highest order nonlinear term. Note, again we use underscores, i.e., `_a[i]` and `_b[i]`, to avoid name conflicts with constants used in the PDEs.

The tanh method for coupled equations then follows the same calculation sequence as for the single equation case, as detailed at the beginning of Section A.2. A Maple procedure that performs this calculation and generates a solution automatically for problems with two coupled equations is included in Listing A.2. It will not solve all problems!

```
tanhMethod2:=proc(M1,M2,pde1,pde2,intFlg1,intFlg2,infoLev)
    local F, FF, G, GG, F11, F12, F21, F22,F31, F32, F41, F42,
          eqns, ode1, ode2, str, tr1, tr2,
          i, j, vars, testFlag;
   global _a, _b, p, q, k, c, sol1,sol2, testSol1,testSol2, N, F5;
   unprotect(_a,_b);
   # Assume a travelling wave solutions of the form
   # U(xi), xi=k*(x-c*t);V(xi), xi=k*(x-c*t);
   tr1:={x=(xi/k+c*tau),t=tau,u=U(xi),v=V(xi)};
   ode1:=dchange(tr1,pde1,[xi,tau,U(xi),V(xi)]);
   ode2:=dchange(tr1,pde2,[xi,tau,U(xi),V(xi)]);
   print('ode1=', ode1); print('ode2= ', ode2);
   if intFlg1 = 1 then
     ode1:=int(lhs(ode1),xi)=0;
     print('After integration, ode1= ', ode1);
   end if;
   if intFlg2 = 1 then
     ode2:=int(lhs(ode2),xi)=0;
     print('After integration, ode2= ', ode2);
   end if;
   tr2:={xi=arctanh(Y),U(xi)=F(Y),V(xi)=G(Y)};
   F11:=dchange(tr2,ode1,[Y,F(Y),G(Y)]); F12:=dchange(tr2,ode2,[Y,F(Y),G(Y)]);
   #print(F11); print(F12);
   F(Y):=add( _a[i]*Y^i, i=0..M1 );
   G(Y):=add( _b[i]*Y^i, i=0..M2 );
   print('F(Y) = ', F(Y)); print('G(Y) = ', G(Y));
   F21:=eval(F11); F22:=eval(F12);
   #print('F21=',F21); print('F22=',F22);
   F31:=collect(expand(F21),{Y}): F32:=collect(expand(F22),{Y}):
   #print('F31=',F31); print('F32=',F32);
   # These lines included incase there are quotient terms
   F31 := numer(lhs(F31))*denom(rhs(F31)) =  numer(rhs(F31))*denom(lhs(F31)):
   F32 := numer(lhs(F32))*denom(rhs(F32)) =  numer(rhs(F32))*denom(lhs(F32)):
   #print('F31=',F31); print('F32=',F32);
   # Get coeff's of Y^m terms
   F41:=CoefficientList(lhs(F31),Y): F42:=CoefficientList(lhs(F32),Y):
   F41:=convert(F41,set): F42:=convert(F42,set):
   eqns:=F41 union F42; #print(eqns);
   # Define variables to be solved for
   vars:={seq(_a[r],r=0..M1),seq(_b[r],r=0..M2),'k','c'};
   if infoLev>0 then print('Unknowns =', vars) end if;
   if infoLev>1 then print('Coefficient List for U =',F41) end if;
   if infoLev>1 then print('Coefficient List for V =',F42) end if;
   # Note: Polynomial coefficients for each power of Y must = 0
   F5:=[solve(eqns,vars)]:
   # This line needed to eliminate 'RootOf' terms
   # from the solve() answer
```

```
    F5:=map(allvalues,F5);
    N:=nops(F5): # No of solutions found by solve()
    print(' Number of solutions found = ', N);
    if infoLev>1 then print(F5) end if;
    #print(F5);
    for i from 1 to N do
      assign(F5[i]):
      FF[i]:=eval(F(Y)): GG[i]:=eval(G(Y)):
      #print('FF',i,'= ', _a[0]); print('GG',i,'= ', GG[i]);
      #sol1[i]:=u=simplify(subs(Y=tanh(k*(x-c*t)),FF[i]),symbolic);
      #sol2[i]:=v=simplify(subs(Y=tanh(k*(x-c*t)),GG[i],symbolic);
      sol1[i]:=u=subs(Y=tanh(k*(x-c*t)),FF[i]);
      sol2[i]:=v=subs(Y=tanh(k*(x-c*t)),GG[i]);
      str:=sprintf('Solution %2d:\n===========', i): printf(str);
      print(sol1[i]); print(sol2[i]);
      # Unassign F(Y) coefficients
      for j from 0 to M1 do
        unassign('_a[j]'):
      end do;
      for j from 0 to M2 do
        unassign('_b[j]'):
      end do;
      unassign('k','c'):
    end do;
    # Test Solutions!!
    testFlag:=0;
    if infoLev > 0 then
      print("About to run solution tests!");
    end if;
    for i from 1 to N do
      if infoLev > 0 then
        print("About to test solution: ", i);
      end if;
      testSol1[i]:=timelimit(20, subs({u=rhs(sol1[i]),v=rhs(sol2[i])},lhs(pde1)) );
        # '0' if true
      testSol2[i]:=timelimit(20, subs({u=rhs(sol1[i]),v=rhs(sol2[i])},lhs(pde2)) );
        # '0' if true
      #print('TS1=', testSol1[i]); print( 'TS2=', testSol2[i]);
      if simplify(testSol1[i]) <> 0 or simplify(testSol2[i]) <> 0 then
        str:=sprintf(' Solution: %d does not pass test() !', i): print(str);
        testFlag:=1;
      end if;
    end do;
    # Final check to see if all solutions found solve 'pde'
    if testFlag = 0 then
      print('All solutions pass test() !');
    end if;
  end proc:
```

LISTING A.2: Maple procedure `tanhMethod2`.

The procedure `tanhMethod2` is applied in a similar way to the `tanhMethod` procedure, which is used for solving single equations. For example, when applied to the coupled KdV

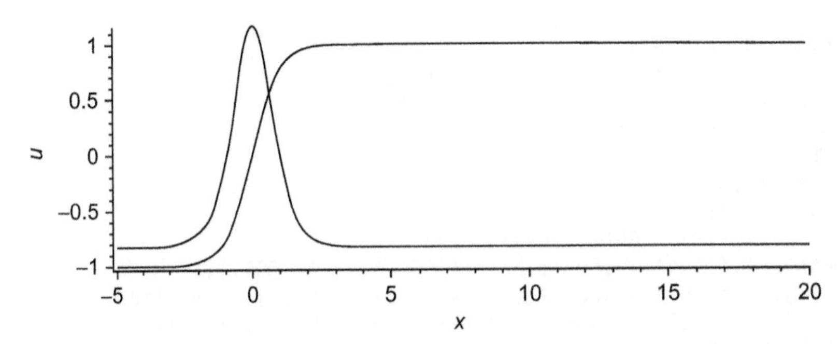

FIGURE A.3: Plot of the solution to the coupled KdV equations for $a = 1$, $b = 1$, $\_b_1 = 1$, and $k = 1$.

equations

$$\frac{\partial u}{\partial t} - a\left(6u\frac{\partial u}{\partial x} + \frac{\partial^3 u}{\partial x^3}\right) - 2bv\frac{\partial v}{\partial x} = 0$$

$$\frac{\partial v}{\partial t} + 3u\frac{\partial v}{\partial x} + \frac{\partial^3 v}{\partial x^3} = 0$$

`tanhMethod2` finds the solution

$$u = -2k^2 \tanh\left[\frac{(2+4a)\,xk^2 - (8ak^4 + \_b_1^2 b)\,t}{(2+4a)\,k}\right]^2 + \frac{4\,(1+4a)\,k^4 + \_b_1^2 b}{(6+12a)\,k^2}$$

$$v = \_b_1 \tanh\left[\frac{(2+4a)\,xk^2 - (8ak^4 + \_b_1^2 b)\,t}{(2+4a)\,k}\right]$$

where u describes a *hump soliton* and v describes a *kink soliton*, both of which propagate left to right in unison. A plot of the output at $t = 0$ is given in Fig. A.3.

The Maple code that derives this solution and generates an animation is detailed in Listing A.3.

```
>╫ Modified KdV Equation
 ╫ Attempt at Malfliet's tanh solution
 ╫ Ref: W. Hereman (?). "SOLITARY WAVE SOLUTIONS OF COUPLED
 ╫      NONLINEAR EVOLUTION EQUATIONS USING MACSYMA",
 ╫      Report, Department of Mathematical and Computer
 ╫      Sciences Colorado School of Mines
 restart; with(PDEtools): with(PolynomialTools):
 with(plots):
>alias(u=u(x,t)): alias(v=v(x,t)):
 ╫ Coupled KdV equations
>pde1:=diff(u,t)-a*(6*u*diff(u,x)+diff(u,x,x,x))
          -2*b*v*diff(v,x)=0;
 pde2:=diff(v,t)+3*u*diff(v,x)+diff(v,x,x,x)=0;
>read("tanhMethod2.txt");
>intFlag1:=0: intFlag2=0: ╫ No integration of U(xi) or V(xi) needed !
 M1:=2: M2:=1: ╫ Set order of approximation
 infoLevOut:=0;
```

```
 tanhMethod2(M1,M2,pde1,pde2,intFlag1,intFlag2,infoLevOut);
>n:=2; # select solution 2
 uSol:= sol1[n];
 vSol:= sol2[n];
># Plot animation
 a:=1;b:=1;_b[1]:=1; k:=1;
 animate({rhs(uSol),rhs(vSol)},x=-5..20,t=0..10,
   axes=framed,thickness=3,frames=50,
   numpoints=300,title="Coupled KdV Equations",
   labels=["x","u"],thickness=3,
   labelfont=[TIMES, ROMAN, 16],axesfont=[TIMES, ROMAN, 16],
   titlefont=[TIMES, ROMAN, 16]);
```

LISTING A.3: Maple code that derives the solution to the above coupled KdV equations.

A.2.4 Extension to higher spatial dimensions

The extension of the tanh method to more than two spatial dimensions is straightforward. We will illustrate this by introducing the traveling wave transformation

$$u(x_1,x_2,x_3,t) \to U(\xi), \quad \xi = (k_1x_1 + k_2x_2 + k_3x_3 - \omega t),$$

where typically, $k_1, k_2, k_3 > 0$ represents *wavenumbers* for the x_1, x_2, and x_3 co-ordinates, respectively, and ω represents *wave frequency*. The wave phase and group velocities are given by

$$c_p = \frac{\omega}{|\mathbf{k}|}$$
$$\mathbf{c_g} = \frac{d\omega}{d\mathbf{k}} = \frac{d\omega}{dk_1}\hat{\mathbf{x}}_1 + \frac{d\omega}{dk_2}\hat{\mathbf{x}}_2 + \frac{d\omega}{dk_3}\hat{\mathbf{x}}_3$$

where $|\mathbf{k}| = \sqrt{k_1^2 + k_2^2 + k_3^2}$ is the absolute value of $\mathbf{k}$, and $\hat{\mathbf{x}}_1$, $\hat{\mathbf{x}}_2$, and $\hat{\mathbf{x}}_3$ are the unit vectors associated with $\mathbf{x}_1$, $\mathbf{x}_2$, and $\mathbf{x}_3$. Therefore as the velocity of a traveling wave is equal to its group velocity (also discussed in the Appendix to Chapter 12—Kawahara equation), we have $\mathbf{c} = \mathbf{c_g}$. For a more detailed discussion refer to [25], p. 236–241.

Partial differentiation operations with respect to t, x_1, x_2, and x_3 are transformed to the following equivalent ordinary differentiation operations

$$\frac{\partial}{\partial t} \to -\omega\frac{d}{d\xi}, \quad \frac{\partial}{\partial x_1} \to k_1\frac{d}{d\xi}, \quad \frac{\partial}{\partial x_2} \to k_2\frac{d}{d\xi}, \quad \frac{\partial}{\partial x_3} \to k_3\frac{d}{d\xi}$$

Problems with higher-order derivatives are handled similarly by a natural extension to the above.

The tanh method for higher spatial dimensions then follows the same calculation sequence as for the one spatial dimension case, as detailed at the beginning of Section A.2. A Maple procedure that performs this calculation and generates a solution automatically for problems with two spatial dimensions is included in Listing A.4. Note, we use k and l

rather than k_1 and k_2 to represent wavenumbers. Also, again, this routine will not solve all problems!

```
tanhMethod3:=proc(M,pde,intFlag,infoLev)
   local F, FF, F1, F2, F3, F4,
         ode1, str, tr1, tr2,
         i, j, vars, testFlag;
   global A, k, l, omega, sol, testSol, N, F5;
   unprotect(_a);
   # Assume a travelling wave solution of the form
   # U(xi), xi=k*x+l*y-omega*t);
   tr1:={x=(xi-l*nu+omega*tau)/k,y=nu,t=tau,u=U(xi)};
   ode1:=dchange(tr1,pde1,[xi,nu,tau,U(xi)]);
   print('ode1=', ode1);
   if intFlag > 0 then
     for i from 1 to intFlag do
       ode1:=eval(int(lhs(ode1),xi))=0;
     end do;
     print('After integration, ode1= ', ode1);
   end if;
   tr2:={xi=arctanh(Y),U(xi)=F(Y)};
   F1:=dchange(tr2,ode1,[Y,F(Y)]);
#print('F1=',F1);
   F(Y):=add( _a[i]*Y^i, i=0..M );
   print('F(Y) = ',F(Y));
   F2:=eval(F1);
#print('F2=',F2);
   F3:=collect(expand(F2),{Y}):
#print('F3=',F3);
   # This line included incase there are quotient terms
   F3 := numer(lhs(F3))*denom(rhs(F3)) =
               numer(rhs(F3))*denom(lhs(F3)):
   # Get coeff's of Y^m terms
   F4:=CoefficientList(lhs(F3),Y):
#print('F4=',F4);
   F4:=convert(F4,set):
#print('F4=',F4);
   # Define variables to be solved for
   vars:={seq(_a[r],r=0..M),'k','l','omega'};
   if infoLev>0 then print('Unknowns =', vars) end if;
   if infoLev>1 then print('Coefficient List =',F4) end if;
   # Note: Polynomial coefficients for each power of Y must = 0
   F5:=[solve(F4,vars)]:
   # This line needed to eliminate 'RootOf' terms
   # from the solve() answer
   F5:=map(allvalues,F5);
   N:=nops(F5): # No of solutions found by solve()
   print(' Number of solutions found = ', N);
   if infoLev>1 then print(F5) end if;
   for i from 1 to N do
     assign(F5[i]): FF[i]:=F(Y):
   #print(F(i));
     sol[i]:=u=eval(simplify(subs(Y=tanh(k*(x+x0)+l*(y+y0)-omega*t),
```

```
                    FF[i]),symbolic));
      str:=sprintf('Solution %2d:\n===========', i): printf(str);
      print(sol[i]); #print(F5[i]);
      #Unassign F(Y) coefficients
      for j from 0 to M do
#print(j);
        unassign('_a[j]');
      end do;
      unassign('k','l','omega');
    end do;
    # Test Solutions!!
    testFlag:=0;
    if infoLev > 0 then
      print("About to run pdetest!");
    end if;
    for i from 1 to N do
      if infoLev > 0 then
        print("About to run pdetest() on solution: ", i);
      end if;
      testSol[i]:=timelimit(20,pdetest(sol[i],pde) ); # '0' if true
      #testSol[i]:=pdetest(sol[i],pde); # '0' if true
      if testSol[i] <> 0 then
        str:=sprintf(' Solution: %d does not pass pdetest() !', i): print(str);
        testFlag:=1;
      end if;
    end do;
# Final check to see if all solutions found solve 'pde'
  if testFlag = 0 then
    print('All solutions pass pdetest() !');
  end if;
end proc:
```

LISTING A.4: Maple procedure `tanhMethod3`.

The procedure `tanhMethod3` is used in a similar way to the `tanhMethod` procedure, which solves equations having one spatial dimension. For example, when applied to the Boussinesq equation

$$\frac{\partial^2 u}{\partial t^2} - \frac{\partial^2 u}{\partial x^2} - \frac{\partial^2 (u^2)}{\partial y^2} + \frac{\partial^4 u}{\partial x^4} = 0, \quad u = u(x,y,t)$$

it finds the solution

$$u = \frac{(4k^4 + \omega^2 - k^2)}{2l^2} - 6\frac{k^2}{l^2}\cosh^2\left[k(x+x_0) + l(y+y_0) - \omega t\right]$$

where u describes a *two-dimensional hump soliton* or a *line soliton*. Plotted output follows as Fig. A.4.

The Maple code that derives this solution is detailed in Listing A.5.

```
># Boussinesq equation - 2D
 # Attempt at Malfliet's tanh solution
```

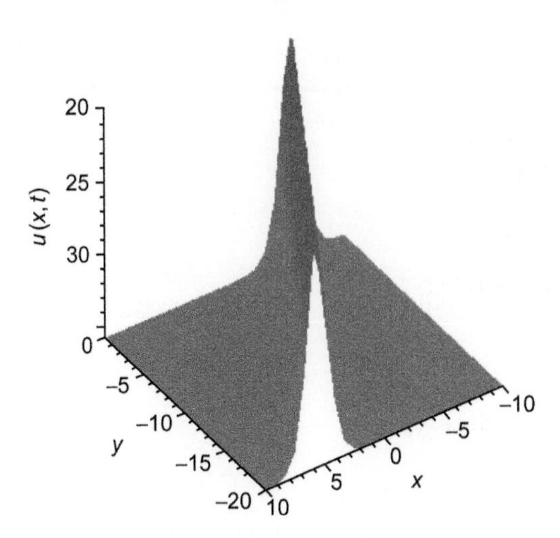

FIGURE A.4: Plot of the derived solution for the two-dimensional Boussinesq equation, with $x_0 = 0$, $y_0 = 10$, $k = 0.9$, $l = 0.5$, and $\omega = -4$, at $t = 0$.

```
# Ref: Tascan, F., & Bekir, A. (2009). Analytic
#      solutions of the (2+1)-dimensional nonlinear
#      evolution equations using the sine?cosine
#      method. Applied Mathematics and Computation,
#      215(8), 3134-3139. doi: 10.1016/j.amc.2009.09.027.
 restart; with(PDEtools): with(PolynomialTools):
 with(plots):
>alias(u=u(x,y,t)): alias(v=v(x,y,t)):
># Set up PDE
pde1:=diff(u,t,t)-diff(u,x,x)-diff(u^2,y,y)+
        diff(u,x,x,x,x)=0;
>read("tanhMethod3.txt");
># Calculate solution
 intFlag:=1: # Integration of U(xi) needed!
 M:=2; # Set order of approximation
 infoLevOut:=2;
 tanhMethod3(M,pde1,intFlag,infoLevOut);
>soln:=simplify(convert(sol[2],sech),size);
 zz1:=rhs(soln);
># Set constants
 const:={x0=0, y0=10,k=0.9,l=0.5,omega=-4};
 zz:=subs(const,zz1);
># Generate 3D animation
 animate(plot3d,[zz,x=-10..10,y=-20..0],t=0..6,
   axes=framed,numpoints=300*300,style=patchnogrid,
   shading=Z,frames=40,orientation=[120,215],
   labeldirections=[HORIZONTAL,HORIZONTAL,VERTICAL],
   title="Boussinesq Equation - 2D",
   labelfont=[TIMES, ROMAN, 16],labels=["x","y","u(x,t)"],
   axesfont=[TIMES, ROMAN, 16], titlefont=[TIMES, ROMAN, 16]);
```

LISTING A.5: Maple code that derives the solution to the above two-dimensional Boussinesq equation.

A.2.5 Coth Method

The *coth method* is effectively the same as the tanh method except that coth is used in lieu of tanh. Recall that if $Y = \coth(\xi)$, then

$$\frac{dY}{d\xi} = 1 - \coth^2(\xi) = 1 - Y^2$$

It therefore follows from the chain rule for differentiation that, for a given function $F(Y)$, we have

$$\frac{dF(Y)}{d\xi} = \frac{dF(Y)}{dY}\frac{dY}{d\xi} = \frac{dF(Y)}{dY}\left(1 - Y^2\right)$$

which is the same result we obtained for $Y = \tanh(\xi)$ in the tanh method.

A.3 Exp Method

The *exp method* is very straightforward to implement and is similar to the tanh method described previously. However instead of using a transformation based on the tanh function to find algebraic functions to represent derivatives, we use a transformation based on the *exponential function*. We start by applying the transformation $U(\xi) = u(x,t)$, $\xi = k(x - ct)$, as for the tanh method. However, instead of assuming a *polynomial* solution in tanh, this method assumes that traveling wave solutions can be expressed in the following *rational* form [7]

$$U(\xi) = \frac{\sum_{n=-N1}^{N2} a_n \exp(n\xi)}{\sum_{m=-M1}^{M2} b_m \exp(m\xi)} \tag{A.21}$$

where $N1$, $N2$, $M1$, and $M2$ are positive integers to be determined, and a_n and b_m are unknown constants. **Note**: Without loss of generality, we can select any constant from a_n, b_m to be unity, and we choose $b_{M2} = 1$ (the same as dividing both the numerator and denominator by a suitable number).

We now proceed by substituting eq. (A.21) into eq. (A.4). The unknowns $N1$ and $M1$ are determined by balancing the *largest linear term of lowest order* with the *largest nonlinear term of lowest order*. Similarly the unknowns $N2$ and $M2$ are determined by balancing the *largest linear term of highest order* with the *largest nonlinear term of highest order*. Depending upon the particular problem, the method does not appear to be overly sensitive to values chosen for $N1$, $N2$, $M1$, and $M2$. With $N1$, $N2$, $M1$, and $M2$ known, we then crossmultiply to eliminate the denominators and collect like terms to obtain an algebraic equation still consisting of exponential terms, i.e.,

$$C_r \exp(-r\xi) + C_{r-1} \exp[-(r-1)\xi] + \cdots C_0 + C_1 \exp(\xi) + \cdots C_s \exp(s\xi) = 0, \tag{A.22}$$

where r and s represent combinations of $N1$, $N2$, $M1$, and $M2$ and are known. $C_{-r} \ldots C_s$ represent expressions consisting of combinations of the known values for $N1, N2, M1$, and $M2$, along with $a_{-N1}, \ldots, a_{N2}$, and $b_{-M1}, \ldots, b_{M2}$, which are unknown. On letting $\exp(\xi) = Y$, we obtain a polynomial of the following form

$$C_{-r}Y^{-r} + C_{-(r-1)}Y^{-(r-1)} + \cdots + C_{-1}Y^{-1} + C_0 + C_1Y^1 + \cdots + C_{s-1}Y^{s-1} + C_sY^s = 0. \tag{A.23}$$

Similarly to the tanh method, for eq. (A.23) to be consistent for all values of Y, the coefficient expressions $C_{-r} \ldots C_s$ must each equate to zero. We therefore use the method of *Lagrange (undetermined) multipliers* to solve the resulting set of equations for the required unknowns, $a_{-N1}, \ldots, a_{N2}$, and $b_{-M1}, \ldots, b_{M2}$. Again for anything other than trivial or simple problems, this is a tedious process that requires the use of a computer algebra system if the solution is to be found in a reasonable time. The final traveling wave solution to the problem is found by substituting $Y = \exp(\xi)$ and $\xi = k(x - ct)$ back into eq. (A.23). For additional information, the reader is referred to [7].

A.3.1 Example - KdV Equation

We will apply the above solution method to the KdV eq. (A.9) (with the analytical solution eq. (A.10) to confirm the analysis).

1. *Apply the traveling wave transformation.*
   Apply the transformation of eq. (A.3) to eq. (A.9) as with the tanh method to obtain eq. (A.4), when we obtain eqns. (A.11) and (A.12).
2. *Apply the exponential approximation.*
   Substitute eq. (A.21) into eq. (A.23) and determine values for $N1, N2, M1$, and $M2$. We now let $\exp(\xi) = Y$ and balance the largest exponent of Y in the highest order nonlinear term with the largest exponent of Y in the highest order linear term. The largest exponent of Y in the highest nonlinear term (the second term) is $2(N2 - M2)$, and the largest exponent of Y in the highest order linear term (the second derivative term) is $N2 - M2$. Therefore on balancing these exponents, we have

   $$2(N2 - M2) = (N2 - M2) \rightarrow N2 = M2$$

   Now balance the largest exponent of Y in the lowest order nonlinear term with the largest exponent of Y in the lowest order linear term. The largest exponent of Y in the lowest order nonlinear term (the second term) is $2(-N1 + M1)$, and the largest exponent of Y in the lowest order linear term (the first term) is $-N1 + M1$. Therefore on balancing these exponents, we have

   $$2(-N1 + M1) = (-N1 + M1) \rightarrow N1 = M1$$

   The summation maximum and minimum index values are not determined uniquely, so we choose $N1 = M1 = N2 = M2 = 1$.

3. *Equate coefficients to zero and solve for unknown parameters.*
After the substitution and on rearranging, eq. (A.23) results in a sixth-order polynomial in Y, which in turn results in seven simultaneous equations (as each coefficient C in equation (A.23) must be equal to zero), i.e.,

$$
\begin{aligned}
0 &= 3ka_1^2 - kca_1 \\
0 &= 6ka_0a_1 - k^3a_1b_0 - 2kca_1b_0 + 3ka_1^2b_0 + k^3a_0 - kca_0 \\
0 &= -2kca_1b_{-1} - 4k^3a_1b_{-1} - k^3a_0b_0 + 3ka_1^2b_{-1} + 6ka_{-1}a_1 \\
&\quad +6ka_0a_1b_0 - 2kca_0b_0 + k^3a_1b_0^2 - kca_1b_0^2 + 3ka_0^2 - kca_{-1} + 4k^3a_{-1} \\
0 &= 6ka_{-1}a_1b_0 - 6k^3a_0b_{-1} - 2kca_0b_{-1} + 3k^3a_{-1}b_0 + 3k^3a_1b_{-1}b_0 \\
&\quad -2kca_1b_{-1}b_0 - kca_0b_0^2 + 6ka_0a_1b_{-1} + 6ka_{-1}a_0 + 3ka_0^2b_0 - 2kca_{-1}b_0 \\
0 &= k^3a_{-1}b_0^2 - 4k^3a_{-1}b_{-1} + 6ka_{-1}a_1a_{-1} - 2kca_0b_{-1}b_0 + 4k^3a_1b_{-1}^2 \\
&\quad -kca_{-1}b_0^2 + 3ka_0^2b_{-1} - 2kca_{-1}b_{-1} + 3ka_{-1}^2 - kca_1b_{-1}^2 \\
&\quad +6ka_{-1}a_0b_0 - k^3a_0b_{-1}b_0 \\
0 &= -kca_0b_{-1}^2 + 6ka_{-1}a_0b_{-1} - 2kca_{-1}b_{-1}b_0 + 3ka_{-1}^2b_0 \\
&\quad -k^3a_{-1}b_{-1}b_0 + k^3a_0b_{-1}^2 \\
0 &= -kca_{-1}b_{-1}^2 + 3ka_{-1}^2b_{-1}
\end{aligned}
\tag{A.24}
$$

Solving these equations using Maple yields the following solutions:

$$
\begin{aligned}
&a_{-1} = a_{-1},\ a_0 = a_0,\ a_1 = a_1,\ b_{-1} = b_{-1},\ b_0 = b_0,\ k = 0,\ c = c && : \text{trivial} \\
&a_{-1} = 0,\ a_0 = 0,\ a_1 = 0,\ b_{-1} = b_{-1},\ b_0 = b_0,\ k = k,\ c = c && : \text{trivial} \\
&a_{-1} = 0,\ a_0 = b_0k^2,\ a_1 = 0,\ b_{-1} = \frac{1}{4}b_0^2,\ b_0 = b_0,\ k = k,\ c = k^2 && : \text{nontrivial} \\
&a_{-1} = 0,\ a_0 = a_1b_0,\ a_1 = a_1,\ b_{-1} = 0,\ b_0 = b_0,\ k = k,\ c = 3a_1 && : \text{trivial} \\
&a_{-1} = b_{-1}a_1,\ a_0 = a_0,\ a_1 = a_1,\ b_{-1} = b_{-1},\ b_0 = \frac{a_0}{a_1},\ k = k,\ c = 3a_1 && : \text{trivial} \\
&a_{-1} = -\frac{3}{16}\frac{a_0^2}{k^2},\ a_0 = a_0,\ a_1 = -\frac{1}{3}k^2,\ b_{-1} = \frac{9}{16}\frac{a_0^2}{k^4},\ b_0 = \frac{3}{2}\frac{a_0}{k^2}. && : \text{nontrivial} \\
&\qquad k = k,\ c = -k^2
\end{aligned}
\tag{A.25}
$$

Recall that b_1 was previously set to 1.

4. *Make back substitutions to obtain final solutions.*
The final results are therefore found by substituting the parameter solution sets (A.25) into eq. (A.23) and then setting $Y = \exp\left[k(x - ct)\right]$ in eq. (A.21), which yields the

following six solutions:

$$
\begin{aligned}
u_1 &= \frac{a_{-1}+a_0+a_1}{b_{-1}+b_0+1} \\
u_2 &= 0 \\
u_3 &= \frac{4b_0k^2}{b_0^2\exp\left(-k\left(x-k^2t\right)\right)+4b_0+4\exp\left(k\left(x-k^2t\right)\right)} \\
u_4 &= a_1 \\
u_5 &= a_1 \\
u_6 &= -\frac{1}{3}\frac{\left\{9a_0^2\exp\left[-k\left(x+k^2t\right)\right]-48k^2a_0+16k^4\exp\left[k\left(x+k^2t\right)\right]\right\}k^2}{9a_0^2\exp\left[-k\left(x+k^2t\right)\right]+24k^2a_0+16k^4\exp\left[k\left(x+k^2t\right)\right]}
\end{aligned}
\tag{A.26}
$$

Where constants a_{-1}, a_0, a_1, b_{-1}, b_0, or k appear in the above solutions, they may take arbitrary values.

We can point out here that Maple may present solutions in a different order than in eqs. (A.26). Also, for difficult problems, Maple may not find all the possible solutions; consequently, it occasionally produces different solutions at different times. For more details on some of the difficulties and possible solutions related to the Maple procedure `solve()`, refer to [8], Chapter 16.

One of the drawbacks with the exp method is that the solutions, at first glance, can appear rather complex. Whilst on further inspection, using hyperbolic relationships such as

$$
\tanh\left[k\left(x+ct\right)\right]=1-\frac{2}{1+\exp\left[2k\left(x+ct\right)\right]}=-\frac{1-\exp\left[2k\left(x+ct\right)\right]}{1+\exp\left[2k\left(x+ct\right)\right]}
$$

we see that they often reduce to simpler forms. This has been highlighted by Kudryashov [11] who illustrates the situation by providing examples of problems for which authors have inadvertently claimed new traveling wave solutions that turned out to be well-known solutions in a different form. Some additional discussion on this subject is given by Parks [16].

A.3.2 A Maple exp Method Procedure

We now present a Maple procedure, `expMethod`, that generates a solution automatically and then illustrate its use by an application example. However, as for the tanh method, we must stress that this procedure will not solve all problems and is only provided to show what can be done with Maple using a fairly short piece of code.

```
expMethod:=proc(Md,Mn,pde,intFlag, infoLev)
   local F, FF, F1, F2, F3, F4, F4_Cs,
         ode1, str, tr1, tr2,
         i, j, vars, testFlag;
   global _a, _b, k, c, sol, testSol, N, Fn, Fd, F5;
   unprotect(_a,_b);
```

```
   # Assume a travelling wave solution of the form
   # U(xi), xi=k*(x-c*t);
   tr1:={x=(xi/k+c*tau),t=tau,u=U(xi)};
   ode1:=dchange(tr1,pde,[xi,tau,U(xi)]);
   print('ode1=', ode1);
   if intFlag > 0 then
     for i from 1 to intFlag do
       ode1:=int(lhs(ode1),xi)=0;
     end do;
     print('After integration, ode1= ', ode1);
   end if;
   tr2:={xi=Y,U(xi)=F(Y)};
   F1:=dchange(tr2,ode1,[Y,F(Y)]);
#print('F1=',F1);
   Fn:=add( _a[i]*exp(i*Y), i=-Mn..Mn ):
   Fd:=add( _b[i]*exp(i*Y), i=-Md..Md ):_b[Md]:=1;
   F(Y):=Fn/Fd;
   print('F(Y) =,',F(Y));
   F2:=simplify(eval(F1)):
#print('F2=',F2);
   # To eliminate exp terms
   F3:=subs(Y=ln(Delta),F2): F3:=simplify(F3):
#print('F3=',F3);
   # This line included incase there are quotient terms
   F3 := numer(lhs(F3))*denom(rhs(F3)) =
         numer(rhs(F3))*denom(lhs(F3)):
   F3:=collect(expand(F3),Delta):F3:=sort(%,Delta):
#print('F3=',F3);
   # Get coeff's of Delta - same as coeff's of exp terms
   F4:=CoefficientList(lhs(F3),Delta): F4_Cs:=convert(F4,set):
   # Define variables to be solved for
   vars:={seq(_a[r],r=-Mn..Mn),seq(_b[r],r=-Md..Md-1),'k','c'};
   if infoLev>0 then print('Unknowns =', vars) end if;
   if infoLev>1 then print('Coefficient List =',F4) end if;
   # Note: Polynomial coefficients for each power of Delta must = 0
   F5:=[solve(F4_Cs,vars)]:
   # This line needed to eliminate 'RootOf' terms
   # from the solve() answer
   F5:=map(allvalues,F5);
   N:=nops(F5): # No of solutions found by solve()
   str:=sprintf(' Number of solution sets found = %d ', N): print(str);
   if infoLev>1 then print(F5) end if;
   for i from 1 to N do
     # print(F5[i]);
     assign(F5[i]): FF[i]:=F(Y):
     sol[i]:=u=eval(simplify(subs(Y=k*(x+x0-c*t),FF[i]),symbolic));
     str:=sprintf('Solution %2d:\n===========', i): printf(str);
     print(sol[i]);
     # Unassign A coefficients
     for j from -Mn to Mn do
        unassign('_a[j]'):
     end do;
     for j from -Md to Md-1 do
       unassign('_b[j]'):
```

```
      end do;
      unassign('k','c'):
    end do;
    # Test Solutions!!
    testFlag:=0;
    for i from 1 to N do
      testSol[i]:=pdetest(sol[i],pde); # '0' if true
      if testSol[i] <> 0 then
        str:=sprintf(' Solution: %d does not pass pdetest() !', i): print(str);
        testFlag:=1;
      end if;
    end do;
    # Final check to see if all solutions found solve 'pde'
    if testFlag = 0 then
      print('All solutions pass pdetest() !');
    end if;
end proc:
```

LISTING A.6: Maple procedure `expMethod()`.

This procedure is saved in the separate file `expMethod.txt` so that it can be reused with any application. Note, because this procedure is general for use with any PDE, we now use `_a[i]` and `_b[i]` rather than simply `a[i]` and `_b[i]` so that there is minimum likelihood of a name conflict with constants in the PDE. Also, `expMethod` introduces the arbitrary constant `x0` into the solution—refer to Section A.7 for an explanation. In addition, because the algorithm appears to be insensitive as to whether the upper and lower summation indices are the same or different, we choose to make them the same. Thus, the numerator indices vary from `-Mn` to `+Mn` and, similarly, the denominator indices vary from `-Md` to `+Md`. It is a simple matter to modify the code to accommodate different upper and lower indices if this is found to be necessary. Also, a number of deactivated (commented) print statements are included; by removing the comment symbol #, they can be used for debugging or to output additional information as the calculations proceed.

We will now use procedure `expMethod()` to solve the Korteweg-de Vries (KdV) eq. (A.9), where we present code segments followed by the associated Maple output.

- *Initialize Maple and load required packages:* `PDEtools, PolynomialTools, plots.` We also have unprotected `_a` and `_b`—see above comments.

```
>#  KdV Equation
 #  Attempt at exp method solution
 restart; with(PDEtools): with(PolynomialTools):
 with(plots): unprotect(_a,_b):
```

- *Define an alias for $u(x, t)$ to simplify writing code and then define the KdV problem PDE.*

```
>alias(u=u(x,t)):
 pde1:=diff(u,t)+6*u*diff(u,x)+diff(u,x,x,x)=0;
```

$$\mathrm{pde1} := \frac{\partial}{\partial t}u + 6u\left(\frac{\partial}{\partial x}u\right) + \frac{\partial^3}{\partial x^3}u = 0$$

- *Read in the procedure* `expMethod` *so that it can be called as required.*

```
> read("expMethod.txt");
```

- *Call the procedure* `expMethod` *with arguments* `Md`, `Mn`, `pde1`, `intFlg`, *and* `infoLevOut`. Note the use of $\_a_i$ and $\_b_i$ rather than a_i and b_i, as mentioned above. Seven solutions are found including two nontrivial and five trivial solutions. All solutions found satisfy the original PDE, eq. (A.9).

```
>intFlg:=1: # Integration of U(xi) needed !
Mn:=1; Md:=1; # Set order of approximation
infoLevOut:=0;
expMethod(Md,Mn,pde1,intFlg,infoLevOut);
```

$$F(Y) := \frac{\_a_{-1}e^{-Y} + \_a_0 + \_a_1 e^{Y}}{\_b_{-1}e^{-Y} + \_b_0 + e^{Y}}$$

$$\text{ode1} =, -ck\left(\frac{d}{d\xi}U(\xi)\right) + 6U(\xi)kd\left(\frac{d}{d\xi}U(\xi)\right) + k^3\left(\frac{d^3}{d\xi^3}U(\xi)\right) = 0$$

$$\text{After integration, ode1} =, -ckU(\xi) + 3kU(\xi)^2 + k^3\left(\frac{d^2}{d\xi^2}U(\xi)\right) = 0$$

"Number of solution sets found = 7"

Solution 1:
=======

$$u = \frac{\_a_{-1} + \_a_0 + \_a_1}{\_b_{-1} + \_b_0 + 1}$$

Solution 2:
=======

$$u = 0$$

Solution 3:
=======

$$u = \frac{1}{3}c$$

Solution 4:
=======

$$u = \frac{4k^2\,\_b_0}{\_b_0^2\,\mathrm{e}^{k(-x-x_0+k^2t)} + 4\,\_b_0 + \mathrm{e}^{-k(-x-x_0+k^2t)}}$$

Solution 5:

=======

$$u = \_a_1$$

Solution 6:

=======

$$u = -\frac{\_a_1}{\_b_1}$$

Solution 7:

=======

$$u = -\frac{1}{3}\frac{k^2\left(\_b_0^2\,\mathrm{e}^{-k(x+x_0+k^2t)} - 8\,\_b_0 + 4\mathrm{e}^{k(x+x_0+k^2t)}\right)}{\_b_0^2\,\mathrm{e}^{-k(x+x_0+k^2t)} + 4\,\_b_0 + 4\mathrm{e}^{k(x+x_0+k^2t)}}$$

This concludes our discussion of the exp method. Additional Maple statements could of course be added to print out additional information and to plot results, as we have done for the tanh method above. Note, `expMethod` also introduces the arbitrary constant x_0 into the solution—refer to Section A.7 for an explanation.

Finally, when using the `expMethod` procedure, care must be taken when selecting variable names in the main body of the application code. It is advisable to avoid using any of the variable names specified as `local` or `global` in the procedure, except as shown in this example. Otherwise, unexpected behavior may result. If in doubt, try to follow the example given here or the additional download examples.

A.4 Riccati Equation Method

A.4.1 Introduction

Consider a differential equation of the form

$$\frac{dy}{d\xi} = f(y,\xi) \tag{A.27}$$

The function f can be extended as a polynomial in y with ξ held constant to give

$$f(y,\xi) = r_0(\xi) + r_1(\xi)y + r_2(\xi)y^2 + \cdots \tag{A.28}$$

Riccati[2] [20] investigated such systems where the approximation was truncated after the second-order term in y to give

$$\frac{dy}{d\xi} = r_0(\xi) + r_1(\xi)y + r_2(\xi)y^2 \tag{A.29}$$

[2]Italian mathematician, *Count Jacopo Francesco Riccati* (1676–1754).

and derived solutions for certain special cases which he published in a paper in 1724. For arbitrary r_0, r_1, and r_2, the Riccati equation cannot be solved by quadrature. However, a characteristic of the Riccati equation is that if a particular solution is known, then this enables us to find another solution. For example, if a solution $y_0(\xi)$ is known, then

$$y(\xi) = y_0(\xi) + \frac{1}{\psi(\xi)} \tag{A.30}$$

when substituted directly into eq. (A.28) gives

$$\frac{d\psi(\xi)}{d\xi} = -\left[r_1(\xi) + 2r_2(\xi)y_0(\xi)\right]\psi(\xi) - r_2(\xi) \tag{A.31}$$

which can be solved as it is a first-order linear differential equation with respect to $\psi(\xi)$. When we solve for ψ, a new solution for y is obtained by back substitution. More information relating to the Riccati equation is given in [5, 27].

A.4.2 ODE Example

Consider the Riccati equation $\frac{dy}{d\xi} = -2 - y + y^2$ for which a known solution is $y_0 = 2$, then on substituting $y = 2 + \frac{1}{\psi}$ into this equation, we obtain after some algebraic manipulation

$$\frac{d\psi}{d\xi} = -(3\psi + 1)$$

$$\Downarrow$$

$$\psi = K\exp(-3\xi) - \frac{1}{3}$$

where K is a constant of integration. After back substitution, we obtain a new solution for y, i.e.,

$$y = 2 + \frac{1}{K\exp(-3\xi) - \frac{1}{3}}$$

A simple Maple code that solves this example by two methods is given below. Method 1 substitutes eq. (A.30) directly into eq. (A.29) and solves the resulting ODE for ψ, althogh Method 2 substitutes the known solution $y_0 = 2$ into eq. (A.31) and solves the resulting ODE for ψ. Both methods yield the same answer, as expected.

```
># Riccati solution - Method 1
  restart;
  alias(y=y(xi),psi=psi(xi));
> # Define ODE
  ode1:=diff(y,xi)=-2-y+y^2;
> # Define Riccati solution
```

```
  f:=2+1/psi;
> # Substitute Riccati solution
  ode2:=simplify(subs(y=f,ode1));
> # Simplify ODE
  ode3 := lhs(ode2)*denom(rhs(ode2))
          = rhs(ode2)*denom(lhs(ode2));
> # Solve for psi
  sol1:=dsolve(ode3);assign(sol1);
> # Find y by back substitution
  y:=f;
> # Riccati solution - Method 2
  restart;
> alias(y=y(xi),psi=psi(xi));
> # Formulate problem
  ode1:=diff(psi,xi)=-(-1+2*(1)*y0)*psi-1;
> # Substitute value for y0
  ode2:=subs(y0=2,ode1);
> # Solve for psi
  sol1:=dsolve(ode2);assign(sol1);
> # Apply Riccati solution
  y:=y0+1/psi;
```

LISTING A.7: Two ODE applications of the Ricatti method.

A.4.3 Application to the Solution of PDEs

We can now use the above idea to help us solve PDEs. We consider again ideas used in the tanh and exp methods. The evolutionary equation for which we wish to find traveling wave solutions is eq. (A.9). The *Riccati method* is very straightforward to implement and is similar to the tanh and exp methods described above. We start by applying the transformation of eq. (A.2), where U has yet to be defined and, typically, $k > 0$ represents *wavenumber* and c represents *wave velocity*. Partial differentiation operations with respect to t and x are, therefore, transformed to the equivalent ordinary differentiation operations of eq. (A.4). Thus, partial differential eq. (A.9) is transformed to ordinary differential eq. (A.11).

We start by setting $U = F(Y)$ and using eq. (A.8) where coefficients a_i are real constants to be determined and M is a positive integer found by balancing (equating) the largest exponent of the highest order linear term with the largest exponent of the highest order nonlinear term.

Here the *Riccati method* deviates from the tanh and exp methods. Instead of using a transformation based on standard transcendental mathematical functions, such as tanh or exp, to find algebraic representations for derivatives, we use the *Riccati differential equation*. We now apply the *Riccati method* [28] and define $r_0(\xi) = A$, $r_1(\xi) = B$, and $r_2(\xi) = C$ of eq. (A.29), where A, B, and C are constants. The Riccati eq. (A.29) therefore becomes

$$\frac{dY}{d\xi} = A + BY + CY^2 \tag{A.32}$$

This Riccati equation has the specific solutions detailed in Table A.1 for $B = 0$ [26]:

Other solutions to eq. (A.32) can be derived for different values of A and C.

Table A.1: Standard known solutions to the Riccati eq. (A.32) where $B = 0$

A	C	Y
$\frac{1}{2}$	$-\frac{1}{2}$	$Y = \tanh\left(\frac{\xi}{2}\right), \coth\left(\frac{\xi}{2}\right)$
$\frac{1}{2}$	$\frac{1}{2}$	$Y = \tan\left(\frac{\xi}{2}\right), -\cot\left(\frac{\xi}{2}\right), \tan(\xi) \pm \sec(\xi)$
1	-1	$Y = \tanh(\xi), \coth(\xi)$
1	1	$Y = \tan(\xi), -\cot(\xi)$
1	-4	$Y = \frac{1}{2}\tanh(2\xi), \frac{1}{2}\coth(2\xi)$
1	4	$Y = \frac{1}{2}\tan(2\xi), -\frac{1}{2}\cot(2\xi)$

From eq. (A.32), we can now proceed to derive algebraic equations for various orders of derivatives of y with respect to ξ using the chain rule for differentiation as follows:

$$\begin{aligned}
\frac{dY}{d\xi} &= A + CY^2 \\
\frac{d^2Y}{d\xi^2} &= 2CY\frac{dY}{d\xi} = 2CY\left(A + CY^2\right) \\
\frac{d^3Y}{d\xi^3} &= 2C\frac{dY}{d\xi}\left(A + CY^2\right) + 4C^2Y^2\frac{dY}{d\xi} \\
&= 2C\left(A + CY^2\right)^2 + 4C^2Y^2\left(A + CY^2\right)
\end{aligned} \tag{A.33}$$

etc.

It is readily seen that there is a simple discernible pattern for higher-order derivatives. Thus, we have found a set of algebraic functions to represent derivatives of various orders. Consequently, after making the substitution $U = F(Y) = \sum_{i=0}^{M} a_i Y^i$, we are now able to convert the ODE of eq. (A.4), to an equivalent algebraic equation by substituting the algebraic relationships of eq. (A.33). This, like the tanh and exp methods, yields a polynomial in Y for which we have to solve for the unknown parameters. The associated coefficient for each power of Y in the polynomial is represented by an expression consisting of parameters A, C and a_i, k and c. Now, for the problem to be consistent for all values of Y, the coefficient expressions must each equate to zero. We therefore use the method of *Lagrange (undetermined) multipliers* to solve the resulting set of equations for the required unknowns a_i, k, and c.

We now have a solution in the form of a polynomial in Y with coefficients where a_i, k, and c are either known or are arbitrary constants. The final solution(s) are obtained by choosing values from Table A.1 for A and C and substituting the appropriate relationship for Y.

For anything other than trivial or simple problems, the process of solving for the unknown coefficients is tedious and usually requires the use of a computer algebra system if the solution is to be found in a reasonable time. In fact, for many problems, it is totally impractical to attempt a manual solution.

For additional information, the reader is referred to [28].

A.4.4 Example—KdV Equation

We will apply the above solution method to the canonical form of the KdV, eq. (A.9), with analytical solution eq. (A.10).

1. *Apply the traveling wave transformation.*
   Apply the transformation of eq. (A.2) to obtain eq. (A.9), which can be integrated once directly with respect to $U(\xi)$ to give eq. (A.12).
   Note: For this method, we set the constant of integration to zero.
2. *Determine the polynomial approximation order, M.*
   We now have to balance the largest exponent of Y in the highest order nonlinear term with the largest exponent of Y in the highest order linear term. When calculating the value of M by balancing exponents, we substitute eq. (A.8) into the problem PDE and then substitute for the derivatives using eq. (A.33). We will illustrate this process by analyzing both eq. (A.11) and eq. (A.12).

   First consider eq. (A.11), i.e., no integration of eq. (A.11). The highest order nonlinear term is the second term. After substitution of the summation into the second term, the largest exponent of Y will occur in the expression $Y^{M-1}Y^{M}\dfrac{dY}{d\xi}$ and, after substitution for the first derivative, this expression becomes $Y^{M-1}Y^{M}Y^{2}$. Thus, the largest exponent of Y in the second term will be equal to $(M-1)+M+2=2M+1$. The highest order linear term is the third term. After substitution of the summation into the third term, the largest exponent of Y, again neglecting coefficients, will occur in one of the expressions $Y^{M-1}\dfrac{d^3Y}{d\xi^3}$, $Y^{M-2}\dfrac{d^2Y}{d\xi^2}\dfrac{dY}{d\xi}$ or $Y^{M-3}\left(\dfrac{dY}{d\xi}\right)^3$ and, after substitution for derivatives, we see on expansion that these expressions become $Y^{M-1}Y^4$, $Y^{M-2}Y^3Y^2$, and $Y^{M-3}\left(Y^2\right)^3$, respectively. Thus, the largest exponent of Y in the third term will be equal to $(M-1)+4=M+3$, $(M-2)+3+2=M+3$, or $(M-3)+2\times 3=M+3$. Each of these alternatives gives the same result, which we would predict from compatibility considerations. Therefore, on balancing these exponents, we have

$$2M+1=M+3\rightarrow M=2$$

   Now consider eq. (A.12), i.e., after integration of eq. (A.11). The highest order nonlinear term is the second term. After substitution of the summation into the second term, the largest exponent in Y will occur in the expression $Y^{M}Y^{M}$. Thus, the largest exponent of Y in the second term will be equal to $2M$. The highest order linear term is the third term. After substitution of the summation into the third term, the largest exponent of Y will occur in either the expression $Y^{M-2}\left(\frac{dY}{d\xi}\right)^2$ or the expression $Y^{M-1}\left(\frac{d^2Y}{d\xi^2}\right)$ and, after substitution for derivatives, we see on expansion that these expressions become $Y^{M-2}Y^4$ and $Y^{M-1}Y^3$, respectively. Thus, the largest exponent of Y in the third term will be equal to either $(M-2)+4=M+2$ or $(M-1)+3=M+2$. Again, each of these alternatives gives the same result, which we would predict from compatibility considerations. Therefore, on balancing these exponents, we have

$$2M = M+2 \rightarrow M=2$$

As to be expected, we obtain $M=2$ from both approaches. It should be stressed that it is easy to make a mistake in calculating M and so use of a symbolic algebra package such as Maple is recommended. However, the Maple procedure `riccatiMethod()`, described below, is sufficiently fast that obtaining M by trial and error is a practical alternative.

3. *Apply the approximating polynomial in Y.*
Substitute $U=F(Y)=\sum_{i=0}^{M} a_i Y^i$ into eq. (A.12) (we choose to use the original equation) and then use eq. (A.33) to substitute for the derivatives. Using a value of $M=2$ and after dividing through by k, we obtain the following fifth-degree polynomial

$$c_0 + c_1 Y + c_2 Y^2 + c_3 Y^3 c_4 Y^4 + c_5 Y^5 = 0 \tag{A.34}$$

where

$$\begin{aligned}
c_0 &= -ca_1A + 2k^2a_1CA^2 + 6a_0a_1A \\
c_1 &= -2ca_2A + 16k^2a_2CA^2 + 12a_0a_2A + 6a_1^2A \\
c_2 &= +8k^2a_1C^2A + 6a_0a_1C + 18a_1a_2A - ca_1C \\
c_3 &= -2ca_2C + 12a_0a_2C + 12a_2^2A + 6a_1^2C + 40k^2a_2C^2A \\
c_4 &= +18a_1a_2C + 6k^2a_1C^3 \\
c_5 &= +12a_2^2C + 24k^2a_2C^3
\end{aligned} \tag{A.35}$$

4. *Solve for unknown parameters.*
For a consistent solution, each of the above coefficient expressions must equate to zero, which results in six simultaneous equations that we solve using Maple. Maple finds the following three parameter solution sets

$$\begin{aligned}
&a_0=a_0,\ a_1=a_1,\ a_2=a_2,\ k=0,\ c=c &&:\ \text{trivial} \\
&a_0=a_0,\ a_1=0,\ a_2=0,\ c=c,\ k=k &&:\ \text{trivial} \\
&a_0=a_0,\ a_1=0,\ a_2=-2k^2C^2,\ c=8k^2CA+6a_0,\ k=k &&:\ \text{nontrivial}
\end{aligned} \tag{A.36}$$

5. *Make back substitutions to obtain final solutions.*
The final results are, therefore, found by substituting the parameter solution sets (A.36) into eq. (A.34) and then using values for A and C from Table A.1 together with the appropriate function for Y. This then, ignoring trivial solutions, yields the following six solutions to eq. (A.9):

$$\begin{aligned}
u_1 &= a_0 - \frac{1}{2}k^2 \tanh\left[\frac{1}{2}k\left(x + 2tk^2 - 6ta_0\right)\right]^2 \\
u_2 &= a_0 - \frac{1}{2}k^2 \tan\left[\frac{1}{2}k\left(x - 2tk^2 - 6ta_0\right)\right]^2 \\
u_3 &= a_0 - 2k^2 \tanh\left[k\left(x + 8tk^2 - 6ta_0\right)\right]^2 \\
u_4 &= a_0 - 2k^2 \tan\left[k\left(x - 8tk^2 - 6ta_0\right)\right]^2 \\
u_5 &= a_0 - 8k^2 \tanh\left[2k\left(x + 32tk^2 - 6ta_0\right)\right]^2 \\
u_6 &= a_0 - 8k^2 \tan\left[2k\left(x - 32tk^2 - 6ta_0\right)\right]^2
\end{aligned} \tag{A.37}$$

Where constants a_0 or k appear in the above solutions, they may take arbitrary values.

We note the following points regarding the above solutions:

- If we set $a_0 = 2k^2$ in the equation for u_3 above, we obtain the known solution, eq. (A.10), for this example.
- For each parameter solution set, we have only included the first of the appropriate solutions given in Table A.1. Additional solutions can be obtained by using the alternative solutions shown. This is left as an exercise for the reader.

A.4.5 A Maple Riccati Method Procedure

We will now present a Maple procedure that generates a solution automatically and then illustrate its use by an application example. However, as for the tanh and exp methods, we must stress that this procedure will not solve all problems and is only provided to show what can be done with Maple using a fairly short piece of code.

```
riccatiMethod:=proc(M,pde,intFlag,infoLev)
    local A, C, F, FF, F1, F2, F3, F4, ZZ,
        d1, d2, d3, d4, d5, d6, d7, eqns, n,
        ode1, ode2, str,  psi, tr1, tr2,
        i, j, vars, testFlag;
    global _a, k, c, xi,
           sol, soln, testSol, N, F5;
    unprotect(_a);
    # Assume a travelling wave solution of the form
    # U(zeta), zeta=k*(x-c*t);
    # Apply travelling wave transformation
    tr1:={x=(xi/k+c*tau),t=tau,u=U(xi)};
```

```
    ode1:=dchange(tr1,pde,[xi,tau,U(xi)]);
    if intFlag > 0 then
      for i from 1 to intFlag do
        ode1:=int(lhs(ode1),xi)=0;
      end do;
      print('After integration, ode1= ', ode1);
    end if;
    # Substitute in polynomial approximation
    F(Y):=add( _a[i]*Y^i, i=0..M );
    print('F(Y) =', F(Y));
    ZZ:=subs(Y=U(xi),F(Y));
    ode2:=expand(subs(U(xi)=ZZ,ode1));
#print('ode2=',ode2);
    # Use Riccati Eqn: Y'=A+B*Y+C*Y^2, where B=0
    # This gives certain solutions to Riccati eqn
    # Define derivatives of Y wrt xi
    d1:=A + C*U(xi)^2;
    d2:=diff(d1,xi):
    d2:=subs(diff(U(xi),xi)=d1,d2);
    d3:=diff(d2,xi):
    d3:=subs(diff(U(xi),xi)=d1,d3);
    d4:=diff(d3,xi):
    d4:=subs(diff(U(xi),xi)=d1,d4);
    d5:=diff(d4,xi):
    d5:=subs(diff(U(xi),xi)=d1,d5);
    d6:=diff(d5,xi):
    d6:=subs(diff(U(xi),xi)=d1,d6);
    d7:=diff(d6,xi):
    d7:=subs(diff(U(xi),xi)=d1,d7);
#print(d1,d2,d3,d4,d5,d6,d7);
    # Substitute d1,..., d7 for derivatives in ode2
    F1:=eval(subs({diff(U(xi),xi)=d1,
                   diff(U(xi),xi,xi)=d2,
                   diff(U(xi),xi,xi,xi)=d3,
                   diff(U(xi),xi,xi,xi,xi)=d4,
                   diff(U(xi),xi,xi,xi,xi,xi)=d5,
                   diff(U(xi),xi,xi,xi,xi,xi,xi)=d6,
                   diff(U(xi),xi,xi,xi,xi,xi,xi,xi)=d7},ode2));
#print('F1=',F1);
F2:=subs(U(xi)=Y,F1);
    F2:=collect(expand(F2),{Y}):
#print('F2=',F2);
    # This line included in case there are quotient terms
    F2 := numer(lhs(F2))*denom(rhs(F2)) =
               numer(rhs(F2))*denom(lhs(F2)):
#print('F2=',F2);
# Make list of coefficient of the powers of Y
    # and convert to a set
    F3:=CoefficientList(lhs(F2),Y):
    F3:=convert(F3,set):
    # Define variables to be solved for
    vars:={seq(_a[r],r=0..M),'c','k'};
    if infoLev>0 then print('Unknowns =', vars) end if;
    if infoLev>1 then print('Coefficient List =',F3) end if;
```

```
    # Solve for unknowns
    F4:=[solve(F3,vars)]:
    # This line needed to eliminate 'RootOf' terms
    # from the solve() answer
    F5:=map(allvalues,F4);
    N:=nops(F5): # No of solutions found by solve()
    str:=sprintf(" Number of solution sets found = %d ", N):
    print(str);
    if infoLev>1 then print(F5) end if;
    # Print out solutions found
    for i from 1 to N do
#print(F5[i]);
      assign(F5[i]): FF[i]:=eval(F(Y)):
      sol[i]:=u=simplify(FF[i],symbolic):
#print(sol[i]);
      for j from 0 to M do
        unassign('_a[j]'):
      end do;
      unassign('c','k'):
    end do:
    # Define the Riccati equation solutions
    psi[1]:=tanh(xi/2);  eqns[1]:=[A=1/2,C=-1/2];
    psi[2]:=tan(xi/2);   eqns[2]:=[A=1/2,C= 1/2];
    psi[3]:=tanh(xi);    eqns[3]:=[A=1  ,C=-1];
    psi[4]:=tan(xi);     eqns[4]:=[A=1  ,C= 1];
    psi[5]:=tanh(2*xi)/2;eqns[5]:=[A=1  ,C=-4];
    psi[6]:=tan(2*xi)/2; eqns[6]:=[A=1  ,C= 4];
    xi:=k*(x+x0-c*t);
    # Check solutions
    for i from 1 to N do
      for n from 1 to 6 do
        assign(F5[i]):
        assign(convert(eqns[n],listlist));
        #print(rhs(sol[i]));
        soln[i,n]:=u=eval(subs(Y=psi[n],rhs(sol[i]))):
        str:=sprintf('Solution %2d,%d:\n=============', i,n): printf(str);
        print(soln[i,n]);
        for j from 0 to M do
          unassign('_a[j]'):
        end do;
        unassign('c','k'):
        unassign('A','C');
      end do:
    end do:
    # Test Solutions!!
    testFlag:=0;
    for i from 1 to N do
      for n from 1 to 6 do
        testSol:=pdetest(soln[i,n],pde);
        if testSol<>0 then
          str:=sprintf(' Solution: %d,%d does not pass pdetest() !', i,n): print(str);
          testFlag:=1;
        end if;
      end do:
```

```
    end do:
    # Final check to see if all solutions found solve 'pde'
    if testFlag = 0 then
      print('All solutions pass pdetest() !');
    end if;
    # Unassign xi so that this proceedure can be reused
    unassign('xi');
end proc:
```

LISTING A.8: Maple procedure `ricattiMethod()`.

This procedure is saved in the separate file `riccatiMethod.txt` so that it can be reused with any application. Note, because this procedure is general for use with any PDE, we now use `_a[i]` rather than simply `a[i]` so that there is minimum likelihood of a name conflict with constants in the PDE. Also, `expMethod` introduces the arbitrary constant `x0` into the solution—refer to section A.7 for an explanation. In addition, a number of deactivated (commented) print statements are included. By removing the comment symbol `#`, they can be used for debugging or to output additional information as the calculations proceed.

We will now use procedure `riccatiMethod()` to solve the Korteweg-de Vries (KdV) eq. (A.9), where we present code fragments followed by the associated Maple output.

- *Initialize Maple and load required packages:* `PDEtools`, `PolynomialTools`, and `plots`.

```
> # The KdV Equation
  # Attempt at Riccati based solution
  restart; with(PDEtools): with(PolynomialTools):
  with(plots):
```

- *Define an alias for* $u(x,t)$ *to simplify writing code and then define the KdV PDE.*

```
> alias(u=u(x,t)):alias(Y=Y(xi)):
  pde1:=diff(u,t)+6*u*diff(u,x)+diff(u,x,x,x)=0;
```

$$\text{pde1} := \frac{\partial}{\partial t}u + 6u\frac{\partial}{\partial x}u + \frac{\partial^3}{\partial x^3}u = 0$$

- *Read in the procedure* `riccatiMethod` *so that it can be called as required.*

```
> read("riccatiMethod.txt");
```

- *Call the procedure* `riccatiMethod` *with arguments* `M`, `pde1`, `intFlg`, *and* `infoLevOut`. Note the use of $\_a_i$ rather than a_i, as mentioned above. This method finds 5×6 solutions. The first 3×6 solutions found are trivial, but the last 2×6 are valid nontrivial solutions. To save space, only a reduced number of solutions are included below. Recall that each solution found is expanded to take on the known solutions to the Riccati equation as detailed in Table A.1.

```
>intFlg:=1: # integration of U(xi) used!
 M:=2; # Set order of approximation
 infoLevOut:=0; # minimum output
 riccatiMethod(M,pde1,intFlg,infoLevOut);
```

$$F(Y) := \_a_0 + \_a_1 Y + \_a_2 Y^2$$

$$\text{ode1} =, -ck\left(\frac{d}{d\xi}U(\xi)\right) + 6U(\xi)kd\left(\frac{d}{d\xi}U(\xi)\right) + k^3\left(\frac{d^3}{d\xi^3}U(\xi)\right) = 0.$$

$$\text{After integration, ode1} =, -ckU(\xi) + 3kU(\xi)^2 + k^3\left(\frac{d^2}{d\xi^2}U(\xi)\right) = 0$$

"Number of solution sets found = 5"

Solution 1,1:
========

$$u = \_a_0$$

Solution 1,2:
========

$$u = \_a_0$$

⋮

Solution 2,1:
========

$$u = 0$$

⋮

Solution 4,1:
========

$$u = \frac{1}{6}k^2 - \frac{1}{2}k^2 \tanh\left[\frac{1}{2}k\left(x + x_0 + k^2 t\right)\right]^2$$

⋮

Solution 5,3:
========

$$u = 2k^2 - 2k^2 \tanh\left[k\left(x + x_0 - 4k^2 t\right)\right]^2$$

⋮

Solution 5,6:
========

$$u = -8k^2 - 8k^2 \tanh\left[2k\left(x + x_0 + 16k^2 t\right)\right]^2.$$

We note that Solution 5,3 is equivalent to the standard solution obtained above using the `tanhMethod` procedure, i.e.,

$$u = \frac{2k^2}{\cosh\left[k\left(-x+4k^2t\right)\right]^2}$$

This concludes our discussion of the Riccati method. Additional Maple statements could of course be added to print out additional information and to plot results, as we have done for the tanh method above.

Finally, when using the `riccatiMethod` procedure, care must be taken when selecting variable names in the main body of the application code. It is advisable to avoid using any of the variable names specified as *local* or *global* in the procedure, except as shown in this example. Otherwise, unexpected behavior may result. If in doubt, try to follow the example given here or the additional download examples.

A.5 Direct Integration

We illustrate this method by application to the KdV eq. (A.9). A closed form *single soliton* solution to the KdV equation can be found using direct integration as follows [22].

Assume a traveling wave solution of the form

$$u(x,t) = U[k(x-ct)] = U(\xi).$$

Then, on substituting into eq. (A.9), the PDE is transformed into the following ODE

$$-ck\frac{dU(\xi)}{d\xi} + 6kU\frac{dU(\xi)}{d\xi} + k^3\frac{d^3U(\xi)}{d\xi^3} = 0$$

Now integrate with respect to ξ and multiply by $\dfrac{dU(\xi)}{d\xi}$ to obtain

$$-ckU(\xi)\frac{dU(\xi)}{d\xi} + 3kU(\xi)^2\frac{dU(\xi)}{d\xi} + k^3\frac{dU(\xi)}{d\xi}\left(\frac{d^2U(\xi)}{d\xi^2}\right) = A\frac{dU(\xi)}{d\xi}$$

Now integrate with respect to ξ once more to obtain

$$-\frac{1}{2}ckU(\xi)^2 + kU(\xi)^3 + \frac{1}{2}k^3\left(\frac{dU(\xi)}{d\xi}\right)^2 = AU(\xi) + B$$

where A and B are arbitrary constants of integration, which we set to zero. We justify this on the assumption that we are modeling a physical system having properties such that U, $\dfrac{dU}{d\xi}$, and $\dfrac{d^2U}{d\xi^2} \to 0$ as $\xi \to \pm\infty$. After rearranging and evaluating the resulting integral, we find

$$U(\xi) = \frac{c}{2}\operatorname{sech}^2\left(\frac{\sqrt{c}}{2k}\xi + \xi_0\right)$$

where ξ_0 is an arbitrary constant of integration - this time not set to zero. On applying the inverse transformation $\xi = k(x - ct)$, we arrive at the final solution

$$U(\xi) = u(x,t) = 2k^2\operatorname{sech}^2\left[k(x + x_0 - ct)\right]$$

where $k = \frac{\sqrt{c}}{2}$ represents *wavenumber*, and we have set the arbitrary constant ξ_0 equal to kx_0. The effect of the new arbitrary constant, x_0, is to shift the solution unchanged to the right or left along the x-axis, according to the sign of x_0. We observe from this solution that the wave travels to the right with a speed that is equal to twice the peak amplitude. Hence, the taller a wave, the faster it travels.

A Maple code that performs the above direct integration calculation for the KdV equation is included with the downloads for this book.

A.6 Factorization

A.6.1 Factoring

The idea of *factoring* is to convert an equation into factors of lower degree. If an ODE can be factored into two or more simple terms, then each factor may be solved independently and each solution is a solution to the original equation. For example, the nonlinear ODE

$$\frac{dy}{dx}\left(\frac{dy}{dx} + y\right) = x(x + y), \quad y = y(x)$$

can be factored into

$$\left(\frac{dy}{dx} + y + x\right)\left(\frac{dy}{dx} - x\right) = 0$$

Equating each of the factors to zero and solving separately yields the following solutions to the original ODE

$$y(x) = \begin{cases} Ae^{-x} + 1 - x \\ B + \frac{1}{2}x^2 \end{cases}$$

where A and B are arbitrary constants. **Note:** The complete solution to the original differential equation may switch from one solution branch to another [32].

A.6.2 Factoring Operators

A more general approach is to employ *factoring operators*, which are used to reduce an ODE or PDE problem to a lower-order problem that can be solved more easily. Here, we concentrate on factoring ODE problems.

For our purposes, this approach is useful for obtaining traveling wave solutions to certain types of PDE. This is achieved by first applying the transformation $u(x,t) = U(\xi)$, $\xi = k(x - ct)$, in order to convert the PDE to an ODE of the form $\mathcal{L}_0[U(\xi)] = 0$, where $\mathcal{L}_0$ is a *differential operator* operating on $U(\xi)$, and then $\mathcal{L}_0$ is factored into two separate operators, i.e., $\mathcal{L}_0 = \mathcal{L}_2\mathcal{L}_1$ and the ODE becomes

$$\mathcal{L}_0[U] = \mathcal{L}_2\mathcal{L}_1[U] = \mathcal{L}_2[\mathcal{L}_1[U]] = 0$$

The above statement should be interpreted as operator $\mathcal{L}_0[U]$ being equivalent to operator $\mathcal{L}_1$ operating on U followed by $\mathcal{L}_2$ operating on the result $\mathcal{L}_1[U]$ (it should be noted that there may be a number of distinct factorizations). The solution is then derived in two steps: $\mathcal{L}_1 = 0 \rightarrow f(\xi)$ followed by $\mathcal{L}_2 = f(\xi) \rightarrow U$. For example, consider the nonlinear ODE

$$\begin{aligned} \mathcal{L}_0[U] &= U_{\xi\xi}^2 - 2U_\xi U_{\xi\xi} + 2UU_\xi - U^2 = 0 \\ &= \left(U_{\xi\xi} - U_\xi\right)^2 - \left(U_\xi - U\right)^2 = 0 \end{aligned}$$

This equation can be factored as $\mathcal{L}_0 = \mathcal{L}_2[\mathcal{U}]\mathcal{L}_1[U] = \mathcal{L}_2[\mathcal{L}_1[U]]$, where [32]

$$\begin{aligned} \mathcal{L}_1[U] &= \left(U_\xi^2 - U^2\right) = 0, \quad \rightarrow f(\xi) \\ \mathcal{L}_2[U] &= \left(U_\xi - U\right) = f(\xi), \quad \rightarrow U \end{aligned}$$

Using the integrating factor method, we first solve $\mathcal{L}_1[U] = 0$ to obtain $f(\xi) = Ce^{\pm\xi}$, where C is an arbitrary constant, then we solve $\mathcal{L}_2[U] = Ce^{\pm\xi}$ to obtain the following two solutions, i.e.,

$$U(\xi) = \begin{cases} (A + C\xi)\,e^{\xi} \\ Ce^{-\xi} + Be^{\xi} \end{cases}$$

where A and B are arbitrary constants.

Care must be taken when using differential operators because in general they will not commute,[3] i.e., $\mathcal{L}_1\mathcal{L}_2$ is not usually the same as $\mathcal{L}_2\mathcal{L}_1$. Also, implicit in this operation is the property that coefficients are differentiable the requisite number of times, although this should not be a problem for our purposes. We observe from the above that the reduced problems $\mathcal{L}_1[U] = 0$ and $\mathcal{L}_2[U] = 0$ are *compatible* with $\mathcal{L}_0[U] = 0$. Thus, if we can solve the reduced problems, then these solutions will also be solutions to the original problem.

[3]Operators $\mathcal{L}_1$ and $\mathcal{L}_2$ will generally only commute if they are both linear.

ASIDE: The same solutions are obtained if we factor the above ODE to obtain

$$\mathcal{L}_0[U] = (U_{\xi\xi} - U)(U_{\xi\xi} - 2U_\xi + U) = 0$$

when the bracketed term on the right yields $U = (A + Cx)e^{\xi}$, and the bracketed term on the left yields $U = Ce^{-\xi} + Be^{\xi}$.

The operator $\mathcal{L}_0$ is said to be reducible if there exists *factoring operators* $\mathcal{L}_1$ and $\mathcal{L}_2$ of lower order such that $\mathcal{L}_0 = \mathcal{L}_2\mathcal{L}_1$; in this case, we say that $\mathcal{L}_1$ is a right factor and $\mathcal{L}_2$ is a left factor of $\mathcal{L}_0$. If an operator is not *reducible*, then it is termed *irreducible*.

For linear systems, we can define a differential operator $\mathcal{L}$ to be a function of $D = d/dx$, $D^2 = d^2/dx^2$, etc. so that an ODE such as $y_x - y = f(x)$ becomes $\mathcal{L}[y] = [D - 1]y = f(x)$. By simple extension, more complex ODEs can be defined in a similar manner. For example, the second-order ODE $y_{xx} + 2y_x - 3 = f(x)$ becomes $\mathcal{L}[y] = [D^2 + 2D - 3]y = f(x)$, which can be factored into $\mathcal{L}[y] = [D - 1][D + 3]y = f(x)$. Similarly, higher-order ODEs can be factored in a similar way. If $f(x)$ can be moved to the left-hand side of the equation and brought into the factoring scheme, then each factor can be set equal to zero and solved. Each solution will then be compatible with the problem equation and will, therefore, be a solution to the original ODE. A computer algorithm for automatic factorization has been developed by Schwartz [23].

For additional discussion relating to factoring, factoring operators, and general factorization methods, readers are referred to [12, 32].

A.6.3 Factorization Method for ODEs with Polynomial Nonlinearity

This method seeks solutions for certain types of PDEs with a *polynomial nonlinearity* by rescaling to eliminate coefficients and assuming a traveling wave solution of the form $u(x,t) = U(\xi)$, where $\xi = k(x - ct)$, c = velocity, and k = wavenumber. The resulting ODE is then factorized and each factor is solved independently [4]. For example, for a PDE of the form

$$\frac{\partial^2 u}{\partial t^2} = \frac{\partial u}{\partial x} + F(u)$$

the traveling wave ODE becomes

$$\frac{d^2U}{d\xi^2} + \gamma\frac{dU}{d\xi} + \hat{F}(U) = 0 \tag{A.38}$$

where $\gamma = \dfrac{c}{k}$ and $\hat{F}(U) = \dfrac{1}{k^2}F(U)$. On factorization, we obtain an equation of the form

$$[D - f_2(U)][D - f_1(U)]U = 0 \tag{A.39}$$

where we define $D = \dfrac{d}{d\xi}$. Care must be taken here as the terms in square brackets are *differential operators* which, may not commute, i.e., the order of evaluation is important. This leads to the equation

$$\frac{d^2U}{d\xi^2} - \frac{df_1}{dU}\frac{dU}{d\xi}U - f_1\frac{dU}{d\xi} - f_2\frac{dU}{d\xi} + f_1f_2U = 0$$

for which there are two possible groupings of terms. The first due to Berkovich [3] is

$$\frac{d^2U}{d\xi^2} - (f_1 + f_2)\frac{dU}{d\xi} + \left(f_1f_2 - \frac{df_1}{dU}\frac{dU}{d\xi}\right)U = 0 \tag{A.40a}$$

and the second due to Cornejo-Perez and Rosu [4] is

$$\frac{d^2U}{d\xi^2} - \left(\frac{df_1}{dU}U + f_1 + f_2\right)\frac{dU}{d\xi} + f_1f_2U = 0 \tag{A.40b}$$

Cornejo-Perez and Rosu recommend the latter grouping as being more convenient. From eq. (A.40b), we have

$$\frac{df_1}{dU}U + f_1 + f_2 = -\gamma \tag{A.41}$$

$$f_1f_2 = \frac{\hat{F}(U)}{U} \tag{A.42}$$

Evaluating eq. (A.41) enables us to determine values for certain unknown coefficients. Also, it follows that the first-order ODE $\left[D - f_1(U)\right]U = 0$ from eq. (A.39) is compatible with eq. (A.38). On solving this ODE, we obtain U and hence u. These steps are illustrated in the following example.

Example: Generalized Fisher Equation

The generalized Fisher equation is

$$\frac{\partial u}{\partial t} - \frac{\partial^2 u}{\partial x^2} - u(1 - u^n) = 0 \tag{A.43}$$

which, on applying the transformation $u(x,t) = U(\xi)$, $\xi = k(x - ct)$, reduces to the traveling wave ODE

$$\frac{d^2U}{d\xi^2} + \gamma\frac{dU}{d\xi^2} + \hat{F}(U) = 0 \tag{A.44}$$

where $\gamma = \dfrac{c}{k}$ and $\hat{F}(U) = \dfrac{1}{k^2}U(1 - U^n)$. We factor the polynomial function as

$$f_1f_2 = \frac{\hat{F}(U)}{U} = \frac{1}{k^2}(1 - U^n) = \frac{1}{k^2}(1 - U^{n/2})(1 + U^{n/2})$$

and choose

$$f_1 = \frac{a_n}{k}(1 - U^{n/2}), \quad f_2 = \frac{1}{a_n k}(1 + U^{n/2}), \quad a \neq 0 \tag{A.45}$$

where we have introduced constant a_n (to be determined). From eqns. (A.41) and (A.45), we obtain

$$\frac{df_1}{dU}U + f_1 + f_2 = -\frac{a_n}{k}\frac{n}{2}U^{n/2} + \frac{a_n}{k}(1 - U^{n/2}) + \frac{1}{a_n k}(1 + U^{n/2}) = -\gamma$$

Equating the coefficients of $U^{n/2}$ to zero (as the ODE is equal to a constant) gives $a_n = \pm\frac{2}{\sqrt{2n+4}}$, $\gamma = -\frac{(n+4)}{\sqrt{2n+4}k}$, and $c = -\frac{(n+4)}{\sqrt{2n+4}}$. Also, as γ is a constant and independent of the value of U, on setting $U = 0$, we find that $f_1 + f_2 = \mp(a_n + a_n^{-1})/k = -\gamma$, from which it follows that the original equation is equal to

$$\frac{d^2U}{d\xi^2} \mp \frac{\left(a_n + a_n^{-1}\right)}{k}\frac{dU}{d\xi^2} + f_1 f_2 U = 0 \tag{A.46}$$

Thus, the corresponding factorization $[D - f_2(U)][D - f_1(U)]U = 0$ becomes

$$\left[D \pm a_n^{-1}(U^{n/2} + 1)/k\right]\left[D \mp a_n(U^{n/2} - 1)/k\right]U = 0 \tag{A.47}$$

Therefore, it follows that eq. (A.46) is compatible with the first-order ODE $[D \mp a_n(U^{n/2} - 1)/k]U = 0$, i.e.,

$$\frac{dU}{d\xi} \mp \frac{a_n}{k}(U^{n/2} - 1)U = 0 \tag{A.48}$$

Integrating eq. (A.48) with $\gamma > 0$, either manually or using, Maple yields

$$U = \left(1 + K\exp\left[\frac{a_n n\xi}{2k}\right]\right)^{-2/n}$$

where K is an arbitrary constant of integration. Substituting back values for ξ, c, and a_n, we find that $k = \frac{n}{\sqrt{(2n+4)}}$, which yields the solution

$$U = \{1 + K\exp[k(x - ct)]\}^{-2/n}$$

where k and c are known.

If we let $K = \pm\exp\left[-kx_0\right]$, we arrive at the standard form of traveling wave solution

$$U_{>}^{\pm} = \left\{1 \pm \exp\left[k\left(x + x_0 - ct\right)\right]\right\}^{-2/n}$$
$$\downarrow$$
$$U_{>}^{+} = \left\{\frac{1}{2} - \frac{1}{2}\tanh\left[\frac{1}{2}k\left(x + x_0 - ct\right)\right]\right\}^{2/n} \tag{A.49}$$
$$U_{>}^{-} = \left\{\frac{1}{2} - \frac{1}{2}\coth\left[\frac{1}{2}k\left(x + x_0 - ct\right)\right]\right\}^{2/n}$$

For $\gamma < 0$, the solution becomes

$$U_{<}^{\pm} = \left\{1 \pm \exp\left[-k\left(x + x_0 - ct\right)\right]\right\}^{-2/n}$$
$$\downarrow$$
$$U_{<}^{+} = \left\{\frac{1}{2} + \frac{1}{2}\tanh\left[-\frac{1}{2}k\left(x + x_0 - ct\right)\right]\right\}^{2/n} \tag{A.50}$$
$$U_{<}^{-} = \left\{\frac{1}{2} + \frac{1}{2}\coth\left[-\frac{1}{2}k\left(x + x_0 - ct\right)\right]\right\}^{2/n}$$

A Maple code that performs the above factorization calculation for the generalized Fisher equation is included with the downloads for this book.

Readers are referred to the papers by Berkovich [3] and Cornejo-Perez and Rosu [4, 21] for more information on this method and additional examples of its use. Additional solutions to eq. (A.43) are given by Polyanin and Zaitsev [17, Chapter 1].

A.7 Additional Solutions by Addition of Arbitrary Constants

All the methods described above derive traveling wave solutions by making the transformation $u(x,t) \to U(\xi)$, $\xi = k(x - ct)$. This effectively transforms the problem from a PDE with dependent variable $u(x,t)$ into an ODE with dependent variable $U(\xi)$. The ODE is then solved to find $U(\xi)$ and, finally, $u(x,t)$ is obtained by applying the inverse transformation $U(\xi) \to u(x,t)$. If the transformation results in an ODE that is autonomous (does not include the independent variable ξ explicitly), with solution $W(\xi)$, it will also admit solutions of the form $W(\xi + \xi_0) \to w(k(x + x_0 - ct))$, where $\xi_0 = kx_0$ is an arbitrary constant. Thus, the number of solutions obtained can be expanded by choosing various values for ξ_0. The effect of ξ_0 is to shift the solution unchanged to the right or left along the x-axis, according to the sign of ξ_0. A positive sign will result in a shift to the left, although a negative sign will result in a shift to the right.

A.8 Other Methods

Although we do not employ these methods here directly, we mention for the interested reader that similar methods to the tanh, exp, and Riccati methods exist based on additional hyperbolic, trigonometrical, elliptic, etc. functions, see survey paper by Baldwin, et al. [2].

In addition to the methods described above, there is a wide range of additional analytical methods that can be employed to find PDE solutions. These include the *inverse scattering method, Bäcklund transformation, Hirota bilinear forms, differential constraints method, Lie group method, pseudo-spectral method,* and others. Interested readers are referred to discussions in [1, 9, 17, 18, 30]. An overview of linear and nonlinear waves is given in [6].

A.9 Maple Built-In Procedure TWSolutions

Maple has a built-in procedure, *TWSolutions,* for obtaining traveling wave solutions to PDE problems. However, as mentioned above, the *tanh-*, *exp-*, and *riccati-* based procedures presented here are not intended to rival this excellent procedure. Rather, they are intended primarily as educational aids for learning the techniques involved. They can also be used as research tools for readers to explore and develop the methods further. For example, it would be a simple matter to modify our procedures to include other mathematical functions or to combine them into a single procedure and to output all the results.

The procedure TWSolutions is generally a more powerful and versatile facility offering additional features, which are described fully in the Maple help system. For example, the Maple help system lists the possible mathematical functions available for TWSolutions as: exp, ln, the trigonometric functions sin, cos, tan, csc, sec, cot, the hyperbolic versions of them, JacobiSN, JacobiCN, JacobiDN, JacobiNS, JacobiNC, JacobiND, and the corresponding InverseJacobi functions and the WeierstrassP function. It also handles systems of equations with the same procedure call and has various options that provide useful insights into traveling wave problems. Nevertheless, the procedures outlined in this book find the same traveling wave solutions to many problems and to some where TWSolutions appears to fail. Consider the following *KdV-Burgers equation* example given in the Maple help:

$$\frac{\partial u}{\partial t}+u\frac{\partial u}{\partial x}-p\frac{\partial^2 u}{\partial x^2}+q\frac{\partial^3 u}{\partial x^3}=0 \tag{A.51}$$

where TWSolutions gives the following two nontrivial solutions:

$$u_1(x,t)=1/25\,\frac{3\,p^3-250\,\_C3\,q^2}{qp}-\frac{6}{25}p^2\tanh\left(\_C1+1/10\,\frac{px}{q}+\_C3\,t\right)q^{-1}-\frac{3}{25}p^2\left[\tanh\left(\_C1+1/10\,\frac{px}{q}+\_C3\,t\right)\right]^2q^{-1}$$

and

$$u_2(x,t)=1/25\,\frac{3\,p^3+250\,\_C3\,q^2}{qp}-\frac{6}{25}p^2\tanh\left(-\_C1+1/10\,\frac{px}{q}-\_C3\,t\right)q^{-1}-\frac{3}{25}p^2\left[\tanh\left(-\_C1+1/10\,\frac{px}{q}-\_C3\,t\right)\right]^2q^{-1}$$

Printed and bound by CPI Group (UK) Ltd, Croydon, CR0 4YY
03/10/2024
01040313-0003

which are the same except for opposite signs for $\_C_1$ and $\_C_3$. Alternatively, if we use our procedure tanhMethod, described above, we obtain the following (repeated) nontrivial solution:

$$u_1(x,t) = \_a_0 - \frac{6}{25}p^2 \tanh\left(\frac{1}{250}\frac{p(25xq + 25x0\,q + 3\,tp^2 - 25\,tq\_a_0)}{q^2}\right)q^{-1}$$
$$- \frac{3}{25}p^2\left[\tanh\left(\frac{1}{250}\frac{p(25xq + 25x0\,q + 3\,tp^2 - 25\,tq\_a_0)}{q^2}\right)\right]^2 q^{-1}$$

It is seen that this solution is equivalent to those obtained using TWSolutions if we define

$$\_a_0 = \frac{3p^3 \pm 250q^2\,\_C_3}{25pq}$$
$$x0 = \pm\frac{10q\_C_1}{p}$$

Many other similar examples could be given where the same solutions are found by our procedures and TWSolutions. But, of course, there are differences. If we repeat with `intFlg:=1` (one integration of the ODE), then tanhMethod finds seven solutions, three trivial, and four nontrivial. Also, for the same example, the expMethod procedure finds 20 exp solutions (2 trivial and 18 nontrivial), whereas TWSolutions appears not to find any exp solutions. On the other hand, TWSolutions finds solutions based on other mathematical functions, such as JacobiSN, that cannot be found using our procedures. However, we must stress that these results are provided by way of example only; they are not intended as a comprehensive comparison.

Maple codes for the above examples are not listed here in order to save space, but they are included with the downloads for this book. Where appropriate, the code is provided in the *mws* file format as well as the *mw* format so that it will also run in early versions of Maple—it has been tested in versions 8 and 13.

References

[1] M.J. Ablowitz, P.A. Clarkson, *Solitons, Nonlinear Evolution Equations and Inverse Scattering*, Cambridge University Press, 1991.

[2] D. Baldwin, U. Goktasa, W. Hereman, L. Hong, R.S. Martino, J.C. Miller, Symbolic computation of exact solutions expressible in hyperbolic and elliptic functions for nonlinear PDEs. *J. Symbolic. Comput.* 37 (2004) 669–705.

[3] L.M. Berkovich, Factorization as a method of finding exact invariant solutions of the Kolmogorov-Petrovskii-Piskunov equation and the related Semenov and Zeldovich equations, *Sov. Math. Dokl.* 45 (1992) 162–167.

[4] O. Cornejo-Perez, H.C. Rosu, Nonlinear second order ODE: Factorizations and particular solutions, *Prog. Theor. Phys.* 114 (3) (2005) 533–538.

[5] E. Fan, Y.C. Hon, Generalized tanh method extended to special types of nonlinear equations, *Zeitschrift fur Naturforschung*, 57a (2002) 692–700.

[6] G.W. Griffiths, W.E. Schiesser, Linear and nonlinear waves, *Scholarpedia*, 4 (7) (2009) 4308. Available online at: <http://www.scholarpedia.org/article/Linear_and_nonlinear_waves>.

[7] J.-H. He, X.-H. Wu, Exp-function method for nonlinear wave equations, *Chaos, Solitons and Fractals*, 30 (2006) 700–708.

[8] A. Heck, *Introduction to Maple*, third ed., Springer, New York, 2000.

[9] E. Infeld, G. Rowlands, *Nonlinear Waves, Solitons and Chaos*, second ed., Cambridge University Press, Cambridge, UK, 2000.

[10] D.J. Korteweg, F. de Vries, On the change of form of long waves advancing in a rectangular canal, and on a new type of long stationary wave, *Phil. Mag.* 39 (1895) 422–433.

[11] N.A. Kudryashov, Seven common errors in finding exact solutions of nonlinear differential equations, *Commun. Nonlinear Sci. Numer. Simulat.* 14 (2009) 3507–3529.

[12] D.A. López Díaz, Symbolic Methods for Factoring Linear Differential Operators, Ph.D. thesis, Research Institute for Symbolic Computation (RISC), Johannes Kepler University, Linz, Austria, 2006.

[13] W. Malfliet, Solitary wave solutions of nonlinear wave equation, *Am. J. of Phys.* 60 (7) (1992) 650–654.

[14] W. Malfliet, W. Hereman, The tanh method: I. Exact solutions of nonlinear evolution and wave equations, *Physica Scripta*, 54 (1996) 563–568.

[15] W. Malfliet, W. Hereman, The tanh method: II. Perturbation technique for conservative systems, *Physica Scripta*, 54 (1996) 569–575.

[16] E.J. Parks, A note on travelling-wave solutions to Lax's seventh-order KdV equation, *Appl. Math. Comput.* 215 (2009) 864–865.

[17] A.D. Polyanin, V.F. Zaitsev, *Handbook of Nonlinear Partial Differential Equations*, Chapman and Hall/CRC Press, Boca Raton, FL, 2004.

[18] A.D. Polyanin, A.c. Manzhirov, *Handbook of Mathematics for Engineers and Scientists*, Chapman and Hall/CRC Press, Boca Raton, FL, 2007.

[19] M.A. Porter, N.L. Zabusky, *Scholarpedia*, 2009; available on-line at: <http://www.scholarpedia.org/article/Soliton>.

[20] J.F. Riccati, Animadversationes in aequationes differentiales secundi gradus, *Acta Eruditorum qae Lipsiae Publicantur Supplementa*, cIII (1724) 67–73.

[21] H.C. Rosu, O. Cornejo-Perez, Supersymmetric pairing of kinks for polynomial nonlinearities. http://arxiv.org/PS_cache/math-ph/pdf/0401/0401040v3.pdf, Accessed 23 April, 2010.

[22] W.E. Schiesser, G.W. Griffiths, *A Compendium of Partial Differential Equation Models: Method of Lines Analysis with Matlab*, Cambridge University Press, Cambridge, UK, 2009.

[23] F. Schwarz, Efficient factorization of linear ODE's. *ACM-SIGSAM Bulletin*, 28 (1994) 9–17.

[24] J. Scott-Russell, *14th Meeting of the British Association for the Advancement of Science*, London, 1844.

[25] G.K. Valis, *Atmospheric and Oceanic Fluid Dynamics: Fundamentals and Large-Scale Circulation*, Cambridge University Press, NY, 2006.

[26] T-Y. Wang, Y-H. Ren, Y-L. Zhao, Exact solutions of $(3+1)$-dimensional stochastic Burgers equation, *Chaos, Solitons and Fractals*, 29 (4) (2006) 920–927.

[27] A.M. Wazwaz, More soliton solutions to the KdV and the KdV-Burgers equations, *Proc. Pakistan Acad. Sci.* 44 (2) (2006) 117–120.

[28] A.M. Wazwaz, The tanh-coth method combined with the Riccati equation for solving the KdV equation, *Arab. J. Math. Math. Sci.* 1 (1) (2007) 27–34.

[29] A.M. Wazwaz, The extended tanh method for new soliton solutions for many forms of the fifth-order KdV equations, *Appl. Math. Comput.* 84 (2) (2007) 1002–1014.

[30] G.B. Whitham, *Linear and Nonlinear Waves*, John Wiley & Sons Inc., New York, 1999.

[31] N.J. Zabusky, M.D. Kruskal, Interaction of 'Solitons' in a collisionless plasma and the recurrence of initial states, *Phys. Rev. Lett.* 15 (1965) 240–243.

[32] D. Zwillinger, *Handbook of Differential Equations*, Academic Press, Chesnut Hill, MA, USA, 1997, 292–302.

Index

A

Adaptive grid, 160
Adaptive mesh refinement (AMR), 160
Advection equation, xii, 7
  analytical solution, 8, 11–12
    routine, 19, 30, 34
  discontinuity, 20, 22, 26, 29
  flux limiter solution, 25–26
  initial condition routine, 18, 20
  linear, 7
  main program, 9–10
  ODE routine, 15–16
  numerical solution, 11–12, 23–24, 33–38
Analytical solution *see also*
  specific PDEs
  discontinuity, 20
  `Maple`, *see* specific PDEs
  PDE, 1
  routine, 19
  smoothness, 5, 7, 20, 35
  verification, 4, 48, 58, 261–262, 309
Antipeakon, 305
Arbitrary constants method, 435

B

Backlund transformation, 436
Benjamin–Bona–Mahony (BBM) equation, 254
Bernoulli equation
see Extended Bernoulli equation
Boundary conditions, 1, 5
  Dirichlet, 49, 60, 67, 70–71
  general, 85
  homogeneous, 70
  infinite domain, 19, 245, 263, 353, 380
  insulated, 104
  Neumann, 67
    analytical, 86, 101
  nonlinear, 67, 84, 98
  periodic, 27
  Robin
    see boundary conditions, third type
  third type, 67, 82, 84, 94, 98
  zero flux, 104
Boundary effects, 269
Boussinesq equation, 339
  analytical solution, 339, 352–353, 358–359, 368–369, 371–374
    direct integration, 372
    `Maple`, 371, 373
    Riccati method, 373
    routines, 348
  boundary conditions, 339
    Dirichlet, 342, 356, 361
    Neumann, 363
  initial conditions, 339
    routines, 346–348
  Jacobian matrix, 350
  main program, 349
  `Maple` procedure, 409
  numerical solution, 351–354, 357–359, 367–369
  ODE routine, 339, 353, 359, 363
    vectorized, 343, 359, 360, 363
  origin, 370
  second order in t, 339
    first order system in t, 342
  stagewise differentiation, 354, 356, 359, 362
    failure, 362
  `weights` (for FDs), 350, 366
  `uxxx7c` (for third derivative), 365
  `u4x11p` (for fourth derivative), 340, 342–346, 350–353
  wavenumber, 372
  wave speed, 372
Burgers–Fisher equation, 123
  analytical solution, 123, 125–130
    `Maple`, 132
    Tanh method, 130
  boundary conditions, 123
  generalized form, 128
  Jacobian matrix, 128
  ODE routine, 123
  initial condition routine, 125
  numerical solution, 126–127
  stagewise differentiation, 127

Burgers–Huxley equation, 111
  analytical solution, 111, 114, 117
  factorization, 118
  Maple, 120–121
  boundary conditions, 112
  generalized form, 118
  initial condition, 111
    routine, 113
  Jacobian matrix, 116
  numerical solution, 114–117
  ODE routine, 111
  wave velocity, 116, 120

C

Cauchy problem, 3, 7
Change of variable PDE, 2
Characteristics, 33
Chromatography, 33
Computer algebra system (CAS), xii, 3, 391
Conservation laws, 37
Convection–diffusion–reaction equation, 57
  analytical solution, 58, 61–65
    routine, 61
    verification, 58
  boundary conditions, 58
  initial condition, 57
    routine, 60
  Jacobian matrix, 64
  nonlinear extension, 63
  numerical solution, 62–65
  ODE routine, 59
Conservation principle, 104, 242
Constraint, 20
Coth method, 411
  see also tanh method
Cubic Klein–Gordon equation, *see* Klein–Gordon equation, cubic

D

d'Alembert solution, 309
Diffusion equation, 47, 67
  analytical solution, 48, 54
    verification, 48
    routine, 51
  boundary conditions, 48
  extension with reaction, 53
  initial conditions, 48
    routine, 50
  Jacobian matrix, 52–53
  main program, 52
  nonlinear
    see nonlinear diffusion equation
  numerical solution, 54
  ODE routine, 48
Diffusion induced chaos, 192
Direct integration method, 2, 372, 429
  general concepts, 429
Dirichlet boundary condition, 49
Discontinuity, 20, 235–236
Dispersion, 234, 236, 257
Dissipation, 234, 236
dss library, 19
dss004, 16, 19, 25, 69, 78, 111, 123, 136, 149, 174, 186, 197, 243, 263
dss044, 49, 51, 59, 62, 277, 294, 296, 311–313, 341–342, 365, 379
dss008, 37, 38

E

Euler equations, 33
Eulerian coordinate, xii, 4, 261, 377
exp method, 3, 195, 389, 391, 411
  applied to KdV, 412
  details of implementation, 412, 416
  general concepts, 411
  Maple procedure, 414
Expansion methods, 3
Exponential time differencing, 194
Extended Bernoulli equation, 261–262, 272
  analytical solution (ODE), 261
    verification, 261
  analytical solution (PDE), 262, 268–269, 272
    Maple, 272
    routine, 264
    verification, 262
  boundary conditions, 264
  initial condition routine, 264
  Jacobian matrix, 267
  main program, 264
  numerical solution, 268–269
  ODE routine, 262

F

Factorization method, 3, 106, 118, 141–145, 179, 391, 430
  algorithm, 432
  applied to generalized Fisher equation, 433
  differential operator, 431

factoring operators, 431–432
commutative, 431–432
general concepts, 430
FD (finite difference), 15
see also, dss004, dss008, dss044
order, 19, 23–25, 35
first derivative, 16, 19, 24
four-point upwind, 27
second derivative, 16
two-point upwind, 23
`u3x9p` (for third derivative), 197–198, 201
exponential differentiation, 230–234
polynomial differentiation, 217–230
`u5x11p` (for fifth derivative), 197–199
polynomial differentiation, 230
Finite difference, *see* FD
Finite volume, 27, 37
Fisher–Kolmogorov equation, 135
analytical solution, 135, 138, 140–144
factorization, 141–145
`Maple`, 145–146
boundary conditions, 136, 138
initial condition routine, 137
Jacobian matrix, 142
main program, 138–139
numerical solution, 140–144
ODE routine, 135
stagewise differentiation, 136
wave velocity, 140–141
Fitzhugh–Nagumo equations, 147
analytical solution, 148, 155–163, 166–169
moving front, 160
routine, 150
boundary conditions, 149–150
dependent variables, 157, 165
excitation threshold, excitability, 165
initial condition routine, 149
Jacobian matrix, 154, 160
main program, 150
`Maple`
exp method, 170
tanh method, 166
numerical solution, 155–163
moving front, 160
ODE routine, 148
ODE version, 165
stagewise differentiation, 149
wavenumber, 164
wave velocity, 164
Flame propagation, 192
Fisher–KPP equation, *see* KPP
Flux limiter, 25, 37–38
Fornberg algorithm, 205
Fortran, 17
Fourier domain analysis, 193

G

Gaussian function, 7, 18
Generalized Fisher equation, 433
factorization solution, 433
Global variables, 10
naming, 402, 429
Godunov barrier theorem, 24, 26
Group velocity, 407

H

Heaviside function, 7, 29
Hirota bilinear forms, 436
Hodgkin–Huxley equation, 147, 164
h refinement, 32, 270
Hump soliton, 406
Hyperbolic Liouville equation, 275
analytical solution, 275, 285–287, 290
routines, 279
Riccati method, 288
transformations, 286, 288, 290
boundary conditions, 275
continuity, 289
initial conditions, 275, 290
routine, 278
Jacobian matrix, 283–284
main program, 279
`Maple`
rational solution, 286–287
transformation, 287, 290
numerical solution, 285–288
ODE routine, 275, 283
second order in t, 275
first order system in t, 277–279
wavenumber, 284
wave velocity, 284
Hyperbolic PDE, 33
Hyperbolic–parabolic PDE, 57

I

Inhomogeneous PDE, 5
Initial condition, 5
Gaussian pulse, 18

Initial condition (*continued*)
  square pulse, 20
  Triangular pulse, 33–34, 36
Initial condition routine, 18
Initial value problem, 3
Integral constraint, 20, 104–106, 242
Invariant, 20, 104–106, 242
Inverse matrix operator, 243
Inverse scattering method, 436

J

Jacobian matrix, 13,
  see also specific PDEs
  banded, 15
  map, 15, 17
  pentadiagonal, 15-16
  routine, 19
  size, 15
  sparse, 15

K

Kawahara equation, 197
  analytical solution, 197, 207–213
    routine, 200
  boundary conditions, 198, 200
  Fornberg algorithm, 205
  initial condition, 199
    routine, 199
Jacobian matrix, 207, 210
  main program, 200
  `Maple`, 237
  numerical solution, 207–213
  ODE routine, 197
  small signal version, 236
  `weights` (for FDs), 201–207
  `u3x9p` (for third derivative), 197–198, 201
  `u5x11p` (for fifth derivative), 197–199
KdV–Burgers equation, 436
  solution by `TWSolutions`, 436
Klein–Gordon equation, 4, 309
  analytical solution, 309, 319, 320, 323, 331, 333–335
    `Maple`, 338
    residual function, 323, 325–328
    routine, 315–316, 321–322, 328
    tanh method, 337
    two forms, 310, 321, 322
    verification, 309, 325–326
  boundary conditions, 310
  cubic, 309, 329–330
  inhomogeneous, 326–328
  initial condition routine, 314
  Jacobian matrix, 317
  linear wave equation, 309
  main program, 316
  Mth-order, 309
  nonhomogeneous, 311, 326–328
  numerical solution, 318–321, 324–327, 330–336
  ODE routine, 310
  quadratic, 309
  second order in t, 309–313, 316
    first order system in t, 310–313, 316
Kolmogorov–Petrovskii–Piskunov equation, *see* KPP
Korteweg–de Vries (KdV) equation, 234, 254, 370, 393–394, 412, 422
  `Maple` solution, 406, 414
  analytical solution, 394–397
KPP, 173
  analytical solution, 173, 177–180
    factorization, 179
    `Maple`, 182
    routine, 176
  boundary conditions, 174
  initial condition, 175
    routine, 175
  Jacobian matrix, 179
  main program, 176
  numerical solution, 177–180
  ODE routine, 173
  stagewise differentiation, 174
  wave velocity, 177–179
Kuramoto-Sivashinsky equation, 185, 192
  analytical solution, 185, 190–194
    `Maple`, 194–195
    routine, 188
  boundary conditions, 185, 188
    Dirichlet, 185
    Neumann, 185, 189
  exponential time differencing
  fourier domain analysis, 193
  initial condition routine, 187
  Jacobian matrix, 192
  main program, 189
  numerical solution, 190–194
  ODE routine, 185
  spectral methods, 194

stagewise differentiation, 187
wavenumber, 191
wave velocity, 191

L

Lagrange multipliers,
see traveling wave methods
Lagrangian coordinate, xii, 4, 139, 261, 377
Lagrangian variable, 115, 164, 177, 191
Level set method, 33
Laboratory coordinates, 302, 304
Lie group methods, 436
Light-cone coordinates, 302, 304
Limiter functions, 40–43
Line soliton, 409

M

Malfliet's tanh method, *see* tanh method
`Maple`, xii, 1
exp method procedure, 414
details of use, 416
plotting, 401
solution check, 401
tanh method procedure, 397
details of use, 399
see also specific PDEs, `TWSolutions`
`Matlab`, 1
see also specific PDEs
inverse matrix operator, 243
vectorized, 17, 243, 341
Method of characteristics, 33
Method of lines, *see* MOL
Mixed partial derivative, 239, 244, 245, 253
FD approximation, 239–241
Mixed-type PDE, 33, 239, 292, 302
Modified wave equation, 377
analytical solution (ODE), 377
analytical solution (PDE), 377, 384–385
exp method, 389
`Maple`, 387
Riccati method, 387
routine, 382
tanh method, 389
verification, 378
Boundary conditions, 380–381, 383, 386
Eulerian coordinate, 377
Initial condition, 381
routine, 381
Jacobian matrix, 383
Lagrangian coordinate, 377
linear wave equation, 377
main program, 382
numerical solution, 384–386
ODE routine, 378
second order in t, 377, 379
first order system in t, 379
MOL (method of lines), xi, xiii, 1
ODE routine, 13
see also specific PDEs
programming, 213–217
stiffness, 251
Moving coordinate, *see* Lagrangian coordinate
Moving front, 153, 160
Mth-order Klein–Gordon equation, *see*
Klein–Gordon equation, mth-order
MUSCL scheme, 37

N

Navier–Stokes equations, 254
Nonhomegenous PDE, 5
Nonlinear diffusion equation, 67
analytical solution, 67, 78, 80, 86–92, 101–103
`Maple`, 108–109
routine, 72
boundary conditions, 67
Dirichlet, 67, 90
nonlinear, 68, 84
Neumann, 67, 69, 79, 87, 92
analytical, 86, 101
third type, 67, 69, 82, 84, 94, 98
initial condition, 71
routine, 71
Jacobian matrix, 79
linear form, 67
main program, 73
numerical solution, 78, 80, 83–103
ODE routine, 67
source terms, 67, 81, 83–84, 86, 88
wavenumber, 67
wave velocity, 67
Nonlinear reaction-diffusion equation, 147,
164
Numerical diffusion, 24, 32
Numerical oscillation, 24, 32
Numerical solution
ODE, 13
computational efficiency, 14
PDE, 1

O

`ode15s`
  sparse option, 13, 74, 76, 151, 265–266, 280
  see also specific PDE, main program
`ode45`, 13, 151, 251, 265, 280
Ordinary differential equation, *see* ODE
ODE, 2
  Jacobian matrix, 13
    see also specific PDEs
  MOL routine, 9
    see also specific PDE, ODE routine
  stiffness, 13
  two-point boundary value, 96, 99

P

Parabolic PDE, 33
Partial differential equation, *see* PDE
PDE (partial differential equation), xi, 1
  analytical solution, 1
  auxiliary conditions, 2
  boundary conditions, 1
  change of variable, 2
  general form, 1
  higher-dimensional, 3
  hyperbolic, 33
  initial conditions, 5
  mixed type, 33, 239, 292, 302
  multi-dimensional, 3
  numerical solution, 1
  parabolic, 33, 47
  residual function method, 1
  second order in t, 3, 275, 293
    first order system in t, 277–279, 295
  smooth solutions, 7, 24, 31–32
  spatial derivatives, 1
  traveling wave solution, 1
Peakon, 305
Peak soliton, 406
Periodic boundary condition, 27
  ghost cell, 27
Phase velocity, 235
p refinement, 32, 270
Procedual programming, 17
Procedure, *see also* `Maple`
  coupled PDEs, 404
  solution check, 401
Pseudo spectral method, 436
Pulse, 20, 33, 36

Q

Quadratic Klein–Gordon equation, *see*
  Klein–Gordon equation, quadratic

R

Regularized long-wave equation, *see* RLW
Residual function method, 1, 4, 323, 326
Riccati method, 3, 195, 288, 373, 388, 391
  applied to
    KdV, 422, 427
    ODE, 419
  details of implementation, 422, 427
  equation, 420–421
    derivatives, 421
    known solutions, 421
  general concepts, 418
  `Maple` procedure, 419, 424
Riemann problem, 7, 25, 29
RLW (regularized long-wave equation), 239
  analytical solution, 239, 253–257
    exp method, 259
    Riccati method, 259
    routine, 245
    tanh method, 258
  boundary conditions, 239
  conservation principles, 242
  initial conditions, 239
    routine, 243
  integral constraints, 242, 252
  invariants, 242, 252
  main program, 247
  mixed partial derivative, 239, 253
  numerical solution, 253–257
    FD matrix approximation, 239–241, 250
  ODE routine, 242

S

Semidiscrete approximation, 39
Shock, 235–236
Simpson's rule, 20, 104, 245, 252
Sine–Gordon equation, 293
  analytical solution, 293, 301–305
    Routine, 298
  antipeakon, 305
  boundary conditions, 294–296
  initial conditions, 293–294, 297
    routine, 296–297
  Jacobian matrix, 298–299
  laboratory coordinates, 302, 304

light-cone coordinates, 302, 304
main program, 298
Maple
tanh method, 305
mixed partial derivative, 302
numerical solution, 300–303
ODE routine, 294
peakon, 305
second order in t, 293
first order system in t, 295
solution sensitivity, 299
Solitary wave, 394
Soliton, 372, 394
hump, 406, 409
peak, 406
Source terms, 67
Spatial derivatives, 1
dss library, 19
FD approximation, 16
linear approximation, 24
nonlinear approximation, 25
Spatial domain, 18
see also Boundary conditions, infinite domain
Spatial error, 270
balanced with temporal error, 270
Spectral methods, 194, 436
Stagewise differentiation, 112, 115, 124, 136, 149, 174, 187, 354, 356, 359
failure, 359–363
Stefan–Boltzmann law, 85
Stiffness, 13, 251
Superbee flux limiter, 25, 27, 41
numerical solution, 28, 30–32, 38–40
routine, 25

T

tanh method, 3, 130, 258, 304, 337, 389, 391
applied to
Boussinesq eqns., 409
KdV, 399, 405
simultaneous (coupled) PDEs, 403
details of implementation, 394
extended form, 393
in higher dimensions, 407
Maple procedure, 397, 404, 408–410
Temporal error, 270
balanced with spatial error, 270
Total variation diminishing, *see* TVD
Traveling wave methods, 1–3
differential operators, 392–393, 407
general concepts, 392, 403, 407
Lagrange multipliers, 391, 412, 421
TVD, 37–38, 42
TWSolutions, 391
applied to KdV–Burgers equation, 436
details of implementation, 437
general features, 436

U

Upwind approximation
four-point, 27
two-point, 23
uxxx7c (for third derivative), 365
u3x9p (for third derivative), 197–198, 201
exponential differentiation, 230–234
polynomial differentiation, 217–230
u4x11p (for fourth derivative), 340, 342–346, 350–353
u5x11p (for fifth derivative), 197–199
polynomial differentiation, 230

V

van Leer flux limiter, 25, 27, 42
routine, 25
numerical solution, 25

W

Water hammer, 33
Wave number, 2, 67, 164, 191, 235, 284, 372, 392, 407, 420, 430, 432
Wave envelope, 235
Wave equation, 309
Wave frequency, 407
Wave packet, 235
Wave speed, 372
Wave velocity, 2, 67, 116, 140–141, 164, 177–179, 191, 284, 392, 420, 432
Weighted essentially non-oscillatory, *see* WENO
weights (for FDs), 201–207
WENO, 33